AF388929

Advanced Deep Learning Strategies for the Analysis of Remote Sensing Images

Advanced Deep Learning Strategies for the Analysis of Remote Sensing Images

Editors

Yakoub Bazi
Edoardo Pasolli

MDPI • Basel • Beijing • Wuhan • Barcelona • Belgrade • Manchester • Tokyo • Cluj • Tianjin

Editors
Yakoub Bazi
King Saud University
Saudi Arabia

Edoardo Pasolli
University of Naples Federico II
Italy

Editorial Office
MDPI
St. Alban-Anlage 66
4052 Basel, Switzerland

This is a reprint of articles from the Special Issue published online in the open access journal *Remote Sensing* (ISSN 2072-4292) (available at: https://www.mdpi.com/journal/remotesensing/special_issues/advanced_deep_learning).

For citation purposes, cite each article independently as indicated on the article page online and as indicated below:

LastName, A.A.; LastName, B.B.; LastName, C.C. Article Title. *Journal Name* **Year**, *Volume Number*, Page Range.

ISBN 978-3-0365-0986-0 (Hbk)
ISBN 978-3-0365-0987-7 (PDF)

Contents

About the Editors

Yakoub Bazi received the State Engineer and M.Sc. degrees in electronics from the University of Batna, Batna, Algeria, in 1994 and 2000, respectively, and the Ph.D. degree in information and communication technology from the University of Trento, Trento, Italy, in 2005. From 2000 to 2002, he was a Lecturer at the University of M'sila, M'sila, Algeria. From January to June 2006, he was a Postdoctoral Researcher at the University of Trento. From August 2006 to September 2009, he was an Assistant Professor at the College of Engineering, Al-Jouf University, Al-Jouf, Saudi Arabia. He is currently a Professor in the Computer Engineering Department, College of Computer and Information Sciences, King Saud University, Riyadh, Saudi Arabia. His research interests include machine learning and pattern recognition methods for signal/image processing and analysis. Dr. Bazi is a referee for several international journals.

Edoardo Pasolli received the M.Sc. degree in Telecommunications Engineering and the Ph.D. degree in ICT from the University of Trento, Trento, Italy, in 2008 and 2011, respectively. He was a postdoctoral fellow at the University of Trento (2011–2012 and 2014–2016); NASA Goddard Space Flight Center, Greenbelt, MD, USA (2012–2013); and Purdue University, West Lafayette, IN, USA (2013–2014). He was a Marie Sklodowska-Curie Individual Fellow at the University of Trento from 2016 to 2018. Since 2018, he has been an Assistant Professor in the Department of Agricultural Sciences, University of Naples Federico II, Naples, Italy. His research interests aim at developing machine learning and data science methodologies for complex ecosystems including the processing of remote sensing images and data.

Preface to "Advanced Deep Learning Strategies for the Analysis of Remote Sensing Images"

The last two decades have unveiled that remote sensing (RS) has become an essential technology in monitoring urban, atmospheric, and ecological changes. The increased availability of satellites and airborne sensors with different spatial and spectral resolutions has made this technology a key component in decision making. In addition to these traditional platforms, a new era has been opened recently by the adoption of UAVs for diverse applications such as policing, precision farming, and urban planning.

The great potential provided in terms of observation capability introduces similarly great challenges in terms of information extraction. However, processing the massive amounts of data collected by these diverse platforms is impractical and ineffective using traditional image analysis methodologies. This calls for the adoption of powerful techniques that can extract reliable and impressive information. In this context, deep learning (DL) strategies have recently been shown to hold the great promise of addressing the challenging needs of the RS community. Indeed, the introduction of DL dates back decades ago, when the first steps towards building artificial neural networks were undertaken. However, due to the limited processing resources, it did not reach a cutting-edge success in data representation and classification tasks until the recent appearance of high-performance computing facilities. This in turn enabled the design of sophisticated deep neural architectures and boosted the precision of many problems to groundbreaking performances. In this context, this book presents several contributions for the analysis of remote sensing imagery, including interesting topics related to scene classification, semantic segmentation, and image retrieval.

Yakoub Bazi, Edoardo Pasolli
Editors

Article

Deep Multi-Scale Recurrent Network for Synthetic Aperture Radar Images Despeckling

Yuanyuan Zhou, Jun Shi *, Xiaqing Yang, Chen Wang, Durga Kumar and Shunjun Wei and Xiaoling Zhang

School of Information and Communication Engineering, University of Electronic Science and Technology of China, Chengdu 611731, China
*   Correspondence: shijun@uestc.edu.cn

Received: 23 September 2019; Accepted: 19 October 2019; Published: 23 October 2019

Abstract: For the existence of speckles, many standard optical image processing methods, such as classification, segmentation, and registration, are restricted to synthetic aperture radar (SAR) images. In this work, an end-to-end deep multi-scale recurrent network (MSR-net) for SAR image despeckling is proposed. The multi-scale recurrent and weights sharing strategies are introduced to increase network capacity without multiplying the number of weights parameters. A convolutional long short-term memory (convLSTM) unit is embedded to capture useful information and helps with despeckling across scales. Meanwhile, the sub-pixel unit is utilized to improve the network efficiency. Besides, two criteria, edge feature keep ratio (EFKR) and feature point keep ratio (FPKR), are proposed to evaluate the performance of despeckling capacity for SAR, which can assess the retention ability of the despeckling algorithm to edge and feature information more effectively. Experimental results show that our proposed network can remove speckle noise while preserving the edge and texture information of images with low computational costs, especially in the low signal noise ratio scenarios. The peak signal to noise ratio (PSNR) of MSR-net can outperform traditional despeckling methods SAR-BM3D (Block-Matching and 3D filtering) by more than 2 dB for the simulated image. Furthermore, the adaptability of optical image processing methods to real SAR images can be enhanced after despeckling.

Keywords: synthetic aperture radar; despeckling; multi-scale; LSTM; sub-pixel

1. Introduction

Synthetic aperture radar (SAR), owing to its all-weather and all-time condition operation, has been widely applied to microwave remote sensing areas, such as topographic mapping, military target reconnaissance, and natural disaster monitoring [1,2]. SAR imaging achieves high range resolution by exploiting pulse compression technique and high azimuth resolution by using radar platform to form a virtual antenna synthetic aperture along track [3,4]. However, speckle noise exists in the imaging results due to the coherent imaging mechanism of SAR, which leads to images quality and readability reduction. Meanwhile, the existence of speckles limits the effectiveness of the application of common optical image processing methods to SAR images [5]. It thus restricts the SAR images to further understanding and interpretation, increasing the difficulty of extracting roads, farmlands, and buildings in the image and the complexity of spatial feature extraction in image registration, and reducing the accuracy of detection and classification of the objects such as vehicles and ships [6]. Speckle suppression is, therefore, an important task in SAR image post-processing.

To improve the quality of SAR images, there have been various speckle suppression methods proposed, including multi-look processing technologies during imaging and image filtering methods after imaging [1,7]. Multi-look processing divides the whole effective synthetic aperture length into multiple segments. The incoherent sub-views are then superimposed to obtain the high signal-to-noise ratio (SNR) images [8]. However, multi-look processing reduces the utilization of Doppler bandwidth, resulting in a decrease of the spatial resolution of the imaging results, which cannot meet the requirements of high resolution [9].

The filtering methods are mainly divided into three categories: the spatial filtering method, the transform domain filtering method, and the non-local mean filtering method. Median filtering and mean filtering are the earliest spatial filtering methods of traditional digital image processing. Although these two methods can suppress speckles to a certain extent, it leads to image blurring and objects edge information loss. Afterward, Lee filter [10], Frost filter [11], and Kuan filter [12] are designed for speckles suppression of SAR images. Based on the coherent speckle multiplier model, Lee filter selects a fixed local window in the image, assuming that the prior mean and variance can be calculated by the local region [10]. This method has a small amount of computation, but the selection of local window size has a great influence on the result, and the details and edge information of the image may be lost [13]. Frost filter assumes that the SAR image is a stationary random process and coherent spot noise is multiplying noise, and uses the least mean square error criterion to estimate the real image [11]. For the reason that the actual SAR image does not fully meet the hypothesis, the SAR image processed by this method will have blurred edges in areas with rich details. Kuan filters to apply sliding windows to estimate the local statistical properties of the image and then replaces the global characteristics of the image with these local statistical properties [12].

The representatives of transform domain filtering methods are threshold filtering method and multi-resolution analysis method based on wavelet transform. Donoho et al. first proposed a hard threshold and soft threshold denoising method based on wavelet transform [14]. After that, Bruce and Gao et al. proposed semi-soft threshold function methods to improve the hard threshold and soft threshold denoising methods [15,16]. This method solved the problems of discontinuity of hard threshold function and constant deviation of reconstructed signal with soft threshold function. He et al. later proposed a wavelet Markov model [17], which achieved a significant result in SAR images denoising. However, the wavelet transform cannot deal with two- and higher-dimensional images well. Because in the case of high-dimensional wavelet basis, one-dimensional wavelet basis cannot obtain the optimal representation of the two-dimensional function. In recent years, the appearance of multi-scale geometric analysis has made up this defect. Also, multi-scale geometric analysis tools are abundant, such as Ridgelets transformation, Curvelets transformation, Brushlets transformation, and Contourlets transformation [18–20]. These transform domain filtering methods have good coherent spots suppression capability, which preserved image details and edge information while speckles are removed. However, the processing of these algorithms is based on local characteristics, which is complex with a large amount of computation and easily producing pseudo-Gibbs stripes.

The non-Local Means (NLM) filtering method [21] proposed by Buades et al. repeatedly searched the whole image with similar image blocks and used similar texture regions instead of noise regions to achieve denoising. The authors of [22,23] applied NLM to SAR image denoising, which can effectively eliminate the speckles. Kervrann et al. [24] further improved NLM by proposing a new adaptive algorithm that modified the similarity measurement parameters of NLM. Dabov et al. [22] proposed a Block-Matching and 3D filtering (BM3D) algorithm which applied the local linear minimum mean variance (MMSE) criterion and wavelet transform, and combined the non-local mean idea with the transform domain filtering method. It is one of the best methods for denoising at present. However, this algorithm needs a large number of search operations at the cost of a large amount of computation and low efficiency.

In recent years, deep convolutional neural network (CNN) has developed rapidly, which provides a new idea for SAR image despeckling. Wang et al. constructed an image despeckling convolutional neural network (ID-CNN) in [25]. ID-CNN can directly estimate the speckle distribution and eliminate the estimated speckles from the image to obtain a clean image. Different from the ID-CNN, Yue et al. in [6], combining the statistical model with CNN, proposed a framework that does not require reference images and could work in an unsupervised way when trained with real SAR images. Bai et al. [26] added fractional total variational loss to the loss function to remove the obvious noise while maintaining the texture details. The authors of [27] proposed a CNN framework based on dilated convolutions called SAR-DRN. This network amplified the receptive field by dilated convolutions and further improved the network by exploiting the skip connections and a residual learning strategy. State-of-the-art results are achieved in both quantitative and visual assessments.

In this study, we design an end-to-end multi-scale recurrent network for SAR image despeckling. Unlike [9,25–27], which only utilized CNN to acquire speckle distribution characteristics and additional division operation or subtraction operation to remove speckle, we use the network to learn the distribution characteristics of speckle noise, meanwhile automatically implementing speckle suppression to output clean images. The proposed network is based on the encoder–decoder architecture. To improve the operation efficiency, in the decoder part, we use the subpixel unit to implement up-sample on the feature maps instead of the deconvolutional layer. Besides, this paper applies a multi-scale recurrent strategy, which inputs the resized images with different scales to the network, and different scale inputs share the same network weights parameters. So, the network performance can be improved without increasing the network parameters and the output which is friendly to the optical image processing algorithm can be obtained. Also, the convolutional LSTM unit is used to implement information transmit among each scale. Although our network is the same as the network based on noise output, i.e., a fully convolutional network, our MSR-net contains the pooling layer that can reduce the dimension of the network and further reduce the amount of computation to a great extent. Lastly, we propose two evaluation criteria based on image processing methods.

The paper consists of 6 parts. In Section 2, we analyze the speckles of SAR images and briefly introduce CNN, and convolutional LSTM. After providing the framework of our proposed MSR-net in Section 3, the result and discussion of the experiment are shown in Sections 4 and 5. The last section will summarize this paper.

2. Review of Speckle Model and Neural Network

2.1. Speckle Model of Sar Images

Multiplicative model is usually used to describe speckle noise [28] and the formula is defined as:

$$I = p_s \cdot n, \tag{1}$$

where I is image intensity, p_s is a constant which denotes the average scattering coefficient of objects or ground, and n denotes the speckle which is independent with p_s statistically.

For the homogeneous SAR image, the single-look intensity I obeys negative exponential distribution [29] and its probability distribution function (PDF) is defined as:

$$p(I) = \frac{1}{p_s} \exp\left(-\frac{I}{p_s}\right). \tag{2}$$

The multi-look processing methods are usually used to improve the quality of SAR images by diminishing the speckle noise. If the Doppler bandwidth is divided into L sub-bandwidths during imaging, and I_i is the single-look intensity image corresponding to each sub-bandwidth, the result of multi-look processing is:

$$I = \frac{1}{L} \sum_{i=1}^{L} I_i, \tag{3}$$

where L is the number of looks. If I_i obeys the exponential distribution in Equation (2), then after multi-look averaging, the L-look intensity image follows the Gamma distribution [1], and the PDF is:

$$p\left(I\right) = \frac{1}{p_s\Gamma(L)}\left(\frac{LI}{p_s}\right)^{L-1}\exp\left(-\frac{LI}{p_s}\right), L \geq 1,\tag{4}$$

where $\Gamma(L)$ denotes the Gamma function. The PDF of L-look speckle n can be obtained by applying the product model on Equations (1) and (4),

$$p(n) = \frac{L^L n^{L-1}\exp\left(-Ln\right)}{\Gamma(L)}, L \geq 1.\tag{5}$$

2.2. Convolutional Long Short-Term Memory

Convolutional neural networks (CNNs) have powerful capabilities of extracting spatial features and can automatically extract universal features through back-propagation algorithms driven by dataset [30,31], however they cannot be used to process sequence signals directly for the reason that the input is independent with each other, and the information flows strictly in one direction from layer to layer.To solve this problem, we introduce convolutional long short-term memory (ConvLSTM) [32] to the network, the inner structure of ConvLSTM is shown in Figure 1. As a special kind of RNN, long short-term memory network (LSTM) has internal hidden memory which allows the model to store information about its past computations and capable of learning long-term dependency [33]. Different than standard LSTM, all of the features variables of ConvLSTM including the input X_t, cell state C_t, the output of the forget gate F_t, input gate i_t, and output gate O_t are three-dimensional tensors, the latter two of which are spatial dimensions width and height. The key equations of ConvLSTM are defined as:

$$\begin{aligned}
F_t &= \sigma(W_f[H_{t-1}, X_t] + b_f),\\
i_t &= \sigma(W_i[H_{t-1}, X_t] + b_i),\\
\tilde{C}_t &= \tanh(W_C[H_{t-1}, X_t] + b_c),\\
C_t &= F_t * C_{t-1} + i_t * \tilde{C}_t,\\
O_t &= \sigma(W_o[H_{t-1}, F_t] + b_o),\\
H_t &= O_t * \tanh(C_t),
\end{aligned}\tag{6}$$

where "$*$" and "σ" denote the Hadamard product and the logistic sigmoid function, respectively. F_t controls the abandoned state information of the last layer and i_t is in charge of current state update, i.e., $\tilde{C}_t$. W_f, W_i, W_c, and W_o represent weights of each neural unit with b_f, b_i, b_c, and b_o denoting the corresponding offsets.

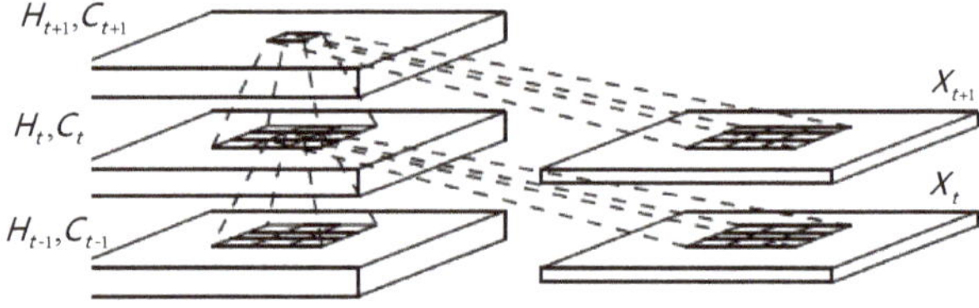

Figure 1. Inner structure of convolutional long short-term memory (ConvLSTM) [32].

3. Proposed Method

An end-to-end network MSR-net for SAR image despeckling is proposed in this paper. Rather than using additional division operation [25,26] or subtraction operation [9,27], our network can automatically perform despeckling and generate a clean image. In this section, we first introduce the multi-scale recurrent architecture and then describe specific details through a single-scale model.

3.1. Architecture

MSR-net is built based on cascaded subnetworks, and each subnetwork contains three parts: encoder, decoder, and ConvLSTM unit, as illustrated in Figure 2. Different levels of subnetworks correspond to different scales of inputs and outputs. The next scale speckled image and the output of current subnetwork are combined as the input of next-level subnetwork. In addition, an LSTM unit with single input and two-output is embedded between encoder and decoder. Specifically, one output is connected to the decoder, and the other output which represents the hidden state is connected to LSTM unit of the next subnetwork.

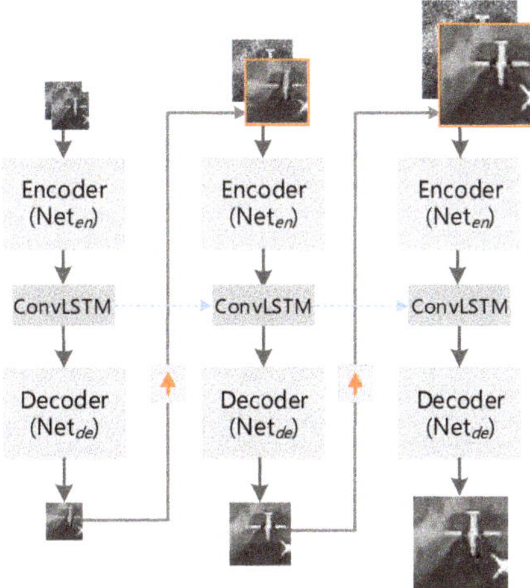

Figure 2. The architecture of Multi-scale Recurrent. The yellow arrow denotes up-sampling by one time.

Different from the general cascaded network like [34], which uses three stages of independent subnetworks, all the state features flow across scales and share the same training parameters in MSR-net. Owing to the multi-scale recurrent and parameter share strategy, the number of parameters that need to be trained in MSR-net is only 1/3 of [34].

For the subnetwork, the output $\mathbf{F}^i$ of encoder $\mathbf{Net}_{en}$, which takes the speckled image and despeckled result up-sampled from the previous scale as input, can be defined as:

$$\mathbf{F}^i = \mathbf{Net}_{en}\left(\mathbf{I}^i_{in}, up(\mathbf{I}^{i+1}_0); \Theta_{en}\right),$$
(7)

where $\mathbf{I}^i_{in}$ is the input image with speckle noise, Θ_{en} is the weights parameters of $\mathbf{Net}_{en}$. $i = 0, 1, 2, ...$ is the scale index. The larger i is, the lower the resolution is. $i = 0$ represents the original resolution and $i = 1$ indicates down sampling once. $\mathbf{I}^{i+1}_0$ is the output of the previous coarse scale. $up(\cdot)$ is the operator that adapts features or images from the (i + 1)-th to the i-th scale, which is implemented by bilinear interpolation.

To exploit the information contained in feature maps of different scales, a convolutional LSTM module is embedded between the encoder and the decoder. The ConvLSTM can be defined as:

$$\mathbf{h}^i, \mathbf{g}^i = \mathbf{ConvLSTM}\left(up(\mathbf{h}^{i+1}), \mathbf{F}^i; \Theta_{LSTM}\right),\qquad(8)$$

where Θ_{LSTM} is the set of parameters in ConvLSTM, $\mathbf{h}^i$ is the hidden state which is passed to the next scale, $\mathbf{h}^{i+1}$ is the hidden state from the previous scales, and $\mathbf{g}^i$ is the output of the current state (scale). Finally, we use Θ_{de} to denote the parameters of the decoder, and the output can be defined as:

$$\mathbf{I}_o^i = \mathbf{Net}_{de}\left(\mathbf{g}^i; \Theta_{de}\right).\qquad(9)$$

3.2. Single Scale Network

Details of the MSR-net are introduced by the single-scale model in this section. As shown in Figure 3, the single-scale model consists of two parts: encoder and decoder. The encoder includes three building blocks: convolutional layer, pooling layer, and Res block.

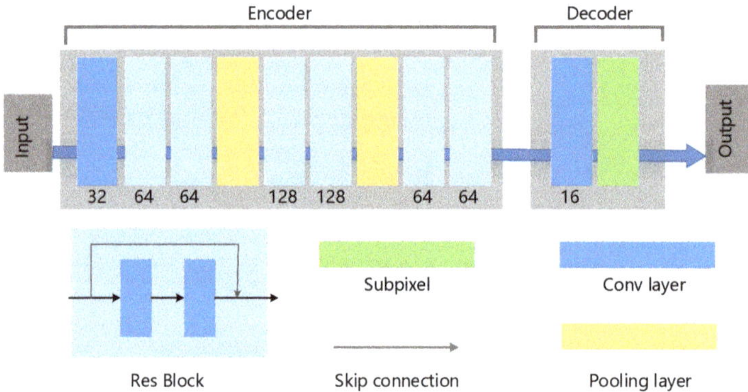

Figure 3. Architecture of Single Scale Modle

The convolution unit performs convolution operation and non-linear activation. Increasing the number of convolutional layers can enhance the feature extraction ability [35,36]. Multiple Res blocks are added after the convolutional layer while designing the network. Unlike the convolution unit, skip connection proposed by He et al. [37] is built into this block, which can effectively avoid gradient explosion or gradient disappearance, as well as increasing the training speed.

The size of the input and output of the convolutional layers keeps the same as the despeckling networks designed in [6,25,27], which increases the amount of computation to a certain extent. We reduce the amount of calculation by decreasing the dimension of the feature maps, i.e., adopting the pooling layer. We choose max pooling operation with the 2×2 pooling kernel in this layer. It should be noted that the pooling layer can also be replaced by strided convolutions [38].

The decoder consists of the convolutional layer and the sub-pixel units. The width and height of the input feature map to the decoder are only 1/4 of the original image after down-sampling twice through the pooling layer. Therefore, the up-sampling operation is required to make the output image of the network the same as the input size. However, an up-sampling operation such as transposed convolution used in [39,40] needs a high amount of computation and causes unwanted checkerboard artifacts [41,42]. A typical checkerboard pattern of artifacts is shown in Figure 4. To reduce the network runtime and avoid the checkerboard pattern of artifacts, the sub-pixel convolution described in Section 3.3 is used to implement the up-sampling operation.

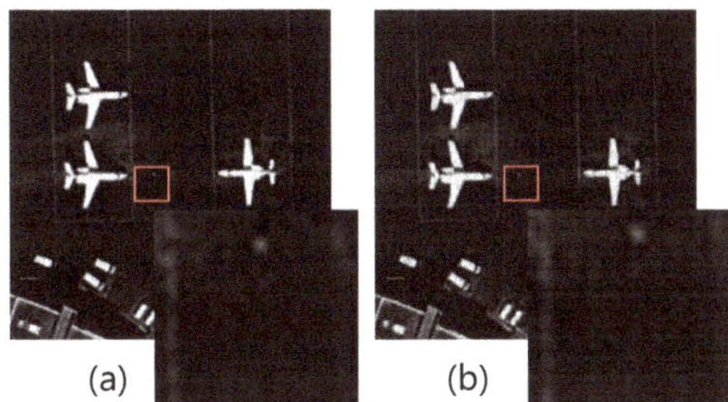

Figure 4. Diagram of checkerboard artifacts. (**a**) original image without noise; (**b**) despeckling image with checkerboard artifacts caused by transposed convolution.

3.3. Sub-Pixel Convolution

Sub-pixel convolution, also called as pixel shuffle, is an upscaling method first proposed in [43] for image super-resolution tasks. Different from the commonly used up-sampling methods in deep learning such as transposed convolution and fractionally strided convolution, sub-pixel convolution adopts channel to space method which achieves spatial scale-up amplification by rearranging pixels in multiple channels of the feature map, as illustrated in Figure 5.

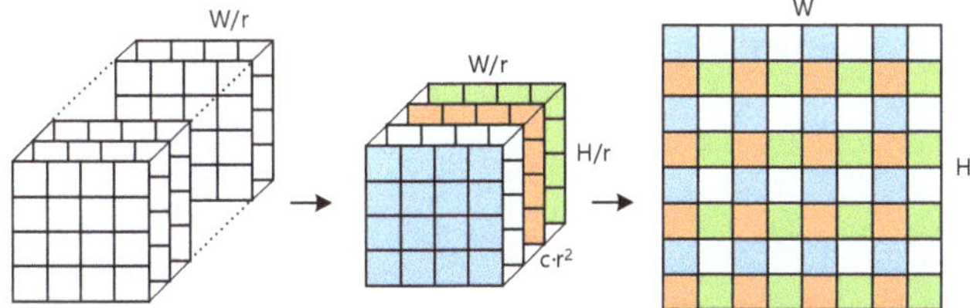

Figure 5. The sub-pixel convolutional operation on the input feature maps with an upscaling factor of r = 2, channel c = 1.

For a sub-pixel unit with r times up-sampling, its output image is defined as $\mathbf{I}^{up}$, and we have $\mathbf{I}^{up} \in \mathbb{R}^{W \times H \times c}$, in which W, H, and c denote the width, height and channels of $\mathbf{I}^{up}$. The sub-pixel convolution operation is defined as:

$$\mathbf{I}^{up}(x,y,c) = \mathbf{F}(\lfloor x/r \rfloor, \lfloor y/r \rfloor, c \cdot r \cdot \mathrm{mod}\,(y,r) + c \cdot \mathrm{mod}\,(x,r) + c) \tag{10}$$

where $\mathbf{I}^{up}(x,y,c)$ is the value of the pixel at the position (x,y) for the cth channel. $\mathbf{F}$ is the input of sub-pixel and $\mathbf{F} \in \mathbb{R}^{W/r \times H/r \times cr^2}$. $\lfloor \cdot \rfloor$ represents floor function that takes as input a real number and gives as output the greatest integer less than or equal to it [43]. After sub-pixel convolution operation, the elements of $\mathbf{F}$ are rearranged to the output $\mathbf{I}^{up}$ by increasing the horizontal and vertical count, and decreasing channel count. For example, when a $64 \times 64 \times 4$ feature map is passed through the sub-pixel unit, an output with shape $128 \times 128 \times 1$ will be obtained.

3.4. Proposed Evaluation Criterion

In this paper, the peak signal to noise ratio (PSNR) [44], structural similarity (SSIM) [45], equivalent number of looks (ENL) [46], and two new proposed evaluation criterions edge feature keep ratio (EFKR) and feature point keep ratio (FPKR) are used to evaluate the performance of despeckling methods.

PSNR is the ratio between the maximum possible power of a signal and the power of corrupting noise that affects the fidelity of its representation, which has been widely used in quality assessment of reconstructed images. SSIM is a metric of image similarity. ENL can describe the smoothness of regions, and no reference image is needed for its calculation, so it can be used to evaluate the performance of despeckling methods for real SAR images.

3.4.1. Edge Feature Keep Ratio and Feature Point Keep Ratio

PSNR and SSIM can effectively evaluate the overall performance of despeckling methods. Specifically, PSNR measures noise level or image distortion, SSIM measures the similarity between two images, and ENL measures the degree of region smoothing. They are not, however, capable of evaluating the edge and typical features retention ability in despeckling tasks directly. In this section, we propose two evaluation criteria that can compensate for the above deficiencies, i.e., edge feature keep ratio (EFKR) and feature point keep ratio (FPKR).

(a) **EFKR:** from the edge detection results shown in Figure 6, we have the following observations. (1) The edge outline of the speckled image is blurred, and there are discrete points in the image; (2) the edge outline is clear after despeckling and there is no discrete point, which is in agreement with the edge detection results of a clean image. Enlightened by this phenomenon, we design a quantitative evaluation criterion EFKR with the ability of edge retention based on counting the number of pixels of edges. The computation steps are as follows:

1. Edge detection processing for clean and test image using edge detection algorithms such as Sobel [47], Canny [48], Prewitt [49], and Roberts [50] methods. After this, two images only with only edge lines are obtained. The values of pixels in the edge position are set to 1, the values of the other position are set to 0.
2. Bit-wise **and** operation on two images from step 1, values at the position where the edges coincide are set to 1, values at other locations are set to 0.
3. Count the number of value 1 in the edge detected the result of a clean image and the number of 1 in step 2, and calculate the ratio of the two numbers.

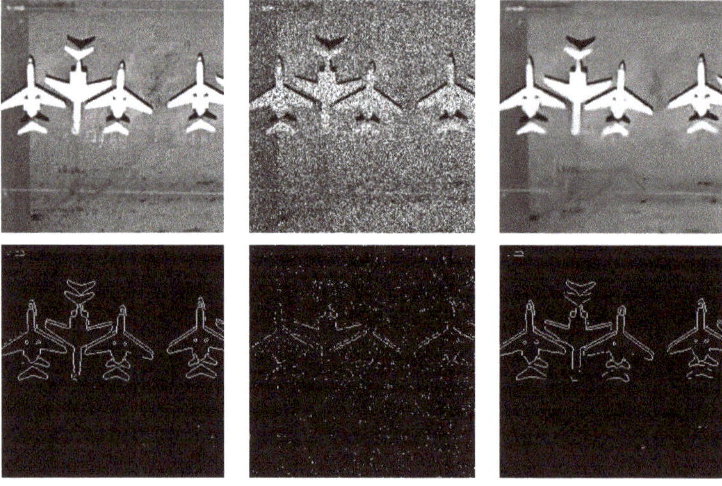

Figure 6. Edge detection results. Images from left to right correspond to a clean image, an image with speckle noise, and image after despeckling.

The ratio of these two factors is the edge feature retention ratio, which is defined as:

$$EFKR = \frac{sum(\mathbf{edge}(X)\&\mathbf{edge}(Y))}{sum(\mathbf{edge}(Y))},$$ (11)

where $\&$ and *sum* denotes the bit-wise conjunction operation and sum operation, and **edge** represents edge detection.

(b) **FPKR:** for real SAR images, ENL is only able to evaluate the smooth level but not the retention ability of typical features such as edges, corners in the image. SIFT [51] can find feature points from different scales and obtain the ultimate descriptor of features. Also, the key points found by SIFT are usually corner points, edge points, bright spots in dark areas, and dark points in bright areas. These points are robust to light, affine transformations, and other transformation. The registration method based on SIFT first uses SIFT to obtain the feature points of the image to be registered, the reference image and their descriptor then matches the feature points according to descriptor and obtains one-to-one corresponding feature point pairs. Finally, the transformation parameters are calculated, and the image registration is carried out.

For SAR images, the registration of feature points and descriptors at the lights spots of speckles are redundant, which also reduce the efficiency and accuracy of subsequent searching of matching points. Based on this phenomenon, we design an evaluation criterion FPKR targeting at key feature points. We first execute an affine transformation to the evaluation image, then use the SIFT algorithm to find the feature points in the two images before and after the transformation, and finally match the feature points. The better the despeckling performs, the more typical features are preserved. The more prominent the feature descriptor obtained by SIFT, the greater the difference of descriptor between different features, so more effective feature point pairs can be searched efficiently. FPKR is defined as:

$$FPKR = \frac{N_{match}(X, X_t)}{\min(N(X), N(X_t))}$$ (12)

where $N(X), N(X_t)$ are the number of key points before and after SIFT, and $N_{match}(X, X_t)$ denotes the number of points for calculating transformation parameters.

4. Experiments and Results

4.1. Dataset

Because it is hard to collect real SAR images without speckle noise, we train networks by using synthetic noise/clean image pairs. Public dataset UC Merced Land Use Dataset (http://weegee.vision.ucmerced.edu/datasets/landuse.html) is chosen as the original clean image for training. The dataset contains 21 scene classes with 100 optical remote sensing images per class. Each image has a size of 256×256 pixels and the pixel resolution is 1 foot [52]. According to [27], we randomly select 400 images from the dataset as the training set and use the remaining images for testing. Some training samples are shown in Figure 7. Finally, after grayscale preprocessing, the speckled images are generated using Equation (1) same as the [25,53]. The noise levels (L = 2, 4, 8, 12) correspond to the number of looks in SAR, and the code of adding speckle noise is available on GitHub (https://github.com/rcouturier/ImageDenoisingwithDeepEncoderDecoder/tree/master/data{_}denoise).

Figure 7. Part of the sample images used to train the network.

4.2. Experimental Settings

All the networks are trained with stochastic gradient descent (SGD) with a mini-batch size of 32. All weights are initialized by a modified scheme of Xavier initialization [54] proposed by He et al. Meanwhile, we use the Adam optimizer [55] with tuned hyper-parameters to accelerate training. The hyper-parameters are kept the same across all layers and all networks. Experiments are implemented on TensorFlow platform with Intel i7-8700 CPU and an NVIDIA GTX-1080(8G) GPU.

The details of the model are specified here. The number of kernels in each unit is shown in Figure 3. The kernel sizes for the first and last convolutional layers are 5×5, while all others are 3×3. Rectified Linear Units (ReLU) are used as the activation function for all layers except for the last convolutional layer before sub-pixel unit. The $\mathcal{L}_1$ loss is chosen to train the network, which is defined as:

$$\mathcal{L}_1(\Theta) = \frac{1}{N} \sum_{y=1}^{W} \sum_{x=1}^{H} |\varphi\left(\mathbf{X}(x,y); \Theta\right) - \mathbf{C}(x,y)|, \tag{13}$$

where Θ is the filter parameters that need to be updated during training, $\mathbf{C}$, $\mathbf{X}$ and $\varphi\left(\cdot\right)$ denote the objective image without noise, the input image with speckle noise, and the output after despeckling, respectively.

4.3. Experimental Results

The test results of our proposed network will be presented in this section. To verify the proposed method, we compare the performance of our MSR-net with other three despeckling methods, SAR-BM3D [22], ID-CNN [25], and Residual Encoder-Decoder network (RED-NET) [53]. The first one is a traditional nonlocal algorithm based on wavelet shrinkage, and the latter two methods are based on deep convolutional neural networks.

4.3.1. Results on Synthetic Images

Building, freeway, and airplane, three classes of synthetic images, are chosen as the test set to evaluate the noise reduction ability of each method. Part of the processing results of different algorithms under different levels of noise are shown in Figures 8 and 9.

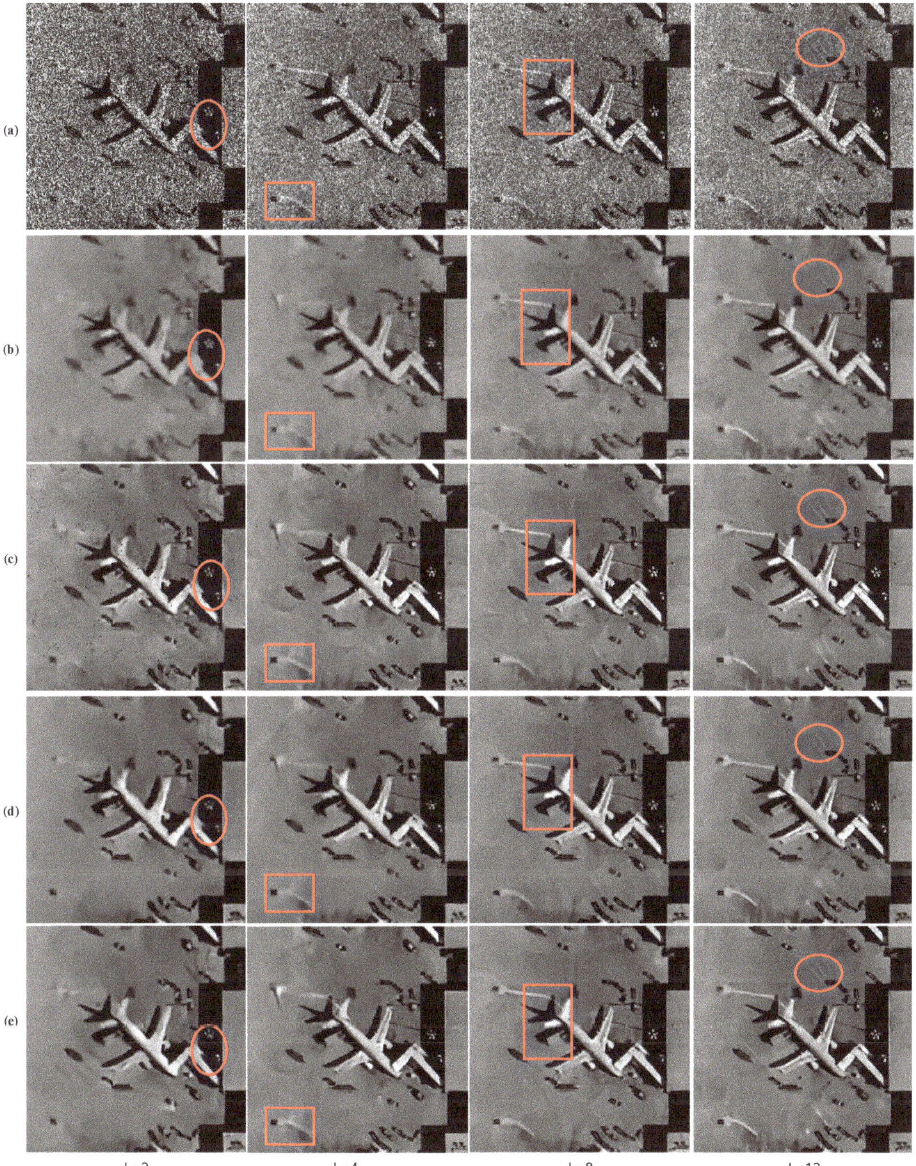

Figure 8. Test results on images of airplanes with four levels of noise. The noise level from left to right is L = 2, L = 4, L = 8, and L = 12. (**a**) Speckled image (**b**) SAR-BM3D (Block-Matching and 3D filtering) (**c**) image despeckling convolutional neural network (ID-CNN) (**d**) RED-Net (**e**) multi-scale recurrent network (MSR-net).

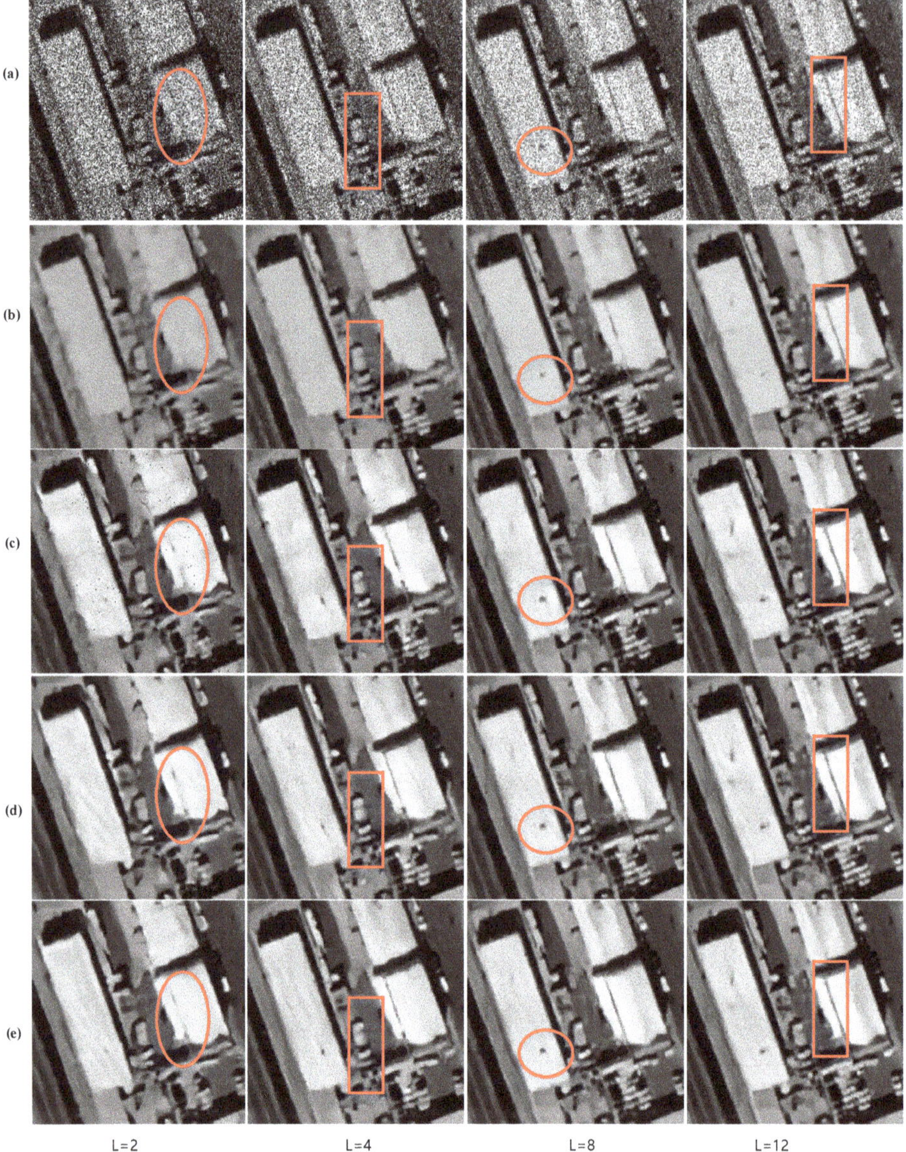

L=2   L=4   L=8   L=12

Figure 9. Test results on images of buildings with four levels of noise. The noise level from left to right is L = 2, L = 4, L = 8, and L = 12. (**a**) Speckled image (**b**) SAR-BM3D (**c**) ID-CNN (**d**) RED-Net (**e**) MSR-net.

From the figures, we can observe that the CNN-based methods, including our MSR-net, can preserve more details like texture features in images than SAR-BM3D after despeckling. When the noise is strong, the SAR-BM3D algorithm will cause blurring at the edge of the objects.

ID-CNN has a good performance on image despeckling, however, after filtering by the network, pepper and salt noise appear in the image, which needs to be processed subsequently by using nonlinear filters such as median filtering and pseudo-median filtering. As the noise intensity increases, the salt and pepper noise increase gradually.

MSR-net has excellent retention performance of spatial geometry features like texture features, lines, and feature points. Compared with the other three algorithms, MSR-net has a higher smoothness of smooth areas as well as a smaller loss of sharpness of edges and details, especially for strong speckle noise. Also, more detail information in image will loose when the speckle noise is strong and more local detail can be preserved in output images when the speckle noise is weak.

When the level of noise added to the test set is small, all the CNN-based approaches can get state-of-art results. Therefore, it is difficult to judge the merits of these algorithms by using visual assessments. Experimental results of evaluation indexes such as PSNR and SSIM are necessary for these circumstances. The PSNR, SSIM, and EFKR evaluation indexes of the above methods are listed in Tables 1–4, respectively. The bold number represents the optimal value in each row, while the underlined number denotes the suboptimal value. We also test MSR-net with only one scale and call it a single scale network (SS-net) during the experiment.

Table 1. The peak signal to noise ratio (PSNR), structural similarity (SSIM), and edge feature keep ratio (EFKR) of test set with noise level L = 2.

Classes	Index	SAR-BM3D	ID-CNN	RED-Net	SS-Net	MSR-Net
building	PSNR	20.343 ± 0.043	21.639 ± 0.073	22.321 ± 0.065	22.467 ± 0.039	**22.768 ± 0.052**
	SSIM	0.579± 0.005	0.599 ± 0.002	0.672 ± 0.005	0.663 ± 0.002	**0.684 ± 0.004**
	EFKR	0.3872± 0.061	0.3888 ± 0.058	0.5131 ± 0.053	0.4950± 0.040	**0.5322 ± 0.049**
freeway	PSNR	22.478 ± 0.081	23.17± 0.046	25.157± 0.033	25.255 ± 0.057	**25.526± 0.064**
	SSIM	0.581 ± 0.002	0.536 ± 0.003	0.662 ± 0.003	0.663 ± 0.048	**0.680 ± 0.002**
	EFKR	0.3046± 0.052	0.3376 ± 0.061	**0.5071 ± 0.061**	0.4929 ± 0.064	0.5065 ± 0.057
airplane	PSNR	21.351± 0.045	24.006 ± 0.058	23.97± 0.062	24.166 ± 0.046	**24.433± 0.049**
	SSIM	0.602 ± 0.004	0.641 ± 0.003	0.638 ± 0.001	0.636 ± 0.002	**0.649 ± 0.005**
	EFKR	0.3048± 0.069	0.2965 ± 0.073	0.4929 ± 0.063	0.4800 ± 0.058	**0.5123 ± 0.065**

The bold number represents the optimal value in each row, while the underlined number denotes the suboptimal value in each row.

Table 2. The PSNR, SSIM, and EFKR of test set with noise level L = 4.

Classes	Index	SAR-BM3D	ID-CNN	RED-Net	SS-Net	MSR-Net
building	PSNR	21.789 ± 0.046	24.319 ± 0.101	24.366 ± 0.059	24.279 ± 0.032	**24.370± 0.038**
	SSIM	0.694 ± 0.003	0.721 ± 0.005	**0.737 ± 0.006**	0.722 ± 0.001	0.735 ± 0.005
	EFKR	0.5029± 0.059	0.5459± 0.057	0.5646± 0.050	0.5742± 0.0423	**0.6021± 0.051**
freeway	PSNR	24.351 ± 0.059	26.008 ± 0.051	26.821 ± 0.063	26.734 ± 0.030	**26.893 ± 0.049**
	SSIM	0.645 ± 0.003	0.670 ± 0.002	0.718 ± 0.005	0.711 ± 0.003	**0.722 ± 0.003**
	EFKR	0.4897± 0.061	0.5019± 0.060	0.5469± 0.057	0.5511± 0.065	**0.5608± 0.051**
airplane	PSNR	23.261 ± 0.041	25.747 ± 0.067	25.849 ± 0.050	25.905 ± 0.046	**26.046 ± 0.064**
	SSIM	0.657 ± 0.003	0.677 ± 0.005	0.691 ± 0.004	0.686 ± 0.002	**0.693 ± 0.005**
	EFKR	0.4976± 0.061	0.5149 ± 0.058	0.5484 ± 0.056	0.5728 ± 0.039	**0.5871 ± 0.055**

The bold number represents the optimal value in each row, while the underlined number denotes the suboptimal value in each row.

Table 3. The PSNR, SSIM, and EFKR of test set with the noise level L = 8.

Classes	Index	SAR-BM3D	ID-CNN	RED-Net	SS-Net	MSR-Net
building	PSNR	23.851 ± 0.071	26.004 ± 0.055	25.685 ± 0.043	25.989 ± 0.064	**26.026 ± 0.048**
	SSIM	0.752 ± 0.003	0.780 ± 0.002	0.769 ± 0.05	0.774 ± 0.004	**0.782 ± 0.004**
	EFKR	0.5992 ± 0.065	0.6191 ± 0.055	0.6305 ± 0.050	0.6261 ± 0.0525	**0.6476 ± 0.060**
freeway	PSNR	26.374 ± 0.081	27.738 ± 0.049	27.785 ± 0.068	28.027 ± 0.053	**28.111 ± 0.057**
	SSIM	0.714 ± 0.003	0.740 ± 0.006	0.751 ± 0.006	0.753 ± 0.004	**0.757 ± 0.003**
	EFKR	0.5432 ± 0.041	0.5809 ± 0.053	**0.5990 ± 0.052**	0.5890 ± 0.057	0.5954 ± 0.048
airplane	PSNR	25.31 ± 0.064	27.438 ± 0.051	27.161 ± 0.044	27.404 ± 0.061	**27.443 ± 0.042**
	SSIM	0.71 ± 0.005	0.731 ± 0.004	0.720 ± 0.004	0.728 ± 0.005	**0.732 ± 0.002**
	EFKR	0.5872 ± 0.057	0.6039 ± 0.047	0.6113 ± 0.053	0.6187 ± 0.038	**0.6293 ± 0.051**

The bold number represents the optimal value in each row, while the underlined number denotes the suboptimal value in each row.

Table 4. The PSNR, SSIM, and EFKR of test set with the noise level L = 12.

Classes	Index	SAR-BM3D	ID-CNN	RED-Net	SS-Net	MSR-Net
building	PSNR	24.979 ± 0.060	26.891 ± 0.043	26.879 ± 0.056	26.929 ± 0.071	**26.942 ± 0.052**
	SSIM	0.781 ± 0.004	0.805 ± 0.0002	0.805 ± 0.004	0.802 ± 0.005	**0.806 ± 0.003**
	EFKR	0.6184 ± 0.051	0.6441 ± 0.049	0.6528 ± 0.047	0.6596 ± 0.054	**0.6664 ± 0.058**
freeway	PSNR	27.444 ± 0.042	28.643 ± 0.059	**28.905 ± 0.047**	28.787 ± 0.066	28.893 ± 0.050
	SSIM	0.747 ± 0.003	0.768 ± 0.006	**0.782 ± 0.007**	0.776 ± 0.004	0.778 ± 0.005
	EFKR	0.5975 ± 0.056	0.6130 ± 0.053	0.6218 ± 0.056	**0.6247 ± 0.067**	0.6197 ± 0.051
airplane	PSNR	26.463 ± 0.053	28.362 ± 0.64	28.352 ± 0.078	28.349 ± 0.059	**28.407 ± 0.065**
	SSIM	0.737 ± 0.005	0.756 ± 0.003	**0.758 ± 0.003**	0.754 ± 0.005	0.756 ± 0.006
	EFKR	0.6008 ± 0.059	0.6275 ± 0.057	0.6432 ± 0.052	**0.6465 ± 0.052**	0.6433 ± 0.053

The bold number represents the optimal value in each row, while the underlined number denotes the suboptimal value in each row.

Consistent with the results shown in Figures 8 and 9, our method has much better speckle-reduction ability than non-learned approach SAR-BM3D at different noise levels. In addition, the advantage of MSR-net will increase as the noise level increases. Taking airplane images for instance, the PSNR/SSIM/EFKR of our proposed MSR-net outperform SAR-BM3D by about 3.082 dB/0.047/0.2075, 2.785 dB/0.036/0.0895, 2.133 dB/0.022/0.0421, 1.944 dB/0.019/0.0425 for L = 2, 4, 8, 12.

Compared with CNN-based methods, MSR-net still has an advantage when the noise is strong. When L = 2, the PSNR/SSIM/EFKR of MSR-net outperform ID-CNN, RED-Net by about 1.129 dB/0.014/0.1434, 2.356 dB/0.144/0.1689, 0.427 dB/0.008/0.2152 and 0.447 dB/0.012/0.0191, 0.369 dB/0.018/-0.006, 0.643 dB/0.011/0.0194 for building, freeway, and airplane, respectively. When L = 4, the PSNR/SSIM/EFKR of MSR-net outperform ID-CNN, RED-Net by about 0.051 dB/0.134/0.0562, 0.885 dB/0.052/0.0589, 0.299 dB/0.016/0.0722 and 0.004 dB/-0.002/0.0375, 0.072 dB/0.004/0.0139, 0.197 dB/0.002/0.0143 for building, freeway and airplane, respectively.

In addition, we can see that our network does not always achieve the best test results. We consider this may have a certain relationship with the feature distribution of the images. Although MSR-net can only get sub-optimal test results for a certain class of images, the difference to the best result is small. For other classes of images, the advantages of MSR-net are more considerable. For example, we can observe from Table 1 that the EFKR of RED-Net only outperforms MSR-net by about 0.006 for freeway images, while MSR-net can outperform RED-Net by about 0.0191 and 0.0194 for building and airplane images. When the noise level L = 12, our network only gets four best test values, which suggests the advantage of the multi-scale network becomes smaller while the noise is weaker, as shown in Table 4.

MSR-net with a single scale (SS-net) also has very good speckle-reduction ability. When the noise intensity is weak, the performance is even better than multi-scale. For example, when L = 12,

PSNR/SSIM/EFKR of the freeway are 28.787 dB/0.776/0.6247 and 28.893 dB/0.778/0.6197 for SS-net and MSR-net, respectively. Ultimately, we can find by comparison that the edge detection effect of the image has been significantly improved after despeckling.

4.3.2. Results on Real Sar Images

To further verify the speckle-reduction ability of our network for real SAR images, two SAR sceneries are selected, as shown in Figures 10a and 11a, and these two images are imaging results of spaceborne SAR RADARSAT-2.

It can be seen by comparing the subgraphs in Figures 10 and 11 that MSR-net generates the visually best output among all the results and retains the edge sharpness as well as the detail information about the structure in the image while removing the speckle noise. After filtering by SAR-BM3D, the loss of edge sharpness of the original SAR image is obvious and most of the lines and texture feature are blurred. ID-CNN and RED-Net can generate smooth results in homogeneous regions while maintaining textural features in the image. However, from the red boxes, we can observe that they are capable of retaining some texture features but not better than MSR-net. Although SS-net performs well in despeckling for real SAR images, it is still worse than multi-scale MSR-net.

The ENL results are shown in Table 5. We can observe from the table that MSR-net has an outstanding performance for real SAR image despeckling. For these four evaluation regions, the three highest scores and one second highest score of the ENL are obtained by our MSR-net.

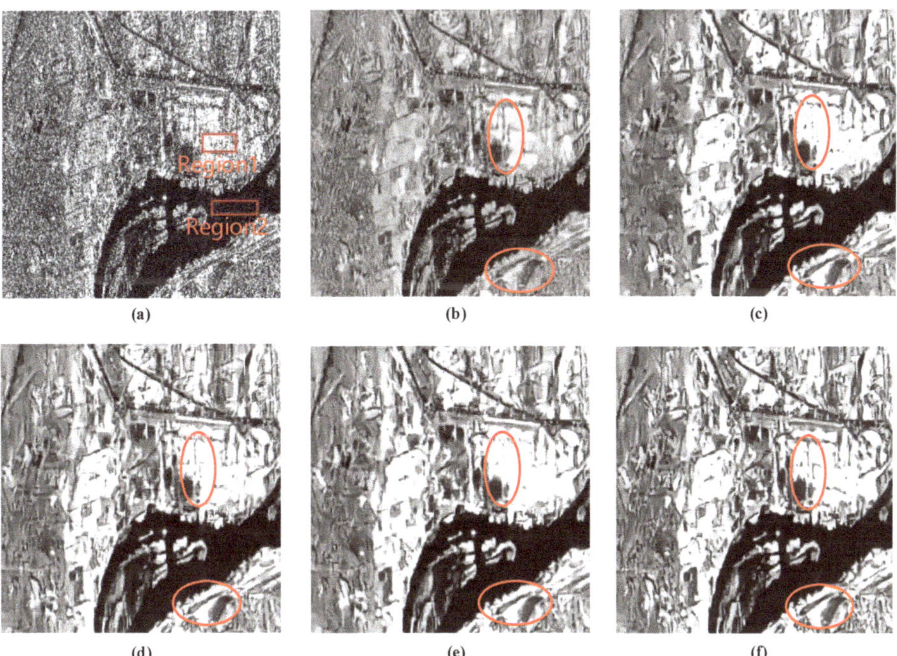

Figure 10. Test results of Real SAR images (RADARSAT-2). (**a**) Original, (**b**) SAR-BM3D, (**c**) ID-CNN, (**d**) RED-Net, (**e**) single scale network (SS-net), (**f**) MSR-net.

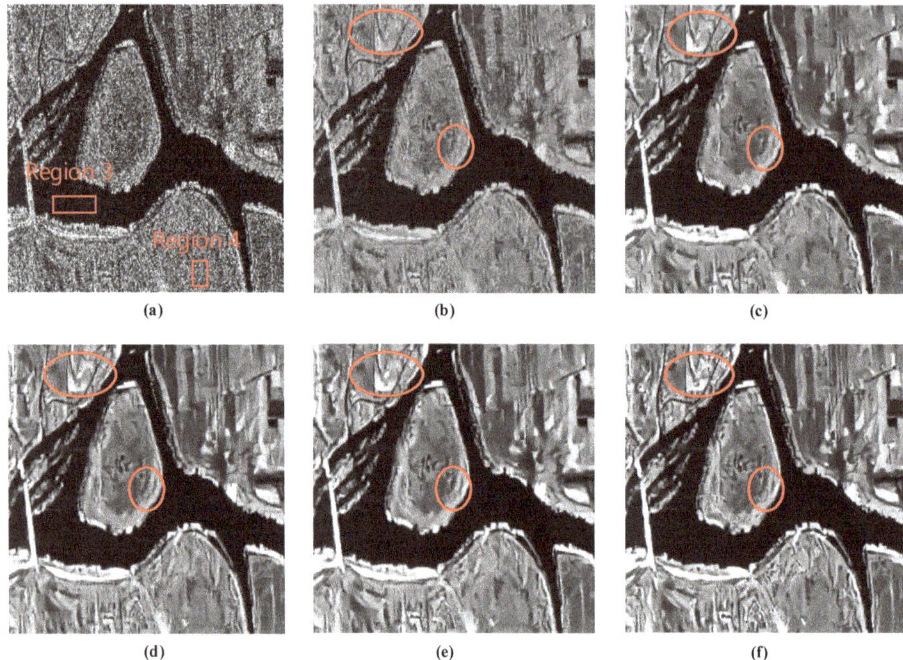

Figure 11. Test results of Real SAR images (RADARSAT-2). (**a**) Original, (**b**) SAR-BM3D, (**c**) ID-CNN, (**d**) RED-Net, (**e**) SS-net, (**f**) MSR-net.

Table 5. The equivalent number of looks (ENL) of real SAR regions

Data	Original	SAR-BM3D	ID-CNN	RED-Net	SS-Net	MSR-Net
region1	3	51	104	317	394	**540**
region2	51	195	173	413	**2906**	644
region3	3	80	101	333	339	**622**
region4	4	99	95	235	369	**560**

The bold number represents the optimal value in each row.

To achieve FPKR results for real SAR data with different methods, we first apply the same affine transformation to each image, as shown in Figure 12. SIFT is then applied to search for the feature points and calculate their descriptors. Matching key points are ultimately conducted by minimizing the Euclidean distance between their SIFT descriptors. Generally, the ratio between distances is used [56] to obtain high matching accuracy. In the experiments, we select three ratios and the FPKR results are shown in Tables 6 and 7. By comparing EFKR of each image, MSR-net performs better than SAR-BM3D. Also, MSR-net shows advantages over other neural network-based algorithms. Specifically, MSR-net achieves the best testing results in five out of six sets of experiments. It also indicates that pre-processing to SAR images by MSR-net can effectively enhance the usefulness of SIFT algorithm to SAR images and improve its performance and efficiency.

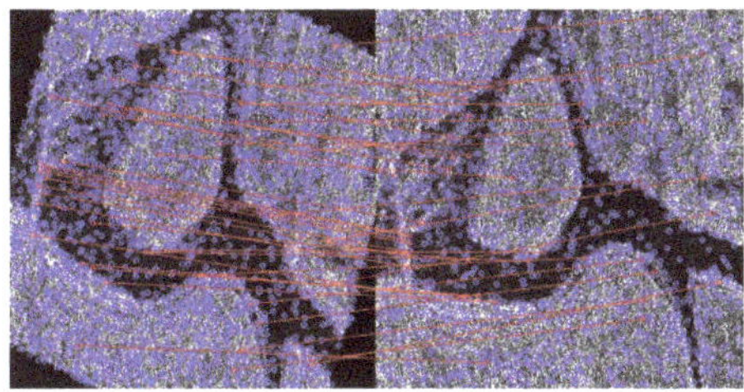 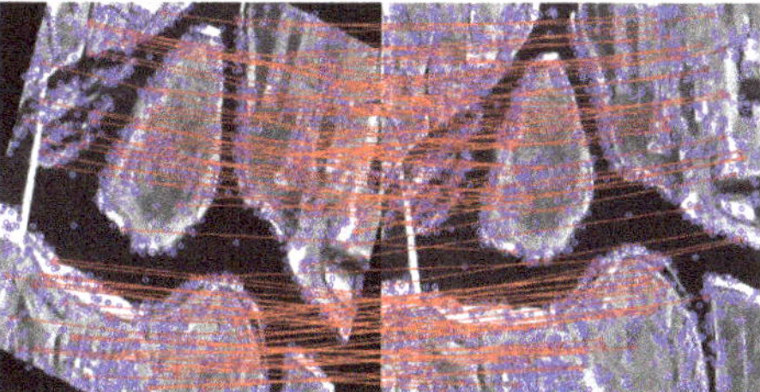

<table>
<tr><td>(a) Before despeckling.</td><td>(b) After despeckling.</td></tr>
</table>

Figure 12. Search results of feature point pairs for synthetic aperture radar (SAR) image before and after despeckling.

Table 6. Feature point keep ratio (FPKR) results before and after despeckling for real SAR image.

	fp1	fp2	FPKR (r = 0.25)	FPKR (r = 0.30)	FPKR (r = 0.35)
original	5637	4531	0.0004 (2)	0.0077 (35)	0.0362 (164)
SAR-BM3D	4554	5947	0.0046 (21)	0.0212 (110)	0.0894 (407)
ID-CNN	3233	3981	0.0049 (16)	0.0294 (95)	0.1148 (371)
RED-Net	3636	4644	0.0055 (20)	0.0292 (106)	0.0976 (355)
SS-Net	3429	4357	0.0052 (18)	0.0286 (98)	**0.1155** (396)
MSR-net	3080	4216	**0.0071** (22)	**0.0425** (106)	0.0981 (302)

The value inside brackets is the number of feature points. The bold number represents the optimal value in each column.

Table 7. FPKR results before and after despeckling for real SAR image.

	fp1	fp2	FPKR (r = 0.25)	FPKR (r = 0.30)	FPKR (r = 0.35)
original	4511	3476	0.0012 (4)	0.0118 (41)	0.0437 (152)
SAR-BM3D	3878	5289	0.0044 (17)	0.0273 (106)	0.0952 (369)
ID-CNN	2021	2533	0.0124 (25)	0.0574 (116)	0.1366 (267)
RED-Net	2137	2797	0.0140 (30)	0.0543 (116)	0.1526 (326)
MSR-net	2021	2533	**0.0166** (33)	**0.0674**(134)	**0.1579** (314)

The value inside brackets is the number of feature points. The bold number represents the optimal value in each column.

4.3.3. Runtime Comparisons

To evaluate the algorithm efficiency, we make statistics of the runtime of each algorithm in CPU implementation. The runtime of different methods on images with different sizes is listed in Table 8. We can see that the proposed denoiser is very competitive although its structure is relatively complex. Such a good compromise between speed and performance over MSR-net is properly attributed to the following two reasons. First, two pooling layers that can achieve spatial dimensionality reduction are embedded in the MSR-net. Each pooling layer with the 2×2 pooling kernel can reduce the amount of data that needs to be processed by the subsequent convolution operation to 25% of before. Second, in contrast to the transposed convolution which increases the resolution of feature maps by padding and complex convolution operation, sub-pixel unit, which up-samples feature maps by a periodic shuffling of pixel values, is adopted to build our network.

Table 8. Runtime (s) of different methods of images with sizes of 256×256 and 512×512.

Size	SAR-BM3D	ID-CNN	RED-Net	SS-Net	MSR-Net
256×256	12.79	0.448	2.241	0.339	0.586
512×512	52.813	1.562	8.943	1.342	2.375

5. Discussion

5.1. Choice of Scales

To select a proper scale, building, freeway, and airplane, three classes of test images are used to analyze. Figure 13 shows the testing results of networks with different scales. In the single-scale network, recurrent modules ConvLSTM are replaced by a convolution layer to keep the same number of convolution layers.

We can observe that as the scale increases, the values of the three evaluation metrics are all improved. It suggests that exploiting the multi-scale information can help with improving network performance. However, we can meanwhile find that the improvement is small while the scale is greater than 3. We thus choose $s=3$ in our network to balance the network performance and complexity.

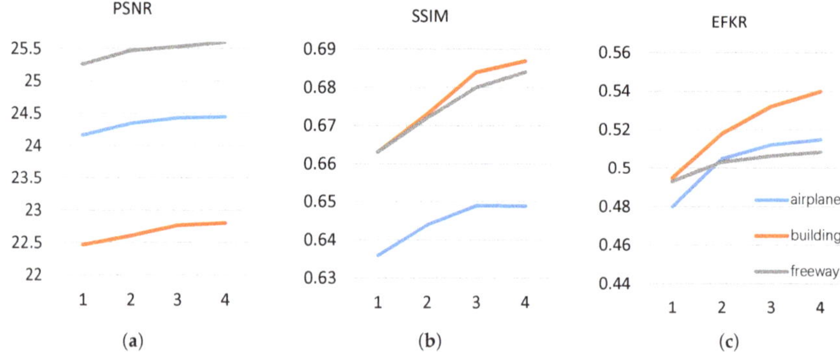

Figure 13. Test results of proposed network with scale 1, 2, 3, 4 when nosie level L = 2.

5.2. Loss Function

The influence of loss function on network performance is also discussed in this paper. Instead of $\mathcal{L}_1$ loss, $\mathcal{L}_2$ loss function is used to train our MSR-net. $\mathcal{L}_2$ norm loss, also called Euclidean loss, is the most commonly used loss function in despeckling tasks. It is defined as:

$$\mathcal{L}_2(\Theta) = \frac{1}{2N} \sum_{y=1}^{W} \sum_{x=1}^{H} \| \varphi\left(\mathbf{X}(x,y); \Theta\right) - \mathbf{C}(x,y) \|_2^2 \tag{14}$$

where Θ is the filter parameter that needs to be updated during the training process, $\mathbf{C}$ is the ground truth image without noise, $\mathbf{X}$ is the input image with speckle noise, and $\varphi\left(\cdot\right)$ is the output after despeckling. The purpose of training network is to minimize the cost. Smaller loss value suggests a smaller error between the network output and its corresponding ground truth.

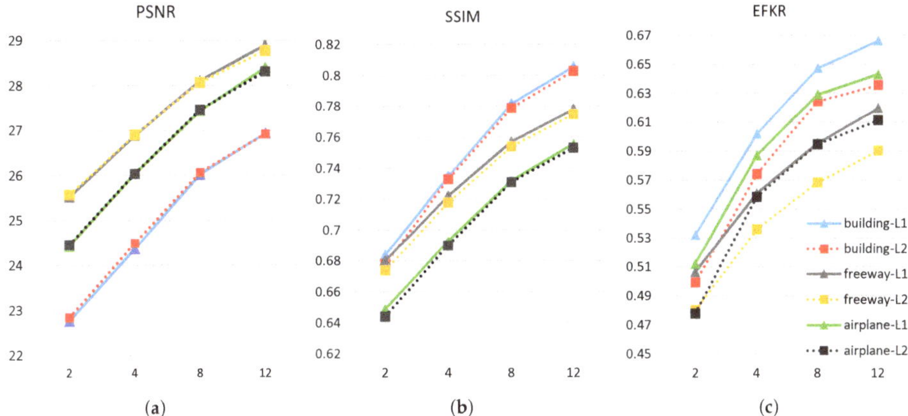

Figure 14. Evaluation index (PSNR, SSIM, and EFKR) values of MSR-net with L1 and L2 Loss.

As shown in Figure 14, the network trained by $\mathcal{L}_2$ loss function is more likely to obtain a higher PSNR only for building images and the network trained by $\mathcal{L}_1$ loss function can obtain both slightly higher PSRN and SSIM with the other images. But for EFKR, the advantage of $\mathcal{L}_1$ loss is significant compared to $\mathcal{L}_2$ loss. Generally speaking, the $\mathcal{L}_1$ loss is more suitable to SAR despeckling task.

6. Conclusions

In summary, different from the existing despeckling network, MSR-net proposed in this paper adopts the coarse-to-fine structure and the convolutional long short-term memory unit that can obtain high-quality despeckling SAR images. During research, we find that the weights sharing strategy of convolutional kernels can reduce network parameters and training complexity, and the sub-pixel unit used in this work can reduce up-sampling complexity, improve network efficiency, and shorten the runtime of the network with respect to the transposed convolutional layer. Meanwhile, new design evaluation metrics EFKR and FPKR are introduced herein to evaluate the compatibility of the despeckling algorithms to the optical image processing algorithms. Experimental results show that our MSR-net has excellent despeckling ability and achieves the state-of-the-art results both for simulated and real SAR images with low computational costs, especially in low signal noise ratio cases. The adaptability of optical image processing algorithms to SAR images can be enhanced after despeckling in our network.

Author Contributions: All of the authors made significant contributions to the work. Y.Z. and J.S. designed the research and analyzed the results. Y.Z. performed the experiments. Y.Z. and X.Y. wrote the paper. C.W., D.K., S.W., and X.Z. provided suggestions for the preparation and revision of the paper.

Acknowledgments: This work was supported in part by the Natural Science Fund of China under Grant 61671113 and the National Key R&D Program of China under Grant 2017YFB0502700, and in part by the Natural Science Fund of China under Grants 61501098, and 61571099.

Conflicts of Interest: The authors declare no conflicts of interest.

References

1. Ranjani, J.J.; Thiruvengadam, S. Dual-tree complex wavelet transform based SAR despeckling using interscale dependence. *IEEE Trans. Geosci. Remote Sens.* **2010**, *48*, 2723–2731. [CrossRef]
2. Ahishali, M.; Kiranyaz, S.; Ince, T.; Gabbouj, M. Dual and Single Polarized SAR Image Classification Using Compact Convolutional Neural Networks. *Remote Sens.* **2019**, *11*, 1340. [CrossRef]

3. Jun, S.; Long, M.; Xiaoling, Z. Streaming BP for non-linear motion compensation SAR imaging based on GPU. *IEEE J. Sel. Top. Appl. Earth Obs. Remote Sens.* **2013**, *6*, 2035–2050. [CrossRef]

4. Tang, X.; Zhang, X.; Shi, J.; Wei, S.; Tian, B. Ground Moving Target 2-D Velocity Estimation and Refocusing for Multichannel Maneuvering SAR with Fixed Acceleration. *Sensors* **2019**, *19*, 3695. [CrossRef]

5. Lee, J.S.; Wen, J.H.; Ainsworth, T.L.; Chen, K.S.; Chen, A.J. Improved sigma filter for speckle filtering of SAR imagery. *IEEE Trans. Geosci. Remote Sens.* **2008**, *47*, 202–213.

6. Yue, D.X.; Xu, F.; Jin, Y.Q. SAR despeckling neural network with logarithmic convolutional product model. *Int. J. Remote Sens.* **2018**, *39*, 7483–7505. [CrossRef]

7. Xie, H.; Pierce, L.E.; Ulaby, F.T. Statistical properties of logarithmically transformed speckle. *IEEE Trans. Geosci. Remote Sens.* **2002**, *40*, 721–727. [CrossRef]

8. Lee, J.S.; Hoppel, K.W.; Mango, S.A.; Miller, A.R. Intensity and phase statistics of multilook polarimetric and interferometric SAR imagery. *IEEE Trans. Geosci. Remote Sens.* **1994**, *32*, 1017–1028.

9. Lattari, F.; Gonzalez Leon, B.; Asaro, F.; Rucci, A.; Prati, C.; Matteucci, M. Deep Learning for SAR Image Despeckling. *Remote Sens.* **2019**, *11*, 1532. [CrossRef]

10. Lopes, A.; Touzi, R.; Nezry, E. Adaptive speckle filters and scene heterogeneity. *IEEE Trans. Geosci. Remote Sens.* **1990**, *28*, 992–1000. [CrossRef]

11. Frost, V.S.; Stiles, J.A.; Shanmugan, K.S.; Holtzman, J.C. A model for radar images and its application to adaptive digital filtering of multiplicative noise. *IEEE Trans. Pattern Anal. Mach. Intell.* **1982**, *PAMI-4l*, 157–166. [CrossRef]

12. Kuan, D.T.; Sawchuk, A.A.; Strand, T.C.; Chavel, P. Adaptive noise smoothing filter for images with signal-dependent noise. *IEEE Trans. Pattern Anal. Mach. Intell.* **1985**, *PAMI-7*, 165–177. [CrossRef]

13. Lopes, A.; Nezry, E.; Touzi, R.; Laur, H. Structure detection and statistical adaptive speckle filtering in SAR images. *Int. J. Remote Sens.* **1993**, *14*, 1735–1758. [CrossRef]

14. Kaur, L.; Gupta, S.; Chauhan, R. Image Denoising Using Wavelet Thresholding. *ICVGIP Indian Conf. Comput. Vision Graph. Image Process.* **2002**, *2*, 16–18.

15. Cohen, A.; Daubechies, I.; Feauveau, J.C. Biorthogonal bases of compactly supported wavelets. *Commun. Pure Appl. Math.* **1992**, *45*, 485–560. [CrossRef]

16. Donovan, G.C.; Geronimo, J.S.; Hardin, D.P.; Massopust, P.R. Construction of orthogonal wavelets using fractal interpolation functions. *Siam J. Math. Anal.* **1996**, *27*, 1158–1192. [CrossRef]

17. Xie, H.; Pierce, L.E.; Ulaby, F.T. SAR speckle reduction using wavelet denoising and Markov random field modeling. *IEEE Trans. Geosci. Remote Sens.* **2002**, *40*, 2196–2212. [CrossRef]

18. Candes, E.J.; Donoho, D.L. *Curvelets: A Surprisingly Effective Nonadaptive Representation for Objects with Edges*; Technical Report; Stanford University, Department of Statistics: Stanford, CA, USA, 2000.

19. Meyer, F.G.; Coifman, R.R. Brushlets: A tool for directional image analysis and image compression. *Appl. Comput. Harmon. Anal.* **1997**, *4*, 147–187. [CrossRef]

20. Do, M.N.; Vetterli, M. The contourlet transform: an efficient directional multiresolution image representation. *IEEE Trans. Image Process.* **2005**, *14*, 2091–2106. [CrossRef]

21. Buades, A.; Coll, B.; Morel, J.M. A non-local algorithm for image denoising. In Proceedings of the 2005 IEEE Computer Society Conference on Computer Vision and Pattern Recognition (CVPR'05), San Diego, CA, USA, 20–25 June 2005; Volume 2, pp. 60–65.

22. Parrilli, S.; Poderico, M.; Angelino, C.V.; Verdoliva, L. A nonlocal SAR image denoising algorithm based on LLMMSE wavelet shrinkage. *IEEE Trans. Geosci. Remote Sens.* **2011**, *50*, 606–616. [CrossRef]

23. Deledalle, C.A.; Denis, L.; Tupin, F.; Reigber, A.; Jäger, M. NL-SAR: A unified nonlocal framework for resolution-preserving (Pol)(In) SAR denoising. *IEEE Trans. Geosci. Remote Sens.* **2014**, *53*, 2021–2038. [CrossRef]

24. Kervrann, C.; Boulanger, J.; Coupé, P. Bayesian non-local means filter, image redundancy and adaptive dictionaries for noise removal. In Proceedings of the International Conference on Scale Space and Variational Methods in Computer Vision, Ischia, Italy, 30 May–2 June 2007; pp. 520–532.

25. Wang, P.; Zhang, H.; Patel, V.M. SAR image despeckling using a convolutional neural network. *IEEE Signal Process. Lett.* **2017**, *24*, 1763–1767. [CrossRef]

26. Bai, Y.C.; Zhang, S.; Chen, M.; Pu, Y.F.; Zhou, J.L. A Fractional Total Variational CNN Approach for SAR Image Despeckling. In Proceedings of the International Conference on Intelligent Computing, Bengaluru, India, 6 July 2018; pp. 431–442.

27. Zhang, Q.; Yuan, Q.; Li, J.; Yang, Z.; Ma, X. Learning a dilated residual network for SAR image despeckling. *Remote Sens.* **2018**, *10*, 196. [CrossRef]
28. Kaplan, L.M. Analysis of multiplicative speckle models for template-based SAR ATR. *IEEE Trans. Aerosp. Electron. Syst.* **2001**, *37*, 1424–1432. [CrossRef]
29. Oliver, C.; Quegan, S. *Understanding Synthetic Aperture Radar Images*; SciTech Publishing: Noida, India, 2004.
30. Wang, C.; Shi, J.; Yang, X.; Zhou, Y.; Wei, S.; Li, L.; Zhang, X. Geospatial Object Detection via Deconvolutional Region Proposal Network. *IEEE J. Sel. Top. Appl. Earth Obs. Remote Sens.* **2019**, *12*, 3014–3027. [CrossRef]
31. Lecun, Y.; Bengio, Y.; Hinton, G. Deep learning. *Nature* **2015**, *521*, 436. [CrossRef]
32. Xingjian, S.; Chen, Z.; Wang, H.; Yeung, D.Y.; Wong, W.K.; Woo, W.C. Convolutional LSTM network: A machine learning approach for precipitation nowcasting. In *Advances in Neural Information Processing Systems*; MIT Press: Cambridge, MA, USA, 2015; pp. 802–810.
33. Hochreiter, S.; Schmidhuber, J. Long short-term memory. *Neural Comput.* **1997**, *9*, 1735–1780. [CrossRef]
34. Chen, Q.; Koltun, V. Photographic image synthesis with cascaded refinement networks. In Proceedings of the IEEE International Conference on Computer Vision, Venice, Italy, 22–29 October 2017; pp. 1511–1520.
35. Srivastava, R.K.; Greff, K.; Schmidhuber, J. Highway networks. *arXiv* **2015**, arXiv:1505.00387.
36. He, K.; Sun, J. Convolutional neural networks at constrained time cost. In Proceedings of the IEEE Conference on Computer Vision and Pattern Recognition, Santiago, Chile, 7–13 December 2015; pp. 5353–5360.
37. He, K.; Zhang, X.; Ren, S.; Sun, J. Deep residual learning for image recognition. In Proceedings of the IEEE Conference on Computer Vision and Pattern Recognition, Las Vegas, NV, USA, 27–30 June 2016; pp. 770–778.
38. Dumoulin, V.; Visin, F. A guide to convolution arithmetic for deep learning. *arXiv* **2016**, arXiv:1603.07285.
39. Long, J.; Shelhamer, E.; Darrell, T. Fully convolutional networks for semantic segmentation. In Proceedings of the IEEE Conference on Computer Vision and Pattern Recognition, Santiago, Chile, 7–13 December 2015; pp. 3431–3440.
40. He, K.; Gkioxari, G.; Dollár, P.; Girshick, R. Mask r-cnn. In Proceedings of the IEEE International Conference on Computer Vision, Venice, Italy, 22–29 October 2017; pp. 2961–2969.
41. DeTone, D.; Malisiewicz, T.; Rabinovich, A. Superpoint: Self-supervised interest point detection and description. In Proceedings of the IEEE Conference on Computer Vision and Pattern Recognition Workshops, Salt Lake City, UT, USA, 18–22 June 2018; pp. 224–236.
42. Odena, A.; Dumoulin, V.; Olah, C. Deconvolution and checkerboard artifacts. *Distill* **2016**, *1*, e3. [CrossRef]
43. Shi, W.; Caballero, J.; Huszár, F.; Totz, J.; Aitken, A.P.; Bishop, R.; Rueckert, D.; Wang, Z. Real-time single image and video super-resolution using an efficient sub-pixel convolutional neural network. In Proceedings of the IEEE Conference on Computer Vision and Pattern Recognition, Las Vegas, NV, USA, 26 June–1 July 2016; pp. 1874–1883.
44. Goodman, J.W. Statistical properties of laser speckle patterns. In *Laser Speckle and Related Phenomena*; Springer: Berlin/Heidelberg, Germany, 1975; pp. 9–75.
45. Wang, Z.; Bovik, A.C.; Sheikh, H.R.; Simoncelli, E.P.; others. Image quality assessment: from error visibility to structural similarity. *IEEE Trans. Image Process.* **2004**, *13*, 600–612. [CrossRef] [PubMed]
46. Anfinsen, S.N.; Doulgeris, A.P.; Eltoft, T. Estimation of the equivalent number of looks in polarimetric synthetic aperture radar imagery. *IEEE Trans. Geosci. Remote Sens.* **2009**, *47*, 3795–3809. [CrossRef]
47. Sobel, I. History and definition of the sobell operator. Retrieved World Wide Web. 2014. Available online: Available online: https://www.researchgate.net/publication/239398674 (accessed on 23 October 2019).
48. Bao, P.; Zhang, L.; Wu, X. Canny edge detection enhancement by scale multiplication. *IEEE Trans. Pattern Anal. Mach. Intell.* **2005**, *27*, 1485–1490. [CrossRef] [PubMed]
49. Dim, J.R.; Takamura, T. Alternative approach for satellite cloud classification: edge gradient application. *Adv. Meteorol.* **2013**, *2013*, 584816. [CrossRef]
50. Davis, L.S. A survey of edge detection techniques. *Comput. Graph. Image Process.* **1975**, *4*, 248–270. [CrossRef]
51. Tareen, S.A.K.; Saleem, Z. A comparative analysis of sift, surf, kaze, akaze, orb, and brisk. In Proceedings of the IEEE 2018 International Conference on Computing, Mathematics and Engineering Technologies (iCoMET), Sukkur, Pakistan, 3–4 March 2018; pp. 1–10.
52. Yang, Y.; Newsam, S. Bag-of-visual-words and spatial extensions for land-use classification. In Proceedings of the 18th SIGSPATIAL International Conference on Advances in Geographic Information Systems, San Jose, CA, USA, 2–5 November 2010; pp. 270–279.

53. Couturier, R.; Perrot, G.; Salomon, M. Image Denoising Using a Deep Encoder-Decoder Network with Skip Connections. In Proceedings of the International Conference on Neural Information Processing, Siem Reap, Cambodia, 13–16 December 2018; pp. 554–565.
54. He, K.; Zhang, X.; Ren, S.; Sun, J. Delving deep into rectifiers: Surpassing human-level performance on imagenet classification. In Proceedings of the IEEE International Conference on Computer Vision, Santiago, Chile, 7–13 December 2015; pp. 1026–1034.
55. Kingma, D.P.; Ba, J. Adam: A method for stochastic optimization. *arXiv* **2014**, arXiv:1412.6980.
56. Kaplan, A.; Avraham, T.; Lindenbaum, M. Interpreting the ratio criterion for matching SIFT descriptors. In Proceedings of the European Conference on Computer Vision, Amsterdam, The Netherlands, 8–16 October 2016; pp. 697–712.

Article

Road Extraction of High-Resolution Remote Sensing Images Derived from DenseUNet

Jiang Xin [1], Xinchang Zhang [2,*], Zhiqiang Zhang [3] and Wu Fang [4]

[1] Department of Geography and Planning, Sun Yat-Sen University, Guangzhou 510275, China; jiangx35@mail2.sysu.edu.cn
[2] School of Geographical Sciences, Guangzhou University, Guangzhou 510006, China
[3] College of Surveying and Geo-informatics, North China University of Water Resources and Electric Power, Zhengzhou 450046, China; zhangzhiqiang@ncwu.edu.cn
[4] College of Surveying and Mapping, Information Engineering University, Zhengzhou 450001, China; xiaoyu_lee@whu.edu.cn
* Correspondence: eeszxc@mail.sysu.edu.cn; Tel.: +86-208-411-5103

Received: 25 September 2019; Accepted: 20 October 2019; Published: 25 October 2019

Abstract: Road network extraction is one of the significant assignments for disaster emergency response, intelligent transportation systems, and real-time updating road network. Road extraction base on high-resolution remote sensing images has become a hot topic. Presently, most of the researches are based on traditional machine learning algorithms, which are complex and computational because of impervious surfaces such as roads and buildings that are discernible in the images. Given the above problems, we propose a new method to extract the road network from remote sensing images using a DenseUNet model with few parameters and robust characteristics. DenseUNet consists of dense connection units and skips connections, which strengthens the fusion of different scales by connections at various network layers. The performance of the advanced method is validated on two datasets of high-resolution images by comparison with three classical semantic segmentation methods. The experimental results show that the method can be used for road extraction in complex scenes.

Keywords: high-resolution remote sensing imagery; multi-scale; road extraction; machine learning; DenseUNet

---

1. Introduction

The traffic road network is one of the essential geographic element of the urban system, which has critical applications in many fields, such as intelligent transportation, automobile navigation, and emergency support [1]. With the development of remote sensing technology and the advancement of remote sensing data processing methods, high temporal and spatial resolution, remote sensing data can provide high-precision ground information and permit the large-scale monitoring of roads. Remote sensing image data has quickly become the primary data source for the automatic extraction of road networks [2]. Automating road extraction plays a vital role in dynamic spatial development. Extracting the road in the urban area is a significant concern for the research on transportation, surveying, and mapping [3]. However, remote sensing images usually have sophisticated heterogeneous regional features with considerable intra-class distinctions and small inter-class distinctions. It is very challenging, especially in the urban area, as many buildings and trees exist, leading to shadow problems and a large number of segmented objects. The shadows of roadside trees or buildings can be observed from high-resolution images. Consequently, it is challenging to obtain high-precision road network information in the automatic extraction of road networks from remote sensing images.

There are many image segmentation methods for these problems by such conventional methods or machine learning algorithms. These methods are mainly divided into two categories: road centerline extraction and road area extraction. This paper focuses on extracting road areas from high-resolution remote sensing images. The road centerline is a linear element, and the spatial geometry is a line formed by a series of ordered nodes, which is an essential characteristic line of the road. The road centerline is generally obtained from the segmented image of road binary map through morphology or Medial Axis Transform (MAT) [4]. The road area is a kind of surface element. The road area is generated by image segmentation. The different spatial shape structure of boundary lines forms a variety of shape structures of surface elements [5]. Road centerline extraction [6,7] is used to detect the skeleton of the road, while road area extraction [8–13] generates the pixel-level label of the road, and there are some methods to extract the road area [14] while obtaining the road centerline. Huang et al. [8] try to extract road networks from the Ranging (LiDAR) data and light detection. Mnih et al. [9] used the Deep Belief Network (DBN) model to identify road targets in airborne images. Unsalan et al. [10] integrated three modules of road shape extraction module, road center probability detection module, and graphics-based module to extract road network from high-resolution satellite images. Cheng et al. [11] automatically extracted the road network information from complex remote sensing images based on the probability propagation method of graph cut. Saito et al. [12], based on the output of the channel function is put forward a new method of CNN's tabbed semantic segmentation. Alshehhi et al. [13] proposed an unsupervised road segmentation method based on the hierarchical graph. Road area extraction can be divided into pixel-level classification or image segmentation problems. Song et al. [15] proposed a method of road area detection based on the shape indexing feature of the support vector machine (SVM). Wang et al. [16] present a road detection method based on salient features and gradient vector flow (GVF) Snake. Rianto et al. [17] proposed a method to detect main roads from SPOT satellite images. The traditional road extraction method depends on the selected features. Zhang et al. [18] selected the seed points on the road, determined the direction, width, and starting point of the road in this section with a radial wheel algorithm, and proposed a semi-automatic method for road network tracking in remote sensing images. Movaghati et al. [19] proposed a new road network extraction framework by combining an extended Kalman filter (EKF) and a special particle filter (PF) to recover road tracks on obstructed obstacles. Gamba et al. [20] used adaptive filtering steps to extract the main road direction, and then proposed a road extraction method based on the prior information of road direction distribution. Li et al. [21] gradually extracted the road from the binary segmentation tree by determining the region of interest of the high-resolution remote sensing image and representing it as a binary segmentation tree.

However, the manually selected set of features is affected by many threshold parameters, such as lighting and atmospheric conditions. This empirical design method only deals with specific data, which limits its application in large-scale datasets. Deep learning is a representation learning method with multiple levels of representation, which is obtained by combining nonlinear but straightforward modules, each module representing a level of representation to a higher, slightly more abstract level. It allows raw data to be supplied to the machine and representations to be automatically discovered. In recent years, the deep convolutional network has been widely used in solving quite complex classification tasks, such as classification [22,23], semantic segmentation [24,25], and natural language processing [26,27].

Most importantly, these methods have proven to be profoundly robust to the appearance of different images, which prompted us to apply them to fully automated road segmentation in high-resolution remote sensing images. Long promoted the fully-convolutional network (FCN) and applied it to the field of semantic segmentation. Likewise, new segmentation methods based on deep neural networks and FCN were developed to extract roads from high-resolution remote sensing images. Mnih [28] put forward a method that combined the context information to detect road areas in aerial images.

He et al. [29] improves the performance of road extraction networks by integrating the spatial pyramid pool (ASPP) with the Encoder–Decoder network to enhance the ability to extract detailed features of the road. Zhang et al. [30] enhanced the propagation efficiency of information flow by fusing dense connections with convolutional layers of various scales. Aiming at the rich details of remote sensing images, Li et al. [31] proposed a Y-type convolutional neural network for road segmentation of high-resolution visible remote sensing images. The proposed network not only avoids background interference but also makes full use of complex details and semantic features to segment multi-scale roads. RSRCNN [32] extracts roads based on geometric features and spatial correlation of roads. Su et al. [33] enhanced the U-Net network model based on available problems. According to the characteristics of a small sample of aerial images, Zhang et al. [34] proposed an improved network-based road extraction design framework. By refining the CNN architecture, Gao et al. [35] proposed the refined deep residual convolutional neural network (RDRCNN) to enable it to detect the road area more accurately. To solve the problems of noise, occlusion, and complex background, Yang et al. [36] successfully designed an RCNN unit and integrated it into the U-Net architecture. The significant advantage of this unit is that it retains detailed low-level spatial characteristics. Zhang et al. [37] proposed the ResU-Net to extract road information by combining the advantages of a residual unit and U-Net. According to the characteristics of the narrow, connected, complex road, Zhou et al. [38] proposed the D-LinkNet model while maintaining the road information, integration of the multi-scale characteristics of the high-resolution satellite images. Based on the iterative search process guided by the decision function of CNN, Bastani [39] proposed RoadTracer, which can automatically construct accurate road network maps directly from aerial images. For irregular footprint problems between road area and image, Li et al. [40] proposed a combining GANs and multi-scale context polymerization of semantic segmentation method, used for road extraction of UAV remote sensing images. Xu et al. [41] put forward a kind of road extraction method based on local and global information, to effectively extract the road information in remote sensing images.

Inspired by the Densely Connected Convolutional Networks and U-Net, we propose the DenseUNet, an architecture that takes advantage of Densely Connected Convolutional Networks and U-Net architecture. The proposed deep convolutional neural network is based on the U-Net architecture. There are three differences between our deep DenseUNet and U-Net.

First, the model used dense units rather than ordinary neural units as the basic building blocks. Second, the proportion of road and non-road in remote sensing images is seriously unbalanced. Thus, this paper tries to analyze and propose ideas in terms of this issue. Finally, the performance of the proposed method is validated by comparison with three classical semantic segmentation methods.

2. Methods

2.1. Encoder–Decoder Architecture

State-of-the-art semantic image segmentation methods are mostly based on Encoder–Decoder architecture such as FCN [42], U-Net [43], SegNet [44]. An end-to-end trainable neural network recognizes the road in images and accurately segmented at the pixel level. Encoders usually use pre-trained models (such as VGG, Inception and ResNet), and each encoding layer includes the convolution, batch normalization (BN), the ReLU function and max pool layer. Each convolutional layer extracts features from all the maps in the previous layer, which has characteristics of simple structure and strong adaptability. Batch normalization [45] normalizes the input of each layer to reduce the internal-covariate-shift problem. It accelerates training and acts as a regularizer. The result shows that estimators based on a connected deep neural network with ReLU activation function and correctly selected the network. Pooling layer aims to compress the input feature map, which reduces the number of parameters in the training process and the degree of overfitting of the model. The main task of the Decoder is to map the distinguishable features to the pixel space for dense classification. Road network density refers to the ratio of the total mileage of the road network to the space of a given areaFor the extraction of relatively dense urban roads (in the same area, there are more roads), especially from high-resolution images, significant obstacles are leading to unreliable extraction results: complex image scenes and road models, as well as occlusion caused by high buildings and their shadows. Because of the above problems, this paper proposes DenseUNet, which is also based on

Encoder–Decoder architectures and designs a more dense connection mechanism for the Encoder layer. Because of the complexity of road scenes, U-Net cannot identify road features at a deeper level, and the generalization ability of multi-scale information is limited, which cannot adequately convey scale information. DenseUNet is a network architecture in which each layer feeds forward (within each dense block) directly to each of the other layers. For each layer, the feature map for all other layers is treated as a separate input, and its feature map is passed as input to all subsequent layers. Additionally, our approach has far fewer parameters due to the intelligent construction of the model. This kind of network design method not only extracts low-level features such as road edges and textures but also identifies the deep contour and location information of the road.

2.2. Backpropagation to Train Multilayer Architectures

Multilayer architectures can be trained by stochastic gradient descent. If only the input function and internal weight of the module are relatively smooth, the gradient can be computed by using the backpropagation process. The backpropagation process used to compute the gradient of the objective function about the weight of stacked multilayer modules is only the practical application of chain rules of derivatives. The significant idea is that the derivative (or gradient) of the module input can be computed by working backward from the gradient of the module output [46].

Figure 1 shows that the input space becomes iteratively warped until the data points become distinguishable through the data flow at various layers of the system. In this way, it can learn highly complex functions. Deep learning is a form of presentation learning—providing the machine with the raw data and developing the representations needed for its pattern recognition—that consists of multiple representation layers. These layers are usually arranged sequentially and consist of a large number of original nonlinear operations, where the representation of such a layer (the original data input) is fed into the next layer and converted to a more abstract representation [47]. The output layer uses softmax activation function to classify the image in one of the classes, and we can use fine-tuned CNNs as feature extractors to achieve better results.

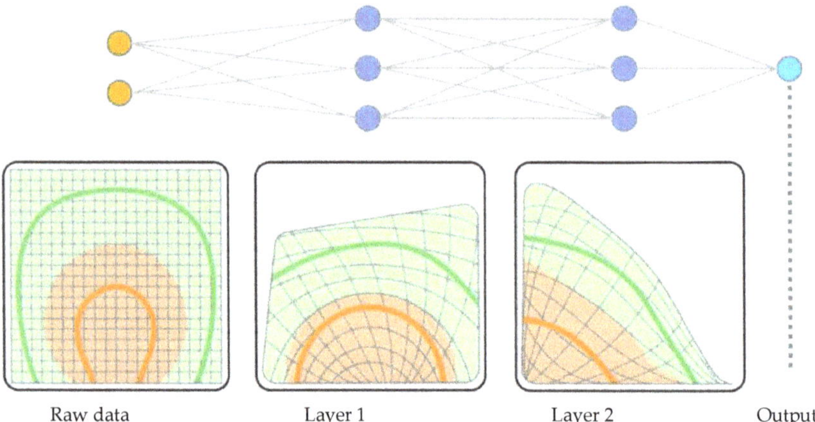

Figure 1. When data flows from one layer to another of the neural network, they are linearly separated by iteratively distorting the data. The final output layer outputs the probabilities of any class. This example illustrates the basic concepts of large-scale network usage.

2.3. DenseUNet

2.3.1. Network Architecture

We chose U-Net as the primary network architecture. In semantic segmentation, in order to achieve better results, it is essential to retain low-level details while acquiring high-level semantic information. The low-level features can be copied to the corresponding high-level to create information transmission paths, allowing signals to propagate naturally between the lower and higher levels, which not only helps the backpropagation in the training process but also compensates for the low-level and details of the high-level semantic features. We show that making use of dense units instead of ugly units can further improve the performance of U-Net. In this paper, the dense block is used as sub-module for feature extraction. By design, DenseUNet allows the layer to access all of its previous feature maps. DenseUNet takes advantage of the potential of the network to efficient compression models through feature reuse. It encourages reuse of features throughout the network and leads to a more compact model.

To restore the spatial resolution, FCN introduces an up-sampling path that includes convolution, up-sampling operations (transpose convolution or linear interpolation), and skip connections. In DenseUNet, we replace the convolution operation with up-sampling operations and transform it. The transition up module consists of a transposed convolution, which upsamples the previous feature mapping. Then the up-sampling feature map is connected to the input from the encoder skip connection to form a new input. We utilize an 11-level deep neural network architecture to extract road areas, as shown in Figure 2.

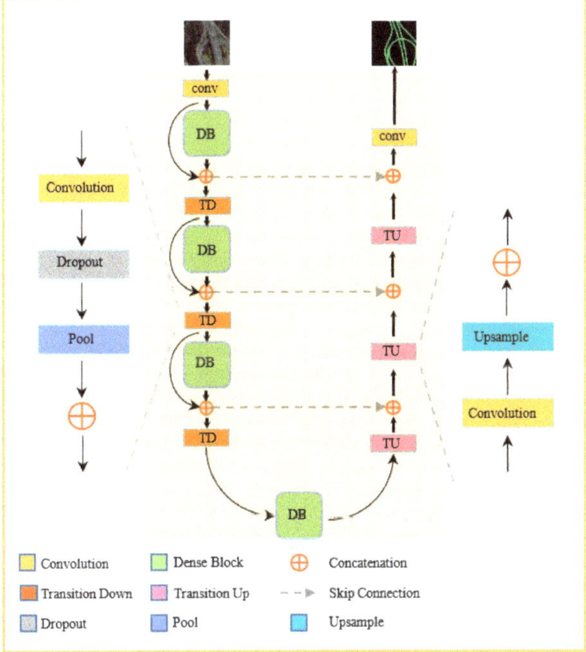

Figure 2. The architecture of the proposed deep DenseUNet. The dense block takes advantage of the potential of the network to efficient compression models through feature reuse.

2.3.2. Dense Block

Deep neural networks extract multi-level features of remote sensing images from low to high by convolution and pooling operations. The first few layers of convolution neural networks mainly extract low-level features such as road edges and textures, while deep-level networks extract features more complete, including road contours and location information. It can improve the performance of multi-layer neural networks and extract higher-level semantic information; however, it may hinder training and cause degradation problems. This is a problem with backpropagation [48]. He et al. [49]

proposed residual neural networks to speed up training and solve degradation problems. The residual neural network consists of a series of residual units. Each unit can be represented in the following form:

$$Z_l = H_l(Z_{l-1}) + Z_{l-1} \qquad (1)$$

Among them, Z_{l-1} and Z_l are the input and output of the lth residual unit, and $H_l(\cdot)$ is the residual function. Therefore, for ResNet model, the output of the lth layer is the composition of the $l-1^{th}$ identity mapping and the $l-1^{th}$ nonlinear transformation. The connection between the low-level and the high-level of the network will facilitate the dissemination of information without degradation. However, this kind of integration destroys the information flow between the layers of the network to a certain extent [50]. Here, we present the DenseUNet, a semantic segmentation neural network that combines the advantages of a densely concatenated convolutional network and U-Net. This architecture can be considered an extension of ResNet, which iteratively sums up the previous feature mappings. However, this small change has some exciting implications: (1) feature reuse, all layers can easily access the previous layer, so that the information in previously computed feature map can be easily reused; (2) parameter efficiency, DenseUNet is more effective in parameter usage; (3) implicit in-depth supervision, because of the short-path of all feature graphs in the architecture, DenseUNet provides deep supervision. Figure 3 is the basic dense network unit in this paper.

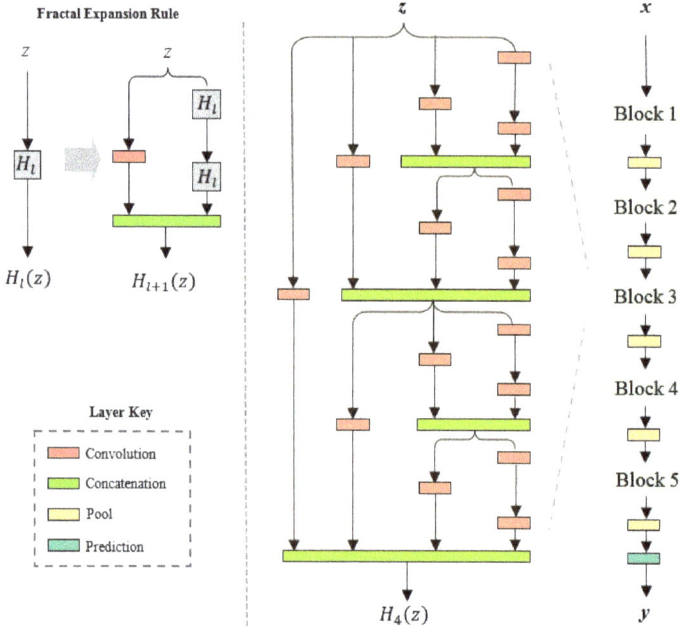

Figure 3. Dense network unit. Fractal structures have statistical or similar self-similar forms.

Dense network elements are fractal architectures. **Dense block layers are connected to each other so that each layer in the network accepts the characteristics of all its previous layers as input.** Left: simple extended rules generate fractal architectures with l intertwined columns. Basically, $H_1(Z)$ has a single layer of the selected type (e.g., convolution) between input and output. The connection layers compute the average value of element-wise. Right: Deep convolution neural network reduces spatial resolution periodically by pooling. A fractal version uses $H_1(Z)$ as the building block between pooling layers. A block such as Stack B produces a network whose total depth (measured as a convolutional layer) is $B \times 2^{C-1}$. Dense units consist of three parts: dense connection, growth rate, and bottleneck layers.

Dense connections. In order to further enhance the transmission of information among network layers, this paper constructs a different connection mode: by introducing direct connections from any layer to all subsequent layers. Figure 3 shows the layout of DenseUNet. Consequently, the Z_l layer receives the feature-maps of all other layers. $Z_0, Z_1, \cdots, Z_{l-1}$, as input:

$$Z_l = H_l([Z_0, Z_1, \cdots, Z_{l-1}]) \tag{2}$$

Among them $[Z_0, Z_1, \cdots, Z_{l-1}]$ refers to the series of features generated in layer $0, \ldots, l-1$. To promote implementation, the multiple inputs of $H_l(\cdot)$ in Equation (2) are concatenated into a single tensor. We define $H_l(\cdot)$ as a composite function of three continuous operations: batch normalization, followed by a 3×3 convolution and a rectified linear unit.

Growth rate. H_l generates G feature-maps, and then the lth layer has $G_0 + G \cdot (l-1)$ input feature maps, where G_0 is the number of channels in the input layer. The difference between DenseUNet and existing network architectures is that DenseUNet can have skinny layers. The hyper-parameter G is called the growth rate of the network.

Bottleneck layers. Although each layer generates only G output element mappings, it usually has more inputs. Literature [51] has noticed that before each 3×3 convolution, 1×1 convolution can be introduced as the bottleneck layer to reduce the number of input feature maps and improve the computational efficiency. We utilize such a bottleneck layer to refer to our network, i.e., the BN-Conv-ReLU version for H_l. Figure 4 shows the operation of dense block layers, transition down and transition up.

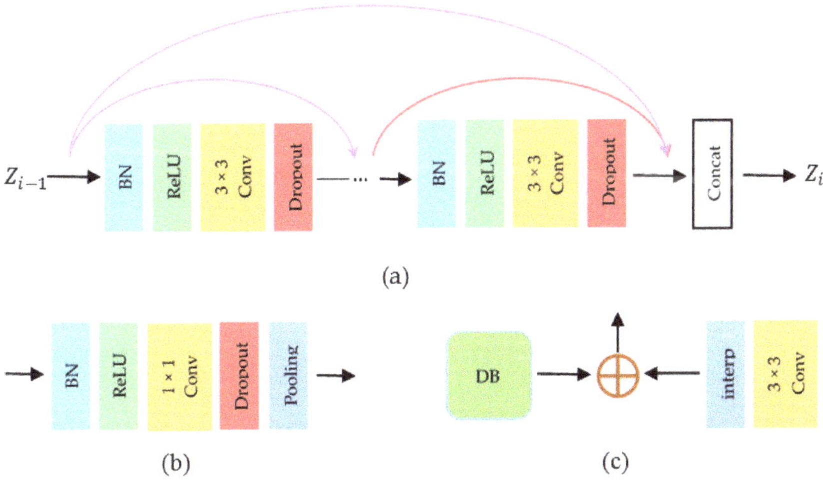

Figure 4. Basic layers of dense block, Transition Down, and Transition Up. (**a**) The dense block layer consists of BN, followed by ReLU and dropout; (**b**) Transition Down consists of BN followed by ReLU, dropout and a max-pooling of size 2×2; (**c**) Transition Up consists of a convolution, using nearest-neighbor interpolation to compensate for the loss of pooling process spatial information.

In our experiments on Conghua roads dataset and Massachusetts roads dataset, we used DenseUNet structure with five dense blocks on 256×256 input images. The number of feature maps in other layers also follows the setting G. In the present study, and we used Adam optimizer to minimize the classification cross-entropy. Let Y be a reference foreground segmentation with values y_i, and X be a prediction probability map of the foreground markers on the N image elements x_i, where the probability of background class is $1 - x_i$. The cross-entropy represents the dissimilarity between the approximate output distribution and the real distribution of the labels. The cross-entropy describes the difference between the true distribution of the input data and the distribution of the model obtained through training. The binary cross-entropy loss function is defined as:

$$loss = -\frac{1}{N} \sum_{i}^{N} (y_i \cdot log x_i + (1 - y_i) \cdot log(1 - x_i)) \tag{3}$$

The reasonable ratio of positive and negative samples is about 1:1 for feature selection in binary classification tasks. However, we find that the serious class imbalance between foreground and background is the central cause of high-resolution remote sensing images in the training process of semantic segmentation.

When the loss function gives equal weight to positive and negative samples, the category with large sample dominates the training process, and the training model is inclined to the category with a large sample, which reduces generalization ability of the model. We suggest reshaping the standard cross-entropy loss to solve the class imbalance problem in order to reduce the loss assigned to large samples. The weighted cross-entropy form of two-class can be expressed as:

$$loss = -\frac{1}{N} \sum_{i}^{N} (\theta_1 \cdot y_i \cdot log x_i + (1 - y_i) \cdot log(1 - x_i)) \tag{4}$$

where θ_1 is attributed to the weight of the foreground class, here defined as:

$$\theta_1 = \frac{N - \sum_{i}^{N} x_i}{\sum_{i}^{N} x_i} \tag{5}$$

By appropriately increasing the loss caused by the fault positive samples, the problem of the vast difference between the positive and negative samples is solved to some extent.

3. Experiments

3.1. Model Preprocessing

3.1.1. Software and Hardware Environment

In order to examine the proposed method, we construct a system platform, which is mainly composed of two parts: the software and hardware environment. The training and testing of deep neural networks require high-performance machines, which consumes a lot of video memory during training. TensorFlow is provided with the advantages of high efficiency, strong expansibility, and high flexibility design, and with the support of TensorFlow researchers, the efficiency of TensorFlow is improved. Based on the above reasons, this paper selects TensorFlow framework for network training. The basic configuration is shown in Table 1:

Table 1. Platform configuration.

Hardware				Software			
memory	hard disk	CPU	GPU	OS	CUDA	TensorFlow	python
16GB	1TB	Core-i7-8700K	GTX1080Ti	Ubuntu16.04	9.0	1.5	2.7.12

3.1.2. Data Augmentation

The deep learning model is trained with sufficient data, with the increase of the input size of the deep neural network, the training parameters after convolution operation also increase. In order to make use of the video memory and increased training efficiency, we utilize a 256 × 256 window to crop image blocks. One of the main problems in such models and signature verification systems is the low number of samples for training the model. Although transfer learning is effective in other domains, the remote sensing images are essentially different from traditional images by rich spectral setting, a wide range of image values, and different color and texture distributions. The image enhancement method is introduced to improve the generalization ability of the model. The deep learning method uses the method to add more data to the training dataset, which is called data augmentation. Data augmentation has already proved to bring many benefits to convolutional neural networks (CNNs) [52]. For example, as a regularizer, it is used to prevent overfitting in neural networks [53] and to improve the performance of unbalanced class problems [54]. As shown in Figure 5, the training set is expanded by six times.

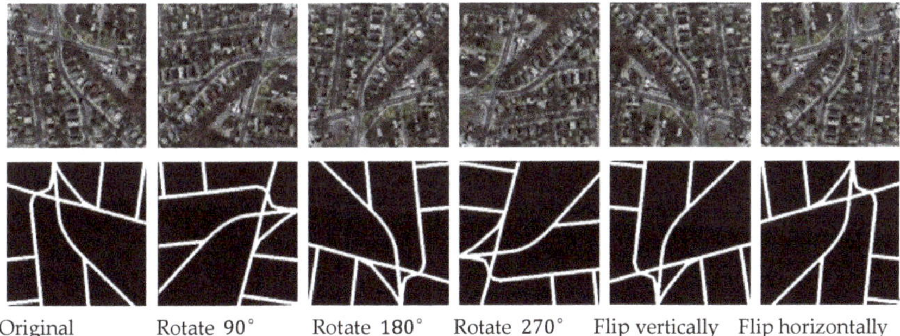

Figure 5. Data augmentation. The method mainly includes rotation, flipping (horizontally and vertically), and cropping operations.

3.1.3. Hyper-Parameters Selection

The process of searching optimal models requires parallel training of multiple models. The selection of learning batch size, learning rate, and optimization algorithm makes the model unique and different from other models. The process of selecting the best model requires the hyper-parameters to be optimized. We use TensorFlow to perform parallel data training with many models. Three hyper-parameters batch size (batch size, learning rate, and epochs) allow parallel training of multiple models, and the accuracy of test datasets determines the best model. We have studied various methods to enable deep learning models to be learned from the training dataset. We studied various methods to learn deep learning models from training data sets. Hyper-parameters can be used to activate the training process. Adam is an adaptive learning method that requires less tuning, is computationally efficient, and is superior to other stochastic optimization methods. The network hyper-parameter settings are shown in Table 2. We chose Adam as the optimization method, and it represents faster convergence than the standard stochastic gradient with momentum. We fix the parameters of Adam as recommended in Reference [55]: $\beta1 = 0.9$ and $\beta2 = 0.999$.

Table 2. Hyper-parameters.

Hyper-Parameter	Grid Search
batch size	(2, 4, 8, 16)
epochs	(50, 100, 150, 200)
learning rates	$(1 \times 10^{-9}, 1 \times 10^{-5}, 1 \times 10^{-3}, 1 \times 10^{-1})$

Compared with the classical U-Net, SegNet, GL-Dense-U-Net, and FRRN-B network, we evaluated the proposed method on two urban scenario datasets: Conghua road dataset, and Massachusetts road dataset. For the sake of quantitatively estimate the performance of the semantic segmentation method, we show the precision, recall, F1-Score, intersection over union (IoU) and kappa as different metrics for performance. The recall rate is defined as the ratio of the correct detection category to the correct detection category and the sum of a false negative, which will be used to assessments of the road integrity. The precision rate is the proportion of successes made by a classifier over the whole instance set, which reflects on the correctness of the road. The F1-score is the harmonic average of precision and recall, computed based on the number of errors detected by computers and manual evaluators. The Intersection over Union (IoU) is only the ratio of the overlap area between the truth and predicted regions of interest on the ground to the area surrounded by them. The kappa coefficient is a statistic which measures inter-rater agreement for specific items, and it is generally used to assess the accuracy of remote sensing image classifications.

3.2. Massachusetts Dataset

The Massachusetts dataset [56] has an image resolution of 1 m, and each image contains $3 \times 1500 \times 1500$ pixels. The open road dataset contains 1711 aerial images with a total area of more than 2600 square kilometers. The dataset is divided into 1108 training images, 14 validation images, and 49 test images. Figure 6 shows that U-Net, SegNet, and FRRN-B models can correctly identify most of the roads. Although these models eliminate the effects of shadows and buildings to a certain extent, the extraction results show that the correctness of intensive road is lower than in other regions. The results of these models are poorly continuous, and the edge of the road is not distinct enough. U-Net and SegNet performed poorly and lack of the necessary connectivity in the intensive road. From the sixth and seventh columns, the performance ability of GL-Dense-U-Net is equal to that of DenseUNet. Both models show good results in both single lane and dual lanes

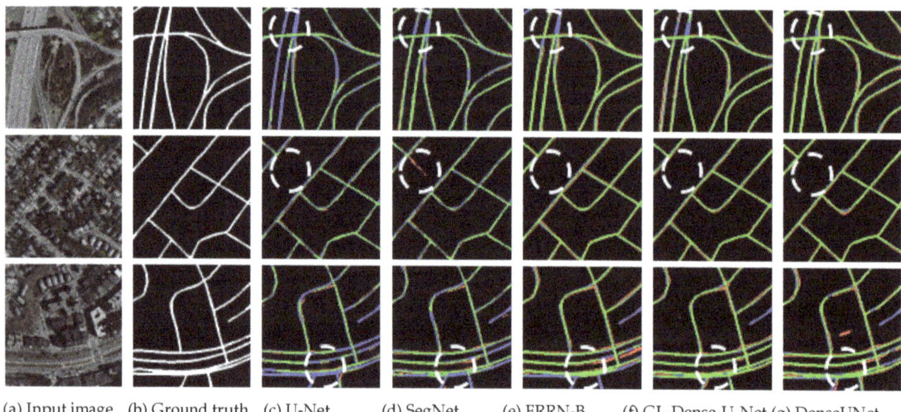

(a) Input image (b) Ground truth (c) U-Net (d) SegNet (e) FRRN-B (f) GL-Dense-U-Net (g) DenseUNet

Figure 6. Images of the original actual color composite image are displayed and classified in three regions using deep learning methods. True positive (TP), false negative (FN) and false positive (FP) were marked as green, blue, and red, respectively.

3.3. Conghua Dataset

The image resolution of Conghua dataset is 0.2 m, which consists of three bands: Red, Green, and Blue (RGB). There are 47 aerial images in this dataset, and each image consists of $3 \times 6000 \times 6000$ pixels. Among these, 80% of the data is used for training, and the remaining 20% data is used for model validation. Figure 7 shows that the white dotted line area is covered with thick trees, especially in urban environments, where model performance is more challenging than other areas, and the road occlusion is more frequent due to trees. The method we propose is hardly affected by shadow occlusion, and the average performance is better than the other three classical semantic segmentation algorithms based on convolution neural network. The performance of the GL-Dense-U-Net model on this data set is comparable to that of DenseUNet, and the extracted road edge information is relatively complete, which maintains functional connectivity. We can extract the local feature information of the image accurately and effectively. Figure 8 shows the details of the shaded area.

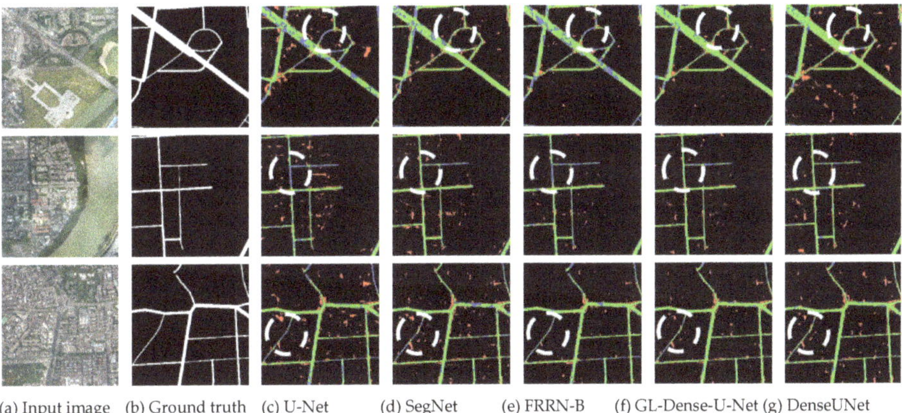

(a) Input image (b) Ground truth (c) U-Net (d) SegNet (e) FRRN-B (f) GL-Dense-U-Net (g) DenseUNet

Figure 7. Images of the original actual color composite image are displayed and classified in three regions using deep learning methods. True positive (TP), false negative (FN) and false positive (FP) were marked as green, blue, and red, respectively. The white dotted line in the images is enlarged for close-up inspection in Figure 7.

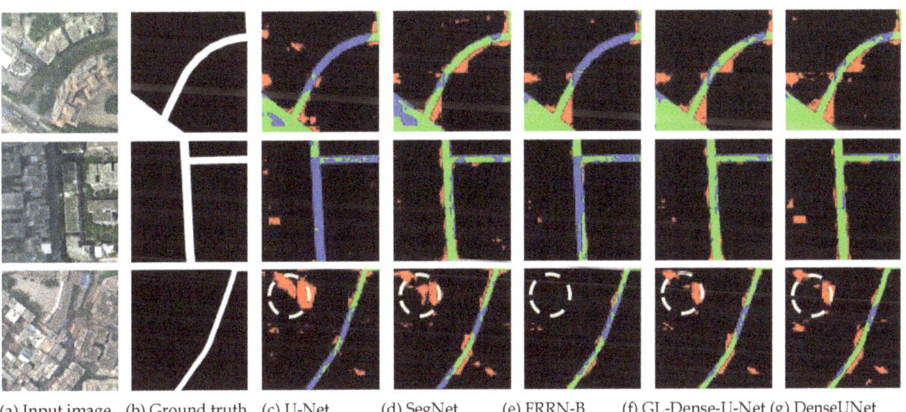

(a) Input image (b) Ground truth (c) U-Net (d) SegNet (e) FRRN-B (f) GL-Dense-U-Net (g) DenseUNet

Figure 8. A close-up view of the original true-color composite image and classification results is displayed across three regions using the deep learning method. The images are the subset from the white dotted line marked in Figure 7. True positive (TP), false negative (FN) and false positive (FP) were marked as green, blue, and red, respectively.

3.4. Accuracy Evaluation

Table 3 shows a comparison of the accuracy of automatic classification. We find that the proposed method achieves the highest accuracy, and both F1-score and kappa are significantly higher than three classical semantic segmentation methods on both datasets. The kappa metrics for the classification results were 0.703 and 0.801, respectively. The proposed method provides the most important value for the F1-score, which involves recall and accurate metrics. The experimental results show that the average performance of the method in recall rate, accuracy, and F1-score is better than the other three classical semantics segmentation methods. In addition, it was found that the method can produce the relatively high average performance of IoU, and kappa over all the images in the test set, which is consistent with the predicted results of Figures 6 and 8.

Table 3. The experimental results of road extraction.

Model	M-Dataset					C-Dataset				
	P (%)	R (%)	F1 (%)	IoU (%)	Kappa	P (%)	R (%)	F1 (%)	IoU (%)	Kappa
U-Net	58.92%	70.81%	60.78%	70.91%	63.65%	84.29%	75.23%	71.83%	77.35%	77.43%
SegNet	61.35%	**71.33%**	62.64%	71.91%	65.41%	85.03%	77.04%	73.94%	78.83%	78.52%
FRRN-B	76.51%	64.87%	66.71%	74.22%	67.70%	83.92%	77.22%	73.62%	78.72%	78.16%
GL- Dense-U-Net	**78.48%**	70.09%	73.98%	72.73%	70.19%	85.33%	**79.07%**	**76.41%**	80.67%	**80.35%**
DenseUNet	78.25%	70.41%	**74.07%**	**74.47%**	**70.32%**	**85.55%**	78.51%	76.25%	**80.89%**	80.11%

Figures 6 and 8 illustrate three example results of U-Net, SegNet, FRRN-B, GL-Dense-U-Net, and the proposed DenseUNet. The results show that compared with the other four methods, our method has the advantages of high accuracy and low noise. Especially in the case of dense roads and shadows, our method can divide each lane with high reliability and get prominent shadows, as shown in the third row of Figures 6 and 8.

3.5. Model Analysis

Road background information is essential when analyzing complex structured objects. Our network takes into account the information around the road to facilitate the distinction between roads and similar objects, such as building roofs and dense trees. The context information is robust when the road is occluded. From the first row of Figure 7, some of the roads in the circle are covered by trees. Three classical semantics segmentation methods cannot detect the road under the tree; however, our method has successfully marked them to some extent. A case of failure is shown in the gold dotted line of Figure 8; the proposed method has a distinct error detection rate in impervious surface. It is mainly because most of the roads in the urban impervious surface are not labeled. Therefore, considering that our network regards them as contextual information of the foreground, these roads share the same characteristics as normal roads. We provide a better insight into the performance of the proposed method. In Figure 9, we show the loss and performance curves during system training. The loss of the four models slowly decreases as the training time increases and eventually stabilizes. Although the U-Net model showed large changes in the initial stage of the model training, it finally reached a convergence state. It can be seen that the improved model ultimately achieves good convergence. The connections in dense units and skipping connections between the lower and higher levels of the network help to spread information without degradation, so that a neural network with fewer parameters can be designed; however, better comparability can be achieved in semantic segmentation performance.

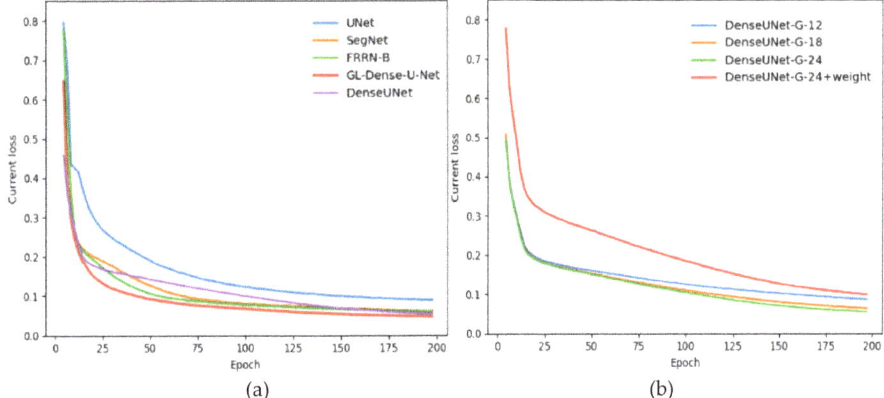

Figure 9. Loss of training. (**a**) The five curves of blue, yellow, green, red and purple represent the losses of U-Net, SegNet, FRRN-B, GL-Dense-U-Net, and DenseUNet; (**b**) The four curves represent models with different growth rates and modified weights

DenseUNet extracts multi-level features from different stages of the dense block, which strengthens the fusion of different scales. We train DenseUNet with different growth rates, G. The main results on two sets of data dataset are shown in Table 4. It can be seen from the accuracy that the model has the best performance (when the parameter G is equal to 24). Besides, Table 4 shows that relatively small growth rates are sufficient to achieve excellent results on the test datasets. The growth rate defines the amount of new information provided by each layer for the global state. It can be accessed from anywhere in the network and does not need to be replicated between layers in traditional network architecture.

Table 4. Results of different growth factors.

G	M-Dataset			C-Dataset		
	OA (%)	IoU (%)	Kappa (%)	OA (%)	IoU (%)	Kappa (%)
12	92.22%	73.24%	69.55%	94.18%	73.24%	69.55%
18	93.13%	74.04%	70.11%	94.87%	74.04%	70.11%
24	**93.93%**	**74.47%**	**70.32%**	**95.02%**	**80.89%**	**70.32%**

4. Discussion

Table 5 shows the statistics of the deep learning model and the variations of DenseUNet. The average running time was calculated by iterating 50 times. U-Net adopts a shallow Encoder–Decoder structure, which requires less computational resources and less reasoning time than other models. However, the road integrity extracted from two sets of data is sparse. DenseUNet adopts a custom encoder–decoder architecture, so it maintains a balance between computing resources and reasoning time. It consumes less computing resources and reasoning time than other models.

Table 5. Compare the network efficiency between the tested deep learning model and DenseUNet.

Model	Inference (ms)	Model Size (MB)	FPS
U-Net	340	106	32
SegNet	204	419	33
FRRN-B	338	297	20
GL-Dense-U-Net	1152	1690	22
DenseUNet-G-12	316	118	23
DenseUNet-G-18	450	279	17
DenseUNet-G-24	472	514	15

GL-Dense-U-Net is equivalent to DenseUNet in terms of various indicators. GL-Dense-U-Net consists of Local Attention Units (LAU) and Global Attention Units (GAU). The $1 \times 1, 3 \times 3, 5 \times 5,$ 7×7 kernels are respectively used for convolution operation by LAU, and finally integrated step by step from the bottom to the top. GAU introduces global average pool (GAP) into the unit to extract comprehensive road information. However, since the GL-Dense-U-Net encoding and decoding layers are composed of dense unit blocks provided by DenseNet, and LAU unit (the feature graph of different scale is fused to realize the attention of pixel-level information) is added in the encoding stage while GAU unit (feature maps from low and high levels are considered, and global information is provided to restore features) is connected later in the decoding stage, the GL-Dense-U-Net model is the largest and the inference time is the longest. DenseUnet adopts dense unit modules in the coding stage, while the sampling stage in the decoding layer adopts the skip connection characteristic of U-Net. Therefore, DenseUNet requires less inference time of 316 ms and a smaller model size of 118 MB than other models. In general, DenseNet is more effective than most models. On the other hand, G feature maps are output after the convolution of all layers in the dense block. The model sets a small growth rate (G = 12) to get good results, as shown in Table 4. The overall accuracy and the mIoU of DenseUNet-G-12 in Massachusetts datasets reached 92.22% and 73.24% respectively

In order to further verify the reliability of the proposed model, two groups of remote sensing image data with different resolutions were selected to compare four classical image segmentation models. In Massachusetts datasets, the overall accuracy and the mIoU of DenseUNet in the Massachusetts dataset achieved 93.93% and 74.47%, respectively. The Conghua dataset achieved 95.02% and 80.89%, respectively. In particular, the classification result of Massachusetts is better than that of GL-Dense-U-Net [41]. In general, DenseUNet performs better than Massachusetts datasets in Conghua datasets, which may be a higher data resolution from the dataset.

The developed DenseUNet has excellent potential for improvement. First, the smoothness of road contour is a key factor that affects the accuracy of road extraction. In the two sets of prediction datasets, we found that compared with the ground truth value, the predicted result road had information loss of edge and contour. Obtaining accurate road profile information is still a challenging task. Second, different network models are suitable for different scenarios, such as PSPNet [57], DeepLabv3+ [58], and BiSeNet [59], etc., which are suitable for real-time segmentation of street view. It is usually necessary to design the network according to specific tasks to obtain the best performance. Neural Architecture Search (NAS) is a kind of automated neural network design technology, which can automatically design high-performance network structure according to the sample set through the algorithm. This architecture can effectively reduce the use and implementation cost of the neural network. Third, we only focused on the performance of different deep learning models during the experiment. Traditional methods, such as threshold-based methods and object-based methods [60], have not been compared, and a more comprehensive comparison of these methods is needed in the future.

5. Conclusions

We propose an efficient road extraction method based on a convolution neural network for high-resolution remote sensing images. The model combines the virtue of dense connection mode and U-Net and solves the problem of tree and shadow occlusion to a certain extent, which we call DenseUNet. In particular, we use a U-Net architecture combined with a suitable weighted loss function to place more emphasis on foreground pixels. Following simple connection rules (fractal extensions), DenseUNet naturally integrates deep supervision, the properties of identity mappings, and diversified depth attributes. The dense connections within dense units and the skip connections between the encoding and decoding paths of the network will help to transfer information and accelerate computation, so they can learn more compactly and get more accurate models.

Although deep neural networks have acquired remarkable success in many fields, there are no sophisticated theories yet. However, one of the critical disadvantages of deep learning models is their limited interpretability, and often these models are described as "black boxes" that do not provide insight into their inner workings. On the other hand, it will be challenging to create a general model through theoretical guidance. Hence, the results obtained from such specific planning problem are difficult to apply to other problems in the same field. We plan to use the trained DenseUNet model to transfer knowledge to improve new tasks in future work.

Author Contributions: J.X. and X.Z. conceived and designed the method; J.X. performed the experiments and wrote the manuscript; Z.Z. provided the Conghua dataset; W.F. revised the manuscript.

Funding: This research was funded by National Key R&D Program of China (Grant No. 2018YFB2100702), National Natural Science Foundation of China (Grant Nos. 41875122, 41431178, 41801351 and 41671453), Natural Science Foundation of Guangdong Province: 2016A030311016, Research Institute of Henan Spatio-Temporal Big Data Industrial Technology: 2017DJA001, National Administration of Surveying, Mapping and Geoinformation of China (Grant No. GZIT2016-A5-147), Hunan Botong Information Co.,ltd.: BTZH2018001.

Acknowledgments: The authors would like to thank the anonymous reviewers for their constructive comments. The authors are also grateful to Hinton et al. for providing the Massachusetts dataset.

Conflicts of Interest: The authors declare no conflicts of interest.

References

1. Zhang, Z.; Zhang, X.; Sun, Y.; Zhang, P. Road Centerline Extraction from Very-High-Resolution Aerial Image and LiDAR Data Based on Road Connectivity. *Remote Sens.* **2018**, *10*, 1284. [CrossRef]
2. Zhou, T.; Sun, C.; Fu, H. Road Information Extraction from High-Resolution Remote Sensing Images Based on Road Reconstruction. *Remote Sens.* **2019**, *11*, 79. [CrossRef]
3. Bong, D.B.; Lai, K.C.; Joseph, A. Automatic Road Network Recognition and Extraction for Urban Planning. *Int. J. Appl. Sci. Eng. Technol.* **2009**, *5*, 209–215.
4. Miao, Z.; Shi, W.; Zhang, H.; Wang, X. Road centerline extraction from high-resolution imagery based on shape features and multivariate adaptive regression splines. *IEEE Geosci. Remote Sens. Lett.* **2012**, *10*, 583–587. [CrossRef]
5. Li, Z.; Huang, P. Quantitative measures for spatial information of maps. *Int. J. Geogr. Inf. Sci.* **2002**, *16*, 699–709. [CrossRef]
6. Liu, B.; Wu, H.; Wang, Y.; Liu, W. Main road extraction from zy-3 grayscale imagery based on directional mathematical morphology and vgi prior knowledge in urban areas. *PLoS ONE* **2015**, *10*, e0138071. [CrossRef]
7. Sujatha, C.; Selvathi, D. Connected component-based technique for automatic extraction of road centerline in high resolution satellite images. *EURASIP J. Image Video Process.* **2015**, *2015*, 8. [CrossRef]
8. Huang, X.; Zhang, L. Road centreline extraction from high—Resolution imagery based on multiscale structural features and support vector machines. *Int. J. Remote Sens.* **2009**, *30*, 1977–1987. [CrossRef]
9. Mnih, V.; Hinton, G.E. Learning to detect roads in high-resolution aerial images. In Proceedings of the European Conference on Computer Vision (ECCV), Crete, Greece, 5–11 September 2010; Daniilidis, K., Maragos, P., Paragios, N., Eds.; Springer: New York, NY, USA, 2010; pp. 210–223.
10. Unsalan, C.; Sirmacek, B. Road network detection using probabilistic and graph theoretical methods. *IEEE Trans. Geosci. Remote Sens.* **2012**, *50*, 4441–4453. [CrossRef]
11. Cheng, G.; Wang, Y.; Gong, Y. Urban road extraction via graph cuts based probability propagation. In Proceedings of the IEEE International Conference on Image Processing (ICIP), Paris, France, 27–30 October 2014; pp. 5072–5076.
12. Saito, S.; Yamashita, T.; Aoki, Y. Multiple object extraction from aerial imagery with convolutional neural networks. *J. Electron. Imaging* **2016**, *2016*, 1–9. [CrossRef]
13. Alshehhi, R.; Marpu, P.R. Hierarchical graph-based segmentation for extracting road networks from high-resolution satellite images. *ISPRS-J. Photogramm. Remote Sens.* **2017**, *126*, 245–260. [CrossRef]
14. Cheng, G.; Wang, Y.; Xu, S. Automatic road detection and centerline extraction via cascaded end-to-end convolutional neural network. *IEEE Trans. Geosci. Remote Sens.* **2017**, *55*, 3322–3337. [CrossRef]

15. Song, M.; Civco, D. Road extraction using SVM and image segmentation. *Photogramm. Eng. Remote Sens.* **2004**, *70*, 1365–1371. [CrossRef]
16. Wang, F.; Wang, W.; Xue, B. Road Extraction from High-spatial-resolution Remote Sensing Image by Combining with GVF Snake with Salient Features. *Acta Geod. Cartogr. Sin.* **2017**, *46*, 1978–1985.
17. Rianto, Y.; Kondo, S.; Kim, T. Detection of roads from satellite images using optimal search. *Int. J. Pattern Recognit. Artif. Intell.* **2000**, *14*, 1009–1023. [CrossRef]
18. Zhang, J.; Lin, X.; Liu, Z.; Shen, J. Semi-automatic road tracking by template matching and distance transformation in urban areas. *Int. J. Remote Sens.* **2011**, *32*, 8331–8347. [CrossRef]
19. Movaghati, S.; Moghaddamjoo, A.; Tavakoli, A. Road extraction from satellite images using particle filtering and extended Kalman filtering. *IEEE Trans. Geosci. Remote Sens.* **2010**, *48*, 2807–2817. [CrossRef]
20. Gamba, P.; Dell'Acqua, F.; Lisini, G. Improving urban road extraction in high-resolution images exploiting directional filtering, perceptual grouping, and simple topological concepts. *IEEE Geosci. Remote Sens. Lett.* **2006**, *3*, 387–391. [CrossRef]
21. Li, M.; Stein, A.; Bijker, W.; Zhang, Q.M. Region-based urban road extraction from VHR satellite images using binary partition tree. *Int. J. Appl. Earth Obs. Geoinf.* **2016**, *44*, 217–225. [CrossRef]
22. Krizhevsky, A.; Sutskever, I.; Hinton, G.E. Imagenet classification with deep convolutional neural networks. In Proceedings of the Twenty-Sixth Annual Conference on Neural Information Processing Systems, Lake Tahoe, NV, USA, 3–8 December 2012; pp. 1097–1105.
23. Farabet, C.; Couprie, C.; Najman, L. Learning hierarchical features for scene labeling. *IEEE Trans. Pattern Anal. Mach. Intell.* **2012**, *35*, 1915–1929. [CrossRef]
24. Noh, H.; Hong, S.; Han, B. Learning deconvolution network for semantic segmentation. In Proceedings of the IEEE international conference on computer vision (ICCV), Santiago, Chile, 11–18 December 2015; pp. 1520–1528.
25. Li, P.; Zang, Y.; Wang, C. Road network extraction via deep learning and line integral convolution. In Proceedings of the IEEE International Geoscience and Remote Sensing Symposium (IGARSS), Beijing, China, 10–15 July 2016; pp. 1599–1602.
26. Kim, Y. Convolutional neural networks for sentence classification. *arXiv* **2014**, arXiv:1408.5882.
27. Tang, D.; Qin, B.; Liu, T. Document modeling with gated recurrent neural network for sentiment classification. In Proceedings of the 2015 Conference on Empirical Methods in Natural Language Processing (EMNLP), Lisbon, Portugal, 17–21 September 2015; Lluís, M., Chris, C.-B., Jian, S., Eds.; Association for Computational Linguistics: Lisbon, Portugal, 2015; pp. 1422–1432.
28. Mnih, V. Machine Learning for Aerial Image Labeling. Ph.D. Thesis, University of Toronto, Toronto, ON, Canada, 2013.
29. He, H.; Yang, D.; Wang, S.; Wang, S.Y.; Li, Y.F. Road Extraction by Using Atrous Spatial Pyramid Pooling Integrated Encoder-Decoder Network and Structural Similarity Loss. *Remote Sens.* **2019**, *11*, 1015. [CrossRef]
30. Zhang, Z.; Wang, Y. JointNet: A Common Neural Network for Road and Building Extraction. *Remote Sens.* **2019**, *11*, 696. [CrossRef]
31. Li, Y.; Xu, L.; Rao, J.; Guo, L.L.; Yan, Z.; Jin, S. A Y-Net deep learning method for road segmentation using high-resolution visible remote sensing images. *Remote Sens. Lett.* **2019**, *10*, 381–390. [CrossRef]
32. Wei, Y.; Wang, Z.; Xu, M. Road structure refined CNN for road extraction in aerial image. *IEEE Geosci. Remote Sens. Lett.* **2017**, *14*, 709–713. [CrossRef]
33. Su, J.; Yang, L.; Jing, W. U-Net based semantic segmentation method for high resolution remote sensing image. *Comput. Appl.* **2019**, *55*, 207–213.
34. Zhang, X.; Han, X.; Li, C.; Tang, X.; Zhou, H.; Jiao, L. Aerial Image Road Extraction Based on an Improved Generative Adversarial Network. *Remote Sens.* **2019**, *11*, 930. [CrossRef]
35. Gao, L.; Song, W.; Dai, J.; Chen, Y. Road Extraction from High-Resolution Remote Sensing Imagery Using Refined Deep Residual Convolutional Neural Network. *Remote Sens.* **2019**, *11*, 552. [CrossRef]
36. Yang, X.; Li, X.; Ye, Y.; Lau, R.Y.K.; Zhang, X.F.; Huang, X.H. Road Detection and Centerline Extraction Via Deep Recurrent Convolutional Neural Network U-Net. *IEEE Trans. Geosci. Remote Sens.* **2019**, *59*, 7209–7220. [CrossRef]

37. Zhang, Z.; Liu, Q.; Wang, Y. Road extraction by deep residual U-Net. *IEEE Geosci. Remote Sens. Lett.* **2018**, *15*, 749–753. [CrossRef]

38. Zhou, L.; Zhang, C.; Wu, M. D-LinkNet: LinkNet With Pretrained Encoder and Dilated Convolution for High Resolution Satellite Imagery Road Extraction. In Proceedings of the IEEE Conference on Computer Vision and Pattern Recognition (CVPR), Salt Lake City, UT, USA, 18–22 June 2018.

39. Bastani, F.; He, S.; Abbar, S.; Alizadeh, M.; Balakrishnan, H.; Chawla, S.; Madden, S.; DeWitt, D. Roadtracer: Automatic extraction of road networks from aerial images. In Proceedings of the IEEE Conference on Computer Vision and Pattern Recognition (CVPR), Salt Lake City, UT, USA, 18–22 June 2018.

40. Li, Y.; Peng, B.; He, L.; Fan, K.; Tong, L. Road Segmentation of Unmanned Aerial Vehicle Remote Sensing Images Using Adversarial Network with Multiscale Context Aggregation. *IEEE J. Sel. Top. Appl. Earth Observ. Remote Sens.* **2019**, *12*, 2279–2287. [CrossRef]

41. Xu, Y.; Xie, Z.; Feng, Y.; Chen, Z.L. Road extraction from high-resolution remote sensing imagery using deep learning. *Remote Sens.* **2018**, *10*, 1461. [CrossRef]

42. Long, J.; Shelhamer, E.; Darrell, T. Fully convolutional networks for semantic segmentation. In Proceedings of the IEEE conference on computer vision and pattern recognition (CVPR), Boston, MA, USA, 7–12 June 2015.

43. Ronneberger, O.; Fischer, P.; Brox, T. U-net: Convolutional networks for biomedical image segmentation. In Proceedings of the International Conference on Medical Image Computing and Computer-Assisted Intervention (MICCAI), Munich, Germany, 5–9 October 2015.

44. Badrinarayanan, V.; Kendall, A.; Cipolla, R. Segnet: A deep convolutional encoder-decoder architecture for image segmentation. *IEEE Trans. Pattern Anal. Mach. Intell.* **2017**, *39*, 2481–2495. [CrossRef] [PubMed]

45. Normalization, B. Accelerating deep network training by reducing internal covariate shift. *arXiv* **2015**, arXiv:1502.03167.

46. LeCun, Y.; Bengio, Y.; Hinton, G. Deep learning. *Nature* **2015**, *521*, 436. [CrossRef] [PubMed]

47. Esteva, A.; Robicquet, A.; Ramsundar, B. A guide to deep learning in healthcare. *Nat. Med.* **2019**, *25*, 24. [CrossRef] [PubMed]

48. Yu, X.; Efe, M.O.; Kaynak, O. A general backpropagation algorithm for feedforward neural networks learning. *IEEE Trans. Neural Netw.* **2002**, *13*, 251–254.

49. He, K.; Zhang, X.; Ren, S. Deep residual learning for image recognition. In Proceedings of the IEEE Conference on Computer Vision and Pattern Recognition (CVPR), Las Vegas, NV, USA, 27–30 June 2016.

50. Huang, G.; Liu, Z.; Van, D.M. Densely connected convolutional networks. In Proceedings of the IEEE Conference on Computer vVision and Pattern Recognition (CVPR), Honolulu, HI, USA, 21–26 July 2017.

51. Sermanet, P.; Chintala, S.; LeCun, Y. Convolutional neural networks applied to house numbers digit classification. *arXiv* **2012**, arXiv:1204.3968.

52. LeCun, Y.; Boser, B.E.; Denker, J.S. Handwritten Digit Recognition with a Back-Propagation Network. In *Advances in Neural Information Processing Systems 2*; Morgan Kaufmann Publishers: San Francisco, CA, USA, 1990; pp. 396–404.

53. Simard, P.Y.; Steinkraus, D.; Platt, J.C. Best practices for convolutional neural networks applied to visual document analysis. In Proceedings of the International Conference on Document Analysis and Recognition (ICDAR), Edinburgh, Scotland, 3–6 August 2003.

54. Chawla, N.V.; Bowyer, K.W.; Hall, L.O. SMOTE: Synthetic minority over-sampling technique. *J. Artif. Intell. Res.* **2002**, *16*, 321–357. [CrossRef]

55. Weinzaepfel, P.; Revaud, J.; Harchaoui, Z. DeepFlow: Large displacement optical flow with deep matching. In Proceedings of the IEEE International Conference on Computer Vision (CVPR), Portland, OR, USA, 23–28 June 2013.

56. Available online: https://www.cs.toronto.edu/~{}vmnih/data/ (accessed on 24 October 2019).

57. Zhao, H.; Shi, J.; Qi, X.; Wang, X.G.; Jia, J.Y. Pyramid scene parsing network. In Proceedings of the IEEE Conference on Computer Vision and Pattern Recognition (CVPR), Honolulu, HI, USA, 21–26 July 2017.

58. Chen, L.C.; Zhu, Y.; Papandreou, G.; Schroff, F.; Adam, H. Encoder-decoder with atrous separable convolution for semantic image segmentation. In Proceedings of the European Conference on Computer Vision (ECCV), Munich, Germany, 8–14 September 2018.

59. Yu, C.; Wang, J.; Peng, C.; Gao, C.X.; Yu, G.; Song, N. Bisenet: Bilateral segmentation network for real-time semantic segmentation. In Proceedings of the European Conference on Computer Vision (ECCV), Munich, Germany, 8–14 September 2018.

60. Baatz, M.; Schäpe, A. Multiresolution segmentation: An optimization approach for high quality multi-scale image segmentation. In *Angewandte Geographische Information—Sverarbeitung*; Strobl, J., Blaschke, T., Griesbner, G., Eds.; Wichmann Verlag: Karlsruhe, Germany, 2000; pp. 12–23.

Article

Lifting Scheme-Based Deep Neural Network for Remote Sensing Scene Classification

Chu He [1,2,*], **Zishan Shi** [1], **Tao Qu** [3], **Dingwen Wang** [3] and **Mingsheng Liao** [2]

[1] Electronic and Information School, Wuhan University, Wuhan 430072, China; shizishan@whu.edu.cn
[2] State Key Laboratory of Information Engineering in Surveying, Mapping and Remote Sensing, Wuhan University, Wuhan 430079, China; liao@whu.edu.cn
[3] School of Computer Science, Wuhan University, Wuhan 430072, China; qutaowhu@whu.edu.cn (T.Q.); wangdw@whu.edu.cn (D.W.)
* Correspondence: chuhe@whu.edu.cn; Tel.: +86-27-6875-4367

Received: 28 September 2019; Accepted: 11 November 2019; Published: 13 November 2019

Abstract: Recently, convolutional neural networks (CNNs) achieve impressive results on remote sensing scene classification, which is a fundamental problem for scene semantic understanding. However, convolution, the most essential operation in CNNs, restricts the development of CNN-based methods for scene classification. Convolution is not efficient enough for high-resolution remote sensing images and limited in extracting discriminative features due to its linearity. Thus, there has been growing interest in improving the convolutional layer. The hardware implementation of the JPEG2000 standard relies on the lifting scheme to perform wavelet transform (WT). Compared with the convolution-based two-channel filter bank method of WT, the lifting scheme is faster, taking up less storage and having the ability of nonlinear transformation. Therefore, the lifting scheme can be regarded as a better alternative implementation for convolution in vanilla CNNs. This paper introduces the lifting scheme into deep learning and addresses the problems that only fixed and finite wavelet bases can be replaced by the lifting scheme, and the parameters cannot be updated through backpropagation. This paper proves that any convolutional layer in vanilla CNNs can be substituted by an equivalent lifting scheme. A lifting scheme-based deep neural network (LSNet) is presented to promote network applications on computational-limited platforms and utilize the nonlinearity of the lifting scheme to enhance performance. LSNet is validated on the CIFAR-100 dataset and the overall accuracies increase by 2.48% and 1.38% in the 1D and 2D experiments respectively. Experimental results on the AID which is one of the newest remote sensing scene dataset demonstrate that 1D LSNet and 2D LSNet achieve 2.05% and 0.45% accuracy improvement compared with the vanilla CNNs respectively.

Keywords: scene classification; lifting scheme; convolution; CNN

---

1. Introduction

1.1. Background

High-resolution remote sensing images contain complex geometrical structures and spatial patterns and thus scene classification is a challenge for remote sensing image interpretation. Discriminative feature extraction is vital for scene classification which aims to automatically assign a semantic label to each remote sensing image to know the category it belongs to. Recently, convolutional neural networks (CNNs) has attracted increasing attention in remote sensing community [1,2], which can learn more abstract and discriminative features and achieve the state-of-the-art performance on classification. A typical CNN framework is shown in Figure 1, where three types of modules are cast into, including feature extraction module, quantization module, and trick module. The quantization module includes spatial quantization, amplitude quantization, and evaluation quantization. Pooling is a typical spatial quantization used for downsampling and dimensionality reduction to reduce the number of parameters and calculations while maintaining significant information. Amplitude quantization is a nonlinear process that compresses real values into a specific range. Examples are sigmoid, tanh, ReLU [3] and its variants [4–6], which are often used to activate neurons because nonlinearity is vital for strengthening the approximation and representation abilities. Evaluation quantization is performed to obtain the outputs that meet the specific requirements. For instance, the softmax function is often used in the final hidden layer of a neural network to output classification probabilities. The trick module contains some algorithms such as dropout [7] and batch normalization [8], for improving training effects and achieving better performance. Among all the modules, the feature extraction module is the most essential, which is based on convolution to extract image features. It invests CNN with some favourable characteristics such as local perception, parameter sharing, and translation invariance.

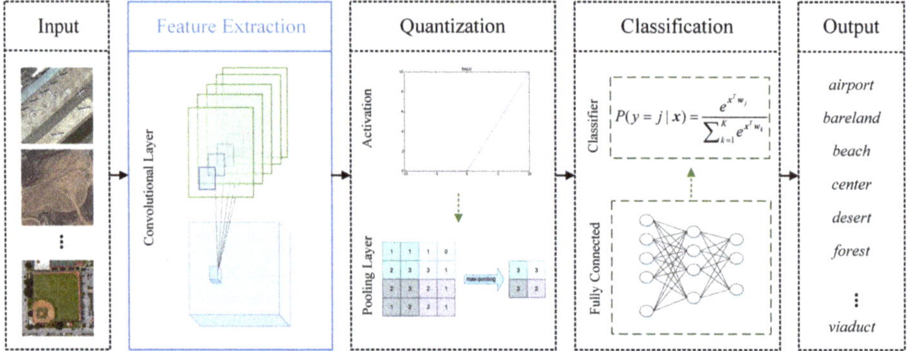

Figure 1. Convolutional neural network (CNN) framework. CNN generally contains three modules, including feature extraction module, quantization module, and tricks module. These three modules are repeatedly stacked to build the deep structure, and finally, a classification module is applied for the specific classification task. The trick module is embedded in the other two modules, which is not displayed in this figure.

The extraction of scene-level discriminative features is the key step of scene classification. In CNN-based methods, convolution is the most critical technology for remote sensing imagery feature extraction. It has the properties of local connection, weight sharing, and translation invariance, which makes full use of the spatial relationship between pixels. However, convolution has some intrinsic drawbacks, such as considerable computational complexity and limitation in transformation capacity due to its linearity. For high-resolution remote sensing images, the computation challenge is severe, and the considerable computational complexity restrains scene classification from that conducted on computationally limited hardware platforms. In addition, it is vital to learn discriminative feature representation of remote sensing images, but the linearity impedes CNNs from extracting more powerful features with better representation and fitting well on complex functions. To extract more representative features, the literature [9] employs collaborative representation after a pre-trained CNN. The literature [10] integrates multilayer features of CNNs. In [11], the last two fully connected layer features of pre-trained VggNet are fused to improve the discriminant ability. In [12], the pre-trained deep CNN is combined with a sparse representation to improve the performance. Most of these methods increase the dimension of features and none of them are proposed to improve the convolutional layer. Therefore, exploration for a better method to extract features is still of great significance for future CNN-based scene classification research.

Convolution is also the paramount operation of the traditional implementation of the wavelet transform (WT). The discrete wavelet transform (DWT) is one branch of WT, which is traditionally implemented by the two-channel filter bank representation [13]. It is widely applied in scientific fields because it has the favourable characteristic of time-frequency localization, but is limited by the considerable computation complexity and the linearity brought by convolution. To overcome those difficulties, W. Sweldens proposed the lifting scheme [14–18], which mainly contains three steps: split, predict, and update. **Split**: The input signal is decomposed into two non-overlapping sample sets. A common method is the lazy wavelet transform, which splits the original signal into an even subset and an odd subset, both of which contain half the number of input samples. **Predict**: The two subsets divided by a signal with strong local correlation are highly correlated in most cases. With one subset known, another subset is reasonably predicted with some predicted error produced. In particular, the predicted errors are all zero for a constant signal. **Update**: One key characteristic of coarse signals is that they have the same average as the input signal, which is guaranteed by this step. The lifting scheme is an efficient algorithm for traditional DWT as Daubechies and W. Sweldens proved that any FIR wavelet transform can be decomposed into a series of prediction and update steps [19]. The form of the lifting scheme makes it easier to perform DWT on a hardware platform, which was adopted as the algorithm used in the wavelet transform module in the JPEG2000 standard [20,21]. Furthermore, the second-generation wavelet, produced directly by designing the prediction and update operators in the lifting scheme [16], can be constructed to perform the nonlinear transformation [22–26]. Briefly, compared with the traditional wavelet transform, the lifting scheme is superior in the following aspects [19]: (1) It leads to a faster implementation of the discrete wavelet transform. (2) It allows a fully in-place implementation of the fast wavelet transform, which signifies that the wavelet transform can be calculated without allocating auxiliary memory. (3) Nonlinear wavelet transforms are easy to build with the lifting scheme. A typical example is the integer wavelet transforms [25].

Therefore, the lifting scheme is a more efficient algorithm for DWT compared with the convolution-based method. In this paper, we prove that the lifting scheme is also an efficient implementation for convolution in the CNN and can substitute the convolutional layer to extract better representations. With the lifting scheme introduced, neural networks share the advantages of nonlinear transformation, all-integral transformation, and the facility of hardware implementation.

1.2. Problems and Motivations

This paper develops the method from the following aspects:

1. Compressing the convolutional layer is critical for easing the computational burden of high-resolution remote sensing imagery analysis and promoting CNN applications on computationally limited platforms. Methods of compressing the convolutional layer are diverse, including utilizing 1×1 convolution to reduce the number of parameters [27,28], depthwise separable convolution [29], and replacing the large size convolutional kernel with several small-size kernels [30]. Most of these methods inevitably result in some accuracy losses. While the lifting scheme is an equivalent implementation for traditional DWT, this paper provides a new perspective on this problem, inspired by the successful application of the lifting scheme on the hardware implementation of the JPEG2000 standard. From this aspect, this paper replaces the vanilla convolutional layer with the lifting scheme, which may be a new direction for better implanting the neural network to computationally limited hardware platforms.

2. Nonlinearity is vital for expanding the function space of CNN and learning discriminative feature representation of remote sensing images. However, the utilization of the convolutional layer limits CNN in extracting favourable representations, where the intrinsic linear characteristic of convolution is responsible. Considering that one of the advantages of the lifting scheme is its extension to nonlinear transformation, this paper proposes a nonlinear feature extractor to enhance the representation ability of neural networks.

3. To insert the lifting scheme into neural networks, for example, to substitute the convolutional layer with the lifting scheme, it is necessary to determine the inner relationship between the lifting scheme and the convolution. Daubechies and W. Sweldens proved that any FIR wavelet transform can be decomposed into a series of lifting and dual lifting steps [19]. In other words, all the first-generation wavelets have the equivalent lifting scheme. However, convolution kernels in CNN are somewhat different from the filters in the two-channel filter bank representation of DWT. In DWT, filter coefficients are pre-designed with some strict restraints and prefixed. In contrast, parameters of convolutional kernels in CNN, which are commonly called weights, are trained and updated ceaselessly in the training stage. Moreover, two-dimensional convolution kernels in the CNN are most non-separable, bringing difficulties for lifting scheme implementation. Therefore, this paper places great emphasis on expanding the few filters that can be replaced with the lifting scheme to convolution kernels with random-valued parameters.
4. The lifting scheme must be compatible with the backpropagation mechanism for optimization. Backpropagation empowers neural networks such that they can learn any arbitrary mapping of input to output [31]. To embed the lifting scheme into neural networks without harming the learning ability, parameters in the lifting scheme must also be trained and updated through backpropagation, which needs to break through some restricts in the original lifting scheme.

1.3. Contributions and Structure

In this paper, the lifting scheme is introduced into neural networks to serve as the feature extraction module, and a lifting scheme-based deep neural network (LSNet) method is proposed to enhance network performance. The main contributions are summarized as follows. (1) This paper introduces the lifting scheme to propose a novel CNN-based method for scene classification, which has potential in easing the computational burden. (2) This paper expands the range of filter bases that can be replaced with the lifting scheme, and the convolution kernel with random-valued parameters are proven to have the equivalent lifting scheme. (3) A learnable lifting scheme block and the backpropagation approach are given. Therefore, any vanilla convolutional layer in CNNs can be replaced with its relative lifting scheme. (4) A novel lifting scheme-based deep neural network (LSNet) model is presented using a nonlinear lifting scheme that is constructed by nonlinear operators. The nonlinear lifting scheme is used as a feature extractor to substitute the vanilla linear convolutional layer to learn discriminative feature representation of remote sensing images. (5) LSNet is validated on the CIFAR-100 and then evaluated on the AID datasets. Experimental results demonstrate that the LSNet outperforms the vanilla CNN and has potential in remote sensing scene classification.

The rest of this paper is organized as follows. Section 2 describes the proposed method and the datasets. Section 3 describes the experimental results on the CIFAR-100 and AID datasets. Section 4 analyzes the results and discusses our future research directions. Section 5 closes with a conclusion.

2. Materials and Methods

In this section, the equivalence between the lifting scheme and the convolution in CNNs is first proven, extending the few wavelet bases that can be replaced with the lifting scheme to convolution kernels with random-valued parameters. With that, the relative lifting scheme is derived for a 1×3 convolution kernel as an example. Finally, we propose a novel lifting scheme-based deep neural network (LSNet), substituting the linear convolutional layers in CNNs with the nonlinear lifting scheme, demonstrating the superiority of the lifting scheme to introduce nonlinearity into the feature extraction module. The datasets used in the experiment are also described in this section.

2.1. Equivalence between the Lifting Scheme and Convolution

In the two-channel filter bank representation of the traditional wavelet transform, a low-pass digital filter h and a high-pass digital filter g are used to process the input signal x, followed by a downsampling operation with base 2, as shown in Equations (1) and (2).

$$a = (x * h) \downarrow 2 \tag{1}$$

$$d = (x * g) \downarrow 2 \tag{2}$$

The filter bank is elaborately designed according to requirements such as compact support and perfect reconstruction, while downsampling is used for redundancy removal. The low-pass filter is highly contacted with the high-pass filter. In other words, with the low-pass filter designed, the high-pass filter is consequently generated. Therefore, the wavelet bases of the first-generation wavelets are finite, restricted, and prefixed. In contrast, parameters of convolution kernels in CNNs are changing ceaselessly during backpropagation, which makes it necessary to expand the lifting scheme to be equivalent to a random-valued filter. In addition, to fit the structure of the convolutional layer, the detailed component d generated by the lifting scheme is removed while the coarse component a is retained in the following proof.

Consider a 1D convolution kernel represented by $h = [h_0, h_1, \ldots, h_{k-1}]$. It is a finite impulse response (FIR) filter from the signal processing perspective, as only a finite number of filter coefficients are non-zero. The z-transform of the FIR filter h is a Laurent polynomial given by

$$H(z) = \sum_{i=0}^{k-1} h_i z^{-i} \tag{3}$$

The degree of the Laurent polynomial $H(z)$ is

$$deg(H(z)) = k - 1 \tag{4}$$

In contrast to the convolution in the signal processing field [32,33], convolution in the CNN omits the reverse operation. The convolution between the input signal x and the convolution kernel h can be written as

$$y = x \odot h = x * \bar{h} \tag{5}$$

where y represents the matrix of output feature maps. Operators "$\odot$" and "$*$" represent the cross-correlation and the convolution in the spotlight of digital signal processing, respectively, while $\bar{h}$ is the reversal signal of h. The z-transform of Equation (5) is

$$Y(z) = X(z)H(z^{-1}) \tag{6}$$

where $H(z^{-1})$ is the z-transform of the reversal sequence $\bar{h} = [h_{k-1}, \ldots, h_1, h_0]$.

In the lifting scheme implementation of traditional wavelets, a common method in the split stage is splitting the original signal $x = [x_0, x_1, x_2, \ldots]$ into an even subset $x_e = [x_0, x_2, \ldots, x_{2k}, \ldots]$ and an odd subset $x = [x_1, x_3, \ldots, x_{2k+1}, \ldots]$. Transforming the signal space to the z-domain, the two-channel filter bank representation shown in Equations (1) and (2) is equivalent to Equation (7), with $A(z)$ and $D(z)$ to represent the z-transform of a and d.

$$\begin{pmatrix} A(z) \\ D(z) \end{pmatrix} = P^T(z^{-1}) \begin{pmatrix} X_e(z) \\ X_o(z) \end{pmatrix} \tag{7}$$

$X_e(z)$ and $X_o(z)$ are the z-transform of x_e and x_o, which are the even subset and odd subset of x, respectively. $P(z)$ is the polynomial matrix of h and g:

$$P(z) = \begin{pmatrix} H_e(z) & G_e(z) \\ H_o(z) & G_o(z) \end{pmatrix} \tag{8}$$

where $H_e(z)$ and $H_o(z)$ are the z-transform of the even subset h_e and the odd subset h_o of h, respectively, while $G_e(z)$ and $G_o(z)$ are the z-transform of the even subset g_e and the odd subset g_o of g, respectively.

Different from the DWT in Equations (1) and (2), there is only one branch in the convolution in CNN, as shown in Equation (6). To adapt to the particular form of the convolution in CNN, we maintain the low-pass filter h while discarding the high-pass filter g, and modify the polynomial matrix to preserve only the relative part of $H(z)$, as in Equation (9).

$$P(z) = \begin{pmatrix} H_e(z) \\ H_o(z) \end{pmatrix} \tag{9}$$

Instead of the lazy wavelet transform, we use the original signal $x = [x_0, x_1, x_2, ...]$ and a time-shift signal $x' = [x_1, x_2, ...]$ as x_e and x_o in the first stage to obtain the same form as Equation (6), as shown in Equation (10).

$$P^T(z^{-2}) \begin{pmatrix} X(z) \\ zX(z)) \end{pmatrix} = (H_e(z^{-2}) + zH_o(z^{-2}))X(z)$$

$$= X(z)H(z^{-1}) \tag{10}$$

Furthermore, the polynomial matrix $P(z)$ can be decomposed into a multiplication form of finite matrices and thus is completed by finite prediction and update steps. As mentioned above, both $H_e(z)$ and $H_o(z)$ are Laurent polynomials. If the following conditions are satisfied:

$$H_e(z) \neq 0, \quad H_o(z) \neq 0$$

$$deg(H_e(z)) \geq deg(H_o(z)) \tag{11}$$

There always exists a Laurent polynomial $q(z)$ (the quotient) with $deg(q(z)) = deg(H_e(z)) - deg(H_o(z))$, and a Laurent polynomial $r(z)$ (the remainder) with $deg(r(z)) < deg(H_o(z))$ so that

$$q(z) = H_e(z)/H_o(z) \tag{12}$$

$$r(z) = H_e(z)\%H_o(z) \tag{13}$$

Iterating the above step, $P(z)$ is then decomposed. As convolution kernels frequently used in CNN are commonly oddly size, such as 1×3, 3×1 [34], 3×3, 5×5, the above conditions are generally satisfied. Comparing Equation (6) with (10), we reach a conclusion that convolution in CNN, which is with random-valued parameters, has equivalent lifting scheme implementation.

2.2. Derivation of the Lifting Scheme for a Relative Convolutional Layer

2.2.1. Lifting Scheme for 1×3 Convolutional Layer

In this subsection, we derive the relative lifting scheme for a 1×3 convolutional layer as an example to illustrate the decomposition of $P(z)$ with the Euclidean algorithm. Given a convolution kernel $h = [h_0, h_1, h_2]$, its polynomial matrix specifically is

$$P(z) = \begin{pmatrix} H_e(z) \\ H_o(z) \end{pmatrix} = \begin{pmatrix} h_0 + h_2 z^{-1} \\ h_1 \end{pmatrix} \tag{14}$$

The first step of decomposition is to divide $H_e(z)$ by $H_o(z)$ and obtain the quotient and remainder.

$$r_1(z) = H_e(z)\%H_o(z) = h_0$$

$$q_1(z) = H_e(z)/H_o(z) = \frac{h_2}{h_1}z^{-1} \tag{15}$$

Then, the divisor $H_o(z)$ in the first step serves as the dividend in the second step, divided by the remainder $r_1(z)$ in the first step.

$$r_2(z) = H_o(z)\%r_1(z) = 0$$
$$q_2(z) = H_o(z)/r_1(z) = \frac{h_1}{h_0} \tag{16}$$

Iteration stops after two steps as the final remainder is 0 and the dividend in the final step $H_o(z) = h_1 = gcd(H_e(z), H_o(z))$, where the operator $gcd(\cdot)$ stands for the greatest common divisor.

With quotients and remainders obtained above, $P(z)$ is decomposed into the form of matrix multiplication, as Equation (17) shows.

$$P(z) = \begin{pmatrix} q_1(z) & 1 \\ 1 & 0 \end{pmatrix} \begin{pmatrix} q_2(z) & 1 \\ 1 & 0 \end{pmatrix} \begin{pmatrix} h_0 \\ 0 \end{pmatrix} = \begin{pmatrix} 1 & \frac{h_2}{h_1}z^{-1} \\ 0 & 1 \end{pmatrix} \begin{pmatrix} 1 & 0 \\ \frac{h_1}{h_0} & 1 \end{pmatrix} \begin{pmatrix} h_0 \\ 0 \end{pmatrix} \tag{17}$$

Therefore,

$$P^T(z^{-2}) = \begin{pmatrix} h_0 & 0 \end{pmatrix} \begin{pmatrix} 1 & \frac{h_1}{h_0} \\ 0 & 1 \end{pmatrix} \begin{pmatrix} 1 & 0 \\ \frac{h_2}{h_1}z^2 & 1 \end{pmatrix} \tag{18}$$

Given the matrix multiplication form in Equation (18), parameters of the lifting scheme, including the prediction and update operators and the scaling factor, are thus obtained. The 1×3 convolutional layer and its equivalent lifting scheme in one spatial plane are shown in Figure 2. In the 1×3 convolution, the convolution kernel moves across the input row by row in the horizontal direction to perform the sliding dot product. The lifting scheme can simultaneously process the entire input image within each of the steps listed in Table 1.

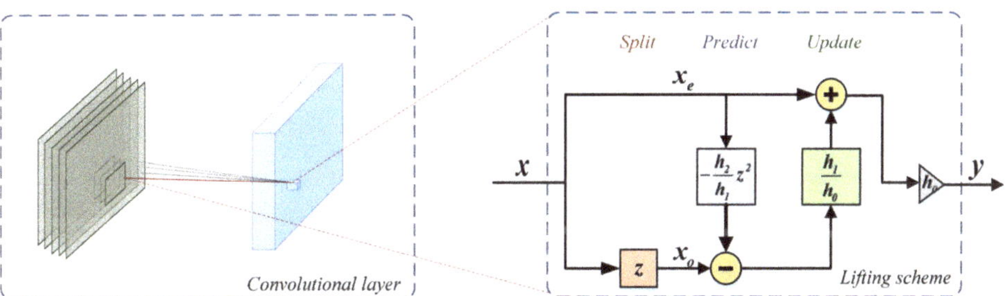

Figure 2. Equivalence between the 1×3 convolution and the lifting scheme. The convolution kernel moves across the entire input image row by row to conduct the sliding dot product and generate the output feature maps. The lifting scheme can process the entire input image in parallel within each step. The lifting scheme implementation with three steps is equivalent to a vanilla 1×3 convolution in one plane.

As illustrated in Table 1, the input image passes through the three steps in the lifting scheme. In the split stage, the input is transformed into 2 branches, x_e and x_o. Different from the lifting scheme in [14], we utilize a sliding window to obtain two branches instead of the lazy wavelet transform. Then, x_o is predicted by x_e with the predict operator $\frac{h_2}{h_1}z^2$, the outcome of which is used to update x_e. The final output is obtained by scaling the updated x_e by the factor h_0.

Table 1. Steps of the lifting scheme equivalent to 1×3 convolutional layer.

Step	Illustration
split	$x_e[i] = x[i]$, $x_o[i] = x[i+1]$
predict	$x_o[i] = x_o[i] + \frac{h_2}{h_1} \times x_e[i+2]$
update	$x_e[i] = x_e[i] + \frac{h_1}{h_0} \times x_o[i]$
scaling	$y[i] = h_0 \times x_e[i]$

Note that notation $x[i]$ represents the ith column in image x.

2.2.2. The Lifting Scheme for the 2D Convolutional Layer

For a convolutional layer with 2D convolutional kernels, the lifting scheme can also be realized through the 1D lifting scheme. The 2D convolution operation is the sum of several 1D convolution operations. For instance, the output of the convolution operation between a $m \times n$ input image and a 3×3 convolutional kernel has a size of $(m-2) \times (n-2)$. Then, the element on the i-th row, j-th column of the output matrix is obtained by

$$y(i,j) = \sum_{u=0}^{2} \sum_{v=0}^{2} x(i+u, j+v) \cdot h(u,v) \tag{19}$$

which can be rewritten as

$$y(i,j) = \sum_{v=0}^{2} x(i,j+v) \cdot h(0,v) + \sum_{v=0}^{2} x(i+1,j+v) \cdot h(1,v) + \sum_{v=0}^{2} x(i+2,j+v) \cdot h(2,v) \tag{20}$$

In other words, the 2D convolution operation is equivalent to the summation of three 1D convolution operations. As the relative lifting scheme of the 1D convolution has been worked out as the right side of Figure 2 shown, the 2D lifting scheme is simply the summation of the relative 1D lifting scheme.

Thus, any convolution kernels with random-valued parameters have equivalent lifting scheme implementation. As the prediction operator and update operator in the lifting scheme can be designed to fit other requirements, which correspond to a larger serving range than vanilla convolution, the lifting scheme is the superset of vanilla convolution. In other words, the vanilla convolutional layer is just a special case of the lifting scheme. To improve the feature extractor in the CNN, a more powerful lifting scheme structure can be designed by choosing other predict operators and update operators.

2.2.3. Backpropagation of the Lifting Scheme

In this part, we propose the backpropagation algorithm to train and update parameters in the lifting scheme. For simplicity, a new list of weights $w = [w_0, w_1, w_2]$ is used to represent the parameters in Figure 2 as

$$w_0 = h_0, w_1 = \frac{h_1}{h_0}, w_2 = \frac{h_2}{h_1} \tag{21}$$

Assume the backpropagation error from the next layer is L, then the gradient with respect to this layer's output y is obtained by

$$\Delta y(i,j) = \frac{\partial L}{\partial y(i,j)} \tag{22}$$

where $y(i,j)$ represents one pixel in the output feature map y.

According to the derivation rule, the gradients with respect to the weights are obtained as follows.

$$\Delta w_0 = \sum_i \sum_j \frac{\partial L}{\partial y(i,j)} \frac{\partial y(i,j)}{\partial w_0} = \Delta \boldsymbol{y} \odot \boldsymbol{x} + w_1 \cdot \Delta \boldsymbol{y} \odot \boldsymbol{x}' + w_1 \cdot w_2 \cdot \boldsymbol{x}'' \tag{23}$$

$$\Delta w_1 = \sum_i \sum_j \frac{\partial L}{\partial y(i,j)} \frac{\partial y(i,j)}{\partial w_1} = w_0 \cdot \Delta \boldsymbol{y} \odot \boldsymbol{x}' + w_0 \cdot w_2 \cdot \Delta \boldsymbol{y} \odot \boldsymbol{x}'' \tag{24}$$

$$\Delta w_2 = \sum_i \sum_j \frac{\partial L}{\partial y(i,j)} \frac{\partial y(i,j)}{\partial w_2} = w_0 \cdot w_1 \cdot \Delta \boldsymbol{y} \odot \boldsymbol{x}'' \tag{25}$$

where $\boldsymbol{x}'$ is the map whose elements in each row are the one-position left shifting elements in the corresponding row of $\boldsymbol{x}$. $\boldsymbol{x}''$ is the map whose elements in each row are the two-position left shifting elements in the corresponding row of $\boldsymbol{x}$. $\boldsymbol{x}'$ and $\boldsymbol{x}''$ maintain the same size as $\boldsymbol{x}$ with some boundary extension methods. The operator "$\odot$" represents the cross-correlation between two matrices. With these gradients, the weights can be updated by stochastic gradient descent.

2.3. Lifting Scheme-Based Deep Neural Network

In this section, the lifting scheme is introduced into the deep learning field, and a lifting scheme-based deep neural network (LSNet) method is proposed to enhance network performance. From Sections 2.1 and 2.2, the lifting scheme can substitute the convolutional layer because it can perform convolution and utilize backpropagation to update parameters. Specifically, operators in the lifting scheme are flexible, which can be designed not only to make the lifting scheme equivalent to vanilla convolutional layers but also extended to meet other requirements. Thus, we develop the LSNet with nonlinear feature extractors utilizing the ability of nonlinear transformation of the lifting scheme.

2.3.1. Basic Block in LSNet

Nonlinearity enables neural networks to fit complex functions and thus strengthens their representation ability. As the lifting scheme is capable of constructing nonlinear wavelets, we introduce nonlinearity into the feature extraction module to build the LSNet. It is realized by designing nonlinear predict and update operators in the lifting scheme, which demonstrates the enormous potential of the lifting scheme to perform nonlinear transformation and enhance the nonlinear representation of the neural network.

We construct the basic block in LSNet based on ResNet34 [35], as shown in Figure 3. The first layer in the basic block is a 3×3 convolutional layer, which is used to change the number of channels and downsampling. The middle layer is the LS block, mainly for feature extraction, with the same number of channels between the input and the output. The plug-and-play LS block is used to substitute the vanilla convolutional layer without any other alterations. In the LS block, the input is split into two parts, including x_e and x_o. The nonlinear predict operator and update operator are constructed by a vanilla convolution kernel followed by a nonlinear function. x_o is then predicted by x_e to gain the detailed component, which is discarded after the update step. x_e is updated by the detailed component, the outcome of which is the coarse component and used as the final output of the LS block. Finally, we add a 1×1 convolutional layer as the third layer to enhance channel-wise communication. Batch normalization and ReLU are followed by the first two layers for overfitting avoidance and activation, respectively. The identity of the input through a shortcut is added to the output of the third layer, followed by batch normalization. The addition outcome is again activated by ReLU, which is the final output of the basic block.

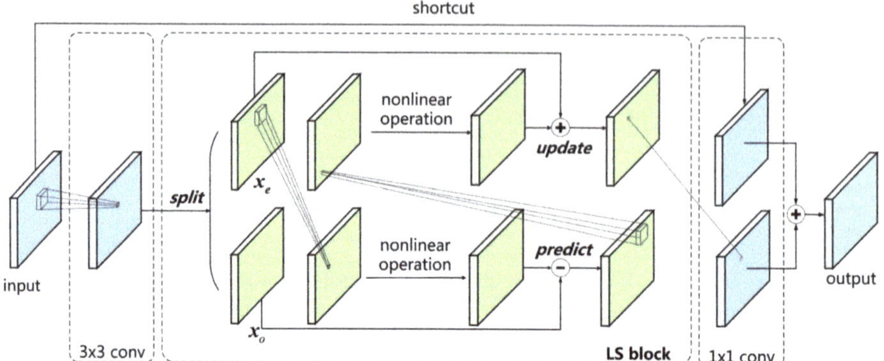

Figure 3. Basic block in LSNet.

As both the 1D and 2D convolutional layers are widely used, we propose the 1D and 2D LS block, named the LS1D block and LS2D block, respectively. The specific processes of which are illustrated in Table 2.

Table 2. Lifting scheme steps of LS1D block and LS2D block.

Step	LS1D Block	LS2D Block
split	$x_e[i] = x[i],\ x_o[i] = x[i+1]$	$x_e[i,j] = x[i,j],\ x_o[i] = x[i+1,j+1]$
predict	$x_o = x_o - \mathcal{N}\{\ \text{Pconv1D}(x_e)\ \}$	$x_o = x_o - \mathcal{N}\{\ \text{Pconv2D}(x_e)\ \}$
update	$x_e = x_e + \mathcal{M}\{\ \text{Uconv1D}(x_o)\ \}$	$x_e = x_e + \mathcal{M}\{\ \text{Uconv2D}(x_o)\ \}$

Note that notation $x[i]$ represents the ith column in image x while the notation $x[i,j]$ represents the ith row, jth column pixel in the image x.

Note that $\mathcal{N}\{\cdot\}$ and $\mathcal{M}\{\cdot\}$ denote the nonlinear transformation in the predict step and the update step, respectively. Pconv$n$D$(\cdot)$ and Uconv$n$D$(\cdot)$ represent the vanilla nD convolution operation in the predict and the update step, respectively. The operators $\mathcal{N}\{\ \text{Pconv}n\text{D}(\cdot)\ \}$ and $\mathcal{N}\{\ \text{Uconv}n\text{D}(\cdot)\ \}$ are the prediction and update operators, respectively, which are changeable to meet other requirements.

2.3.2. Network Architecture and Settings

The network architecture of LSNet is shown in Figure 4. The input of LSNet passes through a single LS block and a 1×1 convolutional layer for initial feature extraction. The feature maps are then processed by stacked basic blocks to obtain low dimension image representation, which is followed by an average pooling for dimension reduction. Finally, the representation is unfolded as a 1D vector, and processed by a fully connected layer and a softmax function to obtain the output.

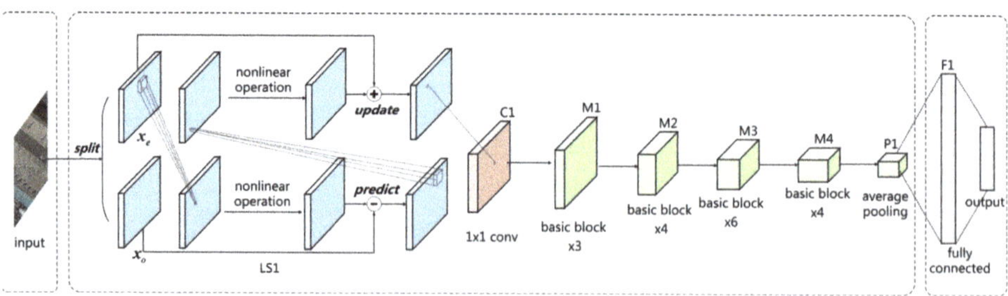

Figure 4. LSNet architecture.

The modified ResNet34 is chosen as the baseline model to evaluate the performance of LSNet. Experiments are separately conducted for the 1D convolution and the 2D convolution, as both are widely used. The setups of each network are listed in Table 3.

In Table 3, the size and the number of convolution kernels in vanilla convolutional layers are listed. For LSNet, the structure of the LS1D block and LS2D block are shown in Table 2, while the number represents the depth of the output. To demonstrate the effect of nonlinearity, we use five different nonlinear functions to construct different LS blocks to compare the performance, which are ReLU [3], leaky ReLU [4], ELU [5], and CELU and SELU [6].

Table 3. Network architectures in contrast experiments.

Layer	ResNet-1d	LSNet-1d	ResNet-2d	LSNet-2d
LS1	$1 \times 3, 3$	LS1D block, 3	$1 \times 3, 3$	LS2D block, 3
C1		$1 \times 1, 64$		
M1	$\begin{bmatrix} 1 \times 3, 64 \\ 1 \times 3, 64 \\ 1 \times 1, 64 \end{bmatrix} \times 3$	$\begin{bmatrix} 1 \times 3, 64 \\ \text{LS1D block}, 64 \\ 1 \times 1, 64 \end{bmatrix} \times 3$	$\begin{bmatrix} 3 \times 3, 64 \\ 3 \times 3, 64 \\ 1 \times 1, 64 \end{bmatrix} \times 3$	$\begin{bmatrix} 3 \times 3, 64 \\ \text{LS2D block}, 64 \\ 1 \times 1, 64 \end{bmatrix} \times 3$
M2	$\begin{bmatrix} 1 \times 3, 128 \\ 1 \times 3, 128 \\ 1 \times 1, 128 \end{bmatrix} \times 4$	$\begin{bmatrix} 1 \times 3, 128 \\ \text{LS1D block}, 128 \\ 1 \times 1, 128 \end{bmatrix} \times 4$	$\begin{bmatrix} 3 \times 3, 128 \\ 3 \times 3, 128 \\ 1 \times 1, 128 \end{bmatrix} \times 4$	$\begin{bmatrix} 3 \times 3, 128 \\ \text{LS2D block}, 128 \\ 1 \times 1, 128 \end{bmatrix} \times 4$
M3	$\begin{bmatrix} 1 \times 3, 256 \\ 1 \times 3, 256 \\ 1 \times 1, 256 \end{bmatrix} \times 6$	$\begin{bmatrix} 1 \times 3, 256 \\ \text{LS1D block}, 256 \\ 1 \times 1, 256 \end{bmatrix} \times 6$	$\begin{bmatrix} 3 \times 3, 256 \\ 3 \times 3, 256 \\ 1 \times 1, 256 \end{bmatrix} \times 6$	$\begin{bmatrix} 3 \times 3, 256 \\ \text{LS2D block}, 256 \\ 1 \times 1, 256 \end{bmatrix} \times 6$
M4	$\begin{bmatrix} 1 \times 3, 512 \\ 1 \times 3, 512 \\ 1 \times 1, 512 \end{bmatrix} \times 3$	$\begin{bmatrix} 1 \times 3, 512 \\ \text{LS1D block}, 512 \\ 1 \times 1, 512 \end{bmatrix} \times 3$	$\begin{bmatrix} 3 \times 3, 512 \\ 3 \times 3, 512 \\ 1 \times 1, 512 \end{bmatrix} \times 3$	$\begin{bmatrix} 3 \times 3, 512 \\ \text{LS2D block}, 512 \\ 1 \times 1, 512 \end{bmatrix} \times 3$
		average pooling, fully connected, Softmax		

In the experiment on the AID dataset, we use the stochastic gradient descent (SGD) as an optimizer, with a momentum of 0.9 and a weight decay of 5×10^{-4}. We train the training set for 100 epochs, setting the mini-batch size as 32. The learning rate is 0.01 initially, which decreases by 5 times every 25 epochs. For the CIFAR-100 dataset, all the settings are the same except that the mini-batch size is 128, and the initial learning rate is 0.1.

2.4. Materials

LSNet is firstly validated using the CIFAR-100 dataset [36], which is one of the most widely used datasets for deep learning reaches. Then, we conduct experiments on the AID dataset [37] to demonstrate the effectiveness of LSNet on the scene classification task.

CIFAR-100 dataset: This dataset contains 60,000 images, which are grouped into 100 classes. Each class contains 600 32×32 colored images, which are further divided into 500 training images and 100 testing images. The 100 classes in the CIFAR-100 dataset are grouped into 20 superclasses. In this dataset, a "fine" label indicates the class to which the image belongs, while a "coarse" label indicates the superclass to which the image belongs.

AID dataset: The AID dataset is a large-scale high-resolution remote sensing dataset proposed by Xia et al. [37] for aerial scene classification. With high intra-class diversity and low inter-class dissimilarity, the AID dataset is suitable as the benchmark for aerial scene classification models. Thirty classes are included, each with 220 up to 420 600 × 600 images. In our experiment, 80% images of each class are chosen as the training data, while the rest 20% are chosen as testing data. Each image is resized to 64 × 64. Some samples of the AID dataset are shown in Figure 5.

Figure 5. Some samples of AID dataset.

3. Results

In this section, LSNet is firstly validated using the CIFAR-100 dataset. Then, we put emphasis on evaluating LSNet on the AID dataset to demonstrate the effectiveness of LSNet on the scene classification task, the performance of which is compared with ResNet34 that utilizes vanilla convolution.

3.1. Results on the CIFAR-100 Dataset

The overall accuracy of the CIFAR-100 test set is used to evaluate the performance of LSNet, as shown in Tables 4 and 5. This metric is defined as the ratio of the number of correctly predicted samples to the number of samples in the whole test set, as shown in Equation (26).

$$p_c = \frac{\sum_{k=1}^{n} P_{kk}}{N} \tag{26}$$

where P_{kk} represents the number of correctly classified samples in the kth class, while n and N are the number of categories and the number of samples in the whole test set, respectively. Compared with metrics that evaluate each category, p_c is more intuitive and straightforward for comparing LSNet and ResNet on the CIFAR-100 dataset, which has 100 categories. The parameter Delta is the accuracy difference between LSNet-1d and ResNet-1d to demonstrate the performance of LSNet more intuitively.

Table 4. Performance of networks on the CIFAR-100 dataset in the 1D experiment.

Method	ResNet-1d	LSNet-1d				
		ReLU	Leaky ReLU	ELU	CELU	SELU
Acc.(test)	73.00%	74.92%	74.71%	**75.48%**	74.98%	75.21%
Delta	–	+1.92%	+1.71%	**+2.48%**	+1.98%	+2.21%

In the 1D experiment, all of the LSNets that utilize different nonlinear functions in the LS1D block outperform the baseline network ResNet-1d by more than 1.5%. In between, LSNet-1d with ELU used in the LS1D block reaches the best test set accuracy, which is superior to the baseline network by 2.48%.

In the 2D experiment, LSNet-2d, whose LS2D block uses SELU as the nonlinear function, is slightly inferior to the baseline network ResNet-2d, while other LSNets all outperform the baseline network. In between, LSNet-2d with leaky ReLU used in the LS2D block performs best, which advantages over ResNet-2d by 1.38%.

Table 5. Performance of networks on the CIFAR-100 dataset in the 2D experiment.

Method	ResNet-2d	LSNet-2d				
		ReLU	Leaky ReLU	ELU	CELU	SELU
Acc.(test)	74.64%	75.52%	76.02%	75.76%	76.00%	74.21%
Delta	−	+0.88%	+1.38%	+1.12%	+1.36%	−0.43%

The experimental results confirm the advantage of introducing nonlinearity into the feature extractor module. With different nonlinear functions, LSNets perform differently, which indicates the importance of choosing suitable nonlinear predict and update operators. For the 1D and 2D experiments, the most appropriate nonlinear functions are different, indicating that they are structure-dependent.

3.2. Results on the AID Dataset

For each evaluated network, the results of the AID test set's overall accuracy are listed in Table 6. For the 1D experiment, all LSNets with different nonlinear functions used in the LS1D block outperform the baseline ResNet-1d. Therefore, LSNet-1d with leaky ReLU performs similarly to ResNet-1d, while LSNet-1d with ELU reaches the highest test set overall accuracy, which is superior to ResNet-1d by 2.05%.

Table 6. Performance of networks on the AID dataset in the 1D experiment.

Method	ResNet-1d	LSNet-1d				
		ReLU	Leaky ReLU	ELU	CELU	SELU
Acc.(test)	82.45%	84.40%	82.55%	83.90%	84.50%	83.50%
Delta	−	+1.95%	+0.10%		1.45%	+2.05%	+1.05%

Confusion matrices for each evaluated network are shown in Figure 6. The probabilities whose values are more than 0.01 are displayed on the confusion matrices. Figure 6b–f are sparser than Figure 6a, which indicates smaller error rates and higher recalls. As Figure 6a shows, ResNet-1d performs well on some classes, such as baseball field, beach, bridge, desert, and forest, the recalls of which are higher than 95%. However, ResNet-1d confuses some classes with other classes. For instance, 20% of the images of the park class are mistaken as the resort class, while approximately 10% of the images of the mountain class are cast incorrectly into the dense residential class.

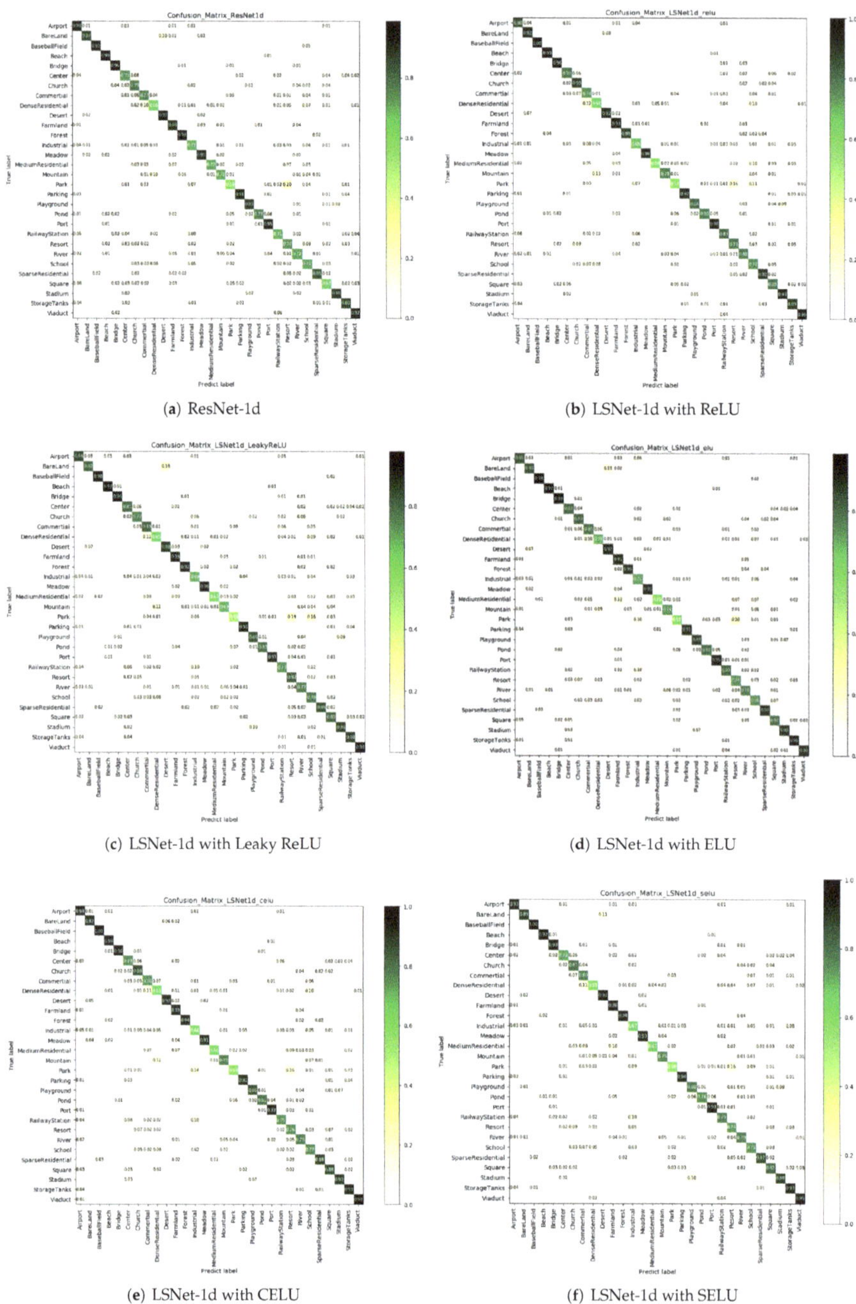

(**a**) ResNet-1d

(**b**) LSNet-1d with ReLU

(**c**) LSNet-1d with Leaky ReLU

(**d**) LSNet-1d with ELU

(**e**) LSNet-1d with CELU

(**f**) LSNet-1d with SELU

Figure 6. Confusion matrices of the AID dataset in the 1D experiment. The number on the *i*th row, *j*th column represents the normalized number of the images in the *i*th class that are classified as the *j*th class. The numbers below 0.01 are not displayed on the confusion matrices.

For further comparison between LSNet-1d and ResNet-1d, we select the classes whose recalls are below 75% in the ResNet-1d experiment and make further analysis. As shown in Figure 7, ResNet-1d surpasses all 1D LSNets on the medium residential class, but it is inferior to all 1D LSNets on five classes including railway station, resort, river, school, and square. All 1D LSNets are well ahead of ResNet-1d on the square class by more than 10%, which demonstrates that features extracted by nonlinear lifting scheme can better distinguish this class from all other classes.

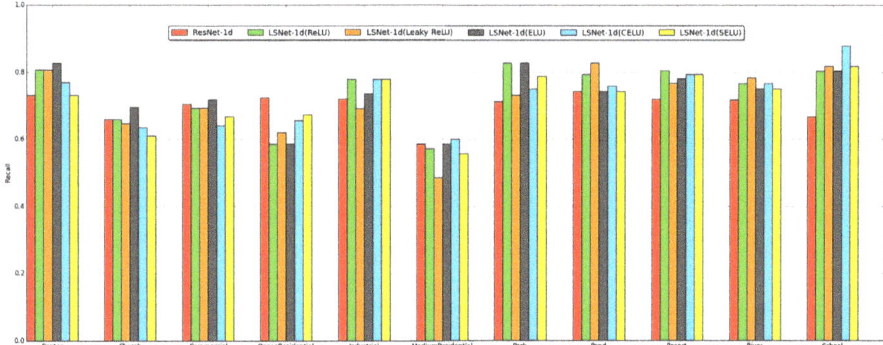

Figure 7. Network performance on partial classes of the AID dataset. The classes whose test set recalls are below 75% in the ResNet-1d experiment are selected for recall comparison between 1D LSNets and ResNet-1d.

In the 2D experiment, LSNets slightly enhance the performance compared with ResNet-2d, as shown in Table 7. In between, LSNet-2d with ReLU used in the LS2D block achieves the highest test set accuracy, which is superior to ResNet-2d by 0.45%. Constructed by different nonlinear predict and update operators, the LS blocks construct different LSNets, which are distinguished in performance. This fact indicates the significance of seeking suitable nonlinear operators.

Table 7. Performance of networks on the AID dataset in the 2D experiment.

Method	ResNet-2d	LSNet-2d				
		ReLU	Leaky ReLU	ELU	CELU	SELU
Acc.(test)	83.30%	**83.75%**	83.20%	83.70%	83.55%	83.65%
Delta	—	**+0.45%**	−0.10%	+0.40%	+0.25%	+0.35%

To explore the comparison between 2D LSNets and ResNet-2d, confusion matrices are drawn to displace the output probabilities whose values are more than 0.01, as shown in Figure 8. Recalls and error rates are displaced for each class. It can be determined that 2D LSNets perform better on some classes, such as the sparse residential class and the viaduct class, where all types of 2D LSNets are superior to ResNet-2d.

Furthermore, we select the classes whose recalls are below 80% in the ResNet-2d experiment and conduct further analysis. As shown in Figure 9, LSNets perform better on most of these classes. For instance, LSNet-2d with ReLU and LSNet-2d with ELU are superior to ResNet-2d by 8.4%, while LSNet-2d with CELU surpasses ResNet-2d by 7.8% on the commercial class. Moreover, all types of LSNets are better than ResNet-2d in river class. This fact indicates that the nonlinear lifting scheme provides an advantage in extracting the features to distinguish the river class from all other classes.

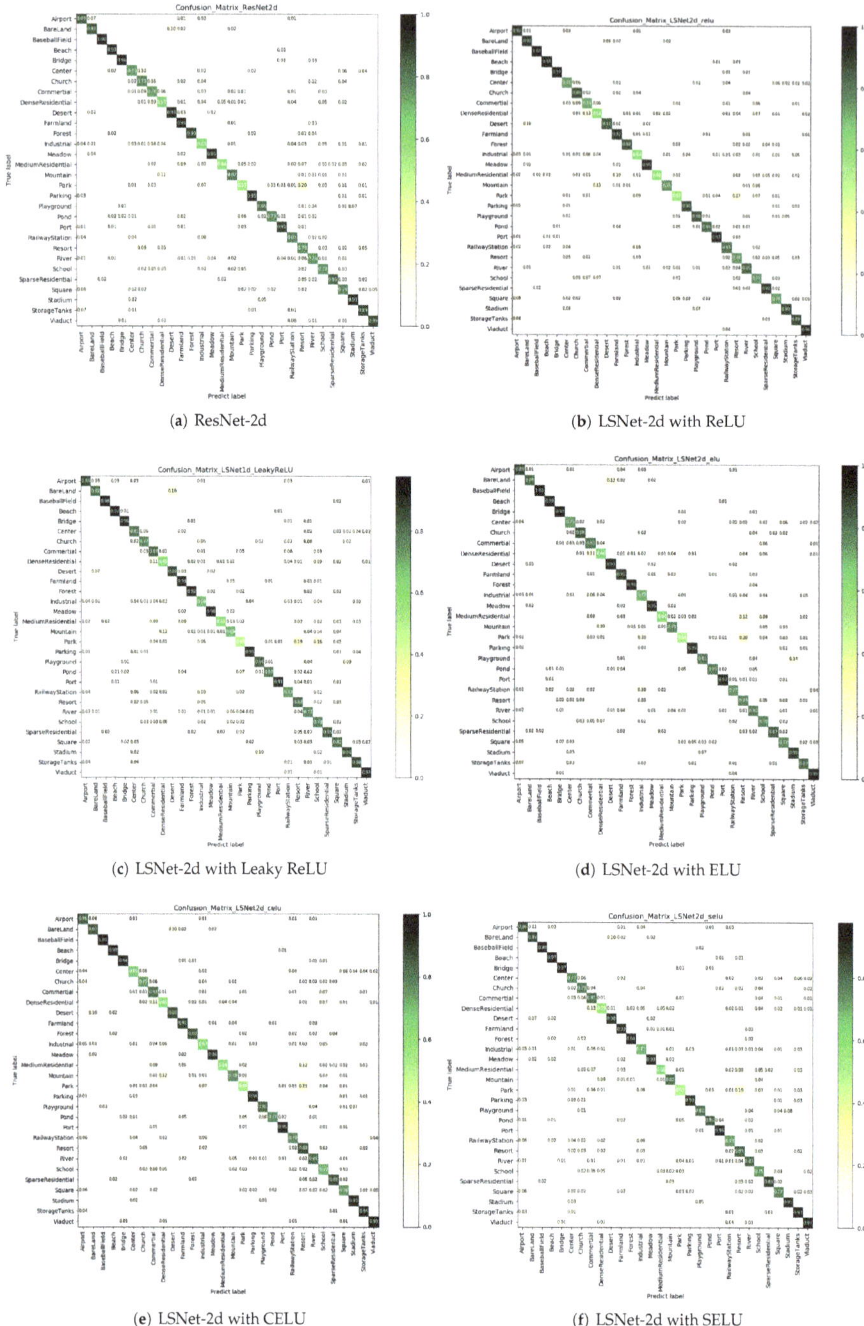

(**a**) ResNet-2d

(**b**) LSNet-2d with ReLU

(**c**) LSNet-2d with Leaky ReLU

(**d**) LSNet-2d with ELU

(**e**) LSNet-2d with CELU

(**f**) LSNet-2d with SELU

Figure 8. Confusion matrices of the AID dataset in the 2D experiment. The number on the ith row, jth column represents the normalized number of the images in the ith class that are classified as the jth class. The numbers below 0.01 are not displayed on the confusion matrices.

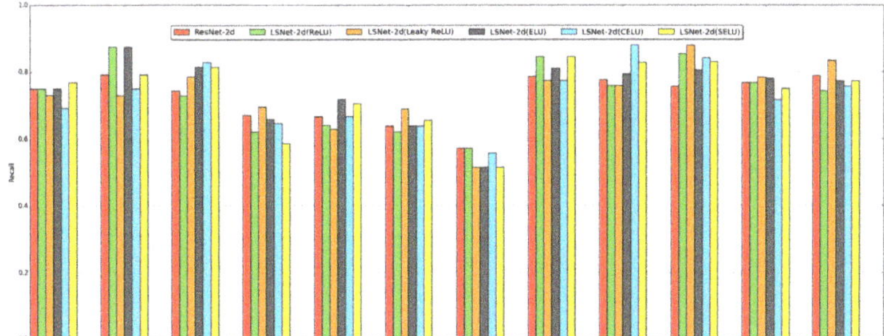

Figure 9. Networks' performance on partial classes of the AID dataset. The classes whose test set recalls are below 80% are selected for recall comparison between 2D LSNets and ResNet-2d.

4. Discussion

Experimental results indicate that the proposed LSNet performs better than ResNet that utilizes the vanilla convolutional layer. An analysis of the improvement is as follows.

1. LSNet can extract more discriminative features with the nonlinearity introduced by the lifting scheme. Corresponding to larger function space, features extracted from the nonlinear module are more distinguishable, which contains the information that cannot be represented by a linear convolutional layer. Therefore, LSNet reaches higher recall in a single class and perform better on the overall accuracy metric.
2. The lifting scheme shown in Figure 2 is more suitable for hardware implementation with the advantages of in-place operation, acceleration, and auxiliary storage free, which can process the entire input image in parallel within each step. In the LSNet, the lifting scheme block performs prediction and update steps without transitional auxiliary storages. This characteristic of the LSNet is expected to provide the potential for easing the computational burden of high-resolution remote sensing imagery analysis and promoting applications on computationally limited hardware platforms.

In the future, we will introduce other nonlinear wavelets into neural networks, such as the morphology wavelet and the adaptive wavelet, to further enhance the nonlinear representation of neural networks. In addition, we will attempt to implant LSNet into some computationally limited platforms, such as FPGA, and compare its performance with existing methods.

5. Conclusions

CNN-based methods for scene classification are restricted by the computational challenge and limited ability to extract discriminative features. In this paper, the lifting scheme is introduced into deep learning and a lifting scheme-based deep neural network (LSNet) is proposed for remote sensing scene classification. The innovation of this approach lies in its capability to introduce nonlinearity into the feature extraction module to extend the feature space. The lifting scheme is an efficient algorithm for the wavelet transform to fit on the hardware platforms, which shows the potential of LSNet to ease the computational burden. Experiments on the AID datasets demonstrate that LSNet-1d and LSNet-2d are superior to ResNet-1d and ResNet-2d by 2.05% and 0.45%, respectively. The method proposed in this paper has room for further improvement, and we will introduce other nonlinear wavelets into neural networks and implant LSNet into some computationally limited platforms in the future.

Author Contributions: Methodology, C.H. and Z.S.; software, T.Q.; writing—original draft preparation, Z.S.; writing—review and editing, C.H.; supervision, D.W. and M.L.

Funding: This research was funded by the National Key Research and Development Program of China (No. 2016YFC0803000), the National Natural Science Foundation of China (No. 41371342 and No. 61331016), and the Hubei Innovation Group (2018CFA006).

Conflicts of Interest: The authors declare no conflict of interest.

Abbreviations

The following abbreviations are used in this manuscript:

CNN	Convolutional Neural Network
WT	Wavelet Transform
LSNet	Lifting Scheme Based Deep Neural Network
JPEG	Joint Photographic Experts Group
CIFAR	Canadian Institute for Advanced Research
AID	Aerial Image Dataset
DWT	Discrete Wavelet Transform
SGD	Stochastic Gradient Descent

References

1. Hu, F.; Xia, G.S.; Hu, J.; Zhang, L. Transferring deep convolutional neural networks for the scene classification of high-resolution remote sensing imagery. *Remote Sens.* **2015**, *7*, 14680–14707. [CrossRef]
2. Castelluccio, M.; Poggi, G.; Sansone, C.; Verdoliva, L. Land use classification in remote sensing images by convolutional neural networks. *arXiv* **2015**, arXiv:1508.00092.
3. Nair, V.; Hinton, G.E. Rectified linear units improve restricted boltzmann machines. In Proceedings of the 27th International Conference on Machine Learning (ICML-10), Haifa, Israel, 21–24 June 2010; pp. 807–814.
4. Maas, A.L.; Hannun, A.Y.; Ng, A.Y. Rectifier nonlinearities improve neural network acoustic models. In Proceedings of the International Conference on Machine Learning (ICML), Atlanta, GA, USA, 16–21 June 2013; Volume 30, p. 3.
5. Clevert, D.A.; Unterthiner, T.; Hochreiter, S. Fast and accurate deep network learning by exponential linear units (elus). *arXiv* **2015**, arXiv:1511.07289.
6. Klambauer, G.; Unterthiner, T.; Mayr, A.; Hochreiter, S. Self-normalizing neural networks. In *Advances in Neural Information Processing Systems, Proceedings of the NIPS 2017, Long Beach, CA, USA, 4–9 December 2017*; Neural Information Processing Systems Foundation: San Diego, CA, USA, 2017; pp. 971–980, arXiv:1706.02515v5.
7. Srivastava, N.; Hinton, G.; Krizhevsky, A.; Sutskever, I.; Salakhutdinov, R. Dropout: A simple way to prevent neural networks from overfitting. *J. Mach. Learn. Res.* **2014**, *15*, 1929–1958.
8. Ioffe, S.; Szegedy, C. Batch normalization: Accelerating deep network training by reducing internal covariate shift. *arXiv* **2015**, arXiv:1502.03167.
9. Liu, B.D.; Meng, J.; Xie, W.Y.; Shao, S.; Li, Y.; Wang, Y. Weighted Spatial Pyramid Matching Collaborative Representation for Remote-Sensing-Image Scene Classification. *Remote Sens.* **2019**, *11*, 518. [CrossRef]
10. Li, E.; Xia, J.; Du, P.; Lin, C.; Samat, A. Integrating Multilayer Features of Convolutional Neural Networks for Remote Sensing Scene Classification. *IEEE Trans. Geosci. Remote Sens.* **2017**, *55*, 5653–5665. [CrossRef]
11. Chaib, S.; Liu, H.; Gu, Y.; Yao, H. Deep Feature Fusion for VHR Remote Sensing Scene Classification. *IEEE Trans. Geosci. Remote Sens.* **2017**, *55*, 4775–4784. [CrossRef]
12. Flores, E.; Zortea, M.; Scharcanski, J. Dictionaries of deep features for land-use scene classification of very high spatial resolution images. *Pattern Recognit.* **2019**, *89*, 32–44. [CrossRef]
13. Mallat, S.G. A theory for multiresolution signal decomposition: The wavelet representation. *IEEE Trans. Pattern Anal. Mach. Intell.* **1989**, *11*, 674–693. [CrossRef]
14. Sweldens, W. Lifting scheme: A new philosophy in biorthogonal wavelet constructions. *Proc. SPIE* **1995**, *2569*, 68–80.
15. Sweldens, W. The lifting scheme: A custom-design construction of biorthogonal wavelets. *Appl. Comput. Harmon. Anal.* **1996**, *3*, 186–200. [CrossRef]
16. Sweldens, W. The lifting scheme: A construction of second generation wavelets. *SIAM J. Math. Anal.* **1998**, *29*, 511–546. [CrossRef]
17. Sweldens, W. Wavelets and the lifting scheme: A 5 minute tour. *ZAMM-Zeitschrift fur Angewandte Mathematik und Mechanik* **1996**, *76*, 41–44.

18. Sweldens, W.; Schröder, P. Building your own wavelets at home. In *Wavelets in the Geosciences*; Springer: Berlin, Germany, 2000; pp. 72–107.

19. Daubechies, I.; Sweldens, W. Factoring wavelet transforms into lifting steps. *J. Fourier Anal. Appl.* **1998**, *4*, 247–269. [CrossRef]

20. Skodras, A.; Christopoulos, C.; Ebrahimi, T. The Jpeg 2000 Still Image Compression Standard. *IEEE Signal Process. Mag.* **2001**, *18*, 36–58. [CrossRef]

21. Lian, C.J.; Chen, K.F.; Chen, H.H.; Chen, L.G. Lifting based discrete wavelet transform architecture for JPEG2000. In Proceedings of the 2001 IEEE International Symposium on Circuits and Systems (ISCAS 2001), Sydney, NSW, Australia, 6–9 May 2001; (Cat. No. 01CH37196); IEEE: Piscataway, NJ, USA, 2001; Volume 2, pp. 445–448.

22. Heijmans, H.J.; Goutsias, J. Nonlinear multiresolution signal decomposition schemes. II. Morphological wavelets. *IEEE Trans. Image Process.* **2000**, *9*, 1897–1913. [CrossRef]

23. Claypoole, R.L.; Baraniuk, R.G.; Nowak, R.D. Adaptive wavelet transforms via lifting. In Proceedings of the 1998 IEEE International Conference on Acoustics, Speech and Signal Processing (ICASSP'98), Seattle, WA, USA, 12–15 May 1998; (Cat. No. 98CH36181); IEEE: Piscataway, NJ, USA, 1998; Volume 3, pp. 1513–1516.

24. Piella, G.; Heijmans, H.J. Adaptive lifting schemes with perfect reconstruction. *IEEE Trans. Signal Process.* **2002**, *50*, 1620–1630. [CrossRef]

25. Calderbank, A.; Daubechies, I.; Sweldens, W.; Yeo, B.L. Wavelet transforms that map integers to integers. *Appl. Comput. Harmonic Analy.* **1998**, *5*, 332–369. [CrossRef]

26. Zheng, Y.; Wang, R.; Li, J. Nonlinear wavelets and bp neural networks adaptive lifting scheme. In Proceedings of the 2010 International Conference on Apperceiving Computing and Intelligence Analysis, Chengdu, China, 17–19 December 2010; IEEE: Piscataway, NJ, USA, 2010; pp. 316–319.

27. Lin, M.; Chen, Q.; Yan, S. Network in network. *arXiv* **2013**, arXiv:1312.4400.

28. Szegedy, C.; Liu, W.; Jia, Y.; Sermanet, P.; Reed, S.; Anguelov, D.; Erhan, D.; Vanhoucke, V.; Rabinovich, A. Going deeper with convolutions. In Proceedings of the IEEE Conference on Computer Vision and Pattern Recognition, Boston, MA, USA, 7–12 June 2015; pp. 1–9.

29. Howard, A.G.; Zhu, M.; Chen, B.; Kalenichenko, D.; Wang, W.; Weyand, T.; Andreetto, M.; Adam, H. Mobilenets: Efficient convolutional neural networks for mobile vision applications. *arXiv* **2017**, arXiv:1704.04861.

30. Szegedy, C.; Vanhoucke, V.; Ioffe, S.; Shlens, J.; Wojna, Z. Rethinking the inception architecture for computer vision. In Proceedings of the 2016 IEEE Conference on Computer Vision and Pattern Recognition (CVPR), Las Vegas, NV, USA, 27–30 June 2016; pp. 2818–2826.

31. Rumelhart, D.E.; Hinton, G.E.; Williams, R.J. Learning representations by back-propagating errors. *Cogn. Modeling* **1988**, *5*, 1. [CrossRef]

32. Oppenheim, A.V. *Discrete-Time Signal Processing*; Pearson Education India: Bengaluru, India, 1999.

33. Jensen, J.R.; Lulla, K. Introductory digital image processing: A remote sensing perspective. *Geocarto Int.* **1987**, *2*, 65. [CrossRef]

34. Romera, E.; Alvarez, J.M.; Bergasa, L.M.; Arroyo, R. Erfnet: Efficient residual factorized convnet for real-time semantic segmentation. *IEEE Trans. Intell. Transp. Syst.* **2017**, *19*, 263–272. [CrossRef]

35. He, K.; Zhang, X.; Ren, S.; Sun, J. Deep residual learning for image recognition. In Proceedings of the IEEE Conference on Computer Vision and Pattern Recognition, Las Vegas, NV, USA, 27–30 June 2016; pp. 770–778.

36. Krizhevsky, A.; Hinton, G. *Learning Multiple Layers of Features from Tiny Images*; Technical report; Citeseer: Princeton, NJ, USA, 2009.

37. Xia, G.S.; Hu, J.; Hu, F.; Shi, B.; Bai, X.; Zhong, Y.; Zhang, L.; Lu, X. AID: A benchmark data set for performance evaluation of aerial scene classification. *IEEE Trans. Geosci. Remote Sens.* **2017**, *55*, 3965–3981. [CrossRef]

Article

Remote Sensing and Texture Image Classification Network Based on Deep Learning Integrated with Binary Coding and Sinkhorn Distance

Chu He [1,2,*,†], Qingyi Zhang [1,†], Tao Qu [3], Dingwen Wang [3] and Mingsheng Liao [2]

1   School of Electronic Information, Wuhan University, Wuhan 430072, China; zhqy@whu.edu.cn
2   State Key Laboratory of Information Engineering in Surveying, Mapping and Remote Sensing,
    Wuhan University, Wuhan 430079, China; liao@whu.edu.cn
3   School of Computer Science, Wuhan University, Wuhan 430072, China; qutaowhu@whu.edu.cn (T.Q.);
    wangdw@whu.edu.cn (D.W.)
*   Correspondence: chuhe@whu.edu.cn; Tel.: +86-27-6875-4367
†   These authors contributed equally to this work.

Received: 7 November 2019; Accepted: 29 November 2019; Published: 3 December 2019

Abstract: In the past two decades, traditional hand-crafted feature based methods and deep feature based methods have successively played the most important role in image classification. In some cases, hand-crafted features still provide better performance than deep features. This paper proposes an innovative network based on deep learning integrated with binary coding and Sinkhorn distance (DBSNet) for remote sensing and texture image classification. The statistical texture features of the image extracted by uniform local binary pattern (ULBP) are introduced as a supplement for deep features extracted by ResNet-50 to enhance the discriminability of features. After the feature fusion, both diversity and redundancy of the features have increased, thus we propose the Sinkhorn loss where an entropy regularization term plays a key role in removing redundant information and training the model quickly and efficiently. Image classification experiments are performed on two texture datasets and five remote sensing datasets. The results show that the statistical texture features of the image extracted by ULBP complement the deep features, and the new Sinkhorn loss performs better than the commonly used softmax loss. The performance of the proposed algorithm DBSNet ranks in the top three on the remote sensing datasets compared with other state-of-the-art algorithms.

Keywords: image classification; deep features; hand-crafted features; Sinkhorn loss

---

1. Introduction

1.1. Background

Image classification has always been an important basic problem in computer vision, and it is also the basis of other high-level visual tasks such as image detection, image segmentation, object tracking, and behavior analysis [1]. To propose an effective method to extract features which can represent the characteristics of the image is always critical in image classification [2]. The methods of extracting features can be divided into hand-crafted feature based methods and deep feature based methods. Before the rise of feature learning, people mostly used hand-crafted feature based methods to extract the essential features of the image such as edge, corner, texture, and other information [3]. For example, Laplacian of Gaussian (LoG) operator [4] and Difference of Gaussian (DoG) operator [5] are designed for detecting blobs in the image, scale invariant feature transform (SIFT) [6] is independent of the size and rotation of the object, local binary pattern (LBP) [7] has rotation invariance and gray invariance, features from accelerated segment test (FAST) operator [8] has high computational performance and high repeatability, bag of visual words model [9] pays more attention to the statistical information of features, Fisher vector [10] expresses an image with a gradient vector of likelihood functions, etc. A few of the most successful methods of texture description are the LBP and its variants such as uniform local binary pattern (ULBP) [11], COVariance and LBP Difference (COV-LBPD) [12], median robust extended LBP (MRELBP) [13], fast LBP histograms from three orthogonal planes (fast LBP-TOP) [14]. The ULBP proposed by Ahonen et al. reduces the number of binary patterns of LBP and is robust for high frequency noise. Hong et al. proposed the LBP difference (LBPD) descriptor and the COV-LBPD descriptor. The LBPD characterizes the extent to which one LBP varies from the average local structure of an image region of interest, and the COV-LBPD is able to capture the intrinsic correlation between the LBPD and other features in a compact manner. The MRELBP descriptor proposed by Liu et al. was computed by comparing image medians over a novel sampling scheme, which can capture both microstructure and macrostructure texture information and has attractive properties of strong discriminativeness, grayscale and rotation invariance, and computational efficiency. Hong et al. proposed the fast LBP-TOP descriptor which fastens the computational efficiency of LBP-TOP on spatial-temporal information and introduced the concept of tensor unfolding to accelerate the implementation process from three-dimensional space to two-dimensional space.

Since the rise of feature learning, deep learning methods have become a research hotspot and have broad application prospects and research value in many fields such as speech recognition and image classification [15]. Deep learning architectures mainly include four types: the autoencoder (AE), deep belief networks (DBNs), convolutional neural network (CNN), and recurrent neural network (RNN) [16]. Among these four deep learning architectures, CNN is the most popular and most published to date. For example, neural networks such as GoogLeNet [17], VGGNet [18], and residual neural network (ResNet) [19] have performed well in the field of image classification. GoogLeNet proposes an inception module, VGGNet explores the effects of the depth of deep neural network, and ResNet solves the problem of degradation of deep networks. These deep learning algorithms build the reasonable model by simulating a multi-layer neural network. High-level layers pay more attention to semantic information and less attention to detail information, while low-level layers are more concerned with detailed information and less with semantic information.

The deep learning algorithms have automatic feature learning capabilities for image data relying on large training sets and large models [20], while traditional methods rely primarily on hand-crafted features. Despite the success of deep features, the hand-crafted LBP texture descriptor and its variants have proven to provide competitive performance compared to deep learning methods in recent texture recognition performance evaluation, especially when there are rotations and multiple types of noise [21]. The LBP method introduces the priori information by presetting thresholds, so it can directly extract useful features through a manually designed algorithm, while the acquisition of deep features with excellent performance requires large training sets and large models. Therefore, there are aspects where hand-crafted features and deep features can learn from each other. For example, Courbariaux et al. proposed the BinaryConnect, which means training a DNN with binary weights during the forward and backward propagations, while retaining precision of the stored weights [22]. Hubara et al. introduced a method to train binarized neural networks (BNNs)—neural networks with binary weights and activations at run-time, and the binary weights and activations are used for computing the parameter gradients at train-time [23]. Inspired by the characteristics of hand-crafted features and deep features, this paper mainly studies the complementary performance between deep features and binary coded features and proposes a more effective feature description method.

During the training of the model, the loss function is used to assess the degree to which a specific algorithm models the given data [24]. The better the loss function, the better the performance of the algorithm. The design of the loss function can be guided by two strategies: empirical risk minimization and structural risk minimization. The average loss of the model on the training data set is called empirical risk, and the strategy of minimizing empirical risk is that the model with the least empirical risk is the best model. The related loss functions include center loss [25] and large-margin softmax loss [26], which are typical improved versions of softmax loss. When the size of training set is small or the model is complex, the model with the least empirical risk makes it easy to overfit the data. The strategy of structural risk minimization adds a regularization term based on the empirical risk minimization strategy. The structural risk minimization strategy means that the model with the least structural risk is the optimal model. The commonly used regularization terms include L1-regularization and L2-regularization. Sinkhorn distance [27] is the approximation of Earth mover's distance (EMD) [28] which can be used as a loss function. Different from other distance functions, EMD solves the correlation between two distributions by a distance matrix and a coupling matrix related to the predicted probability distribution and the actual probability distribution. The presetting distance matrix can increase the influence of the inter-class distance on the loss value, thereby improving the performance of the model. However, when EMD is used as loss function, there will be the problem of excessive computational complexity. Thus, as an approximate representation of EMD, the Sinkhorn distance which adds an entropic regularization term based on EMD is introduced as the loss function of the proposed model. The added entropic constraint turns the transport problem between distributions into a strictly convex problem that can be solved with matrix scaling algorithms and avoids the overfitting program.

This paper mainly verifies the performance of the proposed image classification algorithm in texture classification and remote sensing scene classification. Texture classification is an important basic problem in the field of computer vision and pattern recognition as well as the basis of other visual tasks such as image segmentation, object recognition, and scene understanding. However, texture is only the feature of the surface of an object, which cannot fully reflect the essential properties of the object. High-level image features cannot be obtained using only texture features [29]. Remote sensing scene classification is challenging due to several factors, such as large intra-class variations, small inter-class differences, scale changes, and illumination changes [30]. With the rise of remote sensing instruments, a large amount of satellite data has appeared in the field of remote sensing. Therefore, deep learning is gradually introduced into the image classification of remote sensing scenes. There are wild applications receiving more and more attention, such as land cover classification and target detection.

1.2. Problems and Motivation

Firstly, the features used for classification in the deep model are often global features extracted by high-level layers near the end of the model, with very few local features of the image. The global features have proven important in classification tasks but the local features can enhance the discriminability of features and are also helpful for image classification. Besides, when there are rotation and noise in the image, the traditional hand-crafted features have proven to provide competitive performance compared to the deep features. These two types of features have their own characteristics and can be used for reference for each other in some aspects.

Secondly, after the feature fusion, not only the diversity of features but also the redundancy of features is increased. When training the model by minimizing the loss function, we would like to remove the redundancy of the fused features and maximize the inter-class distance to improve classification performance of the model. However, the common used softmax loss in deep learning usually has insufficient feature distinguishing ability [31]. The loss function that is more suitable for the algorithm needs to be proposed.

1.3. Contributions and Structure

This paper presents a remote sensing and texture image classification network, which is based on deep learning integrated with binary coding and Sinkhorn distance. The general framework of the proposed algorithm is shown in Figure 1.

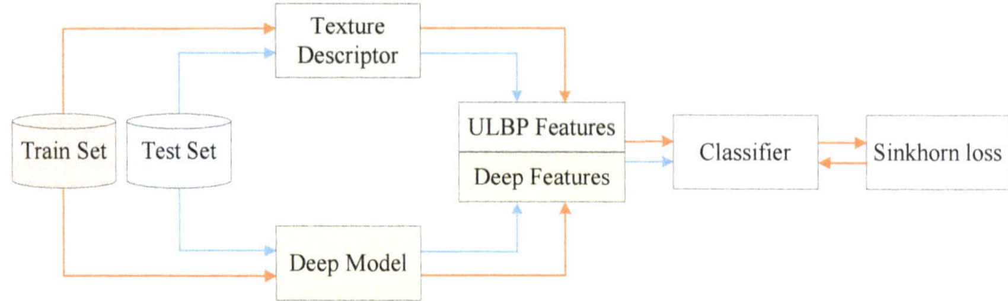

Figure 1. The general framework of the proposed algorithm, deep learning integrated with binary coding and Sinkhorn distance (DBSNet).

The contributions of this paper are summarized as follows.

Firstly, since the deep feature based methods and hand-crafted feature based methods are complementary in some aspects, we combine these two kinds of features to obtain better features to characterize the image. Specifically, the hand-crafted binary coding features extracted by ULBP are introduced to supplement the deep features extracted by representative ResNet-50 in classification, which makes the image features more accurate and comprehensive.

Secondly, a new loss function is proposed, which combines the score function with the Sinkhorn distance to predict the class. Sinkhorn loss analyzes the loss value between distributions from the perspective of doing work and removes the redundancy of the fused features with an entropic regularization term. Since the Sinkhorn distance is bounded from below by the distance between the centers of mass of the two signatures when the ground distance is induced by a norm, we can increase the impact of the inter-class distance on the loss value by presetting the distance matrix to guide the optimization process of the model.

The following sections are arranged as follows: Section 2 introduces the related work; Section 3 shows the proposed image classification algorithm; Section 4 introduces the experiments on texture datasets and remote sensing scene datasets; Section 5 introduces the summary of this paper.

2. Preliminaries

2.1. Deep Feature for Image Classification

Deep learning was successfully applied to image classification, and it is possible to approximate the complex functions of human visual perception by rationally combining several basic modules. The basic modules of the deep model are mainly information extraction module, activation and pooling module, and tricks module. The information extraction module extracts features from the input. The activation and pooling module is mainly devoted to nonlinear transformation and dimensionality reduction. The tricks module can speed up the training procedure and avoid overfitting [32]. The information extraction module of CNN is mainly composed of convolutional layers. Receptive fields are used to describe the area of the input image which can affect the features of the CNN. Figure 2 shows that as the depth of the network deepens, the receptive field of the posterior neurons increases, and the extracted features also change from low-level features such as edge information to mid-level features such as texture information and high-level features such as structural information.

The high-level features are often used for classification in CNN, losing a lot of detailed information, such as edge features and texture features, which may lead to poor performance in image classification requiring more detailed information. In order to improve the discriminative ability of the model, texture features can be used to complement the discriminability of deep features.

Deep models are widely used in texture classification and remote sensing scene classification. ResNet is one of the best deep models for image classification. ResNet was proposed by Kaiming He et al. in 2015. A 152 layer deep neural network was trained with a special network structure and won the championship on the ImageNet competition classification task (top-5 error rate: 3.57%). ResNet overcame the difficulty that the deep network could not be trained, and not only the accuracy of classification was improved, but also the parameter quantity was less than the VGG model. Consequently, ResNet-50 is used as the deep feature extractor in the proposed algorithm.

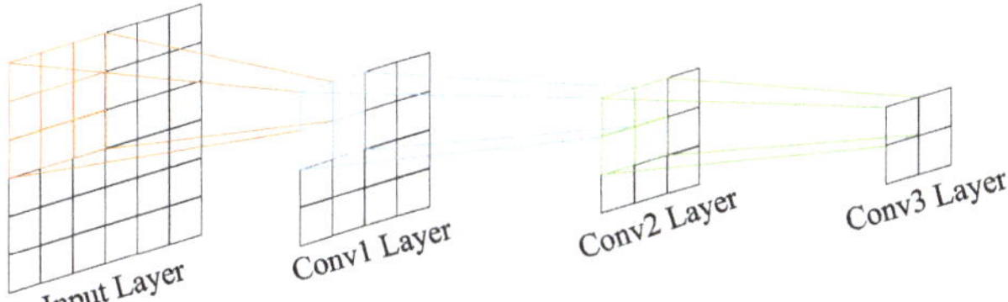

Figure 2. The relationship between receptive field and network depth.

ResNet deepens the network without reducing the accuracy of classification by residual learning. Based on the existing design ideas (batch normalization, small convolution kernel, and fully convolution network), the residual module is introduced. Each residual module contains two paths, one of which performs two or three convolution operations on the input feature to obtain the residual of the feature; the other path is the direct path of the input feature. The outputs of these two paths are finally added together to be the output of the residual module. There is an example of residual module shown in Figure 3. The first 1×1 convolution in the module is used to reduce the dimension (from 256 to 64), and the second 1×1 convolution is used to upgrade the dimension (from 64 to 256). Consequently, the number of input and output channels of the intermediate 3×3 convolution is small (from 64 to 64), and the parameters to be learned can be significantly reduced.

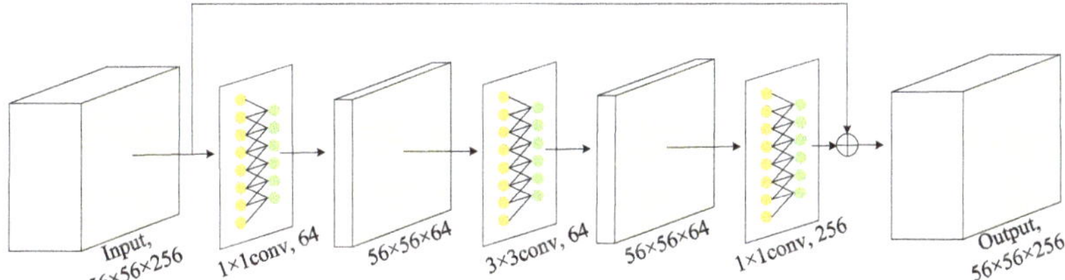

Figure 3. Residual module in ResNet-50.

Figure 4 shows the framework of ResNet-50, in which the residual modules are repeated 3 times, 4 times, 6 times, and 3 times respectively. The deep features of the image are extracted using the fine-tuned ResNet-50.

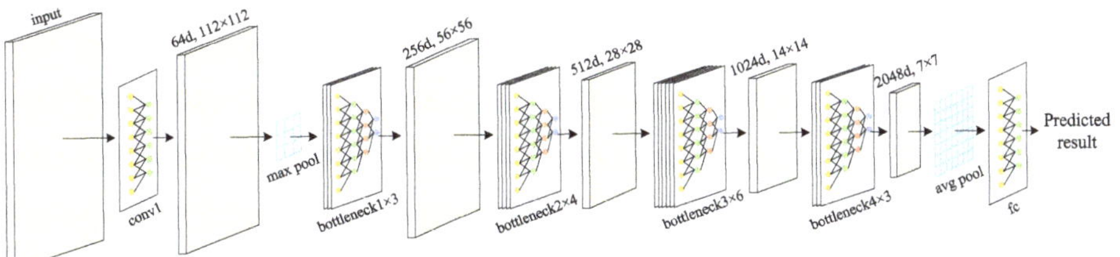

Figure 4. The framework of ResNet-50.

2.2. Binary Coded Feature for Image Classification

In the field of texture recognition, LBP is one of the most commonly used texture description methods, which was firstly proposed by T. Ojala et al. in 1994. It was originally developed as a method of describing texture images and later improved for image feature analysis. LBP has significant advantages such as gray invariance and rotation invariance [29]. LBP has been extensively researched in many fields and has demonstrated outstanding performance in several comparative studies [33,34]. The LBP descriptor works by thresholding the values of its neighborhood pixels, while the threshold is set as the value of the center pixel. The LBP descriptor is capable of detecting local primitives, including flat regions, edges, corners, curves, and edge ends, and it was later extended to obtain multi-scale, rotational invariant, and uniform representations and has been successfully applied to other tasks, such as object detection, face recognition [11], and remote sensing scene classification. The framework of traditional LBP can be presented by Figure 5, which can be divided into three steps. Firstly, the binary relationship between each pixel in the image and its local neighborhood is calculated in grayscale. Then, the binary relationship is weighted into an LBP code according to certain rules. Finally, the histogram sequence obtained by statistics in the LBP image is described as image features.

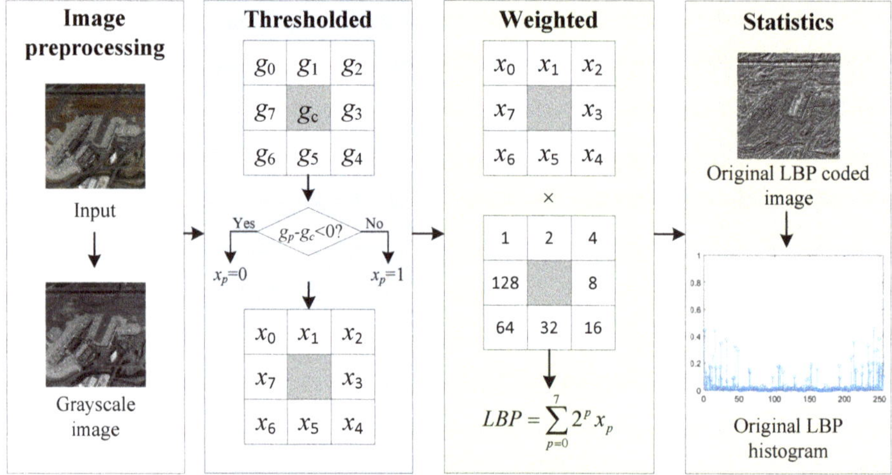

Figure 5. Calculation of the local binary pattern (LBP).

The traditional LBP algorithm can be expressed by Equations (1) and (2):

$$LBP_{P,R} = \sum_{p=0}^{P-1} s(g_p - g_c)2^p,$$

(1)

where

$$s(x) = \begin{cases} 1, & x \geq 0 \\ 0, & x < 0 \end{cases}.$$

P is the number of neighborhoods, R represents the distance between the central pixel and the neighborhood pixel, and we set $R = 1$ here. g_c is the grayscale value of the center pixel, and g_p is the grayscale value of the neighborhood pixel. Compared with the value of center pixel, the value of neighborhood pixel is set as 1 when it is greater or 0 when it is less. Then binary encoding is performed in a certain order. For those pixels which are not at the center of the neighborhood pixels, the grayscale values of their neighborhood pixels can be estimated by linear interpolation. Finally, by traversing all LBP pixel values, a histogram is created to represent the texture features of the image. Assuming the image size is $I \times J$, the histogram is expressed as:

$$H(k) = \sum_{i=1}^{I} \sum_{j=1}^{J} f(LBP_{P,R}(i,j),k),$$

(2)

where

$$k \in [0, 255],$$

$$f(x, y) = \begin{cases} 1, & x = y \\ 0, & otherwise \end{cases}.$$

As one of the binary coded feature extractors, LBP can effectively deal with illumination changes and is widely used in texture analysis and texture recognition. However, with the increase of the variety of patterns, the computational complexity and the data volume of the traditional LBP method will increase sharply. LBP will also be more sensitive to noise, and the slight fluctuations of the central pixel may cause the coded results to be quite different. In order to solve the problem of excessive binary patterns and make the process of statistic more concise, Ojala et al. proposed a uniform pattern to reduce the dimension of the patterns. Ojala believed that most patterns only contain up to two jumps from 1 to 0 or from 0 to 1 in the natural image. Therefore, Ojala defined the "Uniform Pattern", that is, when a loop binary pattern has a maximum of two jumps from 0 to 1 or from 1 to 0, the pattern is called a uniform pattern, such as 00000000 (0 jump), 10001111 (first jump from 1 to 0, then jump from 0 to 1). Except for the uniform pattern, other patterns are classified as mixed pattern, such as 10010111 (totally 4 jumps). The framework of ULBP is shown in Figure 6. $U(LBP_{P,R})$ can be used to indicate the number of jump in the code, which is calculated as follows:

$$U(LBP_{P,R}) = |s(g_{P-1} - g_c) - s(g_0 - g_c)| + \sum_{p=1}^{P-1} |s(g_p - g_c) - s(g_{p-1} - g_c)|. \tag{3}$$

Through such an improvement, the number of patterns is reduced from the original 2^P to $P \times (P-1) + 3$, where P represents the number of sampling points in the neighborhood. For 8 sampling points in the 3×3 neighborhood, the number of binary pattern is reduced from 256 to 59. The values of uniform pattern are assigned from 1 to 58 in ascending order, and the mixed pattern is assigned 0. Since the range of grayscale value is 0–58, the ULBP feature image is entirely dark, which makes the feature vector less dimensional and less impacted by high frequency noise.

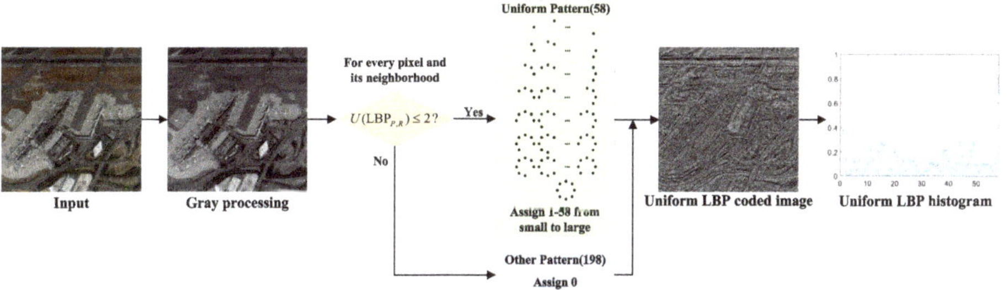

Figure 6. The framework of uniform local binary pattern (ULBP).

3. Methodology

3.1. Sinkhorn Loss

The Sinkhorn loss consists of softmax function and Sinkhorn distance. When the score vector is output from the fully connected layer, we convert it into a probability distribution by the softmax function and then calculate distance between the actual distribution and the predicted distribution using Sinkhorn distance. The approximate solution of optimal transport problem between two distributions can be determined by iterative learning. The advantages of Sinkhorn distance in calculating the distance between two distributions will be introduced next.

Two signatures, P and Q, are defined to represent the predicted distribution and the actual distribution, with m classes respectively. These two signatures can be represented by Equations (4) and (5), where p_i is the label in P, w_{p_i} is the probability of p_i in P, q_j is the label in Q, and w_{q_j} is the probability of q_j in Q. Here we set $p_i = i$ and $q_j = j$ to represent the different labels. The value of w_{p_i} is determined by the output of softmax function. The value of w_{q_i} is determined based on the real class. For a specific sample, if i is the real class of it, we set $w_{q_i} = 1$, otherwise we set $w_{q_i} = 0$.

$$P = \left\{ \left(p_1, w_{p_1}\right), \ldots, \left(p_m, w_{p_m}\right)\right\}, \tag{4}$$

$$Q = \left\{ \left(q_1, w_{q_1}\right), \ldots, \left(q_m, w_{q_m}\right)\right\}. \tag{5}$$

In order to measure the work of transforming one distribution into another, two matrices are introduced: the distance matrix D and the coupling matrix F. Each element d_{ij} in the distance matrix D represents the distance of moving p_i to q_j. Here we set $d_{ij} = 1$ when $i \neq j$ and $d_{ij} = 0$ when $i = j$. Each element f_{ij} in the coupling matrix F indicates the probability quality that needs to be assigned when moving from p_i to q_j. According to the above definition, the total cost $t(P, Q)$ can be calculated by the Frobenius inner product between F and D:

$$t(P, Q) = \langle D, F \rangle = \sum_{i=1}^{m} \sum_{j=1}^{m} d_{ij} f_{ij}. \tag{6}$$

The goal is to find an optimal coupling matrix F^* that minimizes the overall cost function, and the least cost function over all coupling functions is the solution to this optimal transport problem, called EMD.

$$F^* = \arg \min_{F} t(P, Q), \tag{7}$$

$$EMD = \frac{\min\limits_{F} t(P, Q)}{\sum\limits_{i=1}^{m} \sum\limits_{j=1}^{m} f_{ij}}, \tag{8}$$

s.t.

$$f_{ij} \geq 0,$$

$$\sum_{j=1}^{m} f_{ij} \leq w_{p_i},$$

$$\sum_{i=1}^{m} f_{ij} \leq w_{q_j},$$

$$\sum_{i=1}^{m} \sum_{j=1}^{m} f_{ij} = \min\left(\sum_{i=1}^{m} w_{p_i}, \sum_{j=1}^{m} w_{q_j}\right).$$

EMD has a complicated calculation method for finding the optimal solution and is not suitable as a loss function. However, when solving the distance between distributions, it can increase the influence of the inter-class distance on the cost function by reasonably presetting the distance matrix. Thus, we introduce the Sinkhorn distance as loss function which is the approximate value of EMD. It smooths the classic optimal transport problem with an entropic regularization term. The solution to the problem can be rewritten as:

$$For \lambda > 0, SD := \left\langle D, F^{\lambda} \right\rangle, \tag{9}$$

where

$$F^{\lambda} = \arg \min_{F} t(P, Q) - \frac{1}{\lambda} h(F),$$

$$h\left(F\right) = -\sum_{ij} F_{ij} \log F_{ij}.$$

λ is the regularization coefficient. When λ grows, a slower convergence can be observed as F^λ gets closer to the optimal vertex F^*, but the computational complexity will also rise at the same time. Thus we take $\lambda = 10$ where the computational complexity and the accuracy of the approximate solution reach the compromise. By introducing entropy regularization term, the transport problem is turned into a strictly convex problem that can be solved with Sinkhorn's matrix scaling algorithm at a speed which is several orders of magnitude faster than that of transport solvers. For $\lambda > 0$, the solution F^λ of the problem is unique and has the form $F^\lambda = \text{diag}(u)K\text{diag}(v)$, where u and v are two non-negative vectors of $\mathbb{R}^m$ and K is the element-wise exponential of $-\lambda D$.

3.2. Integrating Deep Learning with Binary Coding for Texture and Remote Sensing Image Classification

Nowadays, the networks used for image classification are generally trained and tested through an end-to-end network, and the classification accuracy is improved by optimizing the parameters of the feature extractor and classifier. However, the features extracted by the deep network have limitations. In order to improve the performance of the classification algorithm, the local texture information obtained by the ULBP of the image is used as the supplementary features. This paper combines it with the deep features as the input of fully connected layer, and the optimization of network parameters is guided by Sinkhorn loss. The framework of the two stream model is shown in Figure 7.

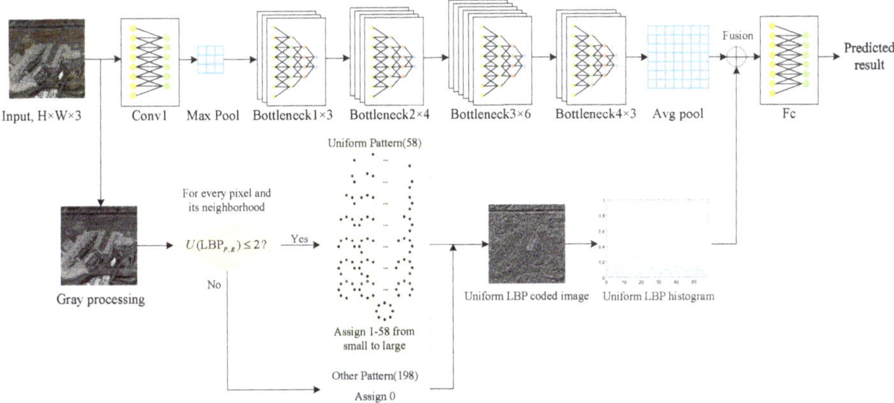

Figure 7. The detailed framework of the proposed algorithm: DBSNet.

The ResNet-50 is pre-trained on the ImageNet 2012 dataset used in the ImageNet Large Scale Visual Recognition Challenge (ILSVRC) [35] and then the original softmax with cross-entropy loss is replaced with the Sinkhorn loss to get a new network (RSNet). Finally, we fine-tuned the RSNet on different datasets and removed the fully connected layer and the classifier to get the deep feature extraction network. The binary coded feature extractor is the ULBP algorithm. The input of the model is an RGB image. Firstly, 2048 dimensional features are extracted through the deep feature extractor. At the same time, the image is grayscale processed and encoded by ULBP to get the 59 dimensional local texture features. After the two sets of features are fused, the class of image is predicted by the output of the fully connected layer.

In order to clearly observe the difference before and after the feature fusion, t-distributed stochastic neighbor embedding (t-SNE) [36] is used to visualize the pre-fusion deep features, ULBP features and the merged DBSNet features extracted on KTH-TIPS2-b texture dataset in the 2D space. The results are shown in Figure 8. As shown in the figure, the deep features have good image characterization capabilities, but the samples of the same class are more scattered. The LBP features have certain image characterization capabilities but the discriminability is not good. The DBSNet features combine the deep features and the LBP features. It can be seen from the reduced-dimensional image features that the image feature representation capability of the DBSNet features is better than the deep features and the ULBP features and the samples of the same class are more compact, indicating that the ULBP features complement deep features.

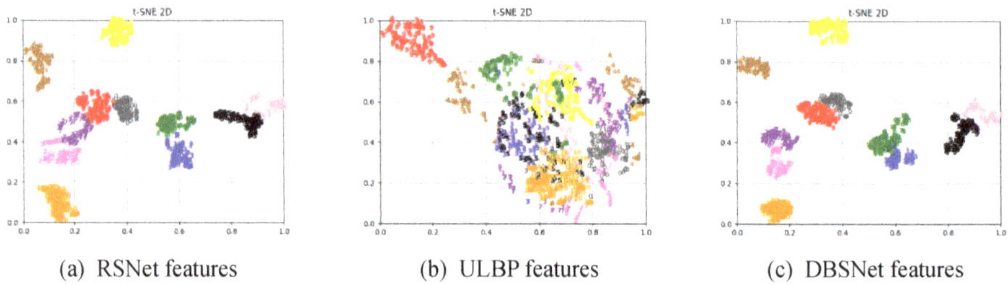

(a) RSNet features   (b) ULBP features   (c) DBSNet features

Figure 8. Comparison of feature maps of RSNet, ULBP, and DBSNet algorithms on KTH-TIPS2-b dataset.

4. Experiment

The performance of the proposed algorithm will be verified on two texture datasets and five remote sensing scene datasets. Firstly, the texture recognition performance of the algorithm is verified on two texture datasets and compared with the ResNet-50, RSNet, and several typical LBP-derived algorithms. Then, the remote sensing scene classification performance of this algorithm is evaluated on five remote sensing scene datasets and compared with the ResNet-50, RSNet, and the representative remote sensing scene classification algorithm.

4.1. Experimental Data

4.1.1. Texture Dataset

The performance of the proposed algorithm is firstly validated on two classic texture datasets: KTH-TIPS2-a dataset and KTH-TIPS2-b dataset.

The KTH-TIPS2-a dataset includes 11 classes of texture images. Most classes of the textures are shot in nine different scales, three poses, and four different lighting conditions, for a total of 4608 images, each with a pixel size of 200×200. We use three sets of samples as the train set and one set of samples as the test set and perform four experiments, with the average of four results as the final result.

The KTH-TIPS2-b dataset includes 11 classes of texture images, each of which is shot in nine different scales, three poses, and four different lighting conditions, for a total of 4752 images, each with a pixel size of 200×200. We use one set of samples as the train set and three sets of samples as the test set and perform four experiments, with the average of four results as the final result.

There are some examples of these texture datasets shown in Figure 9.

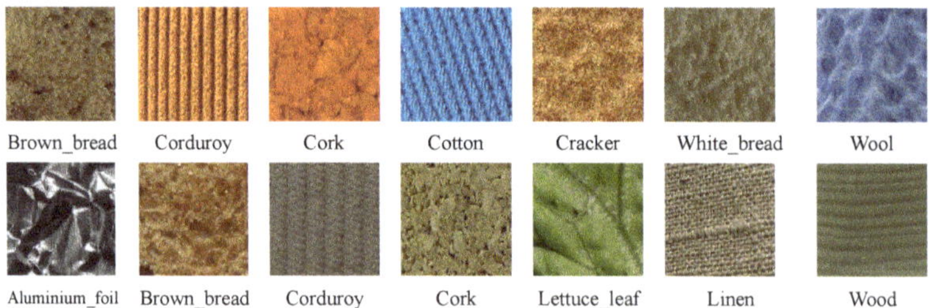

Figure 9. Example images of two texture datasets from top to bottom: KTH-TIPS2-a and KTH-TIPS2-b.

4.1.2. Remote Sensing Scene Dataset

Besides the texture image classification, the performance of the algorithm is also validated on five remote sensing scene datasets: AID dataset, RSSCN7 dataset, UC Merced Land-Use dataset, WHU-RS19 dataset, and OPTIMAL-31 dataset.

AID dataset [37] contains 30 classes of scene images, each class has about 200 to 400 samples, a total of 10,000, and each image has a pixel size of 600×600. Each class of images is randomly selected with ratio of 20:80 to obtain the train and test set.

RSSCN7 dataset [38] contains seven classes of scene images, each with 400 samples, a total of 2800, and each image has a pixel size of 400×400. Each class of images is randomly selected with ratio of 50:50 to obtain the train and test set.

UC Merced Land-Use dataset [39] contains 21 classes of scene images, each with 100 samples, a total of 2100, and each image has a pixel size of 256×256. Each class of images is randomly selected with ratio of 50:50 to obtain the train and test set.

WHU-RS19 dataset [40] contains 19 classes of scene images, each with about 50 samples, a total of 1005, and each image has a pixel size of 600×600. Each class of images is randomly selected with ratio of 60:40 to obtain the train and test set.

OPTIMAL-31 dataset [41] contains 31 classes of scene images, each with 60 samples, a total of 1860, and each image has a pixel size of 256×256. Each class of images is randomly selected with ratio of 80:20 to obtain the train and test set.

There are some examples of these remote sensing scene datasets shown in Figure 10.

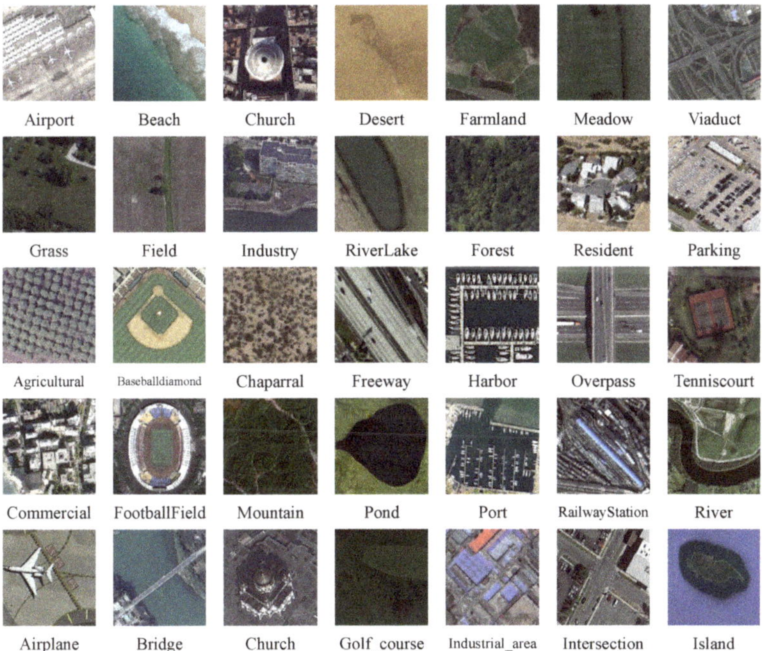

Figure 10. Example images of five remote sensing scene classification datasets from top to bottom: AID, RSSCN7, UC-Merced, WHU-RS19, and OPTIMAL-31.

4.2. Experimental Setup

Performance of the algorithms in the experiments is measured by the overall accuracy (OA) and the confusion matrix (CM) on the test set. The classification accuracy over all scene categories in a dataset is calculated according to $\frac{S_P}{S_T}$, where S_P is the number of correct predictions in the test set and S_T is the total number of images in the test set. The CM allows us to clearly see the classification accuracy of the algorithm for each type of image in the dataset.

In order to verify the performance of the proposed algorithm, we compare the proposed algorithm DBSNet with several representative algorithms on texture datasets and remote sensing datasets. For texture datasets, we compare the DBSNet with the hand-crafted texture feature descriptors ULBP and some efficient and recently proposed LBP derived algorithms such as COV-LBPD, MRELBP, and fast LBP-TOP. We experiment using the source code on the texture datasets. After the feature extraction by the texture feature descriptors, classification using nearest neighbors is conducted. In the proposed algorithm DBSNet, ResNet-50 is one solution for deep feature extractors. Because the ResNet-50 model is complex and the dimension of extracted features is large, we replace ResNet-50 with a shallow CNN model shown in Figure 11 and do the classification experiments on texture datasets to further verify the complementary effect of the hand-crafted texture features on the deep features. The network is trained and tested on four different train-test sets respectively, and then four feature extractors are obtained after removing the fully connected layer. For each feature extractor, we extract the deep features and classify them by the fully connected layer. The deep features fused with ULBP features are also classified by the fully connected layer to obtain the performance of the fused features.

For remote sensing datasets, we compare the proposed method DBSNet with the classic image classification algorithms IFK-SIFT [10], CaffeNet [42], VGG-VD-16 [18], GoogLeNet, ARCNet-VGGNet16 [41], and GBNet + global feature [43]. In addition to comparing the original results in the references [37,41,43], we do experiments on OPTIMAL-31 dataset with IFK-SIFT, CaffeNet, VGG-VD-16, and GoogLeNet referring to the parameter settings in reference [37]. We extract the deep features using the pretrained models without the fully connected layers on ImageNet and the IFK-SIFT features and then classify them respectively by the liblinear [44] for 10 times and take the mean accuracy as the result. Considering that the ResNet-50 used in the proposed methods are fine-tuned for better performance, we fine-tune the deep models CaffeNet, VGG-VD-16, and GoogLeNet for further comparison. We change the output channels of the last fully connected layer and optimize the parameters of deep models with the stochastic gradient descent (SGD). The detailed parameter settings are listed in Table 1.

Table 1. Parameter settings of the deep models.

Parameter	Batch Size	Learning Rate	Momentum	Weight Decay
Value	60	0.0001	0.9	0.0001

Besides the comparison methods mentioned above, three different algorithms are to be compared based on the difference of feature extraction method and the loss function on both texture datasets and remote sensing datasets, which are the fine-tuned ResNet-50, RSNet, and DBSNet algorithms. These three comparison algorithms are respectively tested to verify whether the deep features extracted by the RSNet and the statistical texture features obtained by the ULBP are complementary and whether the proposed Sinkhorn loss has robust performance.

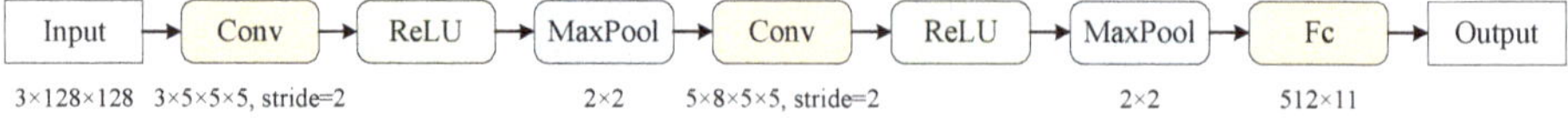

Figure 11. The framework of the shallow convolutional neural network (CNN).

4.3. Experimental Results and Analysis

In this section, we report the classification performance of the proposed DBSNet and other methods for comparison on challenging texture datasets and remote sensing scene classification datasets respectively.

4.3.1. Experiments on Texture Dataset

For the texture recognition, the classification results given in Table 2 show the performance comparison of the different algorithms on KTH-TIPS2-a and KTH-TIPS2-b texture datasets. The accuracy of the best performing algorithm is bolded for different databases. It can be seen that on the KTH-TIPS2-a and KTH-TIPS2-b datasets, the traditional hand-crafted methods are not competitive, and the ResNet-50, the RSNet, and the DBSNet provide incremental performance, which proves that the performance of the Sinkhorn loss is excellent and the features obtained by the ULBP are complementary to the deep features.

Table 2. Classification accuracy of different algorithms on KTH-TIPS2-a and KTH-TIPS2-b texture datasets.

	ULBP	COV-LBPD	MRELBP	Fast LBP-TOP	ResNet-50 (Fine-Tuning)	RSNet	DBSNet
KTH-TIPS2-a	0.6014	0.6291	0.6342	0.6058	0.8247	0.8321	**0.8359**
KTH-TIPS2-b	0.2628	0.5588	0.5475	0.2499	0.7379	0.7458	**0.7511**

Tables 3 and 4 are the confusion matrices of the RSNet algorithm and the DBSNet algorithm on KTH-TIPS2-b texture dataset which clearly reflect the classification performance on each category in the dataset. We compare these two confusion matrices and find that among the 11 classes, DBSNet algorithm outperforms RSNet algorithm in seven classes, which are aluminium foil, brown bread, cork, cracker, lettuce leaf, linen, and wood, and is inferior to the RSNet algorithm in three classes, which are corduroy, cotton, and wool. The overall classification performance of DBSNet is better than the RSNet algorithm, which proves the superiority of the proposed feature extraction method over the normal deep feature based method.

To further verify the complementary effect of the hand-crafted texture features on the deep features, we replace the RSNet feature extractor with a shallow CNN feature extractor. In Table 5, the accuracy of the best performing algorithm is bolded for different databases. It can be seen that the classification performance of the fused features is better than the deep features on four train-test sets of both KTHTIPS2-a and KTHTIPS2-b datasets. Consequently, the ULBP features complement the low-dimensional deep features of shallow CNN in classification task and even though the dimensions of deep features increase, the complement of ULBP features still exists, which has been proved in Table 2.

Table 3. Confusion matrix (CM) of RSNet algorithm on KTH-TIPS-2b dataset.

	Aluminium Foil	Brown Bread	Corduroy	Cork	Cotton	Cracker	Lettuce Leaf	Linen	White Bread	Wood	Wool
aluminium foil	0.9846	0	0	0	0	0	0	0.0154	0	0	0
brown bread	0	0.8549	0	0	0	0.0494	0	0	0.0957	0	0
corduroy	0.0123	0.0031	0.8117	0.0802	0.0062	0.0123	0	0.0463	0.0031	0.0031	0.0216
cork	0	0	0	0.8549	0	0.1204	0	0	0.0247	0	0
cotton	0	0	0.1358	0	0.2531	0	0	0.3827	0.0031	0.0463	0.1790
cracker	0	0.4846	0	0.0710	0	0.4414	0	0.0031	0	0	0
lettuce leaf	0	0	0	0	0.0062	0	0.9938	0	0	0	0
linen	0	0	0.0093	0	0.1790	0	0	0.8117	0	0	0
white bread	0	0	0	0	0	0	0	0	0.9877	0.0123	0
wood	0	0	0	0	0.0247	0	0	0	0.0278	0.9475	0
wool	0.0062	0.0031	0.0309	0.1636	0.0123	0	0	0.5216	0	0	0.2623

Table 4. CM of DBSNet algorithm on KTH-TIPS-2b dataset.

	Aluminium Foil	Brown Bread	Corduroy	Cork	Cotton	Cracker	Lettuce Leaf	Linen	White Bread	Wood	Wool
aluminium foil	1	0	0	0	0	0	0	0	0	0	0
brown bread	0	0.9352	0	0	0	0.0062	0	0	0.0586	0	0
corduroy	0.0741	0.0062	0.7130	0.0401	0.0031	0.0525	0.0031	0.0617	0.0062	0.0401	0
cork	0	0.0062	0	0.9475	0	0.0432	0	0	0.0031	0	0
cotton	0	0	0.2222	0	0.1265	0	0	0.4043	0.0031	0.0617	0.1821
cracker	0.0062	0.3457	0.0031	0.0432	0	0.5988	0	0	0	0.0031	0
lettuce leaf	0	0	0	0	0.0031	0	0.9969	0	0	0	0
linen	0	0	0.0031	0	0.1204	0.0031	0.0031	0.8673	0	0.0031	0
white bread	0	0.0062	0	0	0	0	0	0	0.9877	0.0062	0
wood	0	0	0	0	0.0062	0.0123	0	0	0.0216	0.9599	0
wool	0.1821	0.0062	0.0648	0.0772	0.0123	0	0	0.4630	0	0.0648	0.1296

Table 5. Classification accuracy of different feature sets on KTHTIPS2-a and KTHTIPS2-b texture datasets.

		Train-Test 1	Train-Test 2	Train-Test 3	Train-Test 4	OA
KTHTIPS2-a	Deep features	0.5097	0.4722	0.4318	0.5892	0.5007
	Deep features+ULBP features	0.5182	0.4882	0.4840	0.6002	**0.5227**
KTHTIPS2-b	Deep features	0.3131	0.3151	0.3361	0.4458	0.3525
	Deep features+ULBP features	0.3527	0.3561	0.4234	0.5208	**0.4133**

4.3.2. Experiments on Remote Sensing Scene Dataset

For the remote sensing scene classification, the results given in Table 6 show the performance comparison of the different algorithms on the five challenging remote sensing datasets. The accuracy of the top three best performing algorithms for different databases is bolded. It can be seen that the DBSNet algorithm provides better performance than the RSNet algorithm and the RSNet algorithm performs better than ResNet-50 algorithm on these five datasets, which demonstrates that the features obtained by ULBP still have the performance complementary to the deep features on remote sensing datasets and the proposed Sinkhorn loss can better guide the learning process of the network than the commonly used softmax loss. Compared with the mid-level method IFK-SIFT, the advanced deep feature based methods CaffeNet, VGG-VD-16, GoogLeNet, ARCNet-VGGNet16, and GBNet + global feature achieve improvement performance but the advanced deep feature based methods still have limitations in feature extraction. Based on the deep features, the algorithm DBSNet adds texture features that are instructive for image classification and uses a more suitable loss function. Compared with these representative methods, DBSNet always ranks in the top three on all five datasets.

Table 6. Classification accuracy of different algorithms on AID, RSSCN7, UC-Merced, WHU-RS19, and OPTIMAL-31 remote sensing scene classification datasets.

	AID	RSSCN7	UC-Merced	WHU-RS19	OPTIMAL-31
IFK-SIFT [37]	0.7192	0.8509	0.7874	0.8742	0.6022
CaffeNet [37]	0.8686	0.8825	0.9398	0.9624	0.8586
CaffeNet (fine-tuning)	0.8953	0.9043	0.9525	0.9550	0.8623
VGG-VD-16 [37]	0.8659	0.8718	0.9414	0.9605	0.8610
VGG-VD-16 (fine-tuning)	0.9036	0.9293	0.9552	0.9651	0.8737
GoogLeNet [37]	0.8344	0.8584	0.9270	0.9471	0.8454
GoogLeNet (fine-tuning)	0.9015	**0.9368**	0.9580	0.9650	0.8900
ARCNet-VGGNet16 [41]	0.8875	-	0.9681	**0.9975**	0.9270
GBNet + global feature [43]	0.9220	-	**0.9705**	**0.9925**	**0.9328**
ResNet-50 (fine-tuning)	**0.9233**	0.9312	0.9622	0.9751	0.9301
RSNet	**0.9281**	**0.9400**	**0.9762**	0.9800	**0.9328**
DBSNet	**0.9293**	**0.9521**	**0.9790**	**0.9875**	**0.9344**

The confusion matrices of RSNet algorithm and DBSNet algorithm are compared on RSSCN7 dataset to analyze the classification performance more carefully. It can be seen from Tables 7 and 8 that among the seven classes, DBSNet algorithm outperforms RSNet algorithm in five classes, which are

Grass, Industry, Forest, Resident, and Parking, and is second to the RSNet algorithm in two classes, which are Field and RiverLake. Generally speaking, the overall classification performance of DBSNet algorithm is better than the RSNet algorithm. As a complement, texture features play a role in the classification task.

Table 7. CM of RSNet algorithm on RSSCN7 dataset.

	Grass	Field	Industry	RiverLake	Forest	Resident	Parking
Grass	0.9150	0.0500	0.0100	0.0100	0	0.0100	0.0050
Field	0.0400	0.9600	0	0	0	0	0
Industry	0	0	0.8800	0.0150	0	0.0350	0.0700
RiverLake	0.0050	0.0150	0	0.9650	0.0150	0	0
Forest	0.0100	0.0150	0	0.0050	0.9700	0	0
Resident	0	0	0.0250	0.0050	0	0.9600	0.0100
Parking	0	0.0050	0.0600	0	0	0.0050	0.9300

Table 8. CM of DBSNet algorithm on RSSCN7 dataset.

	Grass	Field	Industry	RiverLake	Forest	Resident	Parking
Grass	0.9550	0.0300	0.0050	0.0100	0	0	0
Field	0.0450	0.9550	0	0	0	0	0
Industry	0	0	0.9000	0.0100	0	0.0350	0.0550
RiverLake	0.0150	0.0100	0.0100	0.9600	0.0050	0	0
Forest	0.0100	0	0	0.0050	0.9850	0	0
Resident	0	0	0.0250	0.0050	0	0.9700	0
Parking	0	0.0050	0.0450	0	0.0050	0.0050	0.9400

5. Conclusions

In this paper we have proposed a robust image classification algorithm based on deep learning integrated with binary coding and Sinkhorn distance. Taking into account the characteristics of hand-crafted features and deep features, we combine their advantages and supplement the deep features with the statistical texture features to fully describe the image. In order to remove redundant information from the fused features and train the model quickly and efficiently, we introduced the Sinkhorn loss where an entropy regularization term plays a key role. In this paper, experiments are carried out on two classic texture datasets and five remote sensing classification datasets. The experimental results show that compared with the ResNet-50, the proposed two stream model DBSNet can improve the overall performance when achieving image classification tasks. In addition, compared with the classic classification algorithms for remote sensing scene classification, the algorithm DBSNet can still provide better results. In the future, we will study how to combine the traditional feature extraction framework with the deep learning framework so that they guide and improve each other.

Author Contributions: Conceptualization, C.H.; Funding acquisition, C.H.; Investigation, Q.Z.; Methodology, C.H. and Q.Z.; Writing—original draft, Q.Z.; Writing—review & editing, T.Q., D.W. and M.L.

Funding: This research was funded by the National Key Research and Development Program of China (No. 2016YFC0803000), the National Natural Science Foundation of China (No. 41371342, No. 61331016), and the Hubei Innovation Group (2018CFA006).

Conflicts of Interest: The authors declare no conflict of interest.

References

1. Lu, D.; Weng, Q. A survey of image classification methods and techniques for improving classification performance. *Int. J. Remote Sens.* **2007**, *28*, 823–870. [CrossRef]

2. Caicedo, J.C.; Cruz, A.; Gonzalez, F.A. Histopathology image classification using bag of features and kernel functions. In Proceedings of the Conference on Artificial Intelligence in Medicine in Europe, Verona, Italy, 18–22 July 2019; Springer: Berlin/Heidelberg, Germany, 2009; pp. 126–135.

3. Szummer, M.; Picard, R.W. Indoor-outdoor image classification. In Proceedings of the 1998 IEEE International Workshop on Content-Based Access of Image and Video Database, Bombay, India, 3 January 1998; pp. 42–51.

4. Marr, D.; Hildreth, E. Theory of edge detection. *Proc. R. Soc. Lond. Ser. B Biol. Sci.* **1980**, *207*, 187–217.

5. Davidson, M.W.; Abramowitz, M. *Molecular Expressions Microscopy Primer: Digital Image Processing-Difference of Gaussians Edge Enhancement Algorithm*; Olympus America Inc. and Florida State University: Tallahassee, FL, USA, 2006.

6. Lowe, D.G. Object recognition from local scale-invariant features. In Proceedings of the International Conference on Computer Vision (ICCV 2019), Seoul, Korea, 27 October –2 November 1999; Volume 99, pp. 1150–1157.

7. Ojala, T.; Pietikäinen, M.; Mäenpää, T. Multiresolution gray-scale and rotation invariant texture classification with local binary patterns. *IEEE Trans. Pattern Anal. Mach. Intell.* **2002**, *24*, 971–987. [CrossRef]

8. Förstner, W.; Gülch, E. A fast operator for detection and precise location of distinct points, corners and centres of circular features. In Proceedings of the ISPRS Intercommission Conference on Fast Processing of Photogrammetric Data, Interlaken, Switzerland, 2–4 June 1987; pp. 281–305.

9. Yang, J.; Jiang, Y.G.; Hauptmann, A.G.; Ngo, C.W. Evaluating bag-of-visual-words representations in scene classification. In Proceedings of the International Workshop on Workshop on Multimedia Information Retrieval, Augsburg, Germany, 24–29 September 2007; ACM: New York, NY, USA, 2007; pp. 197–206.

10. Perronnin, F.; Sánchez, J.; Mensink, T. Improving the fisher kernel for large-scale image classification. In Proceedings of the European Conference on Computer Vision, Crete, Greece, 5–11 September 2010; Springer: Berlin/Heidelberg, Germany, 2010; pp. 143–156.

11. Ahonen, T.; Hadid, A.; Pietikainen, M. Face description with local binary patterns: Application to face recognition. *IEEE Trans. Pattern Anal. Mach. Intell.* **2006**, *28*, 2037–2041. [CrossRef]

12. Hong, X.; Zhao, G.; Pietikäinen, M.; Chen, X. Combining LBP difference and feature correlation for texture description. *IEEE Trans. Image Process.* **2014**, *23*, 2557–2568. [CrossRef]

13. Liu, L.; Lao, S.; Fieguth, P.W.; Guo, Y.; Wang, X.; Pietikäinen, M. Median robust extended local binary pattern for texture classification. *IEEE Trans. Image Process.* **2016**, *25*, 1368–1381. [CrossRef] [PubMed]

14. Hong, X.; Xu, Y.; Zhao, G. Lbp-top: A tensor unfolding revisit. In Proceedings of the Asian Conference on Computer Vision, Taipei, Taiwan, 20–24 November 2016; Springer: Cham, Switzerland, 2016; pp. 513–527.

15. Druzhkov, P.; Kustikova, V. A survey of deep learning methods and software tools for image classification and object detection. *Pattern Recognit. Image Anal.* **2016**, *26*, 9–15. [CrossRef]

16. Ball, J.E.; Anderson, D.T.; Chan, C.S. Comprehensive survey of deep learning in remote sensing: theories, tools, and challenges for the community. *J. Appl. Remote Sens.* **2017**, *11*, 042609. [CrossRef]

17. Szegedy, C.; Liu, W.; Jia, Y.; Sermanet, P.; Reed, S.; Anguelov, D.; Erhan, D.; Vanhoucke, V.; Rabinovich, A. Going deeper with convolutions. In Proceedings of the IEEE Conference on Computer Vision and Pattern Recognition, Boston, MA, USA, 7–12 June 2015; pp. 1–9.

18. Simonyan, K.; Zisserman, A. Very deep convolutional networks for large-scale image recognition. *arXiv* **2014**, arXiv:1409.1556.

19. He, K.; Zhang, X.; Ren, S.; Sun, J. Deep residual learning for image recognition. In Proceedings of the IEEE Conference on Computer Vision and Pattern Recognition, Las Vegas, NV, USA, 27–30 June 2016; pp. 770–778.

20. LeCun, Y.; Kavukcuoglu, K.; Farabet, C. Convolutional networks and applications in vision. In Proceedings of 2010 IEEE International Symposium on Circuits and Systems, Paris, France, 30 May–2 June 2010; pp. 253–256.

21. Liu, L.; Fieguth, P.; Wang, X.; Pietikäinen, M.; Hu, D. Evaluation of LBP and deep texture descriptors with a new robustness benchmark. In Proceedings of the European Conference on Computer Vision, Amsterdam, The Netherlands, 8–16 October 2016; Springer: Cham, Switzerland, 2016; pp. 69–86.

22. Courbariaux, M.; Bengio, Y.; David, J.P. Binaryconnect: Training deep neural networks with binary weights during propagations. In *Advances in Neural Information Processing Systems*; Curran Associates: New York, NY, USA, 2015; pp. 3123–3131.

23. Hubara, I.; Courbariaux, M.; Soudry, D.; El-Yaniv, R.; Bengio, Y. Binarized neural networks. In *Advances in Neural Information Processing Systems*; Curran Associates: New York, NY, USA, 2016; pp. 4107–4115.

24. Aly, M. Survey on multiclass classification methods. *Neural Netw.* **2005**, *19*, 1–9.

25. Wen, Y.; Zhang, K.; Li, Z.; Qiao, Y. A discriminative feature learning approach for deep face recognition. In Proceedings of the European Conference on Computer Vision, Amsterdam, The Netherlands, 8–16 October 2016; Springer: Cham, Switzerland, 2016; pp. 499–515.

26. Liu, W.; Wen, Y.; Yu, Z.; Yang, M. Large-margin softmax loss for convolutional neural networks. In Proceedings of the 2016 International Conference on Machine Learning (ICML 2016), New York, NY, USA, 19–24 June 2016; Volume 2, p. 7.

27. Cuturi, M. Sinkhorn distances: Lightspeed computation of optimal transport. In *Advances in Neural Information Processing Systems*; Curran Associates: New York, NY, USA, 2013; pp. 2292–2300.

28. Rubner, Y.; Tomasi, C.; Guibas, L.J. A metric for distributions with applications to image databases. In Proceedings of the Sixth International Conference on Computer Vision (IEEE Cat. No. 98CH36271), Bombay, India, 7 January 1998; pp. 59–66.

29. Nanni, L.; Lumini, A.; Brahnam, S. Survey on LBP based texture descriptors for image classification. *Expert Syst. Appl.* **2012**, *39*, 3634–3641. [CrossRef]

30. Cheng, G.; Han, J.; Lu, X. Remote sensing image scene classification: Benchmark and state of the art. *Proc. IEEE* **2017**, *105*, 1865–1883. [CrossRef]

31. Wang, H.; Wang, Y.; Zhou, Z.; Ji, X.; Gong, D.; Zhou, J.; Li, Z.; Liu, W. Cosface: Large margin cosine loss for deep face recognition. In Proceedings of the 2018 IEEE Conference on Computer Vision and Pattern Recognition, Salt Lake City, UT, USA, 18–23 June 2018; pp. 5265–5274.

32. He, C.; He, B.; Liu, X.; Kang, C.; Liao, M. Statistics Learning Network Based on the Quadratic Form for SAR Image Classification. *Remote Sens.* **2019**, *11*, 282. [CrossRef]

33. Heikkila, M.; Pietikainen, M. A texture-based method for modeling the background and detecting moving objects. *IEEE Trans. Pattern Anal. Mach. Intell.* **2006**, *28*, 657–662. [CrossRef]

34. Zhao, G.; Pietikainen, M. Dynamic texture recognition using local binary patterns with an application to facial expressions. *IEEE Trans. Pattern Anal. Mach. Intell.* **2007**, *29*, 915–928. [CrossRef]

35. Deng, J.; Dong, W.; Socher, R.; Li, L.J.; Li, K.; Fei-Fei, L. Imagenet: A large-scale hierarchical image database. In Proceedings of the 2009 IEEE Conference on Computer Vision and Pattern Recognition, Miami, FL, USA, 20–25 June 2009; pp. 248–255.

36. Maaten, L.V.D.; Hinton, G. Visualizing data using t-SNE. *J. Mach. Learn. Res.* **2008**, *9*, 2579–2605.

37. Xia, G.S.; Hu, J.; Hu, F.; Shi, B.; Bai, X.; Zhong, Y.; Zhang, L.; Lu, X. AID: A benchmark data set for performance evaluation of aerial scene classification. *IEEE Trans. Geosci. Remote Sens.* **2017**, *55*, 3965–3981. [CrossRef]

38. Zou, Q.; Ni, L.; Zhang, T.; Wang, Q. Deep learning based feature selection for remote sensing scene classification. *IEEE Geosci. Remote Sens. Lett.* **2015**, *12*, 2321–2325. [CrossRef]

39. Yang, Y.; Newsam, S. Bag-of-visual-words and spatial extensions for land-use classification. In Proceedings of the 18th SIGSPATIAL International Conference on Advances in Geographic Information Systems, San Jose, CA, USA, 2–5 November 2010; ACM: New York, NY, USA, 2010; pp. 270–279.

40. Sheng, G.; Yang, W.; Xu, T.; Sun, H. High-resolution satellite scene classification using a sparse coding based multiple feature combination. *Int.l J. Remote Sens.* **2012**, *33*, 2395–2412. [CrossRef]

41. Wang, Q.; Liu, S.; Chanussot, J.; Li, X. Scene classification with recurrent attention of VHR remote sensing images. *IEEE Trans. Geosci. Remote Sens.* **2018**, *57*, 1155–1167. [CrossRef]

42. Jia, Y.; Shelhamer, E.; Donahue, J.; Karayev, S.; Long, J.; Girshick, R.; Guadarrama, S.; Darrell, T. Caffe: Convolutional architecture for fast feature embedding. In Proceedings of the 22nd ACM International Conference on Multimedia, Orlando, FL, USA, 3–7 November 2014; ACM: New York, NY, USA, 2014; pp. 675–678.

43. Sun, H.; Li, S.; Zheng, X.; Lu, X. Remote Sensing Scene Classification by Gated Bidirectional Network. *IEEE Trans. Geosci. Remote Sens.* **2019**. [CrossRef]

44. Fan, R.E.; Chang, K.W.; Hsieh, C.J.; Wang, X.R.; Lin, C.J. LIBLINEAR: A library for large linear classification. *J. Mach. Learn. Res.* **2008**, *9*, 1871–1874.

Article

TextRS: Deep Bidirectional Triplet Network for Matching Text to Remote Sensing Images

Taghreed Abdullah [1], Yakoub Bazi [2,*], Mohamad M. Al Rahhal [3], Mohamed L. Mekhalfi [4], Lalitha Rangarajan [1] and Mansour Zuair [2]

[1] Department of Studies in Computer Science, University of Mysore, Manasagangothri, Mysore 570006, India; taghreed@compsci.uni-mysore.ac.in (T.A.); lalithar@compsci.uni-mysore.ac.in (L.R.)

[2] Computer Engineering Department, College of Computer and Information Sciences, King Saud University, Riyadh 11543, Saudi Arabia; zuair@ksu.edu.sa

[3] Information System Department, College of Applied Computer Science, King Saud University, Riyadh 11543, Saudi Arabia; mmalrahhal@ksu.edu.sa

[4] Department of Information Engineering and Computer Science, University of Trento, Disi Via Sommarive 9, Povo, 28123 Trento, Italy; mohamed.mekhalfi@alumni.unitn.it

[*] Correspondence: ybazi@ksu.edu.sa; Tel.: +966-1014696297

Received: 15 December 2019; Accepted: 23 January 2020; Published: 27 January 2020

Abstract: Exploring the relevance between images and their respective natural language descriptions, due to its paramount importance, is regarded as the next frontier in the general computer vision literature. Thus, recently several works have attempted to map visual attributes onto their corresponding textual tenor with certain success. However, this line of research has not been widespread in the remote sensing community. On this point, our contribution is three-pronged. First, we construct a new dataset for text-image matching tasks, termed TextRS, by collecting images from four well-known different scene datasets, namely AID, Merced, PatternNet, and NWPU datasets. Each image is annotated by five different sentences. All the five sentences were allocated by five people to evidence the diversity. Second, we put forth a novel Deep Bidirectional Triplet Network (DBTN) for text to image matching. Unlike traditional remote sensing image-to-image retrieval, our paradigm seeks to carry out the retrieval by matching text to image representations. To achieve that, we propose to learn a bidirectional triplet network, which is composed of Long Short Term Memory network (LSTM) and pre-trained Convolutional Neural Networks (CNNs) based on (EfficientNet-B2, ResNet-50, Inception-v3, and VGG16). Third, we top the proposed architecture with an average fusion strategy to fuse the features pertaining to the five image sentences, which enables learning of more robust embedding. The performances of the method expressed in terms Recall@K representing the presence of the relevant image among the top K retrieved images to the query text shows promising results as it yields 17.20%, 51.39%, and 73.02% for K = 1, 5, and 10, respectively.

Keywords: remote sensing; text image matching; triplet networks; EfficientNets; LSTM network

---

1. Introduction

The steady accessibility of remote sensing data, particularly high resolution images, has animated remarkable research outputs in the remote sensing community. Two of the most active topics in this regard refer to image classification and retrieval [1–5]. Image classification aims to assign scene images to a discrete set of land use/land cover classes depending on the image content [6–10]. Recently, with rapidly expanded remote sensing acquisition technologies, both quantity and quality of remote sensing data have been increased. In this context, content-based image retrieval (CBIR) has become a paramount research subject in order to meet the increasing need for the efficient organization and management of massive volumes of remote sensing data, which has been a long lasting challenge in the community of remote sensing.

In the last decades, great efforts have been made to develop effective and precise retrieval approaches to search for interest information across large archives of remote sensing. A typical CBIR system involves two main steps [11], namely feature extraction and matching, where the most relevant images from the archive are retrieved. In this regard, both extraction of features as well as matching play a pivotal role in controlling the efficiency of a retrieval system [12].

Content-based remote sensing image retrieval is a particular application of CBIR, in the field of remote sensing. However, the remote sensing community seems to put the emphasis more on devising powerful features due to the fact that image retrieval systems performance relies greatly on the effectiveness of the extracted features [13]. In this respect, remote sensing image retrieval approaches rely on handcrafted features and deep-learning.

As per handcrafted features, low-level features are harnessed to depict the semantic tenor of remote sensing images, and it is possible to draw them from either local or global regions of the image. Color features [14,15], texture features [2,16,17], and shape features [18] are widely applied as global features. On other hand, local features tend to emphasize the description on local regions instead of looking at the image as a whole. There are various algorithms for describing local image regions such as the scale-invariant feature transform (SIFT) and speed up robust features (SURF) [19,20]. The bag-of-word (BOW) model [21], and the vector of aggregated local descriptors (VLAD) [22] are generally proposed to encode local features into a fixed-size image signature via a codebook/dictionary of keypoint/feature vectors.

Recently, remote sensing images have been witnessing a steady increase due to the prominent technological progress of remote sensors [23]. Therefore, huge volumes of data with various spatial dimensions and spectral channels can be availed [24]. On this point, handcrafted features may be personalized and successfully tailored to small chunks of data; they do not meet, however, the standards of practical contexts where the size and complexity of data increases. Nowadays, deep learning strategies, which aim to learn automatically the discriminative and representative features, are highly effective in large-scale image recognition [25–27], object detection [28,29], semantic segmentation [30,31], and scene classification [32]. Furthermore, recurrent neural networks (RNNs) have achieved immense success with various tasks in sequential data analysis as recognition of action [33,34] and image captioning [35]. Recent research shows that image retrieval approaches work particularly well by exploiting deep neural networks. For example, the authors in [36] introduced a content-based remote sensing image retrieval approach depending on deep metric learning using a triplet network. The proposed approach has shown promising results compared to prior state-of-the-art approaches. The work in [37] presented an unsupervised deep feature learning method for the retrieval task of remote sensing images. Yang et al. [38] proposed a dynamic kernel with a deep convolutional neural network (CNN) for image retrieval. It focuses on matching patches between the filters and relevant images and removing the ones for irrelevant pairs. Furthermore, deep hashing neural network strategies are adopted in some works for large-scale remote sensing image retrieval [39]. Li et al. [40] presented a new unsupervised hashing method, the aim of which is to build an effective hash function. In another work, Li et al. [41], investigated cross-source remote sensing image retrieval via source-invariant deep hashing CNNs, which automatically extract the semantic feature for multispectral data.

It is worthwhile mentioning that the aforementioned image retrieval methods are single label retrieval approaches, where the query image and the images to be retrieved are labelled by a single class label. Although these approaches have been applied with a certain amount of success, they tend to abstract the rich semantic tenor of a remote sensing image into a single label.

In order to moderate the semantic gap and enhance the retrieval performance, recent remote sensing research proposed multi-label approaches. For instance, the work in [12] presented a multi-label method, making use of a semi-supervised graph-theoretic technique in order to improve the region-based retrieval method [42]. Zhou et al. [43] proposed a multi-label retrieval technique by training a CNN for semantic segmentation and feature generation. Shao et al. [11] constructed a dense labeling remote sensing dataset to evaluate the performance of retrieval techniques based on traditional handcrafted feature as well as deep learning-based ones. Dai et al. [44] discussed the use of multiple hyperspectral image retrieval labels and introduced a multi-label scheme that incorporates spatial and spectral features.

It is evident that the multi-label scenario is generally favored (over the single label case) on account of its abundant semantic information. However, it remains limited due to the discrete nature of labels pertaining to a given image. This suggests a further endeavor to model the relation among objects/labels using an image description. With the rapid advancement of computer vision and natural language processing (NLP), machines began to understand, slowly but surely, the semantics of images.

Current computer vision literature suggests that, instead of tackling the problem from an image-to-image matching perspective, cross-modal text-image learning seems to offer a more concrete alternative. This concept has manifested itself lately in the form of image captioning, which stems as a crossover where computer vision meets NLP. Basically, it consists of generating a sequential textual narration of visual data, similar to how humans perceive it. In fact, image captioning is considered as a subtle aid for image grasping, as a description generation model should capture not only the objects/scenes presented in the image, but it should also be capable of expressing how the objects/scenes relate to each other in a textual sentence.

The leading deep learning techniques, for image captioning, can be categorized into two streams. One stream adopts encoder–decoder, an end-to-end fashion [45,46] where a CNN is typically considered as the encoder and an RNN as the decoder, often a Long-Short Term Memory (LSTM) [47]. Rather than translating between various languages, such techniques translate from a visual representation to language. The visual representation is extracted via a pre-trained CNN [48]. Translation is achieved by RNNs based language models. The major usefulness of this method is that the whole system adopts end to end learning [47]. Xu et al. [35] went one step further by introducing the attention mechanism, which enables the decoder to concentrate on specific portions of the input image when generating a word. The other stream adopts a compositional framework, such as [49] for instance, which divided the task of generating the caption into various parts: detection of the words by a CNN, generating the caption candidates, and re-ranking the sentence by a deep multimodal similarity model.

With respect to image captioning, the computer vision literature suggests several contributions mainly based on deep learning. For instance, You et al. [50] combined top-down (i.e., image-to-words) and bottom-up (i.e., joining several relevant words into a meaningful image description) approaches via CNN and RNN models for image captioning, which revealed interesting experimental results. Chen et al. [51] proposed an alternative architecture based on spatial and channel-wise attention for image captioning. In other works, a common deep model called a bi-directional spatial–semantic attention network was introduced [52,53], where an embedding and a similarity network were adopted to model the bidirectional relations between pairs of text and image. Zhang and Lu [54] proposed a projection classification loss that classified the vector projection of representations from one form to another by improving the norm-softmax loss. Huang et al. [52] addressed the problem of image text matching in bi-direction by making use of attention networks.

So far, it can be noted that computer vision has been accumulating a steady research basis in the context of image captioning [47,50,55]. In remote sensing, however, contributions have barely begun to move in this direction, often regarded as the 'next frontier' in computer vision. Lu et al. [56] for instance, proposed a similar concept as in [51] by combining CNNs (for image representation) and LSTM network for sentence generation in remote sensing images. Shi et al. [57] leveraged a fully convolutional architecture for remote sensing image description. Zhang et al. [58] adopted an attribute attention strategy to produce remote sensing image description, and investigated the effect of the attributes derived from remote sensing images on the attention system.

As we have previously reviewed, the mainstream of the remote sensing works focuses mainly on scenarios of single label, whereas in practice images may contain many classes simultaneously.

In the quest for tackling this bottleneck, recent works attempted to allocate multiple labels to a single query image. Nevertheless, coherence among the labels in such cases remains questionable since multiple labels are assigned to an image regardless of their relativity. Therefore, these methods do not specify (or else model) explicitly the relation between the different objects in a given image for a better understanding of its content. Evidently, remote sensing image description has witnessed rather scarce attention in this sense. This may be explained by the fact that remote sensing images exhibit a wide range of morphological complexities and scale changes, which render text to/from image retrieval intricate.

In this paper we propose a solution based DBTN for solving the text-to-image matching problem. It is worth mentioning that this work is inspired from [53]. The major contributions of this work can be highlighted as follows:

- Departing from the fact that the task of text-image retrieval/matching is a new topic in the remote sensing community, we deem it necessary to build a benchmark dataset for remote sensing image description. Our dataset will constitute a benchmark for future research in this respect.
- We propose a DBTN architecture to address the problem of text image matching, which to the best of our knowledge, has never been posed in remote sensing prior-art thus far.
- We tie the single models into fusion schemes that can improve the overall performance through adopting the five sentences.

The paper includes five sections, where the structure of the paper is as follows. In Section 2, we introduce the proposed DBTN method. Section 3 presents the TextRS dataset and the experimental results followed by discussions in Section 4. Finally, Section 5 provides conclusions and directions for future developments.

2. Description of the Proposed Method

Assume a training set $\mathcal{D} = \{X_i, Y_i\}_{i=1}^{N}$ composed of N images with their matching sentences. In particular, to each training image X_i we associated a set of M matching sentences $Y_i = \{y_i^1, \ldots, y_i^K\}$. In the test phase, given a query sentence t_q, we aimed to retrieve the most relevant image in the training set $\mathcal{D}$. Figure 1 shows a general description of the proposed DBTN method composed of image and text encoding branches that aimed to learn appropriate image and text embeddings $f(X_i)$ and $g(T_i)$, respectively, by optimizing a bidirectional triplet loss. Detailed descriptions are provided in the next sub-sections.

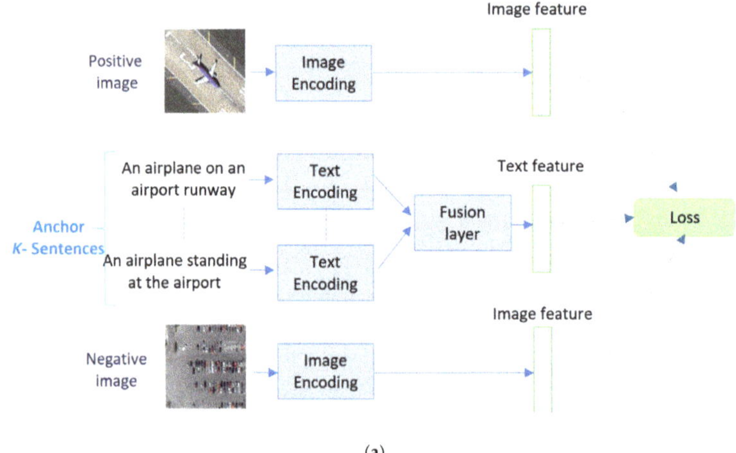

(a)

Figure 1. *Cont.*

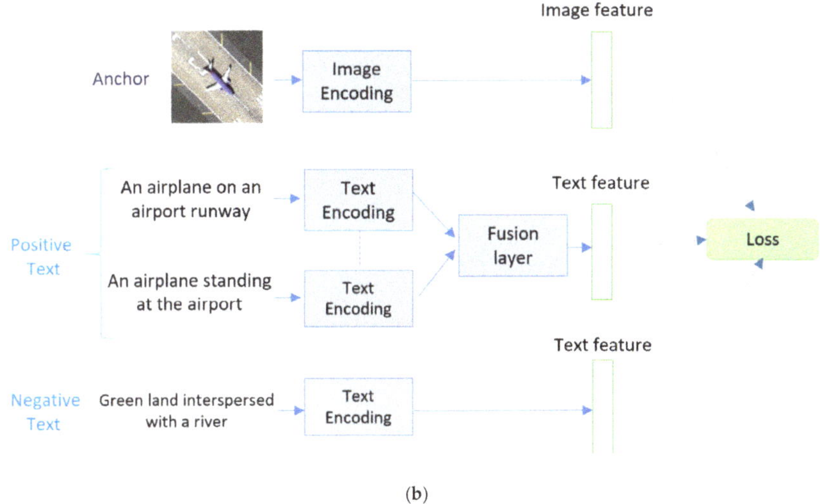

(b)

Figure 1. Flowchart of the proposed Deep Bidirectional Triplet Network (DBTN): (**a**) text as anchor, (**b**) image as anchor.

2.1. Image Encoding Module

The image encoding module uses a pre-trained CNN augmented with an additional network to learn the visual features $f(X_i)$ of the image (Figure 2). To learn informative features and suppress less relevant ones, this extra network applies a channel attention layer termed squeeze excitation (SE) to the activation maps layer obtained after the 3×3 convolution layer. The goal is to enhance further the representation of the features by grasping the significance of each feature map among all extracted feature maps. As illustrated in Figure 2, the squeeze operation produces features of dimension (1,1,128) by means of global average pooling (GAP), which are then fed to a fully connected layer to reduce the dimension by $1/16$. Then the produced feature vector s calibrates the feature maps of each channel (V) by channel-wise scale operation. SE works as shown below [59]:

$$s = Sigmoid(W_2(ReLU(W_1(V))))$$ (1)

$$V_{SE} = s \odot V$$ (2)

where s is the scaling factor, $\odot$ refers to the channel-wise multiplication, and V represents the feature maps obtained from a particular layer of the pre-trained CNN. Then the resulting activation maps V_{SE} are fed to a GAP followed by a fully connected and l_2-normalization for feature rescaling yielding the features $f(X_i)$.

As pre-trained CNNs, we adopted in this work different CNNs including VGG16, inception_v3, ResNet50, and EfficientNet. The VGG16 was proposed in 2014 and has 16-layers [27]. Such network was trained on the imagenet dataset to classify 1.2 million RGB images of size 224×224 pixel into 1000 classes. The inception-v3 network [60], introduced by Google, contains 42 layers as well as three kinds of inception modules, which comprise convolution kernels with sizes of 5×5 to 1×1. Such modules seek to reduce the parameters number. The Residual network (ResNet) [25] is a 50-layer network with shortcut connection. This network was proposed for deeper networks to solve the problem of vanishing gradients. Finally, EfficientNets, which are new state-of-the-art models with up to 10 times better efficiency (faster as well as smaller), were developed recently by a research team from Google [61] to scale up CNNs using a simple compound coefficient. Differently from traditional approaches that scale network dimensions (width, depth, and resolution) individually, EfficientNet tries to scale each dimension in a balanced way using a stationary set of scaling coefficients evenly. Practically, the performance of the model can be enhanced by scaling individual dimensions. Further, enhancing the entire performance can be achieved through scaling each dimension uniformly, which leads to higher accuracy and efficiency.

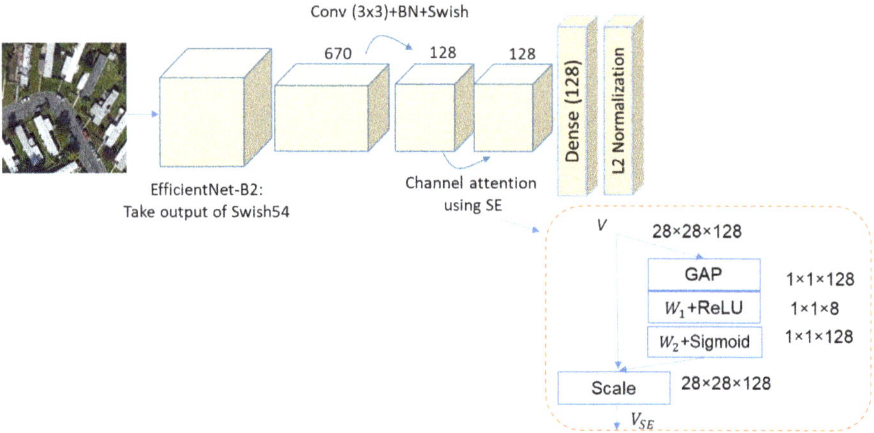

Figure 2. Image encoding branch for extracting the visual features.

2.2. Text Encoding Module

Figure 3 shows the text encoding module, which is composed of K symmetric branches, where each branch is used to encode one sentence describing the image content. These sub-branches use a word embedding layer followed by LSTM, a fully-connected layer, and l_2-normalization.

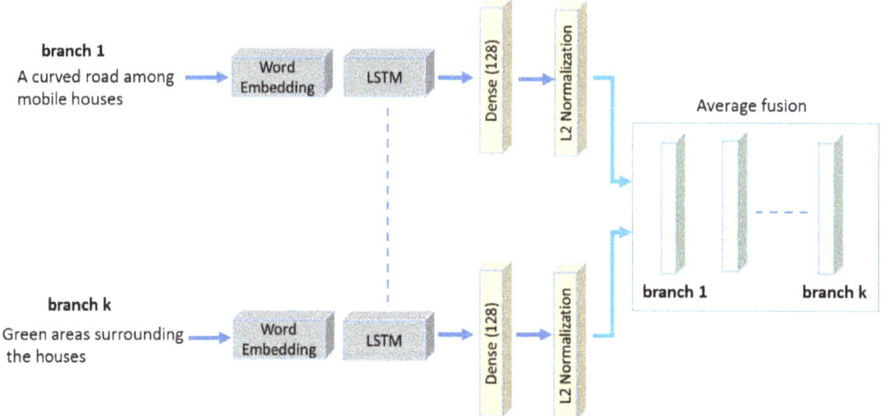

Figure 3. Text embedding branch: The five sentences describing the content of an image are aggregated using an average fusion layer. LSTM is Long-Short Term Memory.

The word embedding layer receives a sequence of integers representing the words in the sentence and transforms them into representations, where similar words should have similar encodings. Then the outputs of this layer are fed to LSTM [62] for modeling the entire sentence based on their long-term dependency learning capacity. Figure 4 shows the architecture of LSTM, with its four types of gates at each time step t in the memory cell. These gates are the input gate i_t, the update gate c_t, the output gate o_t, and the forget gate f_t. For each time step, these gates receive as input the hidden state h_{t-1} and the current input y_t. Then, the cell memory recursively updates itself based on its previous values and forget and update gates.

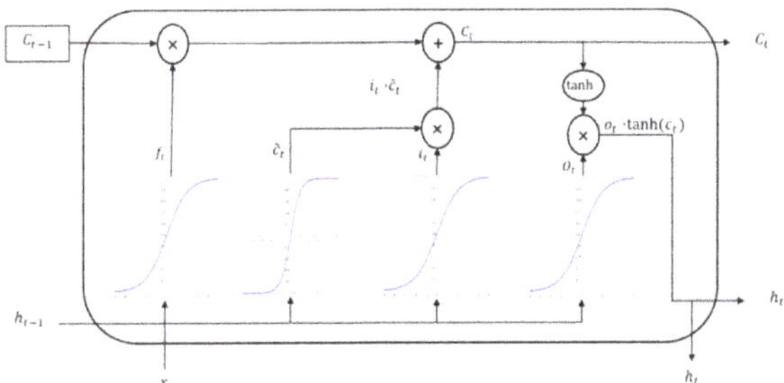

Figure 4. LSTM structure.

The working mechanism of LSTM is given below (for simplicity, we omit the image index *i*) [62]:

$$i_t = sigmoid(W_i.[h_{t-1}, y_t]) \tag{3}$$

$$f_t = sigmoid\big(W_f.[h_{t-1}, y_t]\big) \tag{4}$$

$$\widetilde{c}_t = tanh\big(W_g\,[h_{t-1}, y_t]\big) \tag{5}$$

$$c_t = f_t * c_{t-1} + i_t * \widetilde{c}_t \tag{6}$$

$$o_t = sigmoid(W_o[h_{t-1}, y_t]) \tag{7}$$

$$h_t = o_t * \tanh(c_t) \tag{8}$$

where * denotes the Hadamard product, and W_i, W_f, W_g, and W_o are learnable weights. In general, we can model the hidden state h_t of the LSTM as follows [62]:

$$h_t = LSTM(h_{t-1}, y_t, r_{t-1}) \tag{9}$$

where r_{t-1} indicates the memory cell vector at time step $t-1$.

For each branch, the output of LSTM is fed to an additional fully-connected layer yielding K feature representation $g\big(y_i^k\big)$, $k = 1, \ldots, K$. Then, the final outputs of different branches are fused using an average fusion layer to obtain a feature of dimension 128 [7]:

$$g(T_i) = \frac{\sum_{k=1}^{K} g(y_i^k)}{K} \tag{10}$$

2.3. DBTN Optimization

Many machine learning and computer vision problems are based on learning a distance metric for solving retrieval problems [63]. Inspired by achievements of deep learning in computer vision [26], deep neural networks were used to learn how to embed discriminative features [64,65]. These methods learn to project images or texts into a discriminative embedding space. The embedded vectors of similar samples are closer, while they are farther to those of dissimilar samples. Then several loss functions were developed for optimization such as triplet [65], quadruplet [66], lifted structure [67], N-pairs [68], and angular [69] losses. In this work, we concentrate on the triplet loss, which aims to learn a discriminative embedding for various applications such as classification [64], retrieval [70–74], and person re-identification [75,76]. It is worth recalling that a standard triplet in image-to-image retrieval is composed of three samples: an anchor, a positive sample (from the same category to the anchor), and a negative sample (from the different category to the anchor). The aim of the triplet loss is to learn an embedding space, where anchor samples are closer to positive samples than to negative ones by a given margin.

In our case, the network is composed of asymmetric branches, unlike standard triplet networks, as the anchor; positive and negative samples are represented in a different way. For instance, triplets can be formed using a text as an anchor, its corresponding image as a positive sample in addition to an image with a different content image as a negative. Similarly, one can use an image as an anchor associated with positive and negative textual descriptions. The aim is to learn discriminative features for different textual descriptions and discriminative features for different visual features as well. In addition, we should learn similar features to each image and its corresponding textual representation. For such purpose, we propose a bidirectional triplet loss as a possible solution to the problem. The bidirectional triplet loss is given as follows:

$$l_{DBTN} = \lambda_1 L_1 + \lambda_1 L_2 \tag{11}$$

$$L_1 = \sum_{i=1}^{N} \left[\left\| g(T_i^a) - f(X_i^p) \right\|_2^2 - \left\| g(T_i^a) - f(X_i^n) \right\|_2^2 + \alpha \right]_+ \tag{12}$$

$$L_2 = \sum_{i=1}^{N} \left[\left\| f(X_i^a) \right\| - g(T_i^p)_2^2 - \left\| f(X_i^a) - g(T_i^n) \right\|_2^2 + \alpha \right]_+ \tag{13}$$

where $|z|_+ = max(z,0),\{\displaystyle A\}$ and α is the margin that ensures the negative is farther away than the positive. $g(T_i^a)$ refers to the embedding of the anchor text, $f(X_i^p)$ is the embedding of the positive image, and $f(X_i^n)$ refers to the embedding of the negative image. On the other side, $f(X_i^a)$ refers to the embedding of the anchor image, $g(T_i^p)$ is the embedding of the positive text, and $g(T_i^n)$ refers to the embedding of the negative text. λ_1 and λ_2 are parameters of regularization controlling the contribution of both terms.

The performance of DBTN heavily relies on triplet selection. Indeed, the process of training is often so sensitive to the selected triplets, i.e., selecting the triplets randomly leads to non-convergence. To surmount this problem, the authors in [77] proposed triplet mining, which utilized only semi-hard triplets, where the positive pair was closer than the negative. Such valid semi-hard triplets are scarce, and therefore semi-hard mining requires a large batch size to search for informative pairs. A framework named smart mining was provided by Harwood et al. [78] to find out hard samples from the entire dataset that suffered from the burden of off-line computation. Wu et al. [79] discussed the significance of sampling and proposed a sampling technique called distance weighted sampling, which uniformly samples negative examples by similarity. Ge et al. [80] built a hierarchal tree of all the classes to find out hard negative pairs, which were collected via a dynamic margin. In this paper, we proposed to use a semi-hard mining strategy, as shown in Figure 5, although other sophisticated selection mechanism could be investigated as well. In particular, we selected triplets in an online mode based on the following constraint [77]:

$$d(g(T^a), f(X^p)) < d(g(T^a), f(X^n)) < d(g(T^a), f(X^p)) + \alpha \tag{14}$$

where $d(\cdot)$ is the cosine distance.

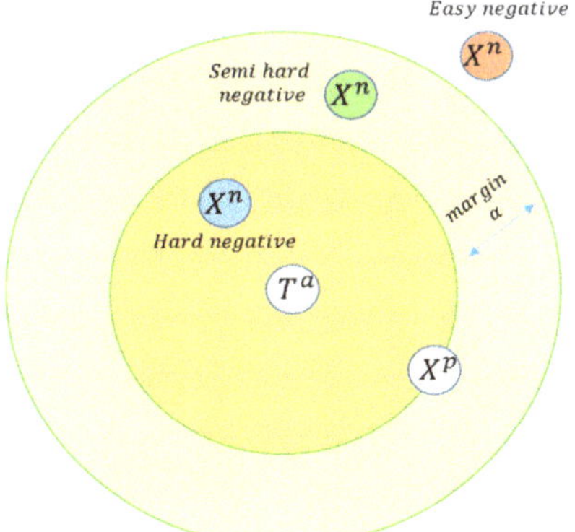

Figure 5. Semi-hard triplet selection scheme.

3. Experimental Results

3.1. Dataset Description

We built a dataset, named TextRS, by collecting images from four well-known different scene datasets, namely the AID dataset, which consists of 10,000 aerial images of size 600×600 pixels within 30 classes collected from Google Earth imagery by different remote sensors. The Merced dataset contains 21 classes; each class has 100 images of size 256×256 pixels with a resolution of 30 cm and RGB color. Such dataset was collected from USGS. The PatternNet was gathered from high-resolution imagery and includes 38 classes; each class contains 800 images of size 256×256 pixels. The NWPU dataset is another scene dataset, which has 31,500 images and is composed of 45 scene classes.

TextRS is composed of 2144 images selected randomly from the above four scene datasets. In particular, 480, 336, 608, and 720 images were selected from AID, Merced, PatternNet, and NWPU, respectively (16 images were selected from each class of such datasets). Then each remote sensing image was annotated by five different sentences; therefore, the total number of sentences was 10,720, and all the captions of this dataset were generated by five people to prove the diversity. It is worth recalling that the choice of the five sentences was mainly motivated by other datasets developed in the general context of computer vision literature [47,81]. During, the annotation we took into consideration some rules that had to be followed during generation of the sentences:

- Focus on the main dominating objects (tiny ones may be useless).
- Describe what exists instead of what does not exist in the scene.
- Try not to focus on the number of objects too much but use generic descriptions such as several, few, many, etc.
- Try not to emphasize the color of objects (e.g., blue vehicles) but rather on their existence and density.
- When mentioning, for instance, a parking lot (in an airport), it is important to mention the word 'airport' as well to distinguish it from any generic parking lot (downtown for example).
- Avoid using punctuation and conjunctions.

Some samples from our dataset are shown in Figure 6.

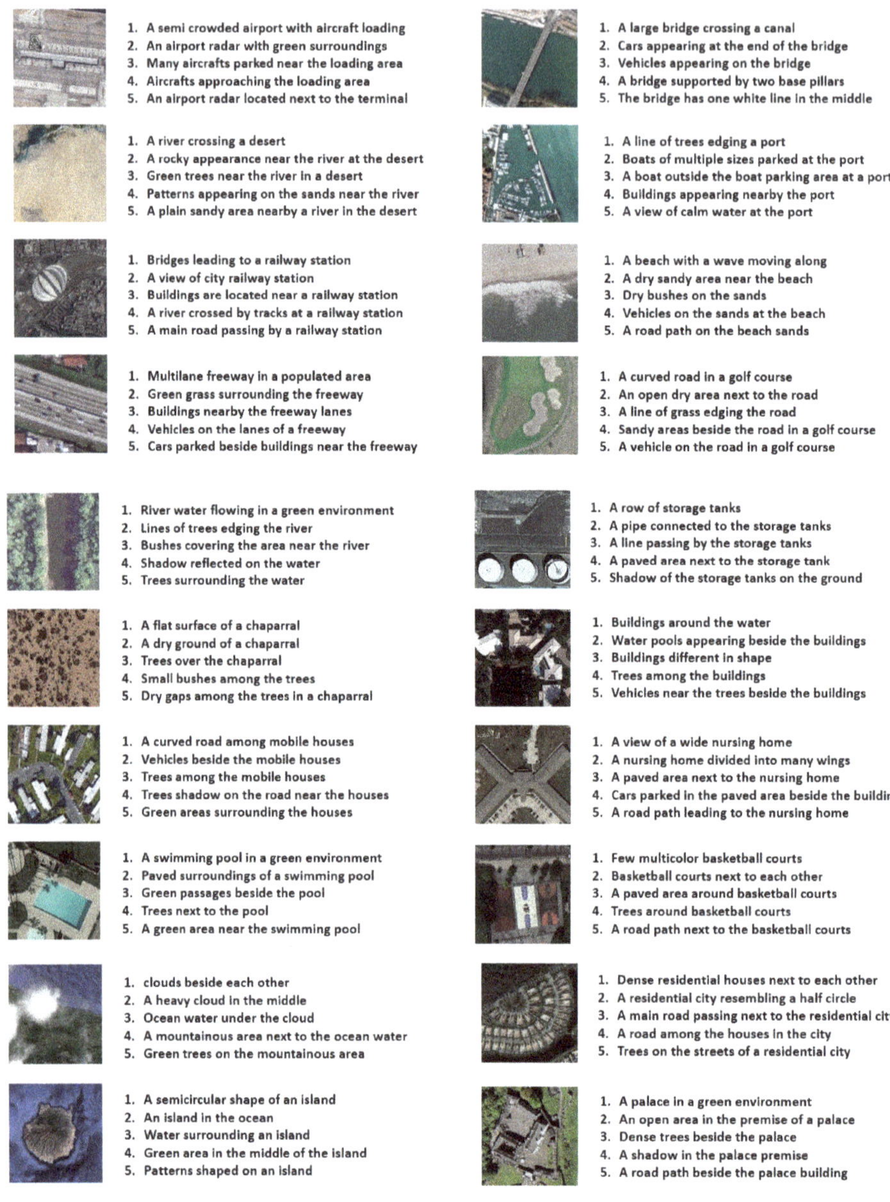

Figure 6. Example of images with the five sentences for each image.

3.2. Performance Evaluation

We implemented the method using the keras open-source library for deep learning written in python. For training the network, we randomly select 1714 images as training and the remaining 430 images as the test corresponding to approximately to 80% for training and 20% for testing. For training the DBTN, we used a mini-batch size of 50 images with the Adam optimization method with a fixed learning rate equal to 0.001 and exponential decay rates for the moment estimates equal to 0.9 and 0.999. Additionally, we set the regularization parameters to the default values of $\lambda_1 = \lambda_2 = 0.5$. To evaluate the performance of the method, we used the wide recall measure, which is suitable for text-to-image retrieval problems. In particular, we presented the results in Recall@K (R@K) terms for different values of K (1, 5, 10), which are the percentage of ground-truth matches shown in the top K-ranked results. We conducted the experiments on a station with an Intel Core i9 processor with a speed of 3.6 GHz and 32 GB of memory, and a Graphical Processing Unit (GPU) with 11 GB of GDDR5X memory.

3.3. Results

As mentioned in the previous sections, we used four different pre-trained CNNs for the image encoding branch, which were EfficientNet, ResNet50, Inception_v3, and VGG16. Figure 7 illustrates the evolution of the triplet loss function during the training phase for these different networks. We can see that the loss function decreased gradually with an increase in the number of iterations. In general, the model reached stable values after 40 iterations. In Figure 8 we show examples of features obtained by the image and text encoding branches at the end of the training process.

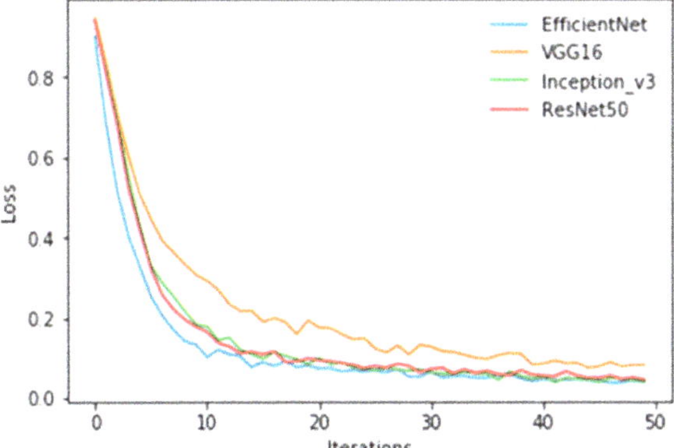

Figure 7. Evolution of loss function for the EfficientNet, ResNet50, Inception_v3, and VGG16.

Table 1 illustrates the performance of DBTN using EfficientNet as a pre-trained CNN for encoding the visual features. It could be observed with one sentence (Sent.1). The method achieved 13.02%, 40%, and 59.30% in R@1, R@5, and R@10, respectively. In contrast, when the five sentences are fused, the performance was further improved to 17.20%, 51.39%, and 73.02% of R@1, R@5, and R@10, respectively. Further, we computed the average of R@1, R@5, and R@10 for each sentence, and for fusion, we observed that the average of fusion had the highest score. Table 2 shows the results obtained using ResNet50 as the image encoder to learn the image features. We can see that the performances in R@1, R@5, and R@10 were 10.93%, 38.60%, and 54.41%, respectively, for Sent.1, while the method achieved 13.72%, 50.93%, and 69.06% of R@1, R@5, and R@10, respectively, with the fusion. Similarly, from Table 3 we observed that with Inception_v3, considering the fusion, the performance was also better than that of individual sentences. Finally, the results of using VGG16 are shown in Table 4. We can see that for Sent.1, our method achieved 10%, 36.27%, and 51.62% of R@1, R@5, and R@10, respectively, whereas the fusion process yielded 11.86%, 44.41%, and 63.72% of R@1, R@5, and R@10, respectively.

According to these preliminary results, one can notice that the fusing of the representations of the five sentences produced better matching results than did using one sentence. Additionally, EfficientNet seemed to be better compared to the other three pre-trained networks. This indicates that learning visual features by EfficientNet was quite effective and allowed better scores to be obtained compared to the other pre-trained CNNs.

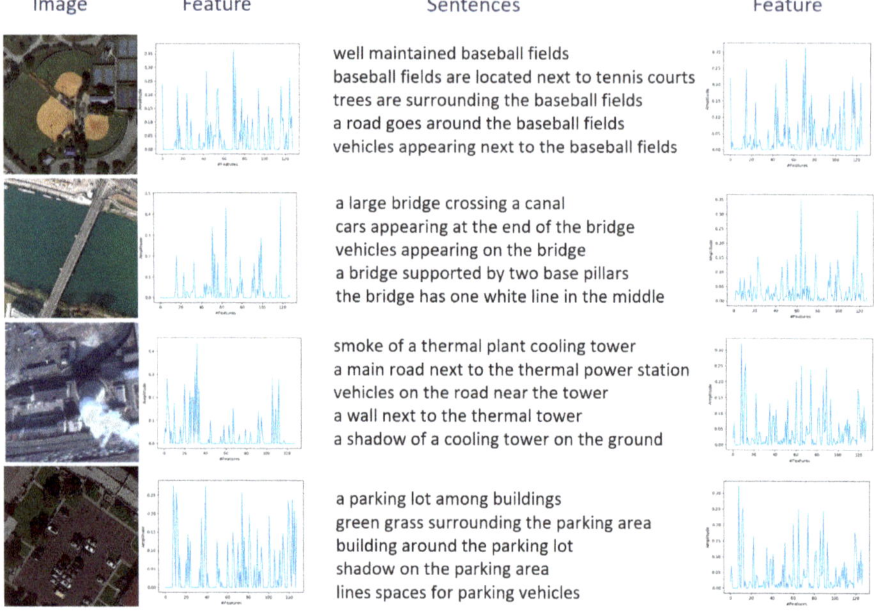

Figure 8. Image and text feature generated by the image and text encoding branches.

Table 1. Bidirectional text image matching results on our dataset by using EfficientNet-B2.

	Sent. 1	Sent. 2	Sent. 3	Sent. 4	Sent. 5	Fusion
R@1	13.02	13.48	14.18	13.02	10.09	17.20
R@5	40.00	44.18	44.18	40.93	38.60	51.39
R@10	59.30	60.46	62.55	57.67	58.37	73.02
Average	37.44	39.37	40.30	37.21	35.69	47.20

Table 2. Bidirectional text image matching results on our dataset by using ResNet50.

	Sent. 1	Sent. 2	Sent. 3	Sent. 4	Sent. 5	Fusion
R@1	10.93	12.79	12.32	12.32	11.86	13.72
R@5	38.60	38.37	42.58	43.02	38.19	50.93
R@10	54.41	56.27	61.16	60.93	55.58	69.06
Average	34.65	35.81	38.69	38.76	35.21	44.57

Table 3. Bidirectional text image matching results on our dataset by using Inception_v3.

	Sent. 1	Sent. 2	Sent. 3	Sent. 4	Sent. 5	Fusion
R@1	8.13	11.86	10.46	10.69	11.16	13.95
R@5	34.88	36.97	36.04	35.58	36.51	46.74
R@10	54.18	55.34	56.27	54.18	55.11	67.44
Average	32.40	34.72	34.26	33.48	34.26	42.71

Table 4. Bidirectional text image matching results on our dataset by using VGG16.

	Sent. 1	Sent. 2	Sent. 3	Sent. 4	Sent. 5	Fusion
R@1	10.00	9.06	11.86	8.13	7.67	11.86
R@5	36.27	35.11	36. 51	34.41	33.25	44.41
R@10	51.62	51.16	56.51	51.16	47.90	63.72
Average	32.63	31.78	38.84	31.23	29.60	40.00

To analyze the performance in detail for image retrieval given a query text, we showed many successful and failure scenarios. For example, we could see (Figure 9) a given query text (five sentences) with its image, and the top nine relevant retrieved images (from left to right); the image in red box is the ground truth image of the query text (true match). We could observe that our method output reasonable relevant images, where all nine images had almost the same content (objects). In these four scenarios, the rank of the retrieved true images was 1, 6, and 1, respectively.

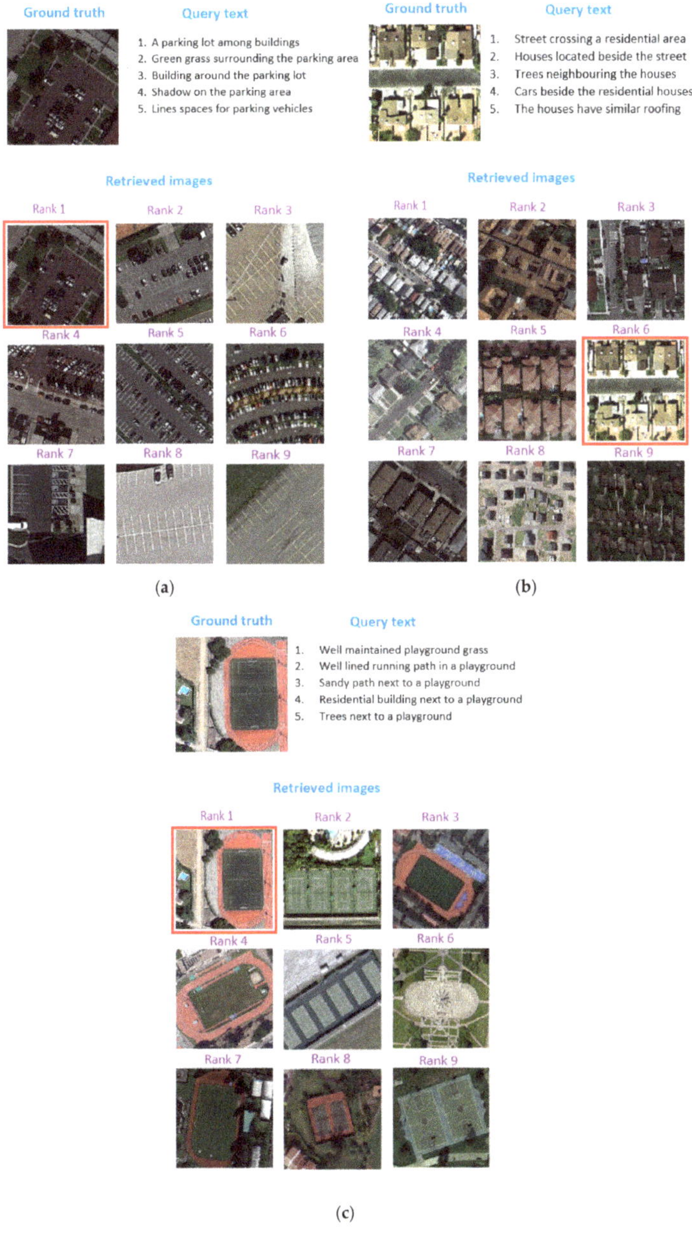

Figure 9. Successful scenarios (**a**, **b** and **c**) of text-to-image retrieval.

In contrast, Figure 10 shows two failure scenarios. In this case, we obtained relevant and irrelevant images, but the true matched image was not retrieved. This gives an indication that the problem was not easy and requires further investigations in improving the alignment of the descriptions to the image content.

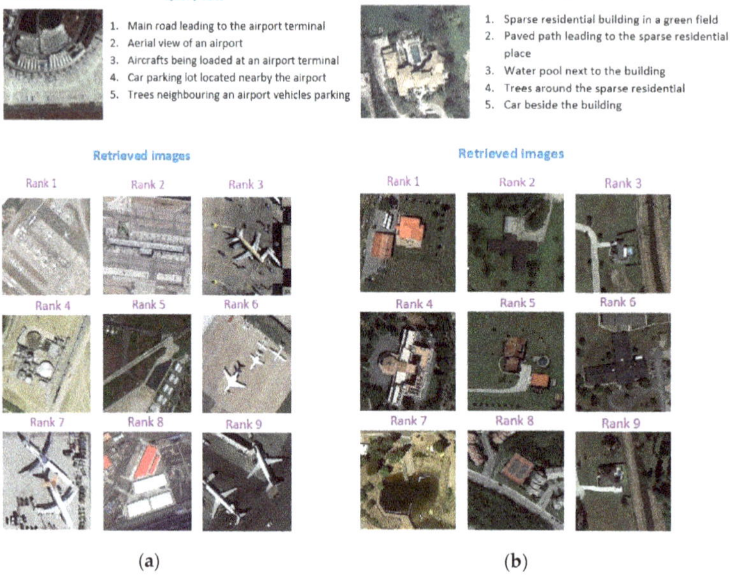

Figure 10. Unsuccessful scenarios (**a** and **b**) of text-to-image retrieval.

4. Discussion

In this section, we analyze further the performances of DBTN using different versions of EfficientNets, which are B0, B3, and B5. The version B0 contains 5.3 M parameters, while B3 and B5 are deeper and have 12M and 30M parameters, respectively. The results reported in Table 5 show that using B2 yields slightly better results compared to the other models. On the other side, B0 seems to be less competing as it provides an average recall of 45.65 compared to 47.20 for B2.

Table 5. Bidirectional text image matching results on our dataset using different EfficientNets.

	B0	B2	B3	B5
R@1	16.74	17.20	16.74	16.51
R@5	51.62	51.39	50.23	51.39
R@10	68.60	73.02	72.09	71.62
Average	45.65	47.20	46.35	46.51

Table 6 shows sensitivity analysis for bidirectional text image matching at multiple margin values. We can observe that setting this parameter to $\alpha = 0.5$ seems to be the most suitable choice. Increasing further this value leads to a decrease in the average recall as the network tends to select easy negative triplets.

In Table 7, we report the recall results obtained by using only one direction instead of bidirectional training. That is, we use text-to-image (Anchor text) and image-to-text (Anchor image). Obviously, the performance with bidirectional achieves the best results where relative similarity in one direction is useful for retrieval in the other direction, in the sense that the model trained with text-to-image triplets obtains a reasonable result in an image-to-text retrieval task and vice-versa. Nevertheless, the model trained with bi-directional triplets achieves the best result, indicating that the triplets organized in bidirectional provide more overall information for text-to-image matching.

Table 6. Sensitivity with respect to the margin parameter α.

	α		
	0.1	**0.5**	**1**
R@1	13	17.20	5
R@5	37.67	51.39	22.09
R@10	54.18	73.02	37.83

Table 7. Comparison between unidirectional and bidirectional loss.

	Anchor Text	**Anchor Image**	**Bidirectional**
R@1	12.55	12.55	17.20
R@5	41.62	39.53	51.39
R@10	62.09	59.53	73.02

5. Conclusions

In this work, we proposed a novel DBTN architecture for matching textual descriptions to remote sensing images. Different from traditional remote sensing image-to-image retrieval, our network seeks to carry out a more challenging problem, which is text-to-image retrieval. Such a network is composed of an image and text encoding branches and is trained using a bidirectional triplet loss. In the experiments, we validated the method on a new benchmark data set termed TextRS. Experiments show in general promising results in terms of the recall measure. In particular, better recall scores were obtained by fusing the textual representations rather than using one sentence for each image. In addition, EfficientNets allows better visual representations to be obtained compared to the other pre-trained CNNs. For future developments, we propose to investigate image-to-text matching and propose advanced solutions based on attention mechanisms.

Author Contributions: T.A., Y.B. and M.M.A.R. designed and implemented the method, and wrote the paper. M.L.M., M.Z. and L.R. contributed to the analysis of the experimental results and paper writing. All authors have read and agreed to the published version of the manuscript.

Funding: This research was funded by the Deanship of Scientific Research at King Saud University through the Local Research Group Program, grant number RG-1435-050.

Acknowledgments: This work was supported by the Deanship of Scientific Research at King Saud University through the Local Research Group Program under Project RG-1435-050.

Conflicts of Interest: The authors declare no conflicts of interest.

References

1. Al Rahhal, M.M.; Bazi, Y.; Abdullah, T.; Mekhalfi, M.L.; AlHichri, H.; Zuair, M. Learning a Multi-Branch Neural Network from Multiple Sources for Knowledge Adaptation in Remote Sensing Imagery. *Remote Sens.* **2018**, *10*, 1890. [CrossRef]
2. Aptoula, E. Remote Sensing Image Retrieval With Global Morphological Texture Descriptors. *IEEE Trans. Geosci. Remote Sens.* **2014**, *52*, 3023–3034. [CrossRef]
3. Paoletti, M.E.; Haut, J.M.; Plaza, J.; Plaza, A. A new deep convolutional neural network for fast hyperspectral image classification. *ISPRS J. Photogramm. Remote Sens.* **2018**, *145*, 120–147. [CrossRef]
4. Schroder, M.; Rehrauer, H.; Seidel, K.; Datcu, M. Interactive learning and probabilistic retrieval in remote sensing image archives. *IEEE Trans. Geosci. Remote Sens.* **2000**, *38*, 2288–2298. [CrossRef]
5. Tuia, D.; Volpi, M.; Copa, L.; Kanevski, M.; Munoz-Mari, J. A Survey of Active Learning Algorithms for Supervised Remote Sensing Image Classification. *IEEE J. Sel. Top. Signal Process.* **2011**, *5*, 606–617. [CrossRef]
6. Cheng, G.; Han, J.; Lu, X. Remote Sensing Image Scene Classification: Benchmark and State of the Art. *Proc. IEEE* **2017**, *105*, 1865–1883. [CrossRef]

7. Mekhalfi, M.L.; Melgani, F.; Bazi, Y.; Alajlan, N. Land-Use Classification With Compressive Sensing Multifeature Fusion. *IEEE Geosci. Remote Sens. Lett.* **2015**, *12*, 2155–2159. [CrossRef]
8. Mekhalfi, M.L.; Melgani, F. Sparse modeling of the land use classification problem. In Proceedings of the 2015 IEEE International Geoscience and Remote Sensing Symposium (IGARSS), Milan, Italy, 26–31 July 2015; pp. 3727–3730.
9. Weng, Q.; Mao, Z.; Lin, J.; Liao, X. Land-use scene classification based on a CNN using a constrained extreme learning machine. *Int. J. Remote Sens.* **2018**, *39*, 6281–6299. [CrossRef]
10. Wu, H.; Liu, B.; Su, W.; Zhang, W.; Sun, J. Deep Filter Banks for Land-Use Scene Classification. *IEEE Geosci. Remote Sens. Lett.* **2016**, *13*, 1895–1899. [CrossRef]
11. Shao, Z.; Yang, K.; Zhou, W. Performance Evaluation of Single-Label and Multi-Label Remote Sensing Image Retrieval Using a Dense Labeling Dataset. *Remote Sens.* **2018**, *10*, 964. [CrossRef]
12. Chaudhuri, B.; Demir, B.; Bruzzone, L.; Chaudhuri, S. Multi-label Remote Sensing Image Retrieval using a Semi-Supervised Graph-Theoretic Method. *IEEE Trans. Geosci. Remote Sens.* **2017**, *56*, 1144–1158. [CrossRef]
13. Shao, Z.; Yang, K.; Zhou, W. Correction: Shao, Z.; et al. A Benchmark Dataset for Performance Evaluation of Multi-Label Remote Sensing Image Retrieval. *Remote Sens.* **2018**, *10*, 1200. [CrossRef]
14. Bosilj, P.; Aptoula, E.; Lefèvre, S.; Kijak, E. Retrieval of Remote Sensing Images with Pattern Spectra Descriptors. *ISPRS Int. J. Geo-Inf.* **2016**, *5*, 228. [CrossRef]
15. Sebai, H.; Kourgli, A.; Serir, A. Dual-tree complex wavelet transform applied on color descriptors for remote-sensed images retrieval. *J. Appl. Remote Sens.* **2015**, *9*, 095994. [CrossRef]
16. Bouteldja, S.; Kourgli, A. Multiscale texture features for the retrieval of high resolution satellite images. In Proceedings of the 2015 International Conference on Systems, Signals and Image Processing (IWSSIP), London, UK, 10–12 September 2015; pp. 170–173.
17. Shao, Z.; Zhou, W.; Zhang, L.; Hou, J. Improved color texture descriptors for remote sensing image retrieval. *J. Appl. Remote Sens.* **2014**, *8*, 083584. [CrossRef]
18. Scott, G.J.; Klaric, M.N.; Davis, C.H.; Shyu, C. Entropy-Balanced Bitmap Tree for Shape-Based Object Retrieval From Large-Scale Satellite Imagery Databases. *IEEE Trans. Geosci. Remote Sens.* **2011**, *49*, 1603–1616. [CrossRef]
19. Lowe, D.G. Distinctive Image Features from Scale-Invariant Keypoints. *Int. J. Comput. Vis.* **2004**, *60*, 91–110. [CrossRef]
20. Bay, H.; Tuytelaars, T.; Gool, L.V. SURF: Speeded Up Robust Features. In Proceedings of the Computer Vision-ECCV 2006, Berlin, Heidelberg, 7–13 May 2006; Springer; pp. 404–417.
21. Yang, J.; Liu, J.; Dai, Q. An improved Bag-of-Words framework for remote sensing image retrieval in large-scale image databases. *Int. J. Digit. Earth* **2015**, *8*, 273–292. [CrossRef]
22. Jégou, H.; Douze, M.; Schmid, C.; Pérez, P. Aggregating local descriptors into a compact image representation. In Proceedings of the 2010 IEEE computer society conference on computer vision and pattern recognition, San Francisco, CA, USA, 13–18 June 2010; pp. 3304–3311.
23. Zhang, F.; Du, B.; Zhang, L. Scene Classification via a Gradient Boosting Random Convolutional Network Framework. *IEEE Trans. Geosci. Remote Sens.* **2016**, *54*, 1793–1802. [CrossRef]
24. Zhang, L.; Zhang, L.; Du, B. Deep Learning for Remote Sensing Data: A Technical Tutorial on the State of the Art. *IEEE Geosci. Remote Sens. Mag.* **2016**, *4*, 22–40. [CrossRef]
25. He, K.; Zhang, X.; Ren, S.; Sun, J. Deep Residual Learning for Image Recognition. In Proceedings of the 2016 IEEE Conference on Computer Vision and Pattern Recognition (CVPR), Las Vegas, NV, USA, 27–30 June 2016; pp. 770–778.
26. Krizhevsky, A.; Sutskever, I.; Hinton, G.E. ImageNet Classification with Deep Convolutional Neural Networks. *Commun ACM* **2017**, *60*, 84–90. [CrossRef]
27. Simonyan, K.; Zisserman, A. Very Deep Convolutional Networks for Large-Scale Image Recognition. *arXiv* **2014**, arXiv:14091556 Cs. Available online: https://arxiv.org/abs/1409.1556 (accessed on 24 January 2020).
28. Girshick, R.; Donahue, J.; Darrell, T.; Malik, J. Region-Based Convolutional Networks for Accurate Object Detection and Segmentation. *IEEE Trans. Pattern Anal. Mach. Intell.* **2016**, *38*, 142–158. [CrossRef] [PubMed]
29. Redmon, J.; Divvala, S.; Girshick, R.; Farhadi, A. You Only Look Once: Unified, Real-Time Object Detection. In Proceedings of the 2016 IEEE Conference on Computer Vision and Pattern Recognition (CVPR), Las Vegas, NV, USA, 27–30 June 2016; pp. 779–788.

30. Long, J.; Shelhamer, E.; Darrell, T. Fully convolutional networks for semantic segmentation. In Proceedings of the 2015 IEEE Conference on Computer Vision and Pattern Recognition (CVPR), Boston, MA, USA, 7–12 June 2015; pp. 3431–3440.

31. Noh, H.; Hong, S.; Han, B. Learning Deconvolution Network for Semantic Segmentation. In Proceedings of the IEEE international conference on computer vision, Santiago, Chile, 7–13 December 2015; pp. 1520–1528.

32. Han, W.; Feng, R.; Wang, L.; Cheng, Y. A semi-supervised generative framework with deep learning features for high-resolution remote sensing image scene classification. *ISPRS J. Photogramm. Remote Sens.* **2018**, *145*, 23–43. [CrossRef]

33. Donahue, J.; Hendricks, L.A.; Rohrbach, M.; Venugopalan, S.; Guadarrama, S.; Saenko, K.; Darrell, T. Long-Term Recurrent Convolutional Networks for Visual Recognition and Description. *IEEE Trans. Pattern Anal. Mach. Intell.* **2017**, *39*, 677–691. [CrossRef] [PubMed]

34. Du, Y.; Wang, W.; Wang, L. Hierarchical recurrent neural network for skeleton based action recognition. In Proceedings of the 2015 IEEE Conference on Computer Vision and Pattern Recognition (CVPR), Boston, MA, USA, 7–12 June 2015; pp. 1110–1118.

35. Xu, K.; Ba, J.; Kiros, R.; Cho, K.; Courville, A.; Salakhutdinov, R.; Zemel, R.; Bengio, Y. Show, Attend and Tell: Neural Image Caption Generation with Visual Attention. In Proceedings of the International Conference on Machine Learning (ICML), Lille, France, 6–11 July 2015; pp. 2048–2057.

36. Cao, R.; Zhang, Q.; Zhu, J.; Li, Q.; Li, Q.; Liu, B.; Qiu, G. Enhancing remote sensing image retrieval using a triplet deep metric learning network. *Int. J. Remote Sens.* **2020**, *41*, 740–751. [CrossRef]

37. Tang, X.; Zhang, X.; Liu, F.; Jiao, L. Unsupervised Deep Feature Learning for Remote Sensing Image Retrieval. *Remote Sens.* **2018**, *10*, 1243. [CrossRef]

38. Yang, J.; Liang, J.; Shen, H.; Wang, K.; Rosin, P.L.; Yang, M.-H. Dynamic Match Kernel With Deep Convolutional Features for Image Retrieval. *IEEE Trans. Image Process.* **2018**, *27*, 5288–5302. [CrossRef]

39. Li, Y.; Zhang, Y.; Huang, X.; Zhu, H.; Ma, J. Large-Scale Remote Sensing Image Retrieval by Deep Hashing Neural Networks. *IEEE Trans. Geosci. Remote Sens.* **2018**, *56*, 950–965. [CrossRef]

40. Li, P.; Ren, P. Partial Randomness Hashing for Large-Scale Remote Sensing Image Retrieval. *IEEE Geosci. Remote Sens. Lett.* **2017**, *14*, 464–468. [CrossRef]

41. Li, Y.; Zhang, Y.; Huang, X.; Ma, J. Learning Source-Invariant Deep Hashing Convolutional Neural Networks for Cross-Source Remote Sensing Image Retrieval. *IEEE Trans. Geosci. Remote Sens.* **2018**, *56*, 6521–6536. [CrossRef]

42. Chaudhuri, B.; Demir, B.; Bruzzone, L.; Chaudhuri, S. Region-Based Retrieval of Remote Sensing Images Using an Unsupervised Graph-Theoretic Approach. *IEEE Geosci. Remote Sens. Lett.* **2016**, *13*, 987–991. [CrossRef]

43. Zhou, W.; Deng, X.; Shao, Z. Region Convolutional Features for Multi-Label Remote Sensing Image Retrieval. *arXiv* **2018**, arXiv:180708634 Cs. Available online: https://arxiv.org/abs/1807.08634 (accessed on 24 January 2020).

44. Dai, O.E.; Demir, B.; Sankur, B.; Bruzzone, L. A Novel System for Content-Based Retrieval of Single and Multi-Label High-Dimensional Remote Sensing Images. *IEEE J. Sel. Top. Appl. Earth Obs. Remote Sens.* **2018**, *11*, 2473–2490. [CrossRef]

45. Wu, Q.; Shen, C.; Liu, L.; Dick, A.; Hengel, A.v.d. What Value Do Explicit High Level Concepts Have in Vision to Language Problems? In Proceedings of the 2016 IEEE Conference on Computer Vision and Pattern Recognition (CVPR), Las Vegas, NV, USA, 27–30 June 2016; pp. 203–212.

46. Cho, K.; van Merrienboer, B.; Gulcehre, C.; Bahdanau, D.; Bougares, F.; Schwenk, H.; Bengio, Y. Learning Phrase Representations using RNN Encoder-Decoder for Statistical Machine Translation. *arXiv* **2014**, arXiv:14061078 Cs Stat. Available online: https://arxiv.org/abs/1406.1078 (accessed on 24 January 2020).

47. Vinyals, O.; Toshev, A.; Bengio, S.; Erhan, D. Show and Tell: Lessons Learned from the 2015 MSCOCO Image Captioning Challenge. *IEEE Trans. Pattern Anal. Mach. Intell.* **2017**, *39*, 652–663. [CrossRef] [PubMed]

48. Krizhevsky, A.; Sutskever, I.; Hinton, G.E. Imagenet classification with deep convolutional neural networks. In *Advances in Neural Information Processing Systems*; MIT Press: Cambridge, MA, USA, 2012; pp. 1097–1105.

49. Fang, H.; Gupta, S.; Iandola, F.N.; Srivastava, R.K.; Deng, L.; Dollar, P.; Gao, J.; He, X.; Mitchell, M.; Platt, J. From captions to visual concepts and back. In Proceedings of the 2015 IEEE Conference on Computer Vision and Pattern Recognition, Boston, MA, USA, 7–12 June 2015; pp. 1473–1482.

50. You, Q.; Jin, H.; Wang, Z.; Fang, C.; Luo, J. Image Captioning with Semantic Attention. In Proceedings of the IEEE Conference on Computer Vision and Pattern Recognition (CVPR), Las Vega, NV, USA, 27–30 June 2016; pp. 4651–4659.

51. Chen, L.; Zhang, H.; Xiao, J.; Nie, L.; Shao, J.; Liu, W.; Chua, T.S. Sca-cnn: Spatial and channel-wise attention in convolutional networks for image captioning. In Proceedings of the 2017 IEEE Conference on Computer Vision and Pattern Recognition (CVPR), Honolulu, HI, USA, 21–26 July 2017; pp. 6298–6306.

52. Huang, F.; Zhang, X.; Li, Z.; Zhao, Z. Bi-directional Spatial-Semantic Attention Networks for Image-Text Matching. *IEEE Trans. Image Process. Publ. IEEE Signal Process. Soc.* **2018**, *28*, 2008–2020. [CrossRef] [PubMed]

53. Wang, L.; Li, Y.; Lazebnik, S. Learning Two-Branch Neural Networks for Image-Text Matching Tasks. *IEEE Trans. Pattern Anal. Mach. Intell.* **2018**, *41*, 394–407. [CrossRef]

54. Zhang, Y.; Lu, H. Deep Cross-Modal Projection Learning for Image-Text Matching. In Proceedings of the European Conference on Computer Vision—ECCV 2018, Munich, Germany, 8–14 September 2018; Springer; pp. 686–701.

55. Yao, T.; Pan, Y.; Li, Y.; Qiu, Z.; Mei, T. Boosting image captioning with attributes. In Proceedings of the 2017 IEEE International Conference on Computer Vision (ICCV), Venice, Italy, 22–29 October 2017; pp. 4904–4912.

56. Lu, X.; Wang, B.; Zheng, X.; Li, X. Exploring Models and Data for Remote Sensing Image Caption Generation. *IEEE Trans. Geosci. Remote Sens.* **2018**, *56*, 2183–2195. [CrossRef]

57. Shi, Z.; Zou, Z. Can a Machine Generate Humanlike Language Descriptions for a Remote Sensing Image? *IEEE Trans. Geosci. Remote Sens.* **2017**, *55*, 3623–3634. [CrossRef]

58. Zhang, X.; Wang, X.; Tang, X.; Zhou, H.; Li, C. Description Generation for Remote Sensing Images Using Attribute Attention Mechanism. *Remote Sens.* **2019**, *11*, 612. [CrossRef]

59. Hu, J.; Shen, L.; Sun, G. Squeeze-and-excitation networks. In Proceedings of the IEEE Conference on Computer Vision and Pattern Recognition, Salt Lake City, UT, USA, 18–22 June 2018; pp. 7132–7141.

60. Szegedy, C.; Vanhoucke, V.; Ioffe, S.; Shlens, J.; Wojna, Z. Rethinking the Inception Architecture for Computer Vision. In Proceedings of the IEEE conference on computer vision and pattern recognition (CVPR), Las Vegas, NV, USA, 27–30 June 2016; pp. 2818–2826.

61. Tan, M.; Le, Q.V. EfficientNet: Rethinking Model Scaling for Convolutional Neural Networks. *arXiv* **2019**, arXiv:190511946 Cs Stat. Available online: https://arxiv.org/abs/1905.11946 (accessed on 24 January 2020).

62. Hochreiter, S.; Schmidhuber, J. Long Short-Term Memory. *Neural Comput.* **1997**, *9*, 1735–1780. [CrossRef]

63. Weinberger, K.Q.; Saul, L.K. Distance Metric Learning for Large Margin Nearest Neighbor Classification. *J. Mach. Learn. Res.* **2009**, *10*, 207–244.

64. Wang, J.; Song, Y.; Leung, T.; Rosenberg, C. Learning Fine-Grained Image Similarity with Deep Ranking. In Proceedings of the IEEE Conference on Computer Vision and Pattern Recognition, Columbus, OH, USA, 23–28 June 2014; pp. 1386–1393.

65. Hoffer, E.; Ailon, N. Deep Metric Learning Using Triplet Network. In Proceedings of the Similarity-Based Pattern Recognition, Copenhagen, Denmark, 12–14 October 2015; Feragen, A., Pelillo, M., Loog, M., Eds.; Springer International Publishing: Cham, Switzerland; pp. 84–92.

66. Law, M.T.; Thome, N.; Cord, M. Quadruplet-Wise Image Similarity Learning. In Proceedings of the 2013 IEEE International Conference on Computer Vision, Sydney, NSW, Australia, 1–8 December 2013; pp. 249–256.

67. Oh Song, H.; Xiang, Y.; Jegelka, S.; Savarese, S. Deep Metric Learning via Lifted Structured Feature Embedding. In Proceedings of the IEEE conference on computer vision and pattern recognition (CVPR), Las Vegas, NV, USA, 27–30 June 2016; pp. 4004–4012.

68. Sohn, K. Improved deep metric learning with multi-class n-pair loss objective. In Proceedings of the Advances in Neural Information Processing Systems (NIPS), Barcelona, Spain, 5–10 December 2016; pp. 1857–1865.

69. Wang, J.; Zhou, F.; Wen, S.; Liu, X.; Lin, Y. Deep Metric Learning with Angular Loss. In Proceedings of the 2017 IEEE International Conference on Computer Vision (ICCV), Venice, Italy, 22–29 October 2017; pp. 2612–2620.

70. Huang, J.; Feris, R.; Chen, Q.; Yan, S. Cross-Domain Image Retrieval with a Dual Attribute-Aware Ranking Network. In Proceedings of the 2015 IEEE International Conference on Computer Vision (ICCV), Santiago, Chile, 7–13 December 2015; pp. 1062–1070.

71. Lai, H.; Pan, Y.; Liu, Y.; Yan, S. Simultaneous Feature Learning and Hash Coding With Deep Neural Networks. In Proceedings of the 2015 IEEE Conference on Computer Vision and Pattern Recognition (CVPR), Boston, MA, USA, 7–12 June 2015; pp. 3270–3278.

72. Zhuang, B.; Lin, G.; Shen, C.; Reid, I. Fast Training of Triplet-Based Deep Binary Embedding Networks. In Proceedings of the 2016 IEEE Conference on Computer Vision and Pattern Recognition (CVPR), Las Vegas, NV, USA, 27–30 June 2016; pp. 5955–5964.

73. Gordo, A.; Almazan, J.; Revaud, J.; Larlus, D. Deep image retrieval: Learning global representations for image search. VI. In Proceedings of the Computer Vision—ECCV 2016, Amsterdam, The Netherlands, 11–14 October 2016; Volume 9910, pp. 241–257.

74. Yuan, Y.; Yang, K.; Zhang, C. Hard-Aware Deeply Cascaded Embedding. In Proceedings of the IEEE international conference on computer vision (ICCV), Venice, Italy, 22–29 October 2017; pp. 814–823.

75. Parkhi, O.M.; Vedaldi, A.; Zisserman, A. Deep Face Recognition. In Proceedings of the British Machine Vision Conference (BMVC); British Machine Vision Association: Swansea, September, 2015; pp. 41.1–41.12.

76. Wang, L.; Li, Y.; Lazebnik, S. Learning Deep Structure-Preserving Image-Text Embeddings. In Proceedings of the 2016 IEEE Conference on Computer Vision and Pattern Recognition (CVPR), Las Vegas, NV, USA, 27–30 June 2016; pp. 5005–5013.

77. Schroff, F.; Kalenichenko, D.; Philbin, J. FaceNet: A Unified Embedding for Face Recognition and Clustering. In Proceedings of the 2015 IEEE Conference on Computer Vision and Pattern Recognition (CVPR), Boston, MA, USA, 7–12 June 2015; pp. 815–823.

78. Harwood, B.; VijayKumar, B.G.; Carneiro, G.; Reid, I.; Drummond, T. Smart Mining for Deep Metric Learning. In Proceedings of the IEEE international conference on computer vision (ICCV), Venice, Italy, 22–29 October 2017; pp. 2821–2829.

79. Wu, C.-Y.; Manmatha, R.; Smola, A.J.; Krähenbühl, P. Sampling Matters in Deep Embedding Learning. In Proceedings of the IEEE international conference on computer vision (ICCV), Venice, Italy, 22–29 October 2017; pp. 2840–2848.

80. Ge, W.; Huang, W.; Dong, D.; Scott, M.R. Deep Metric Learning with Hierarchical Triplet Loss. In Proceedings of the Computer Vision—ECCV 2018, Munich, Germany, 8–14 September 2018; pp. 269–285.

81. Plummer, B.A.; Wang, L.; Cervantes, C.M.; Caicedo, J.C.; Hockenmaier, J.; Lazebnik, S. Flickr30k Entities: Collecting Region-to-Phrase Correspondences for Richer Image-to-Sentence Models. In Proceedings of the 2015 IEEE International Conference on Computer Vision (ICCV), Santiago, Chile, 13–16 December 2015; IEEE; pp. 2641–2649.

Article

Water Identification from High-Resolution Remote Sensing Images Based on Multidimensional Densely Connected Convolutional Neural Networks

Guojie Wang [1,*], **Mengjuan Wu** [1], **Xikun Wei** [1] **and Huihui Song** [2]

[1] Collaborative Innovation Center on Forecast and Evaluation of Meteorological Disasters, School of Geographical Sciences, Nanjing University of Information Science and Technology (NUIST), Nanjing 210044, China; wu.mj@nuist.edu.cn (M.W.); 20181211018@nuist.edu.cn (X.W.)

[2] School of Automation, Nanjing University of Information Science and Technology (NUIST), Nanjing 210044, China; songhuihui@nuist.edu.cn

[*] Correspondence: gwang@nuist.edu.cn; Tel.: +86-025-5873-1418

Received: 30 January 2020; Accepted: 26 February 2020; Published: 2 March 2020

Abstract: The accurate acquisition of water information from remote sensing images has become important in water resources monitoring and protections, and flooding disaster assessment. However, there are significant limitations in the traditionally used index for water body identification. In this study, we have proposed a deep convolutional neural network (CNN), based on the multidimensional densely connected convolutional neural network (DenseNet), for identifying water in the Poyang Lake area. The results from DenseNet were compared with the classical convolutional neural networks (CNNs): ResNet, VGG, SegNet and DeepLab v3+, and also compared with the Normalized Difference Water Index (NDWI). Results have indicated that CNNs are superior to the water index method. Among the five CNNs, the proposed DenseNet requires the shortest training time for model convergence, besides DeepLab v3+. The identification accuracies are evaluated through several error metrics. It is shown that the DenseNet performs much better than the other CNNs and the NDWI method considering the precision of identification results; among those, the NDWI performance is by far the poorest. It is suggested that the DenseNet is much better in distinguishing water from clouds and mountain shadows than other CNNs.

Keywords: convolutional neural network; water identification; water index

---

1. Introduction

Water is an indispensable resource for a sustainable ecosystem on earth. It contributes significantly to the balance of ecosystems, the maintenance of climate change and the carbon cycle [1]. The formation, expansion, shrinkage and disappearance of surface water are important factors influencing the environment and regional climate changes. Water is also an important factor in socioeconomic development, because it affects many agricultural, environmental and ecological issues over time [2,3]. Hence the rapid and accurate extraction of water resource information can provide necessary data, which is of great significance for water resource investigation [4–6], flood monitoring [7,8], wetland protection [9,10] and disaster prevention and reduction [11,12].

In recent years, a lot of research has been done on image foreground extraction and segmentation [13]. This study proposed an Alternating Direction of Method of Multipliers (ADMM) approach to separate the foreground information from the background, and it has a great effect upon the separation of text, moving objects and so on. There are also many algorithms for extracting water from remote sensing images, including spectral classification [14], the threshold segmentation method [7,15] and machine learning [16–18].

However, the accurate identification of water is always a difficult problem because of the complicated terrain, classification methods and remote sensing data itself. Because of its simplicity and convenience, the water index is the most commonly used water identification method. Among them, the Normalized Difference Water Index (NDWI) [19], Modified NDWI (MNDWI) [20] and the Automated Water Extraction Index (AWEI) [21], are most representative methods. The NDWI normalized green and near-infrared bands to enhance the water information to separate the water better, but it had a large error in urban areas [20]. MNDWI ameliorates this problem by using mid-infrared bands [20]. What these water indices have in common, is that they all use differences in the reflectivity of water at different wavebands to enhance water information. The water is then classified by setting a threshold.

There are two problems with the water index approaches, and one of them is that every water index has its drawbacks. For example, the NDWI was poor at distinguishing between water and buildings, and the MNDWI was poor at distinguishing water from snow and mountain shadows. More sophisticated methods for high-precision water maps require auxiliary data, such as digital elevation models and complex rule sets to overcome these problems [22–24]. Another problem is that the optimal threshold to extract water is not only highly subjective, but also varies with region and time. By adopting the method of the Automatic Water Extraction Index [21], the extraction result was improved, but the threshold still changes with the change of time and area.

Statistical models are also used for identifying the water bodies, which can be divided into unsupervised and supervised classifications. It is generally more accurate than other methods, because it does not require an empirical threshold. No prior knowledge is applied in the unsupervised classification, while supervised classification makes classifications by learning from given samples. There are many popular supervised methods, like maximum likelihood [14] and the decision tree [25,26]. Most methods require additional inputs for more accurate results, such as slope, and mountain shadow [25,26], in the original band, and so on. All of these increase the data volume and calculation difficulty.

In recent years, the recognition algorithm based on artificial intelligence has been developing rapidly. Different from the traditional methods, deep learning can adapt learning from a large number of samples with flexibility and universality [27]. The convolutional neural network is one of the commonly used models of deep learning, which greatly reduces the number of parameters, enhances the generalization ability, and realizes the qualitative leap of image recognition by its features of local connection and weight sharing [17]. As part of the study of neural networks, the recent popularity of neural networks has revitalized the research field. As the number of network layers increases, the differences between different structures are also enlarging, which has stimulated the exploration of different network structures [28–32]. Many different network structures have been proposed to realize the semantic segmentation of images. One is the encoder–decoder structure, such as Unet [33], SegNet [34] and RefineNet [35]. The encoder is used to extract image features and reduce image dimensions. The decoder is used to restore the size and the detail of the image. The other is to use the dilated convolutions, such as DeepLab v1 [36], v2 [37], v3 [38], v3+ [39] and PSPNet [40]. They can increase the input field without pooling, so that each convolution contains a larger range of information in the output. In addition, networks that have been proven to be effective in object detection applications were also applied to the instance segmentation field and showed good efficiency. For instance, the regional convolutional network (R-CNN) [41], Fast R-CNN [42], Faster R-CNN [43], Mask R-CNN [44], etc. A new framework has also been proposed called the Hybrid Task Cascade (HTC), which combined cascade architecture with R-CNN for better results [45]. Attention mechanisms have also been applied to segmentation networks by many researchers. Chen et al. [46] showed that the attention mechanism outperforms average and max pooling. More recently, a Dual Attention Network (DANet) [47] has been proposed which appended two types of attention modules on top of dilated FCN, and achieved some new state-of-the-art results on multiple popular benchmarks.

Besides those networks mentioned above, there are many other types of depth model applied to image segmentation, like applying active contour models to convolutional neural networks (CNNs) [48], and so on. Shervin et al. [49] have made a thorough network summary for image segmentation.

The corresponding features from the image of target detection and classification can be extracted by the deep convolutional neural network. It is reported to perform well in image classification and target detection, and there are already some models developed, such as LeNet [50] in 1998, AlexNet [28] in 2012, GoogLeNet [29] and VGG [30] in 2014 and ResNet [31] in 2015. With the technical development, the complexity of these models is increasing. The VGG network uses only a 3×3 convolution kernel and 2×2 pooling kernel [30]. The use of a smaller convolution kernel can increase the linear transformation and improve the classification accuracy. It also shows that the increase of network depth has a great effect on the improvement of the final classification results of the network. However, simply increasing the network depth will lead to gradient vanishing or gradient explosion. ResNet solves this problem by introducing a residual block [31]. It passes information direct to output to protect the integrity of the information. The whole network just needs to know the difference between the input and output, simplifying the learning process. Recent research on ResNet shows that many of its middle layers contribute little to the actual training process, and can be randomly deleted, which makes ResNet similar to the recurrent neural networks [32]; but, since ResNet has its own weight every layer, it has a larger number of parameters. The multidimensional densely connected convolutional neural network (DenseNet) [51] proposed in 2016 does not have the above problems. It gives full play to the idea of a residual block in ResNet, and each layer of its network is directly connected to its previous layer to achieve the reuse of features. This enables the network to be easy to train by improving the flow of information and gradient throughout the network. At the same time, it has a regularization effect, and can prevent the overfitting effect for small data sets. Besides, each layer of the network is very narrow, leading to reduced redundancy. Crucially, unlike ResNet, the DenseNet combines features, not by summing them before passing them to the next layer, but through concatenation instead. Compared to ResNet, the number of its parameters is greatly reduced. The experimental result has shown that the DenseNet has fewer parameters, faster convergence speed and shorter training time under the premise of ensuring the training accuracy [51].

So far, Landsat is one of the most commonly used data satellites in water extraction research, the spatial resolution is 30 meters, and the temporal resolution is 16 days [52]. The GF-1 satellite was launched in April 2013 by China, which was equipped with two full-color cameras with a resolution of 2 m, and a multi-spectral camera with a resolution of 16 m. Since the revisit period of the GF-1 satellite is about four days, it has apparent advantages regarding its spatial and temporal resolutions. However, there are still rare cases using GF-1 satellite images for water body extraction, especially with the deep learning algorithms.

In this paper, we use the convolutional neural network (CNN) to extract water bodies from GF-1 images. We borrowed the idea of DenseNet and added the up-sampling process to form a fully convolutional neural network. At the same time, the skip layer connection was added in the up-sampling and down-sampling processes to improve the efficiency of feature utilization. This paper compares this model with the two segmentation networks of SegNet and DeepLab v3+, two feature extraction networks of ResNet and VGG, and also the traditional water index method to understand their efficiencies in water body identification.

2. Materials and Methods

2.1. Study Area

The Poyang Lake (28°22′–29°45′N, 115°47′–116°45′E), is located in the north of the Jiangxi province. It is the largest freshwater lake in China. In the rainy summer season, the area of lake can exceed 4000 km^2; in the relatively dry autumn and winter, the lake area will typically shrink by more than 1000 km^2. The lake is mainly fed by precipitation, and sometimes the Yangtze River flux. Rainy season in the Jiangxi province usually begins in April, and lasts for about three months.

The increase in precipitation causes the water level of the Poyang Lake to rise. The precipitation amount decreases after July. However, the water level of the Yangtze River rises due to the water supply from precipitation and snowmelt in its upper reaches, which feeds the Poyang Lake and makes the water level of this Poyang Lake continue to rise [53] under the continuous influence of human activities and the Yangtze River water diversion and a large amount of sediment deposits, which has an important influence on the area of Poyang Lake.

Figure 1 shows the river networks in the Poyang Lake basin. Since most of the water bodies in the Poyang Lake basin are distributed in the northern region, we have selected an area of interest to compare the water identification effects of different methods. Due to the influence of monsoon precipitation, the spatial coverage of Poyang Lake changes significantly during the wet and dry seasons. Therefore, we select images in summer and winter, respectively, to evaluate the water body recognition effect of the used models.

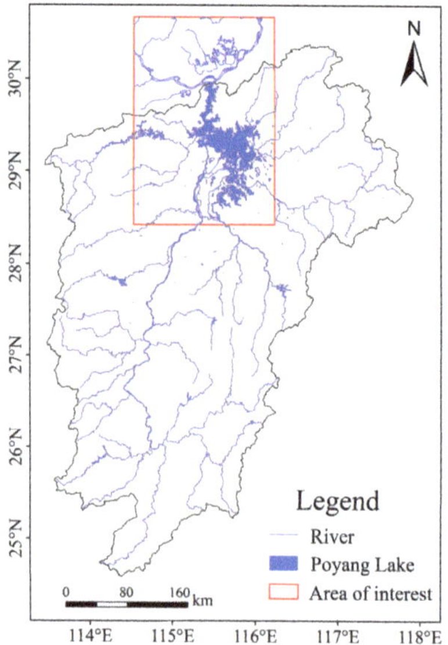

Figure 1. The river networks in Poyang Lake basin.

2.2. Data

The GF-1 satellite was launched in April 2013 and obtained a large amount of data since then. It carries two panchromatic/multi-spectral (P/MS) and four wide-field of view (WFV) cameras. Within the spectral range of the GF-1 WFV sensor (450–890 nm), there are four spectral channels to observe the reflected solar radiation from the earth. It has a spatial resolution of 16 m, a stripe width of 800 km, and consists of four cameras. The temporal resolution is four days. Therefore, it has the characteristics of high frequency revisit time, high spatial resolution and wide coverage, and is an ideal data for large-scale land surface monitoring.

The GF-1 satellite images are provided by the China Resource Satellite Application Center (http://www.cresda.com/CN/). In this study, our model also increases the input channels compared with the conventional neural network, and all the four spectral channels of GF-1 images are used.

2.3. Methods

To produce a water map from high-resolution satellite images, a DenseNet-based water mapping method was proposed. To verify the effectiveness of the proposed method, we compared this method with both methods of water index and classical convolutional neural network.

We select the method of the water index because it is the most widely used and representative method in the field of remote sensing image water extraction. Using the water index, we want to show that the proposed method has better performance than the traditionally used water index in water extraction, and in order to avoid the influence of subjective factors above the threshold selection of water index on the results, we used the Otsu's threshold segmentation method [54,55] to find the optimal threshold. Due to the limitation of GF-1 spectral bands, we choose the NDWI to extract water.

2.3.1. The Normalized Difference Water Index (NDWI)

The GF-1 images only contain four bands, hence NDWI can only be used to identify the water area. The optimal threshold of NDWI is determined using Otsu's method. The NDWI is a widely-used method for water identification based on the green band and near-infrared band. Using GF-1 spectral bands, the NDWI is computed as follows:

$$\text{NDWI} = \frac{b_{\text{green}} - b_{\text{near-infrared}}}{b_{\text{green}} + b_{\text{near-infrared}}} \tag{1}$$

where b_{green} represents the reflectivity of green band, $b_{\text{near-infrared}}$ represents the reflectivity of near-infrared band. Ideally, a positive NDWI value indicates the ground is covered with water, rain or snow; a negative NDWI value indicates vegetation coverage; and the ground is covered by rocks or bare soil if the NDWI is equal to 0. The threshold value is always not 0, due to various influences such as vegetation on the water surface. The selection of threshold is a key and difficult problem for accurate water body identification, and we use Otsu's method to determine it.

This Otsu's method is a classical algorithm in the image segmentation field which was proposed by the Japanese Nobuyuki Otsu in 1979 [54,55]. It is an adaptive threshold determination method. For a color image, it converts the image into a grayscale image and then distinguishes the target from the background according to the grayscale characteristics. The larger the variance of the gray value between target and the background, the greater the difference between these two parts. So, it calculates the maximum value of the class variance between target and background to find the optimal threshold. Among them, the definition of inter-class variance is as follows:

$$e^2(T) = P_O \left(\mu - \mu_O' \right)^2 + P_b \left(\mu - \mu_b' \right)^2 \tag{2}$$

where μ is the grayscale mean of the image, μ_o and μ_b are the means of the target and background, P_o and P_b are the proportion of grayscale of target and background, and T is the threshold. When T is the maximum value of $e^2(T)$, it is the optimal threshold.

In this study, as the pixel-wise NDWI values are derived, it is necessary to stretch them to the gray value from 0 to 256, from which Otsu's threshold is then calculated to segment the water body from the background.

2.3.2. Evolution of Convolutional Neural Network

With the development of technology and the optimization of hardware facilities, many classical networks have emerged after numerous updates of the convolutional neural network. In 2014, researchers developed the new deep convolutional neural network, VGG [30]. They discussed the relationship between the depth and the performance of neural network. VGG [30] successfully constructed the deep layer of 16–19 convolutional neural networks, and it proves that the increase of the network depth affects the performance of the network to some extent. It was once widely used as a backbone feature extraction network for various detection network frameworks [42,56] until the ResNet was proposed.

As a neural network with more than 100 layers, the ResNet's biggest innovation lies in that it solves the problem of network degradation through the introduction of a residual block. The traditional convolutional network has problems such as information loss during information transmission, and leads to the disappearance of gradient or gradient explosion, which makes the deep network unable to train. ResNet passes the input information directly to the output, thus solving this problem to some extent. It simplifies the difficulty of learning by learning the difference between input and output, instead of all input characteristics. DenseNet was proposed based on ResNet, but with considerable improvement.

As shown in Figure 2, the inputs of each layer of DenseNet are the outputs of all previous layers. The information transmission between different layers of the network is guaranteed to be maximized. Instead of connecting layers over summation such as the ResNet, the DenseNet connects the features through concatenating to achieve feature reuse. Meanwhile, a small growth rate is adopted, and the feature graph of each layer is relatively small; thus, to achieve the same accuracy, the computation required by DenseNet is only about half that of the ResNet. Therefore, this study chooses DenseNet as the backbone to extract features.

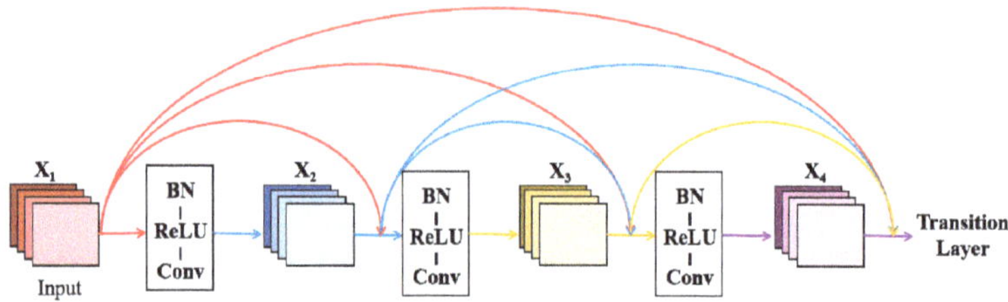

Figure 2. Multi-dimensional Dense Connection Module. (BN refers to Batch Normalization, ReLU refers to Rectified Linear Unit, Conv refers to Convolution.)

For a standard CNN, the output of the layer is the input of the next layer. The ResNet simplifies the training of the deep network by introducing the residual block, of which the output of the layer is the sum of the output of the previous layer and its nonlinear transformation. As for a DenseNet, the input of the l layer is the concatenation of the output characteristic map from 1 to $l - 1$ layer, and then makes nonlinear changes, that is:

$$x_l = K_l([x_{l-1}, x_{l-2}, \ldots, x_1]),\qquad(3)$$

here K is made up of batch normalization, activation functions, convolution and dropout. DenseNet's dense connections increase the utilization of features, make the network easier to train, and has the effect of regularization.

Fully convolutional networks (FCNs) [57,58], as a convolutional neural network, can segment images at pixel scale; therefore, it solves the problem of semantic segmentation. The classic CNN uses the fully connected layers after the convolution layer to obtain the feature vector for classification (fully connected layer + SoftMax output) [59–62]. Unlike the classic CNN, FCN uses deconvolution to return the reduced feature map to the original size after feature extraction. In this way, while preserving the spatial information of the input, the output with the same size of the input is gradually obtained, so as to achieve the purpose of pixel classification. It can accept input images of any size. Many networks have been proposed for image segmentation after FCN. SegNet [34] was proposed as an encoder–decoder network which uses the first 13 layers of VGG16 as encoders, and the max pooling indices as decoders to improve the segmentation resolution. DeepLab v3+ [39] was proposed in 2018, and it is the latest version of DeepLab series. It uses deep convolutional neural network with atrous convolution in the decoder part. Then the Atrous Spatial Pyramid Pooling (ASPP) is used to introduce multiscale information. Compared with DeepLab v3, v3+ introduces the decoder module, which further integrates the low-level features and high-level features to improve the accuracy of segmentation boundary.

2.3.3. Model-Based on DenseNet

Figure 3 shows the architecture of the network we have proposed for water body identification. Our model is a fully convolutional neural network with the fusion of multiscale features. The model chooses DenseNet as the backbone for feature extraction. The DenseNet we use contains four dense blocks. The transition block makes the connection between each dense block. The transition block consists of a 1×1 convolution and a 2×2 pooling operation. It can reduce the spatial dimensionality of feature maps.

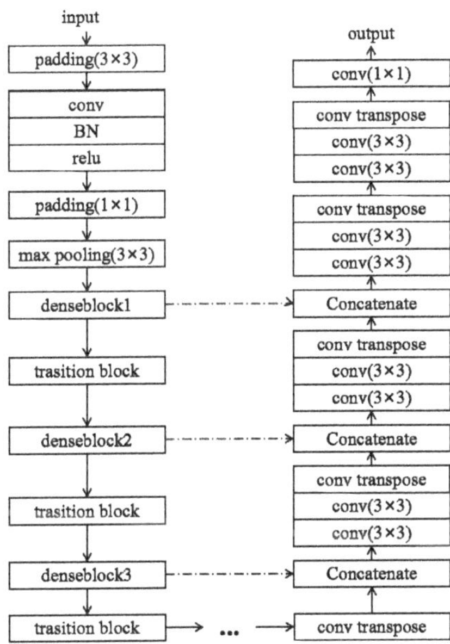

Figure 3. Proposed network architecture for semantic identification based on the DenseNet model.

In our network, in order to recover from the input spatial resolution, the upsampling layer is implemented by the transpose convolution. The feature map of the upsampling is then concatenated to the feature map from the dense block in the down-sampling process. The batch normalization (BN) and the Rectified Linear Unit (ReLU) are performed before the convolution of the image.

Our model can input images of arbitrary size during inference. But for the convenience of training, and to ensure that there is sufficient memory for training, we unified all input images into the size of 224 × 224 pixels. We cut out images of uniform size from GF-1 images, and screened out images containing both water and non-water as effective training data. At the same time, to ensure that the model can directly extract useful features from the original data, we did not do any preprocessing of the input image. We used the Adam optimization algorithm to optimize the weight. Hyperparameters $\beta_1 = 0.9$ and $\beta_2 = 0.999$ are selected as recommended by the algorithm. We trained our model in stages with the initial learning rate $\lambda = 10^{-4}$, which was reduced by 10 times after 30 epochs. The initial learning rate here is the best result from multiple trials. The growth rate of the network is set as 32, weight decay is 10^{-4}, and the Nesterov momentum is 0.9, which remain the same as the classic DenseNet.

In order to determine the number of network layers, we experiment with the number of convolutions in each dense block to find the optimal result. The DenseNet proposed by Huang et al. [51] designed three network layers for different tasks, i.e., Densenet121, Densenet169 and Densenet201. In addition to testing the above-mentioned three networks, we also adjust the number of layers to find the most suitable result for this task. We first halve the convolution layers of first three dense blocks of DenseNet121, the fourth block is unchanged, which is DenseNet79.

Then we tried to halve the convolution layers of four blocks and it became DenseNet63. We trained five DenseNets with different network layers to compare which is the best.

In order to make an effective comparison of the results, we use training time as one indicator to determine which network is faster and more convenient. We use the precision (P), recall (R), F1 score (F1) and mean Intersection over Union (mIoU) to quantitatively measure the performance of the network, which are all based on the confusion matrix. The same indicators were used to evaluate the performances of NDWI, VGG, ResNet, SegNet, DeepLab v3+ and DenseNet. As an evaluation index, the confusion matrix evaluates the performance of a classifier, and it is more accurate for the identification results of unbalanced categories. The confusion matrix divides the image identification results into four parts: true positive (TP), true negative (TN), false positive (FP) and false negative (FN). The specific calculation formula of evaluation index is as follows [63]:

$$P = \frac{TP}{TP + FP} \tag{4}$$

$$R = \frac{TP}{TP + FN} \tag{5}$$

$$F_\alpha = \frac{(1 + \alpha^2)P \times R}{\alpha^2(P + R)} \tag{6}$$

$$mIoU = \frac{TP}{TP + FP + FN} \tag{7}$$

where P means the precision, and R means the recall. MIoU is the intersection of two sets of ground truth and predicted results. Precision is the fraction of correctly identified water pixels (TP) among the predicted water pixels (TP + FP) by the model. Recall is the fraction of correctly identified water pixels (TP) among the actual water pixels (TP + FN). Since precision and recall are sometimes contradictory, we further introduce the F1 score to measure the accuracy of a binary model, which simultaneously takes precision and recall into consideration [64]:

$$F1 = \frac{2(P + R)}{P + R} \tag{8}$$

The comparison results of five networks are shown in Table 1. The best results of all indicators are displayed in bold fonts. We can see that with the increase of network layers, the training time also increases; however, the performance does not become better with layer increase. This may be because the input samples of the network are not enough, and the characteristics of the water are easier to identify, so too many layers will not contribute to the results. Among these five networks, DenseNet79 has the best performance in recall, F1 score and mIoU. Its precision is lower than DenseNet169, but the training time is almost two hours less than DenseNet169. Therefore, DenseNet79 is most suitable for the task of water recognition in this study.

Table 1. Comparison of DenseNets with different layers. The optimal value for each metric is shown in bold. (P refers to Precision; R refers to Recall; F1 refers to F1 score and mIoU refers to mean Intersection over Union).

Network	Time	P	R	F1	mIoU
DenseNet63	**15,463 s**	0.959	0.900	0.928	0.867
DenseNet79	16,377 s	0.961	**0.904**	**0.931**	**0.872**
DenseNet121	20,611 s	0.957	0.901	0.928	0.866
DenseNet169	24,018 s	**0.964**	0.896	0.928	0.867
DenseNet201	27,121 s	0.960	0.899	0.929	0.867

To verify the performance of our implementations, VGG, ResNet, SegNet and DeepLab v3+ were selected to make comparisons. VGG and ResNet were selected, respectively, as representatives of the neural network with less than 100 layers, and the neural network with more than 100 layers. SegNet and DeepLab v3+ were selected as representatives of two segmentation network structures: encoder–decoder structure and Atrous convolution. Also, due to the limitation of computation resources and the number of training datasets, it is not necessary to use powerful and complicated networks as our exception, since, as the backbone of DeepLab v3+, we chose MobileNet [65] as the backbone, which has much less parameter, and can achieve good results on our task in shorter time.

3. Results

To see if the DenseNet is more suitable and efficient than the other methods, we first compare the result of the proposed network with the ground truth to evaluate its effectiveness in identifying the water bodies. Then we compare with the results derived from the NDWI index and four other deep neural networks of VGG, ResNet, SegNet and DeepLab v3+. Finally, we chose the best model and made a simple analysis of the changes in water areas in Poyang Lake area in winter and summer from 2014 to 2018.

3.1. The Image Preprocessing

The dataset contains GF-1 images from the middle and lower reaches of the Yangtze River basin in different periods. The corresponding labels were binary classifications of the water–nonwater area by expert visual interpretation. To improve the efficiency of the model training, we clipped the input data to 224 × 224 pixels. We have deliberately selected some labels with both land and water bodies as training samples. Finally, we have selected 5558 water bodies samples. Of these, 4446 images were used as training sets, while the remaining 1112 images were used as test sets. This data is only used for model training and quantitative evaluation. Since the samples are cut into small pieces, and the selection of training set and test set are random, the recognition efficiency of the model on a large range of images cannot be seen from the existing data. To qualitatively evaluate the performance of different models in different ground object types, we also applied the model to other GF-1 images in different periods.

3.2. Water Identification Result of DenseNet

Figure 4 is the recognition result of 12 remote sensing images selected from the validation dataset, the corresponding ground truths, and the DenseNet result. The sizes of images are all 224 × 224. These 12 images contain water bodies of various shapes and colors, from which the recognition effectiveness of this model on different water bodies can be understood.

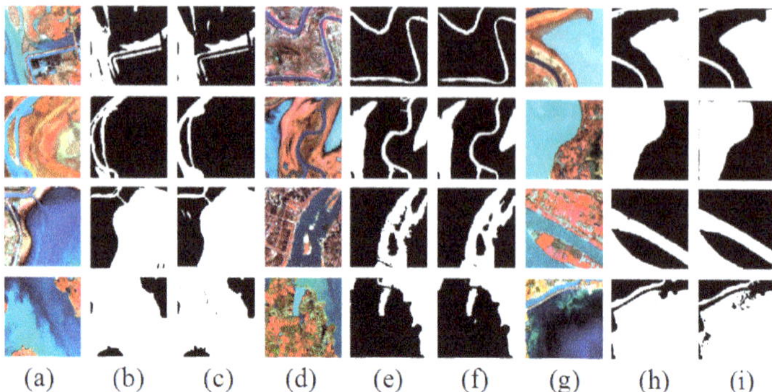

(a)   (b)   (c)   (d)   (e)   (f)   (g)   (h)   (i)

Figure 4. Result of the water body identifications using the DenseNet model. The figure is divided into four rows and nine columns, showing the recognition effects in 12 different regions. Column (**a**), (**d**), (**g**) are the false-color remote sensing images; column (**b**), (**e**), (**h**) are the corresponding ground truths; column (**c**), (**f**), (**i**) are the corresponding model (DenseNet) recognition results.

It can be seen from Figure 4 that the recognition result of DenseNet is consistent with the ground truth. Although this model failed to identify some small water bodies, the error areas are generally very small, and such small errors have little influence on the overall distribution of water bodies, which can be ignored. In addition, the network can accurately identify the water bodies in different forms and regions, and accurately separate small rivers in the towns, and even small barriers such as bridges in the water can be correctly separated. The boundaries between water and land were identified, partly because of the fine resolution of the GF-1 images, and partly because of the efficiency of the proposed DenseNet model.

3.3. Working Efficiency of DenseNet, ResNet, VGG, SegNet and DeepLab v3+ Models

Figure 5 shows the training losses of DenseNet, ResNet, VGG, SegNet and DeepLab v3+. In the convolutional neural networks, the loss function is used to calculate the difference between the output of the model and the ground truth, so as to better optimize the model. The smaller the loss is, the better the robustness of the model is. In Figure 5, one epoch represents 1000 iterations. For the initial epochs, the loss value of VGG is by far the highest, which is two or three times higher than those of ResNet and DenseNet; and it remains the highest until 30 epochs. The initial loss value of SegNet is close to VGG, followed by DeepLab v3+. The DenseNet has a higher initial loss value than the ResNet, but then it declines faster than the ResNet and continues to be lower than the ResNet after five epochs. The loss of DenseNet maintains the lowest after five epochs, indicating the fastest convergence speed compared to the other four models.

Table 2 shows the training time of the five networks. Among them, the VGG has the longest training time. The DeepLab v3+'s training time is the shortest, and DenseNet is next to it. The DenseNet saves about 80 min compared to the VGG model, about 50 min compared to the ResNet model, and 40 min compared to the SegNet model.

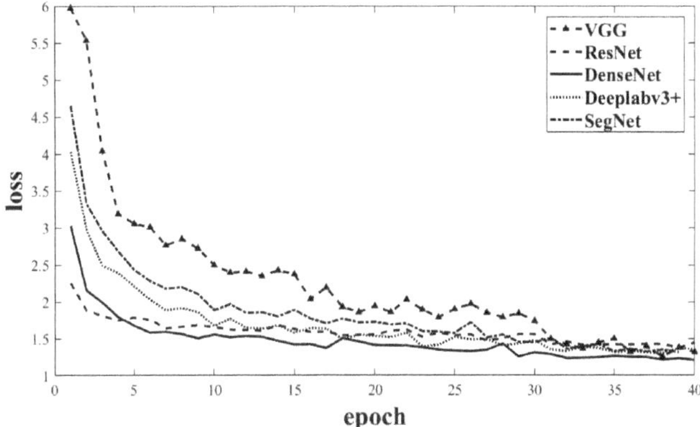

Figure 5. Training losses of the DenseNet, ResNet, VGG, SegNet and DeepLab v3+ models. One epoch represents 1000 iterations.

Table 2. Training time of the DenseNet, ResNet, VGG, SegNet and DeepLab v3+ models.

Network	Time
DenseNet	16,377 s
ResNet	19,436 s
VGG	21,471 s
SegNet	19,021 s
DeepLab v3+	11,924 s

But it takes more than 70 min compared to DeepLab v3+. This indicates that under the same training environment, the DeepLab v3+ requires the least training time; it is easier to train and use the lowest resource consumption capacity. The reason why we did not compare the time consumption of NDWI with these networks is that the NDWI method does not need a lot of time to process, and the required time can be ignored.

3.4. Comparison of Identification Results

The derived P, R, F1 score and mIoU of the VGG, the ResNet, the DenseNet, the SegNet, the DeepLab v3+ and the NDWI models are shown in Table 3. All values in the table were calculated by the prediction results of 1112 images in the test set, and their corresponding ground truth. Given the limited number of samples, we reported the 95% confidence interval of the metrics to see if the result is statistically significant. The best result of each indicator is in bold. We can see from the results that all neural networks' results are much better than the NDWI index. For each network model, the DenseNet result, with a narrower interval, appears more stable than the other methods.

Table 3. The derived P, R, F1score and mIoU of the VGG, ResNet, DenseNet, SegNet, DeepLab v3+ and NDWI models with 95% confidence interval. The optimal value for each metric is shown in bold.

	DenseNet	ResNet	VGG	SegNet	DeepLab v3+	NDWI
P	**0.961 ± 0.011**	0.936 ± 0.014	0.914 ± 0.016	0.911 ± 0.017	0.922 ± 0.016	0.702 ± 0.027
R	0.904 ± 0.017	0.902 ± 0.017	0.915 ± 0.016	0.934 ± 0.015	0.917 ± 0.016	0.983 ± 0.007
F1	**0.931 ± 0.015**	0.919 ± 0.016	0.914 ± 0.016	0.922 ± 0.016	0.919 ± 0.016	0.819 ± 0.023
mIoU	**0.872 ± 0.020**	0.850 ± 0.021	0.842 ± 0.021	0.856 ± 0.021	0.850 ± 0.021	0.767 ± 0.025

Among six models, the DenseNet appears to have the highest precision of 0.961, meaning that 96.1% of the water bodies are correctly predicted among the predicted water bodies by the model. The precisions of ResNet, VGG, SegNet and DeepLab v3+ are 0.936, 0.914, 0.911 and 0.922, respectively.

Such a rank of this precision is as expected, considering the pathway of theoretical improvements of these deep neural network models. However, the NDWI model based on the spectral bands appears to have a rather reduced prediction precision, which is only 0.702, although an adaptive threshold from the Otsu method is employed. Hence, the DenseNet appears to perform the best among the three deep neural networks regarding prediction precision; particularly, such a neural network, at least in this case, is by far the better than normally used NDWI method for water body identification in the remote sensing community.

Among the three deep neural networks, SegNet shows the highest recall value of 0.934. The ResNet shows the lowest recall, which is 0.902. The DenseNet is only 0.02 higher than ResNet. VGG and DeepLab v3+ have a recall of 0.915 and 0.917, respectively. The NDWI model shows the highest recall value of 0.983 among all the six methods, indicating it has successfully identified most of the water body samples in the training dataset. However, its precision value is the lowest, indicating that there are still serious ill predictions from this method. As can be seen, the matrices of recall and precision have given contrary indications of the model performances. To make a comprehensive evaluation of these two indicators, we investigate the F1 score considering both the precision and the recall values. We also use mIoU to evaluate the accuracy of model segmentation results. A higher F1 score and mIoU indicates a better performance. The F1 scores of the DenseNet, ResNet, VGG, SegNet and DeepLab v3+ models are 0.931, 0.919, 0.914, 0.922 and 0.919, respectively, and the mIoUs of them are 0.872, 0.850, 0.842, 0.856 and 0.850, respectively. We can see from the results that the performance of DenseNet is better than ResNet, VGG, SegNet and DeepLab v3+. This may be due to the dense connection, which increases the utilization efficiency of the features. As for the result of DeepLab v3+, the training efficiency is much better than DenseNet. This is because the backbone of DeepLab v3+ we chose is MobileNet, which is a lightweight network using the depth-wise separable convolution to reduce the number of parameters and the amount of calculation. The F1 score and the mIoU of the NDWI index are as low as 0.819 and 0.767, showing that all the deep neural networks have much better performance than the traditional NDWI method from a comprehensive viewpoint.

The recalls of DenseNet and ResNet are not very good in these models, meaning that these networks are not good at capturing all the water areas. Figure 6 shows some examples of this disadvantage. The third column is the result of DenseNet, and the fourth column is the result of ResNet; this figure shows that the water area which DenseNet recognized is the smallest in all six models, and it distributes in small rivers and intertidal zones. Column (h) is the result of NDWI. NDWI recognized the biggest water area, which is consistent with its highest recall value. However, with the increase of identified water area, the probability of recognition error is also increasing, meaning that the precision is more likely to drop with it. To increase the recall value of DenseNet, it may cost a sharp drop of precision. It has good results of F1 score and mIoU, meaning that the overall performance of this network is very good. Therefore, we decided not to further optimize the recall of DenseNet.

In order to further understand the performance of each method in different regions, we selected two GF-1 images of the Poyang Lake during the wet and the dry seasons, respectively, to evaluate the performance of different models, i.e., 29 July and 31 December 2016. Figure 7 shows the results from the image on 31 December 2016, when the Poyang Lake basin was dry with a complex distribution of water area.

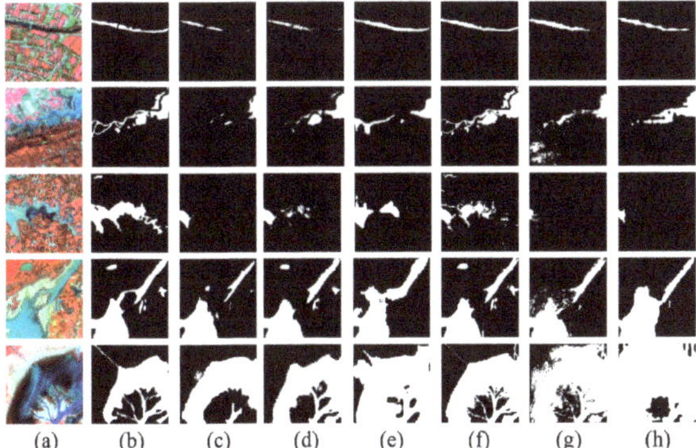

(a)   (b)   (c)   (d)   (e)   (f)   (g)   (h)

Figure 6. Examples of the recognition effect of different models which shows the high false negative (FN) of DenseNet. (**a**) False color composite remote sensing image, and water body identification result by (**b**) the ground truth, (**c**) the DenseNet, (**d**) the ResNet, (**e**) the VGG, (**f**) the SegNet, (**g**) the DeepLab v3+ and (**h**) the NDWI models. White color indicates the identified water bodies.

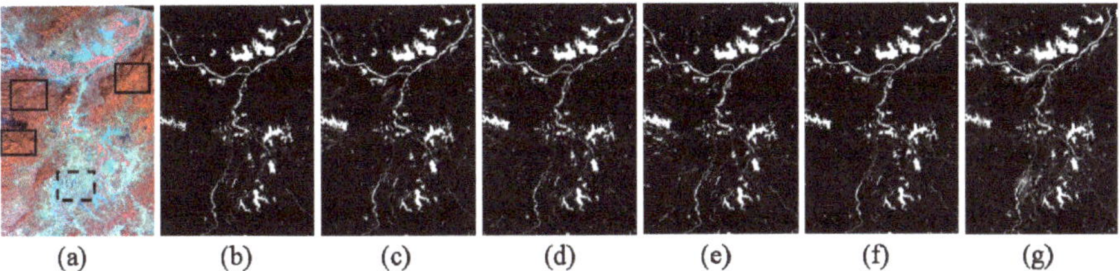

(a)   (b)   (c)   (d)   (e)   (f)   (g)

Figure 7. Comparison of the water identification effect of different models in Poyang Lake on 31 December 2016. (**a**) False color composite remote sensing image, and water body identification result by (**b**) the DenseNet, (**c**) the ResNet, (**d**) the VGG, (**e**) the SegNet, (**f**) the DeepLab v3+ and (**g**) the NDWI models. White color indicates the identified water bodies, solid line depicts mountain area, dashed line depicts urban area.

In the false-color image, the blue area is mostly water body, and the red area is mostly vegetated. The other colored areas include bare land, buildings and other nonwater areas. The mountain area is depicted with a solid line frame, while the urban area is marked with a dashed line frame. In the prediction results of ResNet, many patches in the corresponding region of the mountains are predicted to be water bodies, which proves that the ResNet model is prone to confuse mountain shadows with water. In the same regions, the VGG and SegNet models have also falsely identified some mountain shadow areas as water bodies. DeepLab v3+ has not confused the mountain shadow with a water body, but the boundary of water area it extracted was not as concise as the other methods. The main water body was correctly identified by the NDWI models, which are however much larger than the actual water bodies, and the NDWI model has also identified too many fine patches. The NDWI result also had false detection of the mountain shadows, which is larger than those from the ResNet model, but smaller than those from the VGG model. Other than the mountain shadow, the biggest problem with the NDWI result is that it falsely identified some bare land and urban construction areas as water bodies. The DenseNet model has successfully identified the small rivers and lakes from the GF-1 image, and the mountain shadows and water bodies are successfully separated. In general, these five deep neural networks have consistently identified the large water bodies in winter, although the ResNet and the VGG models show a false identification of mountain shadows. These neural networks have performed much better than the traditionally used NDWI water body index.

Figure 8 shows the identified water bodies from the GF-1 image on 29 July 2016. In the false-color image, the white area in the dotted line indicates the cloud, and the dashed line depicts the urban area. In summer, the Poyang Lake is in a season with abundant water, and its water area reaches its peak within a year. It is found that the VGG, SegNet and DeepLab v3+ models have falsely identified the cloud as water bodies, and the DenseNet also has a small amount of false identification. We can see that the NDWI index can better identify the bulk of the water body, but there is much noise in the boundary areas; besides, it has falsely identified the urban buildings, bare ground and most clouds as water bodies. It is the ResNet model that completely distinguishes between the cloud and the water bodies, which however has some false identification of some water bodies. As for the DenseNet result, it shows a relatively accurate identification of water bodies with clear boundary separation for the transitional areas between land and water. The DenseNet method partially falsely identified cloud as water bodies, but it has filtered out most of it compared to the NDWI result.

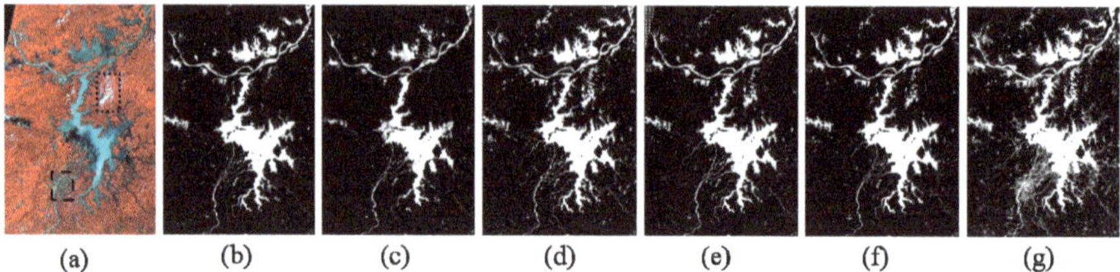

Figure 8. Comparison of water identification effect in Poyang Lake on 29 July 2016. (**a**) False color composite remote sensing image, and water body identification result by (**b**) the DenseNet, (**c**) the ResNet, (**d**) the VGG, (**e**) the SegNet, (**f**) the DeepLab v3+ and (**g**) the NDWI models. White color indicates the identified water bodies, dashed line depicts urban area, dotted line depicts cloud area.

Therefore, for the image of 29 July 2016, these five deep networks have their advantages and disadvantages for the water body identification, but overall show better performances than the NDWI method.

3.5. Interannual Variations of the Water Areas

It can be concluded from the above results that the DenseNet model we proposed has higher accuracy, and can be used for water body identification. Therefore, we have used this model to understand the interannual changes of water areas of Poyang Lake. Since GF-1 was successfully launched in late 2013, we could only study the water area changes from 2014 to 2018. The water areas of Poyang Lake change significantly among seasons, and there is a huge difference between the wet and the dry seasons. The first row of Figure 9 shows the spatial distribution of Poyang Lake in summer from 2014 to 2018. The water area in 2016 was the largest, when there was a flooding disaster event in the Yangtze River basin, and the area in 2018 was the smallest when there was a summer drought due to the reduced precipitation. The second row shows the lake areas in winter. The water areas of Poyang Lake decrease sharply in winter, and the main lake body shrinks to only tributaries and smaller lakes. The disappearance of Poyang Lake is mainly concentrated in the central and southern parts of the lake, leaving only a small part of the water body in the north and northeast. This is principally due to the climatic conditions but is also partly related to the topography, the Yangtze River runoff and the three gorges dam [66,67].

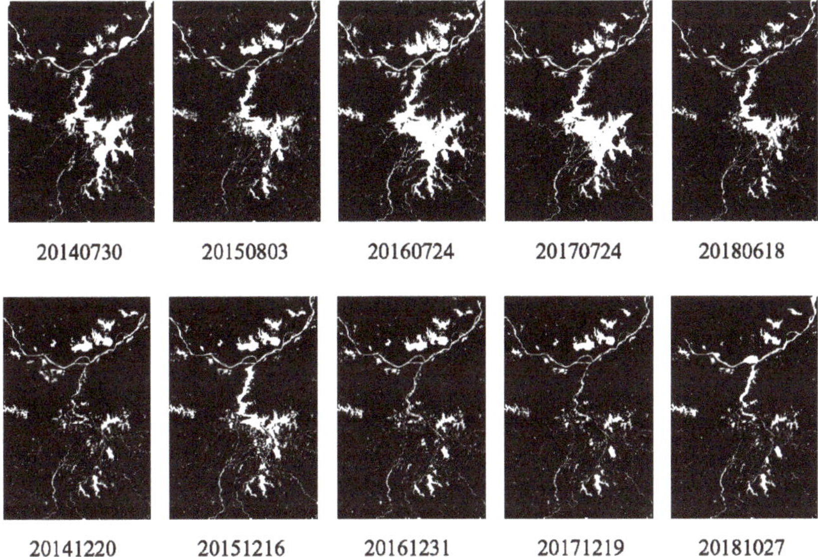

Figure 9. The spatial variations of water area in summer and winter of 2014–2018 in Poyang Lake area based on DenseNet. The first row shows the lake areas in summer and the second row shows those in winter. White color indicates the identified water bodies.

Figure 10 shows the interannual variations of water areas of the Poyang Lake in summer and winter respectively, which were derived from GF-1 images from 2014 to 2018 based on the DenseNet model. The water areas in summer season are generally much larger than those in winter; this is not surprising, because summer is the rainy season in the Poyang Lake basin. The difference in the lake areas in winter and summer is about 2000 km^2 on average. The water area in 2014 summer is about 5200 km^2 and that in winter is about 3200 km^2. In 2015, the water areas in summer and winter are equivalent, amounting to about 4300 km^2; this is because of the increased winter precipitation and reduced summer precipitation contrasting to the normal years. In 2016, the water area in winter is about 3250 km^2 and that doubles in summer, reaching 7000 km^2 due to a severe flooding. It appears clearly that the water areas in summer are decreasing rapidly from 2016 on; however, those in winter show relatively small changes.

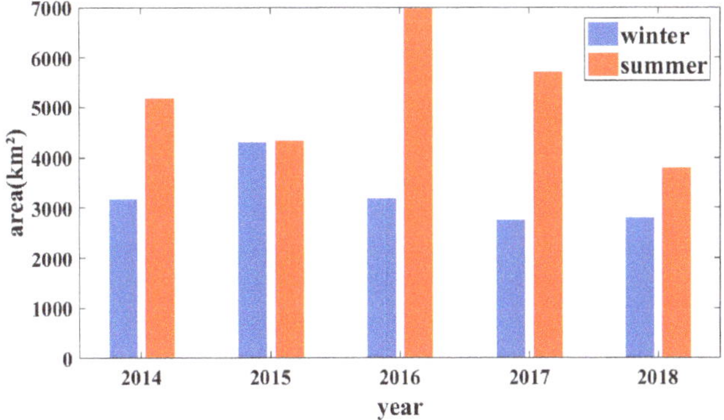

Figure 10. Statistics on the change of water area in summer and winter from 2014 to 2018 in Poyang Lake area, derived from GF-1 images.

4. Discussion

It can be seen from the above results that the performance of a traditionally used water index method is not satisfying, especially in urban areas. This indicates the common problems of water index which are, at least partly, based on thresholds: the thresholds change largely with time and space; the determination of threshold is highly subjective and contains a lot of background information [20,52]. The biggest advantage of NDWI lies in that it is simple and can generate a water map in a very short time. The proposed DenseNet-based water identification method can extract water bodies from the GF-1 images with high accuracy, but it needs hours of training time. However, considering the improvement it has made in recognition accuracy, and once the network is trained, the time to use this network is comparable to NDWI.

So, this network is still a better tool compared to the water index method. Meanwhile, the comparison of the proposed method with other four neural networks shows that it is a more powerful tool for water body recognition.

There are more and more studies using the deep convolution neural network to classify remote sensing images [68]. Our results have approved that, for big remote sensing data like GF-1 images with high spatial and temporal resolutions, the deep learning method can be used to extract water bodies with accurate results efficiently. It can be seen from water area changes in the recent years that the derived water areas from the deep learning method can well reflect the local drought or flooding conditions. Therefore, using the proposed method, the changes of water bodies, such as river and lakes, and wetland as well, can be timely and effectively monitored [69].

The algorithm proposed in this study shows a certain deviation in distinguishing water bodies and clouds, which can be further improved by modifying the model structure and parameters. Also, the cloud area can be removed using image preprocessing to avoid such misjudgment. In this study, we did not preprocess to remove the cloud, such that the original information of the input images are kept. In addition, we use the cloud as one of the indicators to evaluate the effect of water recognition algorithm. When a flooding event occurs, the cloud is always a barrier for water body monitoring with optical remote sensing image. In such a case, the identification results can be improved by removing clouds first or adding samples containing clouds. For cloud removal, it is a solution to integrate optical with microwave remote sensing images. The deficiency of optical remote sensing can be made up by combining with the advantage of microwave remote sensing to penetrate clouds and fog [70,71].

5. Conclusions

This study presents a new multidimensional, densely connected, convolutional network for water identification from high spatial resolution multispectral remote sensing images. It uses DenseNet as the feature extraction network to carry out image downs-sampling, then uses trans-convolution for image upsampling. On this basis, multiscale fusion is added to fuse features of different scales in the down-sampling process into the upsampling process. Compared with the traditionally used water index method, the deep convolutional neural network does not need to find the index threshold, leading to reduced errors, and thus higher accuracy. Meantime, comparing the proposed DenseNet with other networks of ResNet, VGG, SegNet and DeepLab v3+, this DenseNet method requires less training time and has the fastest convergence speed besides DeepLab v3+. The overall performance of DenseNet is still much better. We also added a 95% confidence interval to the evaluation results to reduce the uncertainty caused by the limited samples. The results from the GF-1 images show that, even though DenseNet cannot identify all of the water areas, but it can identify water with great precision, and has much better performance in identifying the boundary between land and water, and can better distinguish the mountain shadows, towns and bare land. Its performance is also better in terms of distinguishing the cloud. Furthermore, the proposed deep learning approach can be easily generalized to an automatic program.

Author Contributions: Conceptualization, G.W. and M.W.; methodology, M.W.; software, M.W.; validation, G.W., M.W. and X.W.; formal analysis, G.W.; investigation, M.W.; resources, G.W.; data curation, M.W.; writing—original draft preparation, M.W.; writing—review and editing, G.W., M.W., X.W. and H.S.; visualization, G.W.; supervision, G.W.; project administration, G.W.; funding acquisition, G.W. All authors have read and agreed to the published version of the manuscript.

Funding: This research was funded by National Key Research and Development Program of China, grant number of 2017YFA0603701; the National Natural Science Foundation of China, grant number of 41875094 and 61872189; the Sino-German Cooperation Group Project, grant number of GZ1447; the Natural Science Foundation of Jiangsu Province of China, grant number of BK20191397 and the APC was founded by Ministry of Science and Technology of China.

Acknowledgments: This work was supported by National Key Research and Development Program of China (2017YFA0603701), the National Natural Science Foundation of China (41875094, 61872189), the Sino-German Cooperation Group Project (GZ1447), and the Natural Science Foundation of Jiangsu Province under Grant nos. BK20191397. All authors are grateful to anonymous reviewers and editors for their constructive comments on earlier versions of the manuscript.

Conflicts of Interest: The authors declare no conflict of interest.

References

1. Tri, A.; Dong, L.; In, Y.; In, Y.; Jae, L. Identification of Water Bodies in a Landsat 8 OLI Image Using a J48 Decision Tree. *Sensors* **2016**, *16*, 1075.
2. Shevyrnogov, A.P.; Kartushinsky, A.V.; Vysotskaya, G.S. Application of Satellite Data for Investigation of Dynamic Processes in Inland Water Bodies: Lake Shira (Khakasia, Siberia), A Case Study. *Aquat. Ecol.* **2002**, *36*, 153–164. [CrossRef]
3. Famiglietti, J.S.; Rodell, M. Water in the Balance. *Science* **2013**, *340*, 1300–1301. [CrossRef] [PubMed]
4. Feng, L.; Hu, C.; Han, X.; Chen, X. Long-Term Distribution Patterns of Chlorophyll-a Concentration in China's Largest Freshwater Lake: MERIS Full-Resolution Observations with a Practical Approach. *Remote Sens.* **2015**, *7*, 275–299. [CrossRef]
5. Ye, Q.; Zhu, L.; Zheng, H.; Naruse, R.; Zhang, X.; Kang, S. Glacier and lake variations in the Yamzhog Yumco basin, southern Tibetan Plateau, from 1980 to 2000 using remote-sensing and GIS technologies. *J. Glaciol.* **2007**, *53*, 183. [CrossRef]
6. Zhang, F.; Zhu, X.; Liu, D. Blending MODIS and Landsat images for urban flood mapping. *Int. J. Remote Sens.* **2014**, *35*, 3237–3253. [CrossRef]
7. Barton, I.J.; Bathols, J.M. Monitoring flood with AVHRR. *Remote Sens. Environ.* **1989**, *30*, 89–94. [CrossRef]
8. Sheng, Y.; Gong, P.; Xiao, Q. Quantitative dynamic flood monitoring with NOAA AVHRR. *Int. J. Remote Sens.* **2001**, *22*, 1709–1724. [CrossRef]
9. Davranche, A.; Lefebvre, G.; Poulin, B. Wetland monitoring using classification trees and SPOT-5 seasonal time series. *Remote Sens. Environ.* **2012**, *114*, 552–562. [CrossRef]
10. Kelly, M.; Tuxen, K.A.; Stralberg, D. Mapping changes to vegetation pattern in a restoring wetland: Finding pattern metrics that are consistent across spatial scale and time. *Ecol. Indic.* **2011**, *11*, 263–273. [CrossRef]
11. Zhao, G.; Xu, Z.; Pang, B.; Tu, T.; Xu, L.; Du, L. An enhanced inundation method for urban flood hazard mapping at the large catchment scale. *J. Hydrol.* **2019**, *571*, 873–882. [CrossRef]
12. Jamshed, A.; Rana, I.A.; Mirza, U.M.; Birkmann, J. Assessing relationship between vulnerability and capacity: An empirical study on rural flooding in Pakistan. *Int. J. Disaster Risk Reduct.* **2019**, *36*, 101109. [CrossRef]
13. Minaee, S.; Wang, Y. An ADMM Approach to Masked Signal Decomposition Using Subspace Representation. *IEEE Trans. Image Process.* **2019**, *28*, 3192–3204. [CrossRef] [PubMed]
14. Frazier, P.S.; Page, K.J. Water body detection and delineation with Landsat TM data. *Photogramm. Eng. Remote Sens.* **2000**, *66*, 1461–1467.
15. Berthon, J.F.; Zibordi, G. Optically black waters in the northern Baltic Sea. *Geophys. Res. Lett.* **2010**, *37*, 232–256. [CrossRef]
16. Karpatne, A.; Khandelwal, A.; Chen, X.; Mithal, V.; Faghmous, J.; Kumar, V. *Global Monitoring of Inland Water Dynamics: State-of-the-Art, Challenges, and Opportunities*; Lässig, J., Kersting, K., Morik, K., Eds.; Computational Sustainability, Springer International Publishing: Cham, Switzerland, 2016; pp. 121–147.
17. Bischof, H.; Schneider, W.; Pinz, A.J. Multispectral classification of Landsat-images using neural networks. *IEEE Trans. Geosci. Remote Sens.* **1992**, *30*, 482–490. [CrossRef]

18. Hughes, M.J.; Hayes, D.J. Automated detection of cloud and cloud shadow in single-date Landsat imagery using neural networks and spatial post-processing. *Remote Sens.* **2014**, *6*, 4907–4926. [CrossRef]
19. Mcfeeters, S.K. The use of the Normalized Difference Water Index (NDWI) in the delineation of open water features. *Int. J. Remote Sens.* **1996**, *17*, 1425–1432. [CrossRef]
20. Xu, H. Modification of normalised difference water index (NDWI) to enhance open water features in remotely sensed imagery. *Int. J. Remote Sens.* **2006**, *27*, 3025–3033. [CrossRef]
21. Feyisa, G.L.; Meilby, H.; Fensholt, R.; Proud, S.R. Automated water extraction index: A new technique for surface water mapping using Landsat imagery. *Remote Sens. Environ.* **2014**, *40*, 23–35. [CrossRef]
22. Zhu, C.; Luo, J.; Shen, Z.; Li, J. River Linear Water Adaptive Auto-extraction on Remote Sensing Image Aided by DEM. *Acta Geodaetica et Cartographica Sinica* **2013**, *42*, 277–283.
23. Tesfa, T.K.; Tarboton, D.G.; Watson, D.W.; Schreuders, K.A.T.; Baker, M.E.; Wallace, R.M. Extraction of hydrological proximity measures from DEMs using parallel processing. *Environ. Modell. Softw.* **2011**, *26*, 1696–1709. [CrossRef]
24. Turcotte, R.; Fortin, J.P.; Rousseau, A.N.; Massicotte, S.; Villeneuve, J.P. Determination of the drainage structure of a watershed using a digital elevation model and a digital river and lake network. *J. Hydrol.* **2001**, *240*, 225–242. [CrossRef]
25. Chen, C.; Fu, J.; Sui, X.; Lu, X.; Tan, A. Construction and application of knowledge decision tree after a disaster for water body information extraction from remote sensing images. *J. Remote Sens.* **2018**, *2*, 792–801.
26. Xia, X.; Wei, Y.; Xu, N.; Yuan, Z.; Wang, P. Decision tree model of extracting blue-green algal blooms information based on Landsat TM/ETM + imagery in Lake Taihu. *J. Lake Sci.* **2014**, *26*, 907–915.
27. Liu, W.; Wang, Z.; Liu, X.; Zeng, N.; Liu, Y.; Alsaadi, F.E. A survey of deep neural network architectures and their applications. *Neurocomputing* **2017**, *234*, 11–26. [CrossRef]
28. Krizhevsky, A.; Sutskever, I.; Hinton, G.E. ImageNet Classification with Deep Convolutional Neural Networks. In Proceedings of the Neural Information Processing Systems, Lake Tahoe, NV, USA, 3–6 December 2012; pp. 1097–1105.
29. Szegedy, C.; Liu, W.; Jia, Y.; Sermanet, P.; Reed, S.; Anguelov, D.; Erhan, D.; Vanhoucke, V.; Rabinovich, A. Going Deeper with Convolutions. In Proceedings of the IEEE Conference on Computer Vision and Pattern Recognition, Boston, MA, USA, 7–12 June 2015; pp. 1–9.
30. Simonyan, K.; Zisserman, A. Very Deep Convolutional Networks for Large-Scale Image Recognition. Available online: https://www.arxiv-vanity.com/papers/1409.1556/ (accessed on 18 December 2019).
31. He, K.; Zhang, X.; Ren, S.; Sun, J. Deep Residual Learning for Image Recognition. In Proceedings of the IEEE Conference on Computer Vision and Pattern Recognition, Las Vegas, NV, USA, 26–30 June 2016; pp. 770–778.
32. Santoro, A.; Faulkner, R.; Raposo, D.; Rae, J.; Chrzanowski, M.; Weber, T.; Wierstra, D.; Vinyals, O.; Pascanu, R.; Lillicrap, T. Relational Recurrent Neural Networks. Available online: https://arxiv.org/abs/1806.01822 (accessed on 18 December 2019).
33. Ronneberger, O.; Fisher, P.; Brox, T. U-Net: Convolutional Networks for Biomedical Image Segmentation. In Proceedings of the Medical Image Computing and Computer Assisted Intervention, Munich, Germany, 5–9 October 2015; pp. 234–241.
34. Badrinarayanan, V.; Kendall, A.; Clipolla, R. A Deep Convolutional Encoder-Decoder Architecture for Image Segmentation. *IEEE Trans. Pattern Anal. Mach. Intell.* **2017**, *39*, 2481–2495. [CrossRef] [PubMed]
35. Lin, G.; Milan, A.; Shen, C.; Reid, I. Refine Net: Multi-Path Refinement Networks for High-Resolution Semantic Segmentation. Available online: https://arxiv.org/abs/1611.06612 (accessed on 18 December 2019).
36. Chen, L.C.; Papandreou, G.; Kokkinos, I.; Murphy, K.; Yuille, A.L. Semantic Image Segmentation with Deep Convolutional Nets and Fully Connected CRFs. Available online: https://arxiv.org/abs/1412.7062 (accessed on 10 February 2020).
37. Chen, L.C.; Papandreou, G.; Kokkinos, I.; Murphy, K.; Yuille, A.L. DeepLab: Semantic Image Segmentation with Deep Convolutional Nets, Atrous Concolution, and Fully Connected CRFs. Available online: https://arxiv.org/abs/1606.00915 (accessed on 10 February 2020).
38. Chen, L.C.; Papandreou, G.; Schroff, F.; Papandreou, G. Rethinking Atrous Convolution for Semantic Image Segmentation. Available online: https://arxiv.org/abs/1706.05587 (accessed on 10 February 2020).
39. Chen, L.C.; Zhu, Y.; Papandreou, G.; Schroff, F.; Papandreou, G. Encoder-Decoder with Atrous Separable Convolution for Semantic Image Segmentation. In Proceedings of the European Conference on Computer Vision, Munich, Germany, 8–14 September 2018; pp. 833–851.

40. Zhao, H.; Shi, J.; Qi, X.; Wang, X.; Jia, J. Pyramid Scene Parsing Network. In Proceedings of the IEEE Conference on Computer Vision and Pattern Recognition, Honolulu, HI, USA, 21–26 July 2017; pp. 6230–6239.

41. Girshick, R.; Donahue, J.; Darrel, T.; Malik, J. Rich feature hierarchies for accurate object detection and semantic segmentation. Available online: https://arxiv.org/abs/1311.2524 (accessed on 10 February 2020).

42. Girshick, R. Fast R-CNN. In Proceedings of the International Conference on Computer Vision, Santiago, MN, USA, 13–16 December 2015; pp. 1440–1448.

43. Ren, S.; He, K.; Girshick, R.; Sun, J. Faster R-CNN: Towards Real-Time Object Detection with Region Proposal Networks. In *Advances in Neural Information Processing Systems 28*; Cortes, C., Lawrence, N.D., Lee, D.D., Sugiyama, M., Garnett, R., Eds.; Curran Associates, Inc.: Red Hook, NY, USA, 2015.

44. He, K.; Gkioxari, G.; Dollar, P.; Girshick, R. Mask R-CNN. Available online: https://arxiv.org/abs/1703.06870 (accessed on 10 February 2020).

45. Chen, K.; Pang, J.; Wang, J.; Xiong, Y.; Li, X.; Sun, S.; Feng, W.; Liu, Z.; Shi, J.; Ouyang, W.; et al. Hybrid Task Cascade for Instance Segmentation. Available online: https://arxiv.org/abs/1901.07518 (accessed on 19 February 2020).

46. Chen, L.; Yang, Y.; Wang, J.; Xu, W.; Yuille, A.L. Attention to Scale: Scale-Aware Semantic Image Segmentation. In Proceedings of the IEEE Conference on Computer Vision and Pattern Recognition, Las Vegas, NV, USA, 26–30 June 2016; pp. 3640–3649.

47. Fu, J.; Liu, J.; Tian, H.; Li, Y.; Bao, Y.; Fang, Z.; Lu, H. Dual attention network for scene segmentation. In Proceedings of the IEEE Conference on Computer Vision and Pattern Recognition, Long Beach, CA, USA, 16–20 June 2019; pp. 3146–3154.

48. Marquez-Neila, P.; Baumela, L.; Alvarez, L. A Morphological Approach to Curvature-Based Evolution of Curves and Surfaces. *IEEE Trans. Pattern Anal. Mach. Intell.* **2014**, *36*, 2–17. [CrossRef]

49. Minaee, S.; Boykov, Y.; Porikli, F.; Plaza, A.; Kehtarnavaz, N.; Terzopoulos, D. Image Segmentation Using Deep Learning: A Survey. Available online: https://arxiv.org/abs/2001.05566 (accessed on 19 February 2020).

50. Lecun, Y.; Bottou, L.; Bengio, Y.; Haffner, P. Gradient-based learning applied to document recognition. *Proc. IEEE* **1998**, *86*, 2278–2324. [CrossRef]

51. Huang, G.; Liu, Z.; van Maaten, L.; Weinberger, K.Q. Densely Connected Convolutional Networks. In Proceedings of the IEEE Conference on Computer Vision and Pattern Recognition, Honolulu, HI, USA, 21–26 July 2017; pp. 2261–2269.

52. Feng, M.; Sexton, J.O.; Channan, S.; Townshend, J.R. A global, 30-m resolution land-surface water body dataset for 2000: first results of a topographic-spectral classification algorithm. *Int. J. Digit. Earth* **2016**, *9*, 113–133. [CrossRef]

53. Guo, H.; Hu, Q.; Zhang, Q. Changes in Hydrological Interactions of the Yangtze River and the Poyang Lake in China during 1957–2008. *Acta Geographica Sinica* **2011**, *66*, 609–618.

54. Otsu, N. A Threshold Selection Method from Gray-Level Histograms. *IEEE Trans. Syst. Man Cybern.* **1979**, *9*, 62–66. [CrossRef]

55. Zhang, C.; Xie, Y.; Liu, D.; Wang, L. Fast Threshold Image Segmentation Based on 2D Fuzzy Fisher and Random Local Optimized QPSO. *IEEE Trans. Image Process.* **2017**, *26*, 1355–1362. [CrossRef] [PubMed]

56. Erhan, D.; Szegedy, C.; Toshev, A.; Anguelov, D. Scalable Object Detection using Deep Neural Networks. In Proceedings of the IEEE Conference on Computer Vision and Pattern Recognition, Columbus, OH, USA, 24–27 June 2014; pp. 2155–2162.

57. Long, J.; Shelhamer, E.; Darrell, T. Fully convolutional networks for semantic segmentation. In Proceedings of the IEEE Conference on Computer Vision and Pattern Recognition, Boston, MA, USA, 7–12 June 2015; pp. 3431–3440.

58. Shelhamer, E.; Long, J.; Darrell, T. Fully convolutional networks for semantic segmentation. *IEEE Trans. Pattern Anal. Mach. Intell.* **2017**, *39*, 640–651. [CrossRef] [PubMed]

59. Srivastava, R.K.; Greff, K.; Schmidhuber, J. Highway Networks. Available online: https://arxiv.org/abs/1505.00387 (accessed on 15 December 2019).

60. Huang, G.; Sun, Y.; Liu, Z.; Sedra, D.; Weinberger, K. Deep Networks with Stochastic Depth. In Proceedings of the European Conference on Computer Vision, Amsterdam, The Netherlands, 10–16 October 2016; pp. 646–661.

61. Larsson, G.; Maire, M.; Shakhnarovich, G. FractalNet: Ultra-Deep Neural Networks without Residuals. Available online: https://arxiv.org/abs/1605.07648 (accessed on 15 December 2019).

Remote Sens. **2020**, *12*, 795

62. Kingma, D.P.; Ba, J. Adam: A Method for Stochastic Optimization. Available online: https://arxiv.org/abs/1412.6980 (accessed on 10 October 2019).

63. Stehman, S.V. Selecting and interpreting measures of thematic classification accuracy. *Remote Sens. Environ.* **1997**, *62*, 77–89. [CrossRef]

64. Powers, D.M.W. Evaluation: From Precision, Recall and F-Measure to ROC, Informedness, Markedness & Correlation. *J. Mach. Learn. Tch.* **2011**, *2*, 37–63.

65. Howard, A.G.; Zhu, M.; Chen, B.; Kalenichenko, D.; Wang, W.; Weyand, T.; Andreetto, M.; Adam, H. MobileNets: Efficient Convolutional Neural Networks for Mobile Vision Applications. Available online: https://arxiv.org/abs/1704.04861 (accessed on 10 December 2019).

66. Hu, Q.; Feng, S.; Guo, H.; Chen, G.; Jiang, T. Interactions of the Yangtze River flow and hydrologic progress of the Poyang Lake, China. *J. Hydrol.* **2007**, *347*, 90–100. [CrossRef]

67. Zhou, Y.; Jeppesen, E.; Li, J.; Zhang, X.; Li, X. Impacts of Three Gorges Reservoir on the sedimentation regimes in the downstream-linked two largest Chinese freshwater lakes. *Sci. Rep.* **2016**, *6*, 35396. [CrossRef]

68. Yang, Z.; Mu, X.D.; Zhao, F.A. Scene Classification of Remote Sensing Image Based on Deep Network Grading Transferring. *Optik* **2018**, *168*, 127–133. [CrossRef]

69. Gao, H.; Birkett, C.; Lettenmaier, D.P. Global monitoring of large reservoir storage from satellite remote sensing. *Water Resour. Res.* **2012**, *48*, W09504. [CrossRef]

70. Liu, P.; Du, P.; Tan, K. A novel remotely sensed image classification based on ensemble learning and feature integration. *J. Infrared Millim. Waves* **2014**, *33*, 311–317.

71. Teimouri, M.; Mokhtarzade, M.; Zoej, M.J.V. Optimal fusion of optical and SAR high-resolution images for semiautomatic building detection. *GISci. Remote Sens.* **2016**, *53*, 45–62. [CrossRef]

Article

An End-to-End and Localized Post-Processing Method for Correcting High-Resolution Remote Sensing Classification Result Images

Xin Pan [1,2], **Jian Zhao** [1] **and Jun Xu** [2,*]

[1] School of Computer Technology and Engineering, Changchun Institute of Technology, Changchun 130012, China; panxin@neigae.ac.cn (X.P.); zhaojian08@mails.jlu.edu.cn (J.Z.)
[2] Jilin Provincial Key Laboratory of Changbai Historical Culture and VR Reconstruction Technology, Changchun Institute of Technology, Changchun 130012, China
* Correspondence: xujun@ccit.edu.cn; Tel.: +86-13844908223

Received: 7 February 2020; Accepted: 5 March 2020; Published: 6 March 2020

Abstract: Since the result images obtained by deep semantic segmentation neural networks are usually not perfect, especially at object borders, the conditional random field (CRF) method is frequently utilized in the result post-processing stage to obtain the corrected classification result image. The CRF method has achieved many successes in the field of computer vision, but when it is applied to remote sensing images, overcorrection phenomena may occur. This paper proposes an end-to-end and localized post-processing method (ELP) to correct the result images of high-resolution remote sensing image classification methods. ELP has two advantages. (1) End-to-end evaluation: ELP can identify which locations of the result image are highly suspected of having errors without requiring samples. This characteristic allows ELP to be adapted to an end-to-end classification process. (2) Localization: Based on the suspect areas, ELP limits the CRF analysis and update area to a small range and controls the iteration termination condition. This characteristic avoids the overcorrections caused by the global processing of the CRF. In the experiments, ELP is used to correct the classification results obtained by various deep semantic segmentation neural networks. Compared with traditional methods, the proposed method more effectively corrects the classification result and improves classification accuracy.

Keywords: semantic segmentation; high-resolution remote sensing image; pixel-wise classification; result correction; conditional random field (CRF)

---

1. Introduction

With the advent of high-resolution satellites and drone technologies, an increasing number of high-resolution remote sensing images have become available, making automated processing technology increasingly important for utilizing these images effectively [1]. High-resolution remote sensing image classification methods, which automatically provide category labels for objects in these images, are playing an increasingly important role in land resource management, urban planning, precision agriculture, and environmental protection [2]. However, high-resolution remote sensing images usually contain detailed information and exhibit high intra-class heterogeneity and inter-class homogeneity characteristics, which are challenging for traditional shallow-model classification algorithms [3]. To improve classification ability, deep learning technology, which can extract higher-level features in complex data, has been widely studied in the high-resolution remote sensing classification field in recent years [4].

Deep semantic segmentation neural networks (DSSNNs) are constructed based on convolutional neural networks (CNNs); the input of these models is an image patch, and the output are category labels for the image patch. With the help of this structure, a DSSNN gains the ability to perform rapid pixel-wise classification and end-to-end data processing—characteristics that are critical in automatically analyzing remote sensing images [5,6]. In the field of remote sensing classification, the most widely used DSSNN architectures include fully convolutional networks (FCNs), SegNETs, and U-Nets [7–9]. Many studies have extended these architectures by adding new neural connection structures to address different remote sensing object recognition tasks [10]. In the process of segmenting high-resolution remote sensing images of urban buildings, due to their hierarchical feature extraction structures, DSSNNs can extract buildings' spatial and spectral building features and achieve better building recognition results [11,12]. When applied to road extraction from images, DSSNNs can integrate low-level features into higher-level features layer by layer and extract the relevant features [13,14]. Based on U-Nets, FCNs, and transmitting structures that add additional spatial information, DSSNNs can improve road border and centerline recognition accuracy obtained from high-resolution remote sensing images [15]. Through the FCN and SegNET architectures, DSSNNs can obtain deep information regarding land cover areas and can classify complex land cover systems in an automatic and end-to-end manner [16–18].

In many high-resolution remote sensing semantic segmentation application tasks, obtaining accurate object boundaries is crucial [19,20]. However, due to the hierarchical structures of DSSNNs, some object spatial information may be lost during the training and classification process; consequently, their classification results are usually not perfect, especially at object boundaries [21,22]. Conditional random field (CRF) methods are widely used to correct segmentation or super-pixel classification results [23,24]. CRFs can also be integrated into the training stage to enhance a CNN's classification ability [25,26]. Due to their ability to capture fine edge details, CRFs are also useful for post-processing the results of remote sensing classification [27]. Using CRFs, remote sensing image semantic segmentation results can be corrected, especially at ground object borders [28–30]. Although CRFs have achieved many successes in the field of computer vision, when applied to remote sensing images where the numbers, sizes, and locations of objects vary greatly, some of the ground objects will be excessively enlarged or reduced and shadowed areas may be confused with adjacent objects [31]. To address the above problem, training samples or contexts can be introduced to restrict the behavior of CRFs, limiting their processes within a certain range, category, or number of iterations [32,33]. Unfortunately, these methods require the introduction of samples to the CRF control process, and these samples cause the corresponding methods to lose their end-to-end characteristics. For classification tasks based on deep semantic segmentation networks, when the end-to-end characteristics are lost, the classification work must add an additional manual sample set selection and construction process, which reduces the ability to apply semantic segmentation to networks automatically and decreases their application value [34]. Therefore, new methods must be introduced to solve this problem.

To address the above problems, this paper proposes an end-to-end and localized post-processing method (ELP) for correcting high-resolution remote sensing image classification results. By introducing a mechanism for end-to-end evaluation and a localization procedure, ELP can identify which locations of the resulting image are highly suspected to have errors without requiring training and verification samples; then, it can apply further controls to restrict the CRF analysis and update areas within a small range and limit the iteration process. In the experiments, we introduce study images from the "semantic labeling contest of the ISPRS WG III/4 dataset" and apply various DSSNNs to obtain classification result images. The results show that compared with the traditional CRF method, the proposed method more effectively corrects classification result images and improves classification accuracy.

2. Methodology

2.1. Deep Semantic Segmentation Neural Networks' Classification Strategy and Post-Processing

Deep semantic segmentation neural networks such as FCNs, SegNET, and U-NET have been widely studied and applied by the remote sensing image classification research community. The training and classification processes of these networks are illustrated in Figure 1.

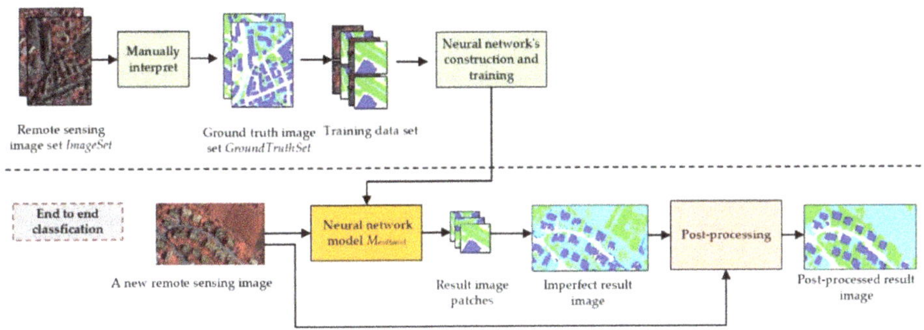

Figure 1. End-to-end classification strategy.

As shown in Figure 1, the end-to-end classification strategy is usually adopted for DSSNNs' training and classification. During the training stage, a set of remote sensing images $ImageSet = \{I_1, I_2, \ldots, I_n\}$ is adopted and manually interpreted into a ground truth set $GroundTruthSet = \{I_{gt1}, I_{gt2}, \ldots, I_{gtn}\}$; then, the $GroundTruthSet$ is separated into patches to construct the training dataset. The classification model $M_{endtoend}$ is obtained based on this training dataset. During the classification stage, the classification model is utilized to classify a completely new remote sensing image I_{new} (not an image from $ImageSet$). This strategy achieves a higher degree of automation; the classification process has no relationship with the training data or the training algorithm, and newly obtained or other images in the same area can be classified automatically with $M_{endtoend}$, forming an input-to-output/end-to-end structure. Thus, this strategy is more valuable in practical applications when massive amounts of remote sensing data need to be processed quickly.

However, the classification results of the end-to-end classification strategy are usually not "perfect", and they are affected by two factors. On the one hand, because the training data are constructed by manual interpretation, it is difficult to provide training ground truth images that are precise at the pixel level (especially at the boundaries of ground objects). Moreover, the incorrectly interpreted areas of these images may even be amplified through the repetitive training process [16]. On the other hand, during data transfer among the neural network layers, along with obtaining high-level spatial features, some spatial context information may be lost [35]. Therefore, the classification results obtained by the "end-to-end classification strategy" may result in many flaws, especially at ground object boundaries. To correct these flaws, in the computer vision research field, the conditional random field (CRF) method is usually adopted in the post-processing stage to correct the result image. The conditional random field can be defined as follows:

$$P(X|F) = \frac{1}{Z(F)} \exp\left(-\sum_{c \in C_g} \log(X_C|F)\right), \tag{1}$$

where F is a set of random variables $\{F_1, F_2, \ldots, F_N\}$; F_i is a pixel vector; X is a set of random variables $\{x_1, x_2, \ldots, x_N\}$, where x_i is the category label of pixel i; $Z(F)$ is a normalizing factor; and c is a clique in a set of cliques C_g, where g induces a potential φ_c [23,24]. By calculating Equation (1), the CRF adjusts the category label of each pixel and achieves the goal of correcting the result image. The CRF is highly effective at processing images that contain only a small number of objects. However, the numbers, sizes, and locations of objects in remote sensing images vary widely, and the traditional CRF tends to perform a global optimization of the entire image. This process leads to some ground objects being excessively enlarged or reduced. Furthermore, if the different parts of ground objects that are shadowed or not shadowed are processed in the same manner, the CRF result will contain more errors [31]. In our previous work, we proposed a method called the restricted conditional random field (RCRF) that can handle the above situation [31]. Unfortunately, the RCRF requires the introduction of samples to control its iteration termination and produce an output integrated image. When integrated into the classification process, the need for samples will cause the whole classification process to lose its end-to-end characteristic; thus, the RCRF cannot be integrated into an end-to-end process. In summary, to address the above problems, the traditional CRF method needs to be further improved by adding the following characteristics:

(1) End-to end result image evaluation: Without requiring samples, the method should be able to automatically identify which areas of a classification result image may contain errors. By identifying areas that are strongly suspected of being misclassified, we can limit the CRF process and analysis scope.

(2) Localized post-processing: The method should be able to transform the entire image post-processing operation into local corrections and separate the various objects or different parts of objects (such as roads in shadow or not in shadow) into sub-images to alleviate the negative impacts of differences in the number, size, location, and brightness of objects.

To achieve this goal, a new mechanism must be introduced to improve the traditional CRF algorithm for remote sensing classification results post-processing.

2.2. End-to-End Result Image Evaluation and Localized Post-Processing

The majority of evaluation methods for classification results require samples with category labels that allow the algorithm to determine whether the classification result is good; however, to achieve an end-to-end classification results evaluation, samples cannot be required during the evaluation process. In the absence of testing samples, although it is impossible to accurately indicate which pixels are incorrectly classified, we can still find some areas that are highly suspected of having classification errors by applying some conditions.

Therefore, we need to establish a relation between the remote sensing image and the classification result image and find the areas where the colors (bands) of the remote sensing image are consistent, but the classification results are inconsistent; these are the areas that may belong to the same object but are incorrectly classified into different categories. Such areas are strong candidates for containing incorrectly classified pixels. Furthermore, we try to correct these errors within a relatively small area.

To achieve the above goals, for a remote sensing image I_{image} and its corresponding classification image I_{cls}, the methods proposed in this paper are illustrated in Figure 2:

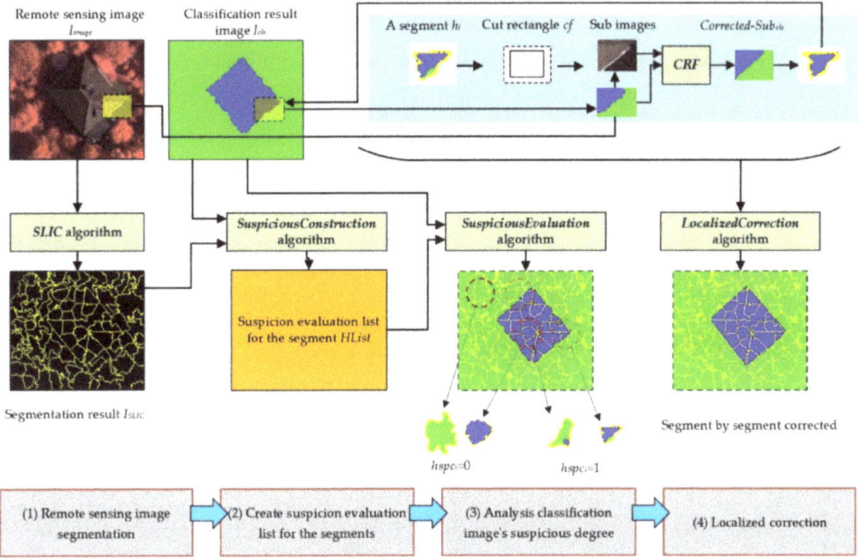

Figure 2. End-to-end result image evaluation and localized post-processing.

As shown in Figure 2, we use four steps to perform localized correction:

(1) Remote sensing image segmentation

We need to segment the remote image based on color (band value) consistency. In this paper, we adopt the simple linear iterative clustering (SLIC) algorithm as the segmentation method. The algorithm initially contains k clusters. Each cluster is denoted by $C_i = \{l_i, a_i, b_i, x_i, y_i\}$, where l_i, a_i, and b_i are the color values of C_i in CIELAB color space, and x_i, y_i are the center coordinates of C_i in the image. For two clusters, C_i and C_j, the SLIC algorithm is introduced to compare color and space distances simultaneously, as follows:

$$distance_{color}(C_i, C_j) = \sqrt{(l_i - l_j)^2 + (a_i - a_j)^2 + (b_i - b_j)^2},$$ (2)

$$distance_{space}(C_i, C_j) = \sqrt{(x_i - x_j)^2 + (y_i - y_j)^2},$$ (3)

where $distance_{color}$ is the color distance and $distance_{space}$ is the spatial distance. Based on these two distances, the distance between the two clusters is:

$$distance(C_i, C_j) = \sqrt{\left(\frac{distance_{color}(C_i, C_j)}{N_{color}}\right)^2 + \left(\frac{distance_{space}(C_i, C_j)}{N_{space}}\right)^2},$$ (4)

where N_{color} is the maximum color distance and N_{space} is the maximum position distance. The SLIC algorithm uses the iterative mechanism of the k-means algorithm to gradually adjust the cluster position and the cluster to which each pixel belongs, eventually obtaining $N_{segment}$ segments [36]. The advantage of the SLIC algorithm is that it can quickly and easily cluster adjacent similar regions into a segment; this characteristic is particularly useful for finding adjacent areas that have a consistent color (band value). For I_{image}, the SLIC algorithm is used to obtain the segmentation result I_{SLIC}. In each segment in I_{SLIC}, the pixels are assigned the same segment label.

(2) Create a list of segments with suspicious degree evaluations

For all the segments in I_{SLIC}, a suspicion evaluation list for the segments $HList = \{h_1, h_2, \ldots, h_n\}$ is constructed, where h_i is a set $h_i = \{hid_i, hpixels_i, hrec_i, hspc_i\}$, hid_i is a segment label, $hpixels_i$ holds the locations of all the pixels in the segment; *hrec_i* is the location and size of the enclosing frame rectangle of *hpixels_i*; and *hspc_i* is a suspicious evaluation value, which is either 0 or 1—a "1" means that the pixels in the segment are suspected of being misclassified, and a "0" means that the pixels in the segment are likely correctly classified. The algorithm to construct the suspicious degree evaluation list is as follows (SuspiciousConstruction Algorithm):

Algorithm SuspiciousConstruction

Input: I_{SLIC}
Output: *HList*
Begin
    HList = an empty list;
    foreach (segment label *i* in I_{SLIC}){
        hid_i = *i*;
        $hpixels_i$ = Locations of all the pixels in corresponding segment *i*;
        $hrec_i$= the location and size of $hpixels_i$'s enclosing frame rectangle;
        $hspc_i$ = 0;
        h_i = Create a set {hid_i, $hpixels_i$, $hrec_i$, $hhyp_i$};
        HList ← h_i;

    }
    return *HList*;
End

In the SuspiciousConstruction algorithm, by default, each segment's spc_i = 0 (the hypothesis is that no misclassified pixels exist in the segment).

(3) Analyze the suspicious degree

As shown in Figure 2, for a segment, the spc_i value can be calculated based on the pixels in I_{SLIC}; h_i's corresponding pixels can be grouped as SP = {$sp_1, sp_2, \ldots , sp_m$}, where sp_i is the pixel number belonging to category *i*, and the inconsistency degree of SP can be described using the following formula:

$$inconsistence(SP) = 1 - \frac{\max_{i=1..m}(sp_i)}{\sum_i^m sp_i}, \tag{5}$$

Based on this formula, the value of hyp_i can be expressed as follows:

$$hspc_i = \begin{cases} 0 & inconsistence < \alpha \\ 1 & inconsistence \geq \alpha \end{cases}, \tag{6}$$

where α is a threshold value (the default is 0.05). When a segment's $hspc_i$ = 0, the segment's corresponding pixels in I_{SLIC} all belong to the same category, which indicates that pixels' features are consistent in both I_{image} (band value) and I_{SLIC} (segment label); in this case, the segment does not need correction by CRF. In contrast, when a segment's $hspc_i$ = 1, the pixels of the segment in I_{SLIC} belong to different categories, but the pixel's color (band value) is comparatively consistent in I_{image}; this case may be further subdivided into two situations:

A. Some type of classification error exists in the segment (such as the classification result deformation problem appearing on an object boundary).

B. The classification result is correct, but the segment crosses a boundary between objects in I_{SLIC} (for example, the number of segments assigned to the SLIC algorithm is too small, and some areas are undersegmented).

In either case, we need to be suspicious of the corresponding segment and attempt to correct mistakes using CRF. Based on Formulas 5 and 6, the algorithm for analyzing I_{cls} using *HList* is as follows (SuspiciousEvaluation Algorithm):

Algorithm SuspiciousEvaluation

Input: *HList*, I_{cls}
Output: *HList*
Begin
    foreach(Item h_i in *HList*){
        SP = new m-element array with zero values
        foreach(Location *pl* in $h_i.hpixels_i$){
            px = Obtain pixel at *pl* in I_{cls};
            $pxcl$ = Obtain category label of *px*;
            $SP[pxcl]$ = $SP[pxcl]$ + 1;
        }
        inconsistence = Use Formula (5) to calculate *SP*;
        $h_i.hyp_i$ = Use Formula (6) with *inconsistence*
    }
    return *HList*;
End

By applying the SuspiciousEvaluation algorithm, we can identify which segments are suspicious and require further post-processing.

(4) Localized correction

As shown in Figure 2, for a segment h_i that is suspected of containing error classified pixels, the post-processing strategy can be described as follows: First, based on $h_i.hrec_i$, create a cut rectangle *cf*, (*cf* = rectangle $h_i.hrec_i$ enlarged by β pixels), where β is the number of pixels to enlarge and the default value is 10. Second, use the *cf* cut sub-images from I_{image} and I_{cls} to obtain a Sub_{image} and a Sub_{cls}. Third, input Sub_{image} and Sub_{cls} to the CRF algorithm, and obtain a corrected classification result $Corrected\text{-}Sub_{cls}$. Finally, based on the pixel locations in $h_i.hpixels_i$, obtain the pixels from $Corrected\text{-}Sub_{cls}$ and write them to I_{cls}, which constitutes localized area correction on I_{cls}. For the entire I_{cls}, the localized correction algorithm on I_{cls} is as follows (LocalizedCorrection Algorithm):

Algorithm LocalizedCorrection

Input: I_{cls}, *HList*
Output: I_{cls}
Begin
    foreach(h_i in *HList*){
        if ($h_i.hspc_i$==0) **then** continue;
        cf = Enlarge rectangle $h_i.hrec_i$ by β pixels;
        Sub_{image}, Sub_{cls} = Cut sub-images from I_{image} and I_{cls};
        $Corrected\text{-}Sub_{cls}$ = Process Sub_{image} and Sub_{cls} by CRF algorithm;
        pixels = Obtain pixels in $h_i.hpixels_i$ from $Corrected\text{-}Sub_{cls}$;
        $I_{cls} \leftarrow pixels$;
    }
    return I_{cls}
End

By applying the LocalizedCorrection algorithm, the I_{CLS} will be corrected segment by segment through the CRF algorithm.

2.3. Overall Process of the End-to-End and Localized Post-Processing Method

Based on the four steps and algorithms described in the preceding subsection, we can evaluate the classification result image without requiring testing samples and correct the classification result image within local areas. By integrating these algorithms, we propose an end-to-end and localized post-processing method (ELP) whose input is a remote sensing image I_{image} and a classification result image I_{cls}, and whose output is the corrected classification result image. Through the iterative and progressive correction process, the goal of improving the quality of the I_{CLS} can be achieved. The process of ELP is shown in Figure 3.

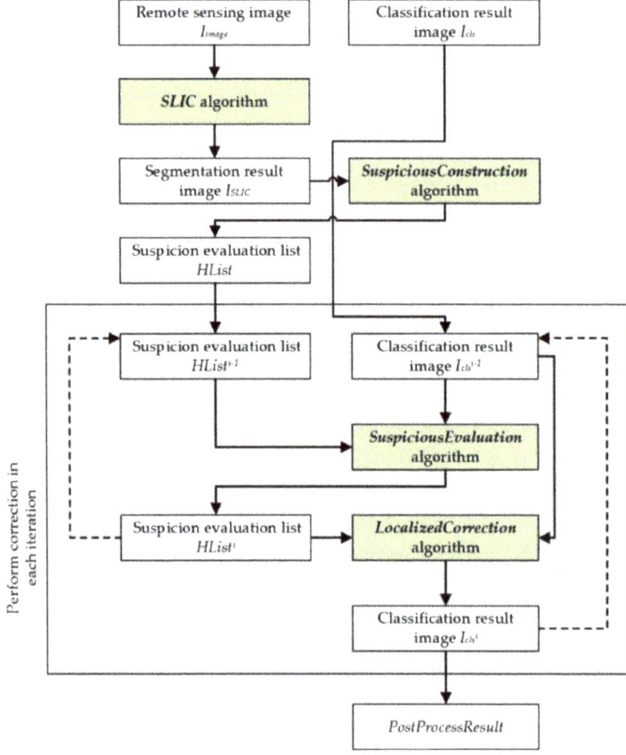

Figure 3. Overall processes of end-to-end and localized post-processing method (ELP).

As Figure 3 shows, the ELP method is a step-by-step iterative correction process that requires a total of γ iterations to correct the I_{cls} content. Before beginning the iteration, the ELP method obtains the segmentation result image I_{SLIC} before iteration:

$$I_{SLIC} = SLIC(I_{image}), \tag{7}$$

Then, it evaluates the segments and constructs the suspicion evaluation list for the segments *HList*:

$$HList = SuspiciousConstruction(I_{SLIC}), \tag{8}$$

In each iteration, ELP updates *HList* to obtain $HList^{\eta}$, and it outputs a new classification result image I_{CLS}^{η}, where η is the iteration value (in the range $[1,\gamma]$). The *i*-th iteration's output is:

$$HList^{i} = SuspiciousEvaluation(HList^{i-1}, I_{cls}^{i-1}), \tag{9}$$

$$I_{cls}^{i} = LocalizedCorrection(I_{cls}^{i-1}, HList^{i}), \tag{10}$$

When $\eta = 1$, $HList^{0} = HList$, and $I_{cls}^{0} = I_{cls}$; when $\eta \geq 2$, the current iteration result depends on the result of the previous iteration. Based on the above two formulas, the ELP algorithm will update *HList* and I_{cls} in each iteration; *HList* indicates suspicious areas, and these areas are corrected and stored in I_{cls}. As the iteration progresses, the final result is obtained:

$$PostProcessResult = I_{cls}^{\gamma}, \tag{11}$$

Through the above process, ELP achieves both the desired goals: end-to-end result image evaluation and localized post-processing.

3. Experiments

We implemented all the codes in Python 3.6; the CRF algorithm was implemented based on the PyDenseCRF package. To analyze the correction effect for the deep semantic segmentation model's classification result images, this study introduces images from Vaihingen and Potsdam in the "semantic labeling contest of the ISPRS WG III/4 dataset" as the two test datasets.

3.1. Comparison of CRF and ELP on Vaihingen Dataset

3.1.1. Method Implementation and Study Images

We introduced five commonly used DSSNN models as testing targets: FCN8s, FCN16s, FCN32s, SegNET, and U-Net. All these deep models are implemented using the Keras package, and all the models take a 224 × 224 image patch as input and output a corresponding semantic segmentation result. The five image files from Vaihingen were selected as the study images and are listed in Table 1.

Table 1. Five study images from the Vaihingen dataset.

Image ID	Filename	Size	Role
1	top_mosaic_09cm_area23.tif	1903 × 2546	Training/Testing image
2	top_mosaic_09cm_area1.tif	1919 × 2569	Testing image
3	top_mosaic_09cm_area3.tif	2006 × 3007	Training/Testing image
4	top_mosaic_09cm_area21.tif	1903 × 2546	Testing image
5	top_mosaic_09cm_area30.tif	1934 × 2563	Testing image

All five images contain five categories: impervious surfaces (I), buildings (B), low vegetation (LV), trees (T), and cars (C). These images have three spatial bands (near-infrared (NIR), red (R), and green (G)). The study images and their corresponding ground truth images are shown in Figure 4.

We selected study images 1 and 3 as training data and used all the images as test data. Study images 1 and 3 and their ground truth images were cut into 224 × 224 image patches with 10-pixel intervals; all the patches were stacked into a training set, and all the deep semantic segmentation models were trained on this training set.

This study used two methods to compare correction ability:

(1) CRF: We compared our method with the traditional CRF method. For each classification result image, the CRF was executed 10 times, and each time, the corresponding correct-distance parameters were set to 10, 20, . . . , 100. Since all the study images have corresponding ground truth images, the CRF algorithm selects the result with the highest accuracy among the 10 executions.

(2) ELP: Using the proposed method, the threshold value parameter α was set to 0.05, and the CRF's correct-distance parameters in the LocalizedCorrection algorithm were set to 10. The number of ELP iterations was set to 10. Since ELP emphasizes "end-to-end" ability, no ground truth is needed to analyze the true classification accuracy during the iteration process; therefore, ELP directly outputs the result of the last iteration as the corrected image.

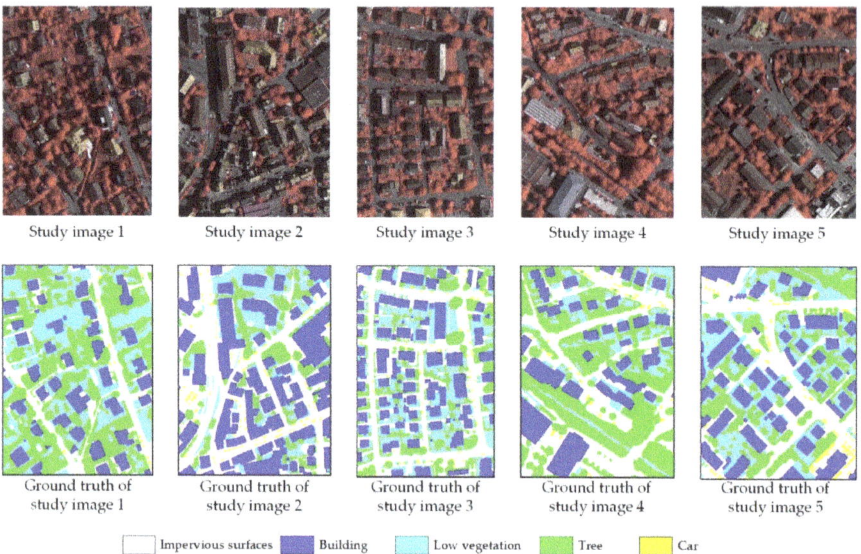

<table>
<tr><td align="center">Study image 1</td><td align="center">Study image 2</td><td align="center">Study image 3</td><td align="center">Study image 4</td><td align="center">Study image 5</td></tr>
<tr><td align="center">Ground truth of
study image 1</td><td align="center">Ground truth of
study image 2</td><td align="center">Ground truth of
study image 3</td><td align="center">Ground truth of
study image 4</td><td align="center">Ground truth of
study image 5</td></tr>
</table>

Figure 4. The study images and their corresponding ground truth from the Vaihingen dataset.

3.1.2. Classification Results of Semantic Segmentation Models

We used all five deep semantic segmentation models as end-to-end classifiers to process five study images. The classification results are illustrated in Figure 5.

As shown in Figure 5, because study images 1 and 3 are used as training data (the deep neural network is sufficiently large to "remember" these images), the classification results by the five models for study images 1 and 3 are close to "perfect": almost all the ground objects and boundaries are correctly identified. However, because just two training images cannot exhaustively represent all the boundaries and object characteristics, these models cannot perfectly process study images 2, 4, and 5. As shown in Figure 5, there are obvious defects and boundary deformations, and many objects are misclassified in large areas. Based on the ground truth images, the classification accuracies of these result images are as follows:

As shown in Figure 5, because study images 1 and 3 are used as training data (the deep neural network is sufficiently large to "remember" these images), the classification results by the five models for study images 1 and 3 are close to "perfect": almost all the ground objects and boundaries are correctly identified. However, because just two training images cannot exhaustively represent all the boundaries and object characteristics, these models cannot perfectly process study images 2, 4, and 5. As shown in Figure 5, there are obvious defects and boundary deformations, and many objects are misclassified in large areas. Based on the ground truth images, the classification accuracies of these result images are as follows:

Table 2 shows that because study images 1 and 3 are training images, all five models' classification accuracies of these two images are above 95%, which is a satisfactory result. However, on study images 2, 4, and 5, due to the many errors on ground objects and boundaries, all five models' classification accuracies degrade to approximately 80%. Therefore, it is necessary to introduce a correction mechanism to correct the boundary errors in these images.

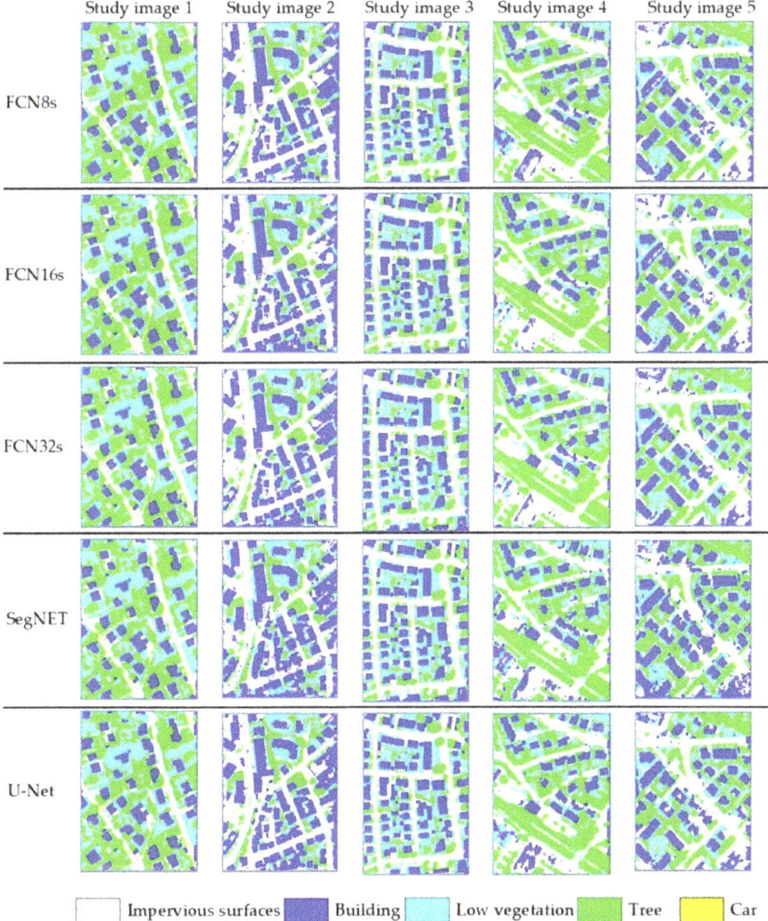

Figure 5. Classification results of five deep semantic segmentation models.

Table 2. Classification accuracy for the study images.

Model	Classification Accuracy of Study Images (%)				
	1	2	3	4	5
FCN8s	97.46	81.92	96.16	79.52	80.37
FCN16s	96.90	79.61	95.04	76.98	80.12
FCN32s	97.22	82.27	95.01	78.59	81.63
SegNET	97.83	78.95	95.39	76.50	77.11
U-Net	97.91	82.02	97.38	79.43	80.07

3.1.3. Comparison of the Correction Characteristics of ELP and CRF

To compare the correction characteristics of the ELP and CRF, this section uses U-Net's classification result for test image 5 and applies ELP and CRF to correct a subarea of the result image. The detailed results of the two algorithms with regard to iterations (ELP) and execution (CRF) are as shown in Figure 6.

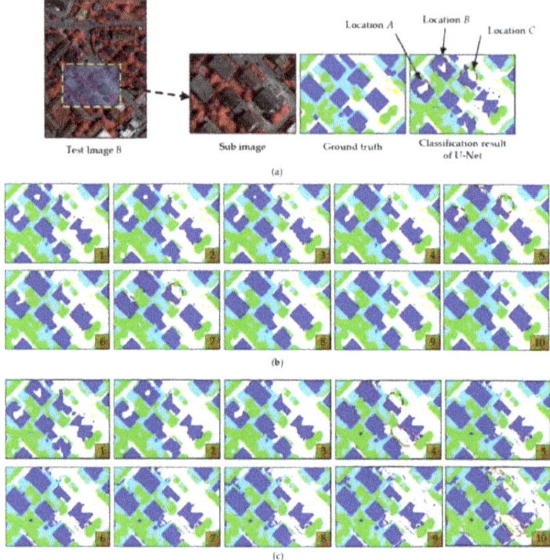

Figure 6. Detailed comparison of the results of the ELP and conditional random field (CRF) algorithms. (**a**) Ground truth and U-Net's classification results of a subarea of test image 5; (**b**) results of ten iterations of ELP; (**c**) results of ten executions of CRF.

As shown in Figure 6a, for the sub-image of test image 5, the classification results obtained by U-Net are far from perfect, and the boundaries of objects are blurred or chaotic. Especially at locations *A*, *B*, and *C* (marked by the red circles), the buildings are confused with impervious surfaces, and the buildings contain large holes or misclassified parts. On this sub-image, the classification accuracy of U-Net is only 79.02%.

Figure 6b shows the results of the 10 ELP iterations. As the method iterates, the object boundaries are gradually refined, and the errors at locations *A* and *B* are gradually corrected. By the 5th iteration, the hole at position *A* is completely filled, and by the 7th iteration, the errors at location *B* are also corrected. For location *C*, because our algorithm follows an end-to-end process, no samples exist in this process to determine which part of the corresponding area is incorrect; therefore, location *C* is not significantly modified during the iterations. Nevertheless, the initial classification error is not enlarged. As the iteration continues, the resulting images change little from the 7th to 10th iterations, and the algorithm's result becomes stable.

Figure 6c shows the results of the CRF. From executions 1 to 3, it can be seen that the CRF can also perform boundary correction. After the 4th iteration, the errors in locations *A* and *B* are corrected. It can also be seen that at position *C*, part of the correctly classified building roof was modified into impervious surfaces, further exaggerating the errors. The reason for this outcome is that at location *C*, for the corresponding roof color, the correctly classified part is smaller than the misclassified part. In the global correction context, the CRF algorithm more easily replaces relatively small parts. At the same time, as the iteration progresses, errors gradually appear due to the CRF's correction process (as marked in orange); some categories that were originally not dominant in an area (such as trees and cars) experience large decreases, and the classification accuracy continues to decrease with further iterations.

Based on the ground truth image, we evaluate the classification accuracy of the two methods after each iteration/execution as shown in Table 3:

Table 3. Comparison of two algorithms by iteration/execution.

Methods	Classification Accuracy at each Iteration/Execution (%)									
	1	**2**	**3**	**4**	**5**	**6**	**7**	**8**	**9**	**10**
ELP	80.93	81.49	82.88	83.11	84.24	84.34	85.76	85.81	85.59	85.67
CRF	79.41	82.55	83.21	83.76	81.45	80.93	80.55	79.65	77.83	73.86

As seen from Table 3, compared to the original classification result, whose accuracy is 79.02%, the ELP's classification accuracy increases to 80.93% after the first iteration—the lowest among its ten iterations, and it reaches 85.81% by the 8th iteration for a classification accuracy improvement of 6.79%. The CRF's classification accuracy is 79.41% after the first executions, and it reaches its highest accuracy of 83.76% in the 4th execution; subsequently, the classification accuracy gradually declines during the remaining iterations, falling to 73.86% by the 10th execution. Overall, the CRF reduced the accuracy by 5.16% compared with the original classification image. A graphical comparison of the classification accuracy of the two methods is shown in Figure 7:

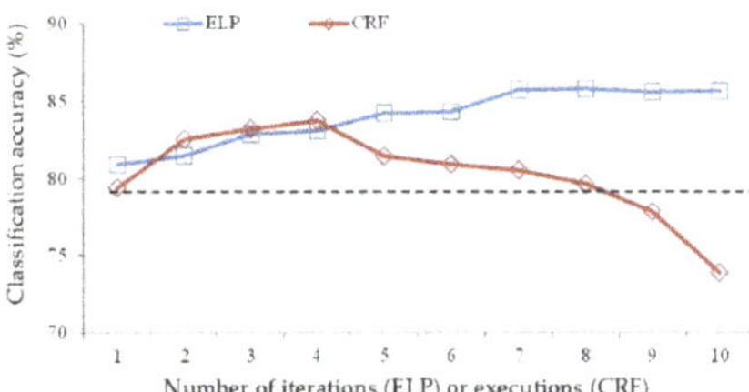

Figure 7. Graphical comparison of the classification accuracy of the two methods.

In Figure 7, the black dashed line indicates the original classification accuracy of 79.02%. The CRF's accuracy improvement was slightly better than that of the ELP algorithm in the second, third, and fourth iterations; however, its classification accuracy decreases rapidly in the later iterations, and by the ninth iteration, the classification accuracy is lower than that of the original classification result image. In contrast, the classification accuracy of the ELP increases steadily, and after approaching its highest accuracy, in subsequent iterations the accuracy remains relatively stable. From the above results, the ELP not only achieves a better correction result but also avoids causing obvious classification accuracy reductions from performing too many iterations. In end-to-end application scenarios where no samples participate in the result evaluation, we cannot know when the highest correction result has been reached; thus, the ideal method of termination conditions are also unknown. Specifying a too-small distance parameter will cause under-correction, while a too-large parameter will cause over-correction. The relatively stable characteristics and greater accuracy of ELP clearly allow it to achieve better processing results than those of CRF.

3.1.4. Correction Results Comparison

The correction results of the CRF method are shown in Figure 8.

As shown in Figure 8, for images 1 and 3, because the classification accuracy is relatively high, less room exists for correction, and the resulting images change only slightly. For images 2, 4, and 5, although the large errors and holes are corrected, numerous incorrect borders are present, pixels appear at shadowed parts of the ground objects, and many small objects (e.g., cars or trees) are erased by larger objects (e.g., impervious surfaces or low vegetation). For ELP, the correction results are shown in Figure 9.

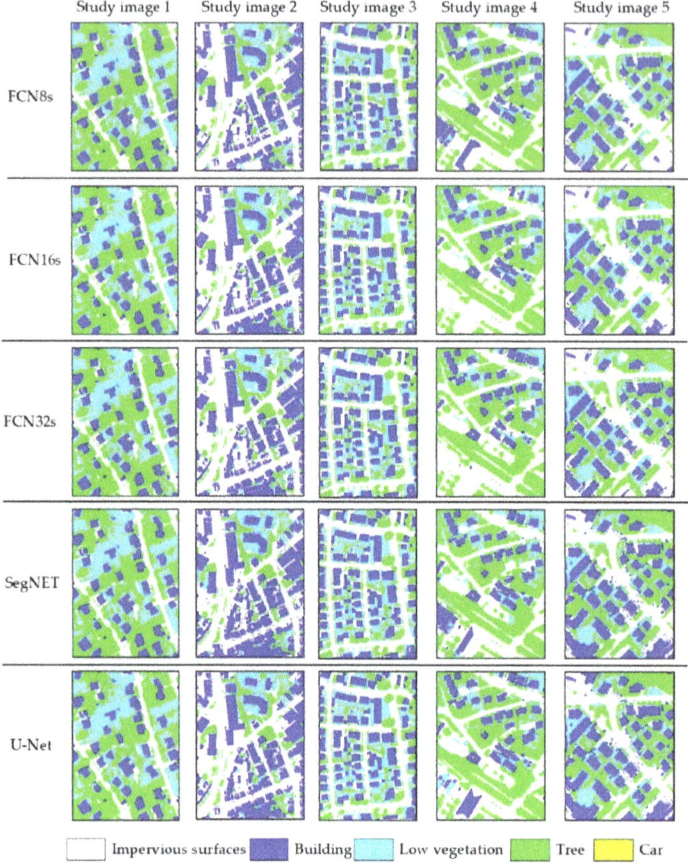

Figure 8. CRF correction results.

As Figure 9 shows, ELP also corrects the large errors and holes but does not produce overcorrection errors in the shadowed parts of the ground objects, and small objects are not erased. Therefore, in general, the correction results of ELP are better than those of CRF. Correction accuracy comparisons of the two algorithms are shown in Table 4.

Table 4. Correction accuracy comparisons of CRF and ELP.

Model	Classification Accuracy of Study Images (%)									
	1		2		3		4		5	
	CRF	ELP	CRF	ELP	CRF	ELP	CRF	ELP	CRF	ELP
FCN8s	97.91	97.88	84.72	88.32	97.31	97.02	81.24	85.47	82.81	85.87
FCN16s	97.41	97.37	83.33	87.61	96.94	97.11	80.57	84.67	84.02	85.82
FCN32s	97.67	97.62	84.90	87.32	96.71	96.31	82.02	84.89	83.02	85.70
SegNET	97.89	97.94	80.98	86.51	96.44	97.03	80.07	84.51	80.05	85.01
U-Net	97.95	97.98	85.07	88.91	97.42	97.47	81.53	86.93	83.12	86.07
Average Improvement	0.30	0.29	2.84	6.78	1.16	1.19	2.88	7.09	2.74	5.83

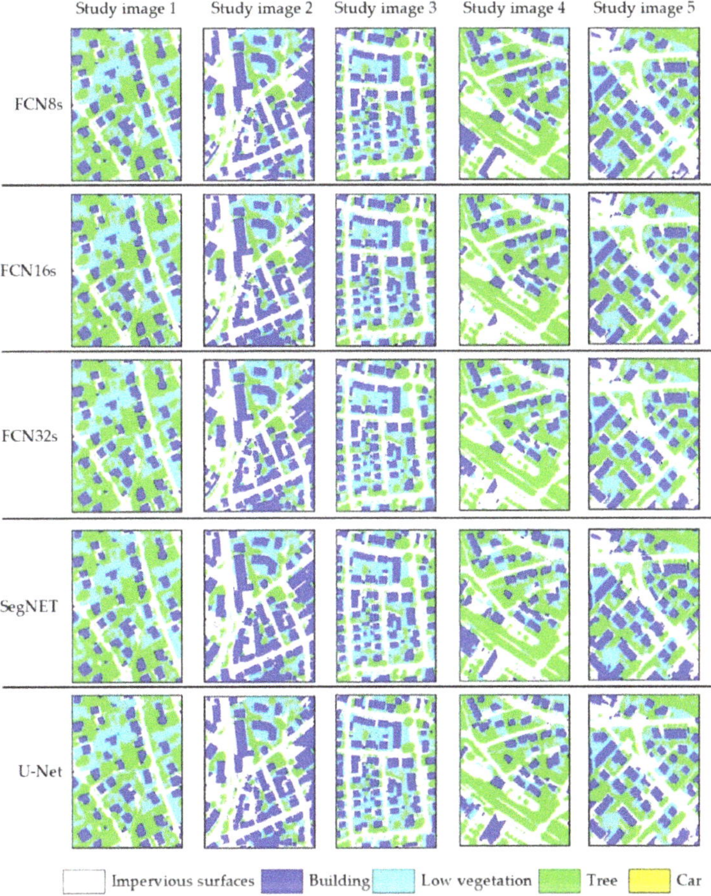

Figure 9. ELP correction results.

As shown in Table 3, for study images 1 and 3, because the original classification accuracy is high, the correction results of CRF and ELP are similar to the original classification result, and the improvements are limited. On study images 2, 4, and 5, the ELP's average improvements are 6.78%, 7.09%, and 5.83%, respectively, while the corresponding CRF improvements are only 2.84%, 2.88%, and 2.74%. Thus, ELP's correction ability is significantly better than that of CRF.

3.2. Comparison of Multiple Post-Processing Methods on Potsdam Dataset

3.2.1. Test Images and Methods

This study introduces four images from the Potsdam dataset, which are listed in Table 5.

Table 5. Four study images from the Potsdam dataset.

Image Name	Filename	Size
Training image 1	top_potsdam_2_10_RGBIR	6000×6000
Training image 2	top_potsdam_3_10_RGBIR	6000×6000
Testing image 1	top_potsdam_2_12_RGBIR	6000×6000
Testing image 2	top_potsdam_3_12_RGBIR	6000×6000

We selected two images as training data and the other two images as test data. Three bands (red (R), green (G), blue (B)) of images were selected. These images contain six categories: impervious surfaces (I), buildings (B), low vegetation (LV), trees (T), cars (C), and clutter/ background (C/B). The study images and their corresponding ground truth images are shown in Figure 10.

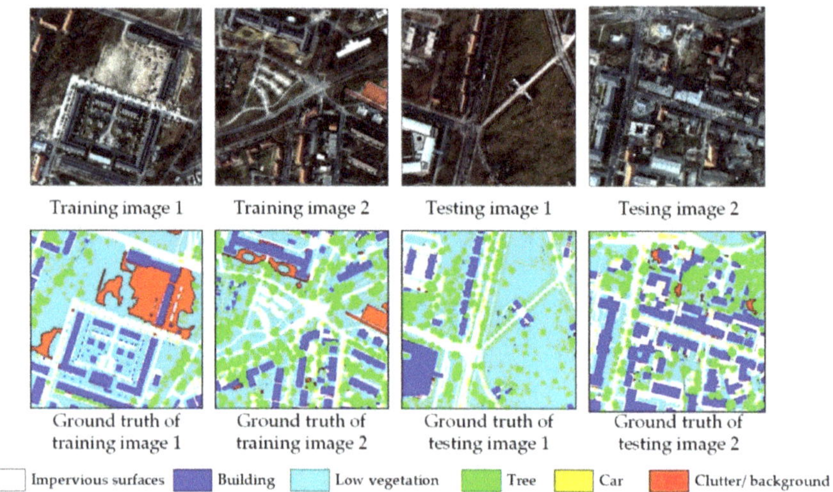

Figure 10. The study images and their corresponding ground truth from the Potsdam dataset.

To further evaluate ELP's ability, this paper compares four methods:

(1) U-Net + CRF: Use U-Net to classify an image and use CRF to perform post-processing.

(2) U-Net + MRF: Use U-Net to classify an image and use the Markov random field (MRF) to perform post-processing.

(3) DeepLab: Adopt DeepLab v1 model; in the DeepLab v1, the model has a built-in CRF as the last processing component, and this model can obtain a more accurate boundary than a model without the CRF component.

(4) U-Net + ELP: Use U-Net to classify an image and use ELP to perform post-processing.

3.2.2. Process Results of Four Methods

For the two testing images, the final process results of the four methods are illustrated in Figure 11.

As can be seen in Figure 11, because U-Net + CRF uses a global CRF processing strategy, there are many overcorrection areas, and some objects in the result image contain chaotic wrongly classified pixels. For the U-Net + MRF, the majority of noise pixels are removed, but the correction effects are not obvious. DeepLab's CRF is performed on image patch, not the whole image, so the overcorrection phenomenon is less than that of U-Net + CRF to some extent. U-Net + ELP obtained the best classification among all of the methods. The classification accuracies of the four methods are presented in Table 6.

Table 6. Classification accuracies of the four methods.

Image File	Classification Accuracy of the Four Methods (%)			
	U-Net + CRF	U-Net + MRF	DeepLab	U-Net + ELP
Testing image 1	81.67	78.45	82.85	85.31
Testing image 2	80.49	76.33	83.57	86.58

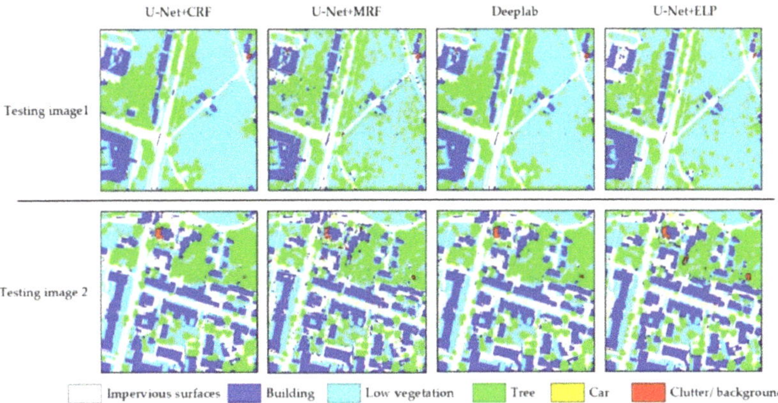

Figure 11. Results of the four methods.

As shown in Table 6, U-Net + MRF achieves the lowest classification accuracy, U-Net + CRF and DeepLab are higher than U-Net + MRF, and U-Net + ELP achieves the best classification accuracy.

3.2.3. Analysis of computational complexity

To analyze the computational complexity of the methods, we use four methods to process the testing image 1 and run the process five times. The experiments are performed on a computer (i9 9900k/64 GB/RTX 2080ti 11G), and the average process times are listed in Table 7.

Table 7. Process time of four methods.

Methods	Process Time of Each Step (Seconds)				Overall Time (Seconds)
	Image Separate	Patches Classification	Patches Combination	Post-Processing	
U-Net + CRF	7.52	45.84	8.32	216.34	278.02
U-Net + MRF	7.43	45.67	8.51	170.97	232.58
DeepLab	7.61	232.98	8.45	\	249.04
U-Net + ELP	7.57	46.02	8.77	506.45	568.81

As shown in Table 7, because the U-Net model can make full use of the graphics processing unit (GPU), and the processing speed of CRF and MRF on the whole image is also fast, U-Net + CRF and U-Net + MRF obtain results in a short time. DeepLab performs CRF after each image patch classification, so it does not need the post-processing stage, but the patch-based CRF process needs additional data access time and duplicate pixels at the patches' border, so its process time is similar to that of U-Net + CRF and U-Net + MRF. Since U-Net + ELP adopts the same deep model, its process time of the first three steps is similar to that of U-Net + CRF and U-Net + MRF, but at the post-processing stage, it needs a much longer time than the other methods.

For the ELP algorithm, the *HList* is updated at each iteration, and the suspicious areas marked by *HList* will change constantly. Each suspicious area needs to be processed by the CRF method, so the processing complexity of the ELP will vary along with the complexity of the image content. Although each suspicious area is small, ELP's greater amount of iterations, localization process, and result image update mechanism will introduce an additional computational burden, so ELP needs more process time than traditional methods.

3.2.4. Analysis of Different Threshold Parameter Values of ELP

The threshold value α of ELP will determine the choice of suspicious areas. To test the influence of this parameter, we chose UNet + ELP to process testing image 1, and set α in the range 0, 0.01, 0.02, ..., 0.09, which can vary from 0 to 0.09 with an interval of 0.01. The classification accuracy is shown in Table 8.

Table 8. Classification accuracy comparison of different threshold values α.

Method	Classification Accuracy at Different Threshold Parameter Value (%)									
	0.00	0.01	0.02	0.03	0.04.	0.05	0.06	0.07	0.08	0.09
U-Net + ELP	84.97	85.40	85.01	85.07	85.37	85.31	84.87	83.13	81.29	80.46

As can be seen from Table 8, when α is less than 0.6, the classification accuracy does not change significantly; when α is larger than 0.7, the classification accuracy is decreased. The main reason for this phenomenon is that when α is small, ELP will be more sensitive to the discovery of suspicious areas; however, too many suspicious areas merely increase the computational burden without contributing to obvious changes in accuracy. In contrast, when α is larger, ELP will have a diminished capability to discover suspicious areas, and many suspicious areas that need correction will be omitted, which will cause a decrease in accuracy. At the same time, we can see from Table 8 that in a large range (0.0 to 0.6), the accuracy of ELP does not change greatly, which reveals that ELP has good stability with the threshold value α.

3.2.5. Analysis of Different Segmentation Number of ELP

The ELP method adopts the SLIC algorithm as the segmentation method, and an important parameter of the SLIC algorithm is $N_{segment}$ which decides the number of segments after the algorithm performed. When the $N_{segment}$ is assigned an overly small value, under-segmentation may appear; conversely, when $N_{segment}$ is assigned an overly large value, over-segmentation may appear. To test the influence of this parameter on the ELP, we set $N_{segment}$ = 1000, 2000, ... , 10,000 and allowed $N_{segment}$ to vary from 1000 to 10,000 with an interval of 1000. The classification accuracy of testing image 1 by U-Net + ELP is shown in Table 9.

Table 9. Classification accuracy comparison of different segment number.

Method	Classification Accuracy at Different Segment Number (%)									
	1000	2000	3000	4000	5000	6000	7000	8000	9000	10,000
U-Net + ELP	81.75	82.02	83.21	85.22	85.31	85.52	85.67	85.23	82.22	80.32

It can be seen from Table 9 that when $N_{segment}$ = 1000 to 3000, because the image is large (6000 × 6000) and the segment number is relatively small, the image is under-segmented, and each segmentation may contain pixels with a different color or brightness. This situation makes it difficult for ELP to focus on suspicious areas, and the classification accuracy is low. When $N_{segment}$ = 9000 to 10,000, the image is obviously over-segmented, and the segmentations are too small. This situation leads ELP to have a small update size in its LocalizedCorrection algorithm and leads to ELP's poor performance. For $N_{segment}$ = 4000 to 8000, the classification accuracy of ELP does not vary greatly, which indicates that ELP does not have restrictive requirements for the segment parameter; as long as the segment method can correctly separate the regions with similar colors/brightness and the segment size is not too small, ELP can achieve satisfactory results.

4. Conclusions

Deep semantic segmentation neural networks are powerful end-to-end remote sensing image classification tools that have achieved successes in many applications. However, it is difficult to construct a training set that thoroughly exhausts all the pixel segmentation possibilities for a specific ground area; in addition, spatial information loss occurs during the network training and inference stages. Consequently, the classification results of DSSNNs are usually not perfect, which introduces a need to correct the results.

Our experiments demonstrate that when faced with complicated remote sensing images, the CRF algorithm often has difficulty achieving a substantially improved correction effect; without restricting the mechanism by using additional samples, the CRF may overcorrect, leading to a decrease in the classification accuracy. Our approach improves on the traditional CRF global processing effects by offering two advantages:

(1) End-to-end: ELP identifies which locations of the result image are highly suspected of containing errors without requiring samples; this characteristic allows ELP to be used in an end-to-end classification process.

(2) Localization: Based on the suspect areas, ELP limits the CRF analysis and update area within a small range and controls the iteration termination condition; these characteristics avoid the overcorrections caused by the global processing of the CRF.

The experimental results also show that the ELP achieves a better correction result, is more stable, and does not require training samples to restrict the iterations. The above advantages ensure that ELP is better able to adapt to correct the classification results of remote sensing images and provides it with a higher degree of automation.

The typical limitation of ELP is that, in comparison with the traditional CRF, the additional iterations, the localization process, and the update mechanism of the result image will introduce an additional computational burden. Consequently, ELP is much slower than the traditional CRF method. Fortunately, the localization process also ensures that different parts of areas do not affect each other, which makes ELP easier to parallelize. In further research, we will adjust the processing structure of ELP, facilitating a GPU implementation that enables ELP to execute faster. For semantic segmentation neural networks, differences in the training set size can cause various degrees of errors in the resulting image, which has an apparent influence on the post-processing task. In future research, to construct a more adaptive post-processing method, we will study the relationship between training/testing dataset size and the post-processing methods used and consider the problems faced by post-processing methods in more complex application scenarios.

Author Contributions: Conceptualization, X.P. and J.X.; data curation, J.Z.; methodology, X.P.; supervision, J.X.; visualization, J.Z.; writing—original draft, X.P. All authors have read and agree to the published version of the manuscript.

Funding: This research was jointly supported by the National Natural Science Foundation of China (41871236; 41971193) and the Natural Science Foundation of Jilin Province (20180101020JC).

Conflicts of Interest: The authors declare no conflict of interest.

References

1. Pan, X.; Zhao, J.; Xu, J. An object-based and heterogeneous segment filter convolutional neural network for high-resolution remote sensing image classification. *Int. J. Remote Sens.* **2019**, *40*, 5892–5916. [CrossRef]
2. Tong, X.-Y.; Xia, G.-S.; Lu, Q.; Shen, H.; Li, S.; You, S.; Zhang, L. Land-cover classification with high-resolution remote sensing images using transferable deep models. *Remote Sens. Environ.* **2020**, *237*, 111322. [CrossRef]
3. Pan, X.; Zhao, J. A central-point-enhanced convolutional neural network for high-resolution remote-sensing image classification. *Int. J. Remote Sens.* **2017**, *38*, 6554–6581. [CrossRef]
4. Chen, Y.Y.; Ming, D.P.; Lv, X.W. Superpixel based land cover classification of VHR satellite image combining multi-scale CNN and scale parameter estimation. *Earth Sci. Inform.* **2019**, *12*, 341–363. [CrossRef]

5.   Long, J.; Shelhamer, E.; Darrell, T. Fully Convolutional Networks for Semantic Segmentation. In Proceedings of the rfgvb 2015 Ieee Conference on Computer Vision and Pattern Recognition (Cvpr), Boston, MA, USA, 7–12 June 2015; pp. 3431–3440.

6.   Zhu, X.X.; Tuia, D.; Mou, L.; Xia, G.S.; Zhang, L.; Xu, F.; Fraundorfer, F. Deep learning in remote sensing: A Comprehensive Review and List of Resources. *IEEE Geosci. Remote Sens. Mag.* **2017**, *5*, 8–36. [CrossRef]

7.   Long, J.; Shelhamer, E.; Darrell, T. Fully convolutional networks for semantic segmentation. In Proceedings of the IEEE Conference on Computer Vision and Pattern Recognition, Boston, MA, USA, 7–12 June 2015; pp. 3431–3440. [CrossRef]

8.   Badrinarayanan, V.; Kendall, A.; Cipolla, R. SegNet: A Deep Convolutional Encoder-Decoder Architecture for Image Segmentation. *IEEE Trans. Pattern Anal. Mach. Intell.* **2017**, *39*, 2481–2495. [CrossRef]

9.   Ronneberger, O.; Fischer, P.; Brox, T. U-Net: Convolutional Networks for Biomedical Image Segmentation. In *Medical Image Computing and Computer-Assisted Intervention, Pt Iii*; Springer: Munich, Germany, 2015; Volume 9351, pp. 234–241. [CrossRef]

10.   Kemker, R.; Salvaggio, C.; Kanan, C. Algorithms for semantic segmentation of multispectral remote sensing imagery using deep learning. *Isprs J. Photogramm.* **2018**, *145*, 60–77. [CrossRef]

11.   Vakalopoulou, M.; Karantzalos, K.; Komodakis, N.; Paragios, N. Building Detection in Very High Resolution Multispectral Data with Deep Learning Features. In Proceedings of the 2015 IEEE International Geoscience and Remote Sensing Symposium (IGARSS), Milan, Italy, 26–31 July 2015; pp. 1873–1876. [CrossRef]

12.   Yi, Y.N.; Zhang, Z.J.; Zhang, W.C.; Zhang, C.R.; Li, W.D.; Zhao, T. Semantic Segmentation of Urban Buildings from VHR Remote Sensing Imagery Using a Deep Convolutional Neural Network. *Remote Sens.-Basel* **2019**, *11*, 1774. [CrossRef]

13.   Zhou, L.C.; Zhang, C.; Wu, M. D-LinkNet: LinkNet with Pretrained Encoder and Dilated Convolution for High Resolution Satellite Imagery Road Extraction. In Proceedings of the 2018 IEEE/CVF Conference on Computer Vision and Pattern Recognition Workshops (CVPRW), Salt Lake City, UT, USA, 18–22 June 2018; pp. 192–196. [CrossRef]

14.   Tao, C.; Qi, J.; Li, Y.S.; Wang, H.; Li, H.F. Spatial information inference net: Road extraction using road-specific contextual information. *Isprs J. Photogramm.* **2019**, *158*, 155–166. [CrossRef]

15.   Zhang, Z.X.; Liu, Q.J.; Wang, Y.H. Road Extraction by Deep Residual U-Net. *IEEE Geosci. Remote Sens.* **2018**, *15*, 749–753. [CrossRef]

16.   Volpi, M.; Tuia, D. Dense Semantic Labeling of Subdecimeter Resolution Images with Convolutional Neural Networks. *IEEE Trans. Geosci. Remote* **2017**, *55*, 881–893. [CrossRef]

17.   Chen, G.Z.; Zhang, X.D.; Wang, Q.; Dai, F.; Gong, Y.F.; Zhu, K. Symmetrical Dense-Shortcut Deep Fully Convolutional Networks for Semantic Segmentation of Very-High-Resolution Remote Sensing Images. *IEEE J.-Stars* **2018**, *11*, 1633–1644. [CrossRef]

18.   Mohammadimanesh, F.; Salehi, B.; Mandianpari, M.; Gill, E.; Molinier, M. A new fully convolutional neural network for semantic segmentation of polarimetric SAR imagery in complex land cover ecosystem. *Isprs J. Photogramm.* **2019**, *151*, 223–236. [CrossRef]

19.   Mboga, N.; Georganos, S.; Grippa, T.; Lennert, M.; Vanhuysse, S.; Wolff, E. Fully Convolutional Networks and Geographic Object-Based Image Analysis for the Classification of VHR Imagery. *Remote Sens.* **2019**, *11*, 597. [CrossRef]

20.   Mi, L.; Chen, Z. Superpixel-enhanced deep neural forest for remote sensing image semantic segmentation. *Isprs J. Photogramm.* **2020**, *159*, 140–152. [CrossRef]

21.   Fu, G.; Liu, C.J.; Zhou, R.; Sun, T.; Zhang, Q.J. Classification for High Resolution Remote Sensing Imagery Using a Fully Convolutional Network. *Remote Sens.* **2017**, *9*, 498. [CrossRef]

22.   Marmanis, D.; Schindler, K.; Wegner, J.D.; Galliani, S.; Datcu, M.; Stilla, U. Classification with an edge: Improving semantic image segmentation with boundary detection. *Isprs J. Photogramm.* **2018**, *135*, 158–172. [CrossRef]

23.   Lafferty, J.D.; McCallum, A.; Pereira, F.C.N. Conditional random fields: Probabilistic models for segmenting and labeling sequence data. In Proceedings of the Eighteenth International Conference on Machine Learning (ICML 2001), Williamstown, MA, USA, 28 June–1 July 2001; pp. 282–289.

24.   Alam, F.I.; Zhou, J.; Liew, A.W.C.; Jia, X.P. Crf learning with cnn features for hyperspectral image segmentation. In Proceedings of the 2016 Ieee International Geoscience and Remote Sensing Symposium (Igarss), Beijing, China, 10–15 July 2016; pp. 6890–6893.

25. Zhang, B.; Wang, C.P.; Shen, Y.L.; Liu, Y.Y. Fully Connected Conditional Random Fields for High-Resolution Remote Sensing Land Use/Land Cover Classification with Convolutional Neural Networks. *Remote Sens.* **2018**, *10*, 1889. [CrossRef]

26. Zheng, S.; Jayasumana, S.; Romera-Paredes, B.; Vineet, V.; Su, Z.Z.; Du, D.L.; Huang, C.; Torr, P.H.S. Conditional Random Fields as Recurrent Neural Networks. In Proceedings of the 2015 IEEE International Conference on Computer Vision (ICCV), Santiago, Chile, 7–13 December 2015; pp. 1529–1537. [CrossRef]

27. Chen, Y.S.; Zhao, X.; Jia, X.P. Spectral-Spatial Classification of Hyperspectral Data Based on Deep Belief Network. *IEEE J.-Stars* **2015**, *8*, 2381–2392. [CrossRef]

28. Marmanis, D.; Wegner, J.D.; Galliani, S.; Schindler, K.; Datcu, M.; Stilla, U. Semantic segmentation of aerial images with an ensemble of cnss. In Proceedings of the ISPRS Annals of the Photogrammetry, Remote Sensing and Spatial Information Sciences, Prag, Tschechien, 12–19 July 2016; Volume 3, pp. 473–480.

29. Maggiori, E.; Tarabalka, Y.; Charpiat, G.; Alliez, P. Fully Convolutional Neural Networks for Remote Sensing Image Classification. In Proceedings of the 2016 Ieee International Geoscience and Remote Sensing Symposium (Igarss), Beijing, China, 10–15 July 2016; pp. 5071–5074. [CrossRef]

30. He, C.; Fang, P.Z.; Zhang, Z.; Xiong, D.H.; Liao, M.S. An End-to-End Conditional Random Fields and Skip-Connected Generative Adversarial Segmentation Network for Remote Sensing Images. *Remote Sens.* **2019**, *11*, 1604. [CrossRef]

31. Pan, X.; Zhao, J. High-Resolution Remote Sensing Image Classification Method Based on Convolutional Neural Network and Restricted Conditional Random Field. *Remote Sens.* **2018**, *10*, 920. [CrossRef]

32. Ma, F.; Gao, F.; Sun, J.P.; Zhou, H.Y.; Hussain, A. Weakly Supervised Segmentation of SAR Imagery Using Superpixel and Hierarchically Adversarial CRF. *Remote Sens.* **2019**, *11*, 512. [CrossRef]

33. Wei, L.F.; Yu, M.; Zhong, Y.F.; Zhao, J.; Liang, Y.J.; Hu, X. Spatial-Spectral Fusion Based on Conditional Random Fields for the Fine Classification of Crops in UAV-Borne Hyperspectral Remote Sensing Imagery. *Remote Sens.* **2019**, *11*, 780. [CrossRef]

34. Parente, L.; Taquary, E.; Silva, A.P.; Souza, C.; Ferreira, L. Next Generation Mapping: Combining Deep Learning, Cloud Computing, and Big Remote Sensing Data. *Remote Sens.* **2019**, *11*, 2881. [CrossRef]

35. Kampffmeyer, M.; Salberg, A.B.; Jenssen, R. Semantic Segmentation of Small Objects and Modeling of Uncertainty in Urban Remote Sensing Images Using Deep Convolutional Neural Networks. In Proceedings of the 29th Ieee Conference on Computer Vision and Pattern Recognition Workshops, (Cvprw 2016), Las Vegas, NV, USA, 26 June–1 July 2016; pp. 680–688. [CrossRef]

36. Achanta, R.; Shaji, A.; Smith, K.; Lucchi, A.; Fua, P.; Susstrunk, S. SLIC Superpixels Compared to State-of-the-Art Superpixel Methods. *IEEE Trans. Pattern Anal.* **2012**, *34*, 2274–2281. [CrossRef] [PubMed]

 *remote sensing*

Article

Vehicle and Vessel Detection on Satellite Imagery: A Comparative Study on Single-Shot Detectors

Tanguy Ophoff [1,*], **Steven Puttemans** [2], **Vasileios Kalogirou** [3], **Jean-Philippe Robin** [3] and **Toon Goedemé** [1]

1   KU Leuven, EAVISE - Jan Pieter De Nayerlaan 5, 2860 Sint-Katelijne-Waver, Belgium;
    toon.goedeme@kuleuven.be
2   Flanders Innovation & Entrepreneurship (VLAIO)—Koning Albert II Laan 35, 1030 Brussel, Belgium;
    steven.puttemans@vlaio.be
3   European Union Satellite Centre—Apdo de Correos 511, Torrejon de Ardoz, 28850 Madrid, Spain;
    vasileios.kalogirou@satcen.europa.eu (V.K.); jean-philippe.robin@satcen.europa.eu (J.-P.R.)
*   Correspondence: tanguy.ophoff@kuleuven.be; Tel.: +32-15-31-69-44

Received: 6 March 2020; Accepted: 2 April 2020; Published: 9 April 2020

Abstract: In this paper, we investigate the feasibility of automatic small object detection, such as vehicles and vessels, in satellite imagery with a spatial resolution between 0.3 and 0.5 m. The main challenges of this task are the small objects, as well as the spread in object sizes, with objects ranging from 5 to a few hundred pixels in length. We first annotated 1500 km^2, making sure to have equal amounts of land and water data. On top of this dataset we trained and evaluated four different single-shot object detection networks: YOLOV2, YOLOV3, D-YOLO and YOLT, adjusting the many hyperparameters to achieve maximal accuracy. We performed various experiments to better understand the performance and differences between the models. The best performing model, D-YOLO, reached an average precision of 60% for vehicles and 66% for vessels and can process an image of around 1 Gpx in 14 s. We conclude that these models, if properly tuned, can thus indeed be used to help speed up the workflows of satellite data analysts and to create even bigger datasets, making it possible to train even better models in the future.

Keywords: satellite; object detection; neural networks; single-shot

---

1. Introduction

Historically, spaceborne remote sensing has been an industry for governments and some heavyweight corporations. However, recent advancements, like the ability to use cost-effective off-the-shelf components (COTS) for cubesats, or the downstream opportunities created by the European Union Copernicus program, have radically changed the industry [1]. This has given rise to new businesses emerging and taking advantage of this geospatial data. To process these huge quantities of data coming from satellites, the space industry needs to speed up and automate workflows, which are traditionally handled by manual operators. Artificial intelligence (AI), being very good at general pattern recognition, lends itself to being a great contender for these tasks.

Imagery intelligence (IMINT) is a discipline which collects information through aerial and satellite means, allowing the monitoring of agricultural crop growth [2], performance of border and maritime surveillance [3,4] and inference of land changes [5] for other applications. Recent advances in computer vision, using deep learning techniques, already allow successful automation of IMINT cases on aerial images [6–8]. Furthermore, locating and segmenting larger objects, e.g., buildings, in satellite imagery is something that is already being used presently [9]. However, detecting smaller objects like vehicles and vessels at these spatial resolutions remains a challenging topic. In this paper, we investigate how different object detection networks handle the task of vehicle and vessel detection and classification in satellite imagery with a spatial resolution between 0.3 and 0.5 m, where objects appear only a few pixels large in the image.

Traditionally, computer vision algorithms used handcrafted features as an input to a machine learning classifier, in combination with a sliding window approach, in order to perform various tasks like object detection and classification [10–12]. However, with the rise of convolutional neural networks (CNN), deep learning has outperformed these traditional techniques by a significant margin. For the specific case of object detection, there are two different and commonly used approaches: two-stage and single-shot detectors. Two-staged methods like the R-CNN detector [13] will first generate several bounding boxes around potential objects, called region proposals. Each of these proposals will then be run through a classifier to determine whether it actually is an object and what class of object it is. Because they need to run each potential object through a classification algorithm, these techniques are quite slow. Therefore, optimizations have been made in fast R-CNN [14] to share computations between both stages. Faster R-CNN [15] improved upon this even further, by using a deep-learned approach to generate the box proposals, reducing the number of false positive boxes and thus increasing the runtime speeds even further. However, these techniques remain orders of magnitude slower than single-shot detectors. Single-shot detectors [3,4,16–18] are faster because they only process the images through a single neural network, detecting and classifying multiple objects at the same time. Because these detectors work on the entire image at once, they should also be able to use contextual visual information, to detect and classify objects. For the specific case of vessel detection, this means that these single-shot detectors can recognize the structures of sea waves, shorelines, ports, etc. and use that contextual information around the vessels to correctly detect and classify them.

In this paper, we will take a look at the YOLO (You Only Look Once) detector, a well-known and high-performing single-shot detector, and assess its performance for our use case of vehicle and vessel detection in satellite imagery. More specifically, we will compare the YOLOV2 [17] and YOLOV3 [18] detectors, as well as some variations of these, YOLT [3] and D-YOLO [4], which were specifically engineered for aerial object detection in remote sensing applications. We will train these four detectors on our custom dataset for vehicle and vessel detection and will perform various experiments, to assess the strengths and weaknesses of each detector.

In the remainder of this paper, we will first discuss the labeling of ground truth data and how we trained and evaluated the different models (Section 2). Afterwards, in Section 3, we will report various experiments we conducted to assess and understand the performance of the different models. Finally, we will conclude our paper by formulating an answer to the following questions (Section 4):

- Which model is best suited for satellite vehicle and vessel detection?
- What are the different trade-offs between the models?
- Can we consider the problem of automatic satellite object detection solved?

2. Materials and Methods

In this section, we will discuss the creation of our dataset (Section 2.1), the different models we used (Section 2.2) and how we trained them (Section 2.3).

2.1. Ground Truth Acquisition

To train and assess the performance of our algorithms, we need labeled data. In this paper, we collected images from four different optical satellites (i.e., WorldView-2 and -3, GeoEye and Pleiades), typically used to perform IMINT tasks for security applications. The images were acquired under various acquisition angles varying from 7 to 36 degrees, resulting in different spatial resolutions between 0.3 m and 0.5 m. The images were delivered in 3-band true-color RGB (see Figure 1). We take three images of each type of satellite (further referred to as WV, GE and PL), totaling in nine annotated images, covering around 1500 km^2. The entire dataset contains around 53% land and 47% water data.

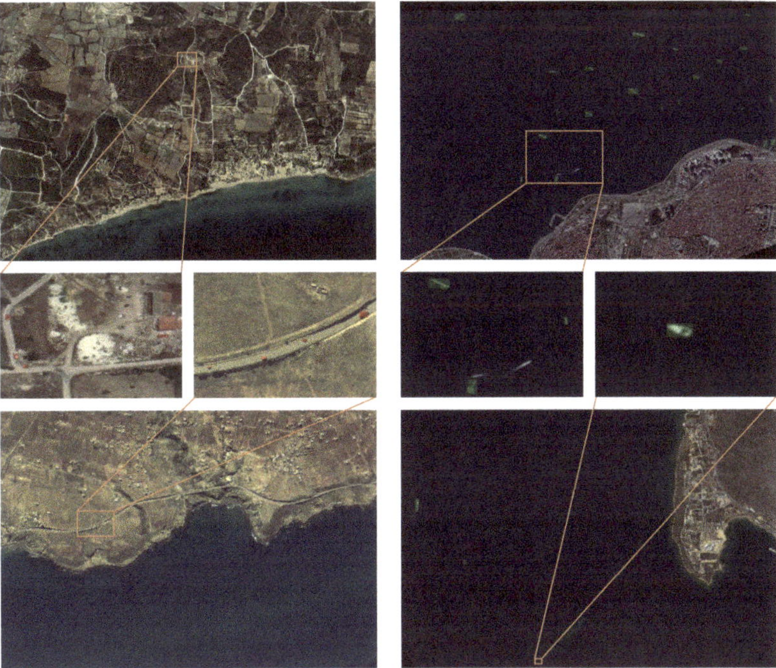

Figure 1. Example image crops taken from the training dataset. Vehicles are shown in red, vessels in green.

As not to miss any object during annotation, we decided to split the images in overlapping patches of around 500 × 500 pixels. Since annotating the image dataset is a time-consuming and demanding task, we first defined land and water regions in our images. This allows us to only process land region patches during vehicle annotation and water region patches for vessels, speeding up the annotation process. These regions later prove to be beneficial for training and testing our detectors as well. There is little use in training and running a vehicle detector on water bodies and vice-versa for vessel detectors. Doing so will only make it harder for the detector to converge towards an optimal solution and with datasets like the Global Surface Water Explorer [19], which provide water coverage maps with a ground resolution of 30 m, it is possible to automatically create these regions, providing a scene constraint that will increase the performance of a detector.

Figure 2 shows the width and length of the manually annotated vehicles and vessels. This graph shows the two main difficulties detectors will have to cope with, namely that vehicles are only around 10 × 10 pixels big and vessels have a huge spread in size, between 5 and almost 500 pixels long. To both train and test the detectors, we split our dataset in a training and testing set, picking two images at random from each type of satellite for training and the remaining image for testing. The number of annotations of each type can be found in Table 1.

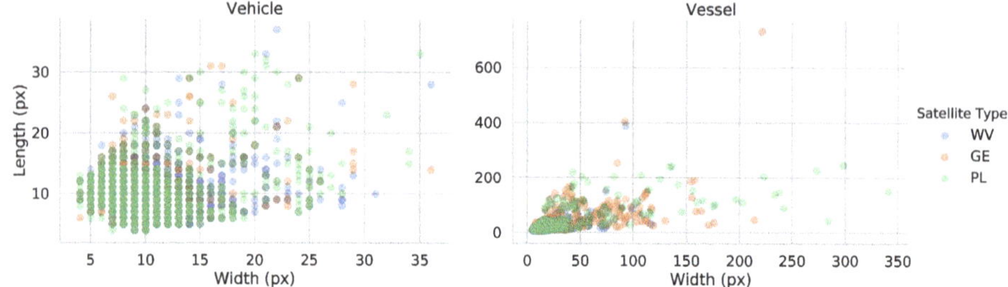

Figure 2. Length and Width properties for vehicles and vessels in the annotated dataset.

Table 1. Number of annotations in our dataset.

Satellite	Vehicles			Vessels		
	Train	**Test**	**Total**	**Train**	**Test**	**Total**
WV	1477	323	1800	252	97	349
GE	413	220	633	301	50	351
PL	1318	358	1676	184	211	395
Total	3208	901	4109	737	358	1095

Detecting vehicles and vessels for operational IMINT purposes can be a useful task, when deployed in large satellite image datasets. However, the value of automation can be considerably higher if the detections can be further classified or labeled on certain categories. The classification scheme adopted in this study, for both vehicles and vessels, considers semantic requirements coming from the IMINT domain (see left columns in Figure 3). For this reason, it includes classes which rarely appear in most vessel detection studies, which focus their research on the detection of larger ships [20]. On the contrary, this study considers also smaller vessels, such as e.g., inflatable rubber boats and skiffs, which can be as small as 3–6 m in length.

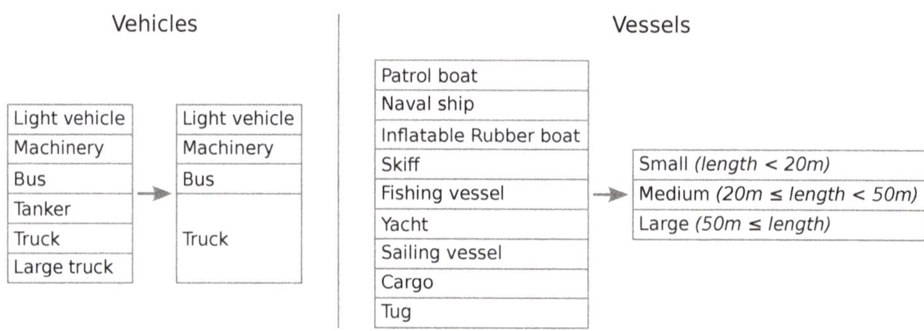

Figure 3. Initially defined labels vs. simplified labels.

However, after labeling the vehicles and comparing the labels of two independent researchers, we concluded that the originally defined labels were too specific compared to the spatial resolution of the data. Table 2 shows that there is a huge annotator bias (e.g., when is something considered a truck or a large truck; is any vehicle located in fields to be considered machinery?).

In some cases, due to the semantic or structural similarities existing between the classes (e.g., a sailing vessel vs. a yacht) or due to the limited spatial resolution of some images, the labeling task becomes challenging even for experienced image analysts. In these cases, analysts usually consult collateral sources of information to assign a label, using Automatic Identification System (AIS) tracking data for vessels or generic contextual information (e.g., port type, or vessel location at port etc.). The scheme was further simplified, to avoid the aforementioned semantic bias and improve the quality of the annotated dataset. Figure 3 shows the original and simplified labels and Table 3 shows the number of objects for each of the simplified labels in our dataset. To decrease the complexity further, we decided to train separate models for the case of vehicle and vessel detection. We will first look at single-class detectors, which are only able to detect vehicles or vessels, but not label them further. Afterwards, we will also train multi-class detectors, which are capable of both detection and classification of our objects.

Table 2. Classification differences between two annotators for the original vehicle classes.

Annotator 1 \ Annotator 2	Light Vehicle	Machinery	Bus	Tanker	Truck	Large Truck
Light vehicle	**1386**	35	9	4	90	25
Machinery	31	**7**	0	1	3	2
Bus	24	1	**4**	1	0	1
Tanker	4	2	0	**0**	1	2
Truck	179	6	1	0	**32**	4
Large truck	47	22	0	8	14	**18**

Table 3. Number of annotations per label in the dataset.

Dataset	Light Vehicle	Machinery	Bus	Truck	Small Vessel	Medium Vessel	Large Vessel
Train	2561	44	106	497	404	258	75
Test	690	26	16	169	246	93	19
Total	3251	70	122	666	650	351	94

2.2. Network Architectures

In this paper, we compare the YOLOV2 [17], YOLOV3 [18], YOLT [3] and D-YOLO [4] network architectures. All networks were re-implemented with PyTorch [21] and are available in our open-source library Lightnet [22]. This section will provide a quick overview of the different architectures and talk about key differences between them. For a more detailed explanation of each network, we refer to the original research papers.

In 2017 Redmon & Farhadi released YOLOV2 [17], a general-purpose single-shot object detector based on the Darknet19 classification network (see Figure 4a). The network is fully convolutional and concatenates fine-grained features from an earlier feature map in the network to increase the detection accuracy on smaller objects. To be able to combine these feature maps with different spatial resolutions, they invented a reorganization scheme that transforms the higher resolution map into a smaller resolution by dividing it and stacking in the depth dimension.

In 2018 they released an improvement of this network, called YOLOV3 [18]. This network is based on the Darknet53 classification network and is thus much deeper (see Figure 4b). Taking inspiration from feature pyramid networks [23], they concatenate fine-grained features twice, upsampling the lower resolution feature map rather than reorganizing the higher resolution one, and perform predictions from different spatial resolutions as well. These changes resulted in a network that performs better—mostly on smaller objects—but takes a longer time to run.

In 2018, Van Etten released the YOLT model [3], a variation of YOLOV2 that is especially engineered to increase the performance on remote sensing object detection. He noted that YOLOV2 had trouble detecting small objects and thus tried to solve this issue by having less maxpool subsampling operations. Van Etten's network is also slightly smaller (see Figure 4c), to make up for the fact that the reduced maxpool operations increase the number of computations and thus slow down the network.

Lastly, in 2018, Acatay et al. released their model D-YOLO [4], which is a variation of YOLOV2 as well. Trying to solve the same issue as YOLT, they instead opt to take inspiration from feature pyramid networks [23], with deconvolution operations instead of upsampling (see Figure 4d). This means that all layers except the last 3 operate at the same spatial resolution as YOLOV2, resulting in runtime speeds much closer to it than YOLT.

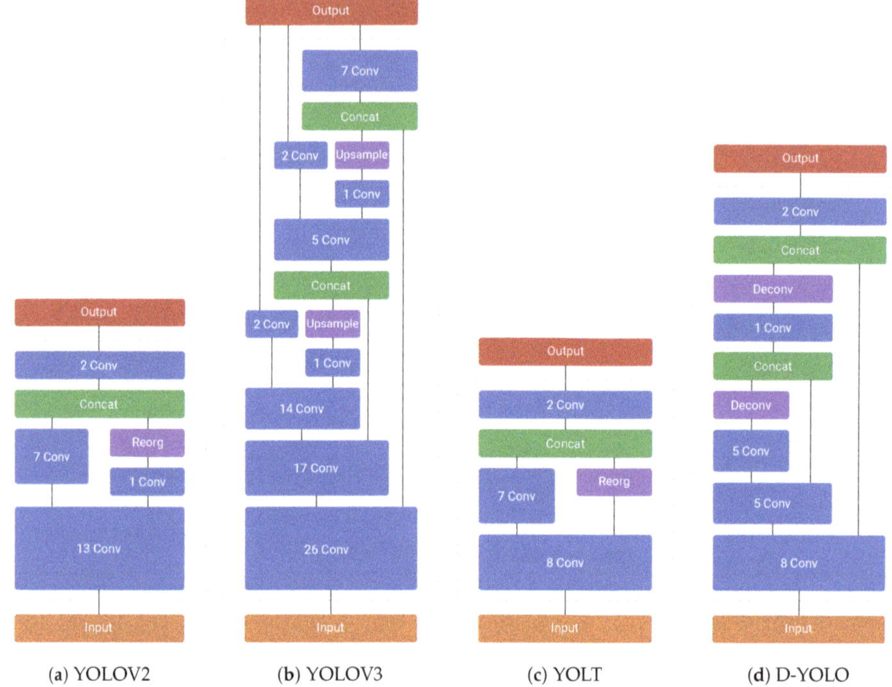

(a) YOLOV2   (b) YOLOV3   (c) YOLT   (d) D-YOLO

Figure 4. Different network architectures.

2.3. Training on Satellite Imagery

Satellite images are considerably large (average width-height in our dataset: 34,060 × 34,877 pixels) and can thus not be processed by off-the-shelf detectors at once, as the computed feature maps would be too big to fit in GPU memory. Therefore, the images were cut into patches of 416 × 416 pixels, with a 10% horizontal and vertical overlap, which get processed individually by the detection networks. Of course, these settings are quite arbitrary and are in fact hyperparameters, which one can tune, depending on the application and the size of the objects that need to be detected. One can then consider the collection of all these patches as the entire training dataset, but filtering the patches and only keeping those which actually contain objects to detect, will increase the performance of the detectors. Indeed, every part of a patch that does not contain an object is already providing a negative example for the detector to train on. This number of negative examples has empirically showed to be enough to successfully train a model and as such we do not need more negative examples by adding these empty patches. However, to correctly evaluate the performance of the different models, we do need to include all patches during the testing phase of the detectors. Because we allow for overlap between the different patches, the non-maximal suppression post-processing step should be used after combining the bounding boxes of the different patches, to remove duplicate detections.

When training a network, one needs to set up some parameters which influence how the model trains. These are called hyperparameters and correctly fine-tuning them makes the difference between a good and a bad model. Moreover, these hyperparameters need to be tweaked for every new use case and can differ wildly, depending on the dataset, model, etc. Hyperparameters like learning rate, momentum, training time, etc. have a big influence on the training outcome and were thus tweaked manually, by trial and error (combined with the researcher's experience and insights), to try and achieve the best performance. However, we observed that some lesser known hyperparameters that influence the yolo-specific loss function, had a significant impact on the achievable accuracy of the models as well. Indeed, if we break down what single-shot detection networks need to predict, we can find three distinct tasks: detecting objects, finding bounding box coordinates and optionally classifying these objects. These three different tasks each have a different loss function, which get combined by taking the weighted sum of these different sub-losses. The task of detecting objects can be further broken down into two parts, namely how important it is to predict a high confidence when there is an object and how important it is to predict a low confidence when there is no object. When training a model, one can change the weight of each of these four sub-losses, effectively modifying the relative importance of these different parts and thus the final behavior of the model.

Care must be taken not to overfit a model on the testing dataset when selecting these hyperparameters, and to that end we chose to tune our parameters on a subset of our data. We thus trained and evaluated the four different models on the GE subset of our data and tweaked the hyperparameters in order to achieve the highest possible average precision at an IoU threshold of 50% (see Table 4). Once we found these hyperparameters, we used these exact values for the remainder of our experiments.

Table 4. Manually tuned hyperparameters for the different models on the GE subset of our data. Please note that not all hyperparameters are shown.

Hyperparameter	Vehicles				Vessels			
	YOLOV2	YOLOV3	YOLT	D-YOLO	YOLOV2	YOLOV3	YOLT	D-YOLO
Learning rate	0.001 with division by factor 10 after 4000, 7000, 10,000 batches							
Momentum	0.9	0.9	0.9	0.9	0.9	0.9	0.9	0.9
Batch size	32	32	32	32	32	32	32	32
Max batches	15,000	15,000	15,000	15,000	15,000	15,000	15,000	15,000
Object Scale	5.0	7.0	6.0	10.0	5.0	5.0	5.0	5.0
No-Object Scale	1.0	3.0	1.5	3.0	1.0	2.0	2.5	2.0
Coordinate Scale	2.0	1.0	2.0	1.0	1.5	1.5	1.0	1.0
Classification Scale	1.0	1.0	1.0	1.0	1.0	1.0	1.0	1.0

3. Results

In this section, we will discuss the experiments we ran with our trained networks, to better understand the strengths and weaknesses of each detection model for satellite object detection.

3.1. Precision–Recall

The first and most obvious way to compare object detection networks is to use Precision (P) Recall (R) curves and the Average Precision (AP) metric. These metrics have been a standard for object detection for a long time and provide an easy way to compare different architectures, as well as giving an insight into different working points for using these detectors.

Looking at the AP of the four detectors on the entire dataset (see Figure 5), we can see that for the case of satellite object detection, YOLOV2 has a 20–30% lower AP than the other detectors. For the specific case of vessel detection, the YOLT detector seems to have a really interesting performance, beating both D-YOLO and YOLOV3 by over 4% and 8% respectively. These results can also be seen on the qualitative examples in Figure 6.

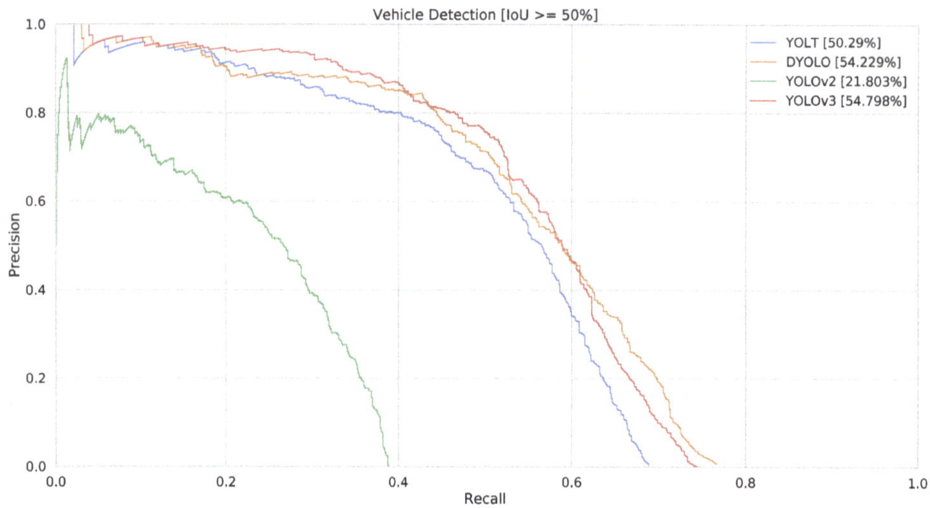

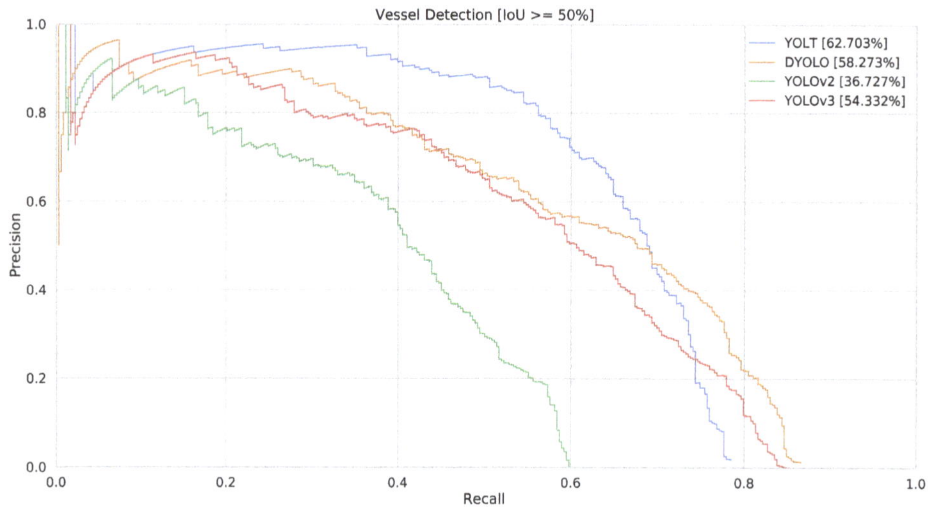

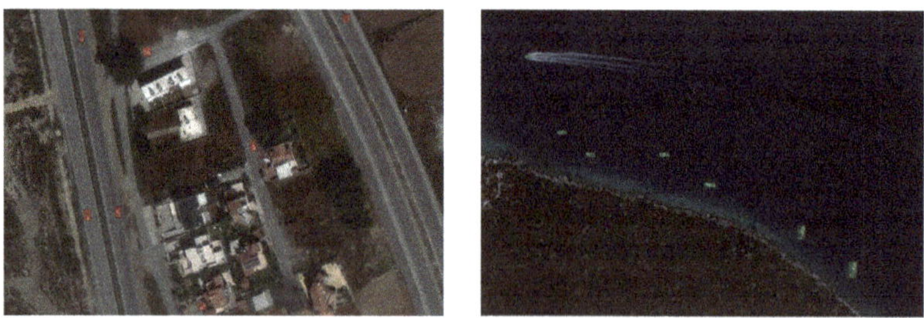

Figure 5. PR-curves of our detectors for vehicles and vessels. The IoU threshold was set at 50%. AP values are written in the legend.

(**a**) Annotation bounding boxes

Figure 6. *Cont.*

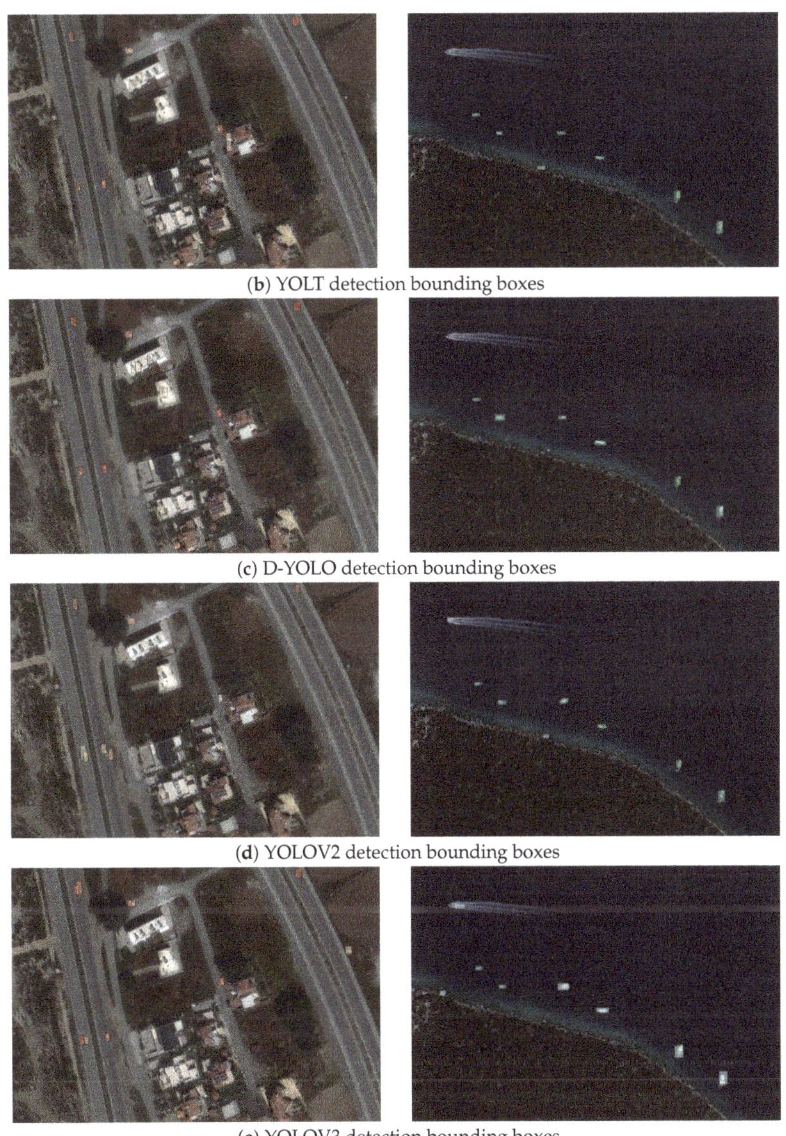

(**b**) YOLT detection bounding boxes

(**c**) D-YOLO detection bounding boxes

(**d**) YOLOV2 detection bounding boxes

(**e**) YOLOV3 detection bounding boxes

Figure 6. Qualitative comparison of our models. Darker colors mean higher confidences. Please note that we only show detections with a confidence higher than 10%. Please note that these images are small crops, taken from the full satellite images, to be able to show the objects that need to be detected. (**a**) Annotation; (**b**) YOLT; (**c**) D-YOLO; (**d**) YOLOV2; (**e**) YOLOV3.

3.2. Localization Error

Traditional PR/AP metrics work at an Intersection over Union (IoU) threshold of 50%, meaning they count a detection as a true positive as long as it overlaps with a ground truth object with an IoU of at least 50%. This means that these metrics do not allow you to assess exactly how accurate the bounding boxes from a detector are aligned with the ground truth. One way to be able to compare the localization accuracy of different detectors is to compute the AP of these detectors at different IoU thresholds, which is what we did in Figure 7.

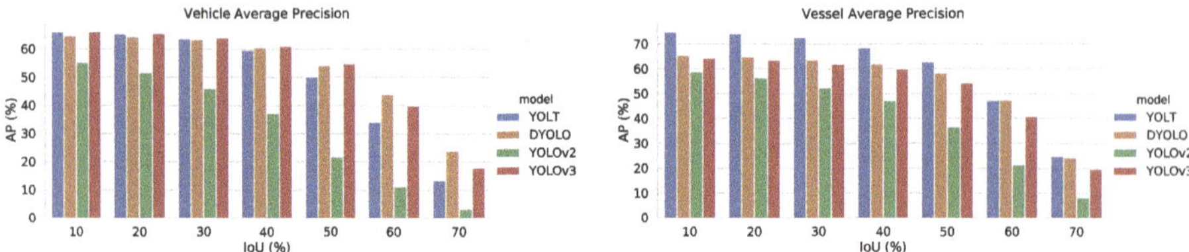

Figure 7. Average Precision of the different detectors at increasing IoU thresholds.

These figures show that D-YOLO and YOLOV3 provide the most accurate bounding boxes, as they relatively maintain their AP levels, for increasing values of IoU threshold. While YOLT seems to be on par with D-YOLO and YOLOV3 for vessels, this graph shows that for vehicles, YOLT has difficulties to accurately define bounding boxes. As vehicles are considerably smaller targets, this probably indicates that we are reaching the limits in terms of bounding box localization for YOLT.

It also shows that YOLOV2 performs considerably good at finding objects, but it demonstrates difficulty to accurately delineate them in bounding boxes. Indeed, when comparing the detectors with a lower IoU threshold of 10%, we find that YOLOV2 only has a 10–15% accuracy drop compared to the other detectors and when comparing at a threshold of 60%, this difference increases to 20–35%.

3.3. Pretrained Weights

When training neural networks for a certain task like object detection, it is known to be beneficial to start from pretrained weights from a similar network, which might have been trained for a different task. This is done to cut the training times and size of the dataset needed to train a network. In our case we start from weights from a classification network (*i.c.* Darknet19, Darknet53) trained on the ImageNet dataset [24].

However, the difference between the ImageNet dataset and the satellite images used in this study is significant, as the first is based on "natural" real-world images, while the latter depicts the world from a different view—from several kilometers above the Earth. Still, certain studies suggest that the low- and mid-level features extracted from ImageNet-trained CNNs have high potential in other tasks [25].

To evaluate the impact of pretrained weights, we compare training the networks from pretrained ImageNet weights and from pretrained weights originating from the Dataset for Object Detection in Aerial Images (DOTA) [6]. While the objects from DOTA and our dataset are not the same and not of the same size, it is highly likely that the pretrained weights from DOTA should be more attuned to satellite footage. Moreover, DOTA being an object detection dataset, these weights should also be more geared towards detection, compared to the weights from ImageNet, which is a classification dataset. However, only the first few layers can be loaded with pretrained weights from the ImageNet-trained classification networks, as they are the only that remain the same between the classification and detection networks (e.g., until the 23th layer for Darknet19/YOLOV2). To be able to compare our results, only those layers were loaded with pretrained weights from DOTA as well.

Comparing the graphs in Figure 8 with the previous ones (Figures 5 and 7), one can infer that using pretrained weights from a similar domain does offer advantages over weights from a different domain like ImageNet. Furthermore, the D-YOLO architecture seems to be outperforming YOLT in these experiments for both tasks of vehicle (+7.3%) and vessel detection (+4%), indicating that this network is more capable than YOLT, but might need more training data when starting with ImageNet weights.

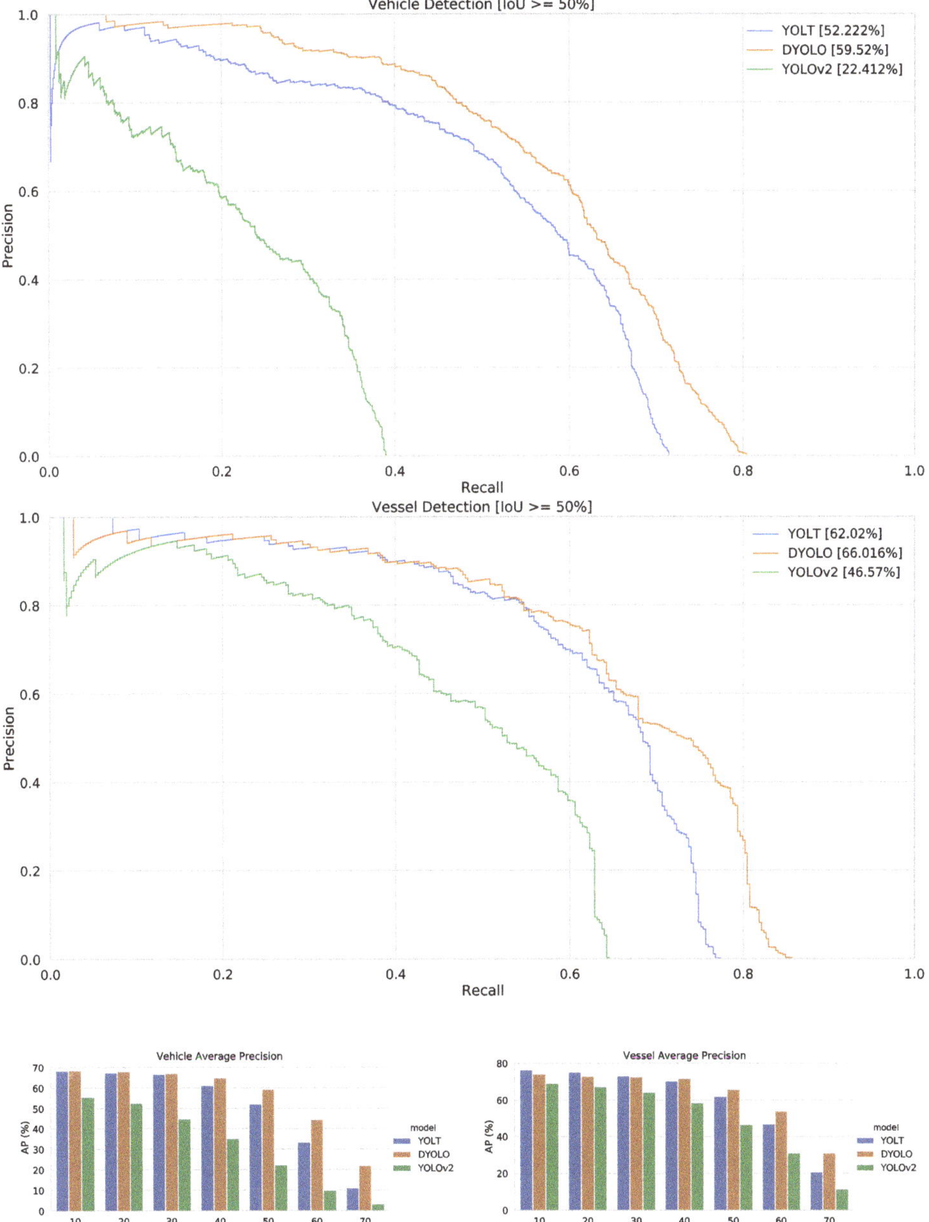

Figure 8. PR-curves and AP-IoU graphs of the different detectors trained starting from DOTA pretrained weights. Please note that the YOLOV3 network is not shown, as we could not find pretrained DOTA weights for this network.

3.4. Image Variation

Our dataset consists of images from 3 different types of satellite: WorldView (WV), GeoEye (GE) and Pleiades (PL). To test whether the different models can cope with the variation in spatial resolution, we performed an ablation study, training and testing on only subsets of our data containing certain types of satellite imagery. Bear in mind that for these training routines, we did not modify any of the hyperparameters, still using the ones adapted on only the images from the GE subset.

The results suggest that our models can cope with imagery from different sources and varying qualities (see Figures 9 and 10). When comparing the AP of our models when using the entire dataset (last row: WV+GE+PL) with the AP of a single type of satellite (first three rows: WV, GE and PL), we can see a general trend where the performance increases when we use the combined data. This indicates that there is a bigger advantage in using more images, compared to the disadvantage of added complexity, brought by combining images from different types of satellites. Thus, training on our entire dataset results in more robust detection models.

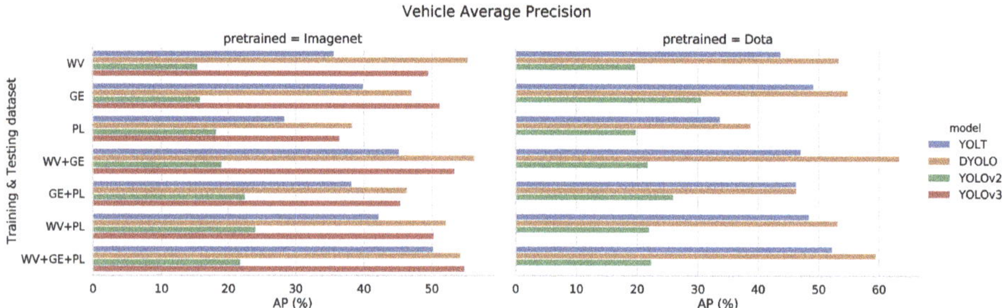

Figure 9. Average Precision of the different vehicle detectors for different subsets of our data.

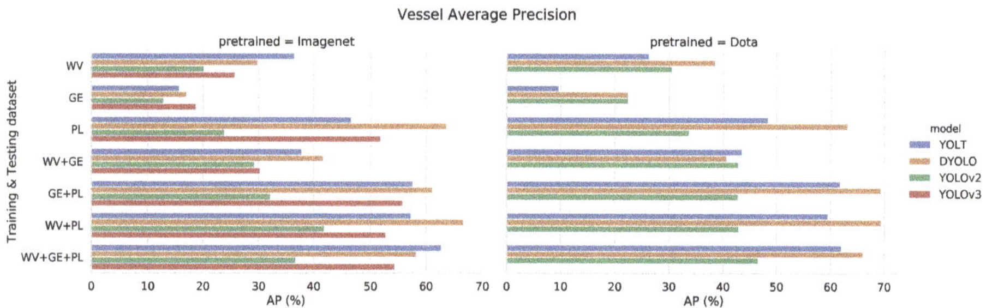

Figure 10. Average Precision of the different vessel detectors for different subsets of our data.

3.5. Multi-Label Detection

Besides being able to detect vehicles and vessels, we are also interested to know what kind of vehicle or vessel is found at a certain location. However, the problem of classifying vehicles at these resolutions remains a very challenging task even for image analysts, who use additional data or context to successfully label several object categories.

Nevertheless, we wanted to investigate whether our chosen algorithms could differentiate between these different types, solely based on visual information. Because single-shot detectors are also capable of performing classification, the most straight-forward approach was to train a multi-class detector. In this experiment, we chose the best performing detectors on the entire dataset (YOLT for vessels and D-YOLO for vehicles) and fine-tuned their training to perform multi-class detection.

Figure 11 shows the PR-curve of our single-class and multi-class detectors, when disregarding the class labels. These graphs clearly demonstrate that adding a classification task to these networks does not deteriorate their detection capabilities, but even marginally increases their detection performance.

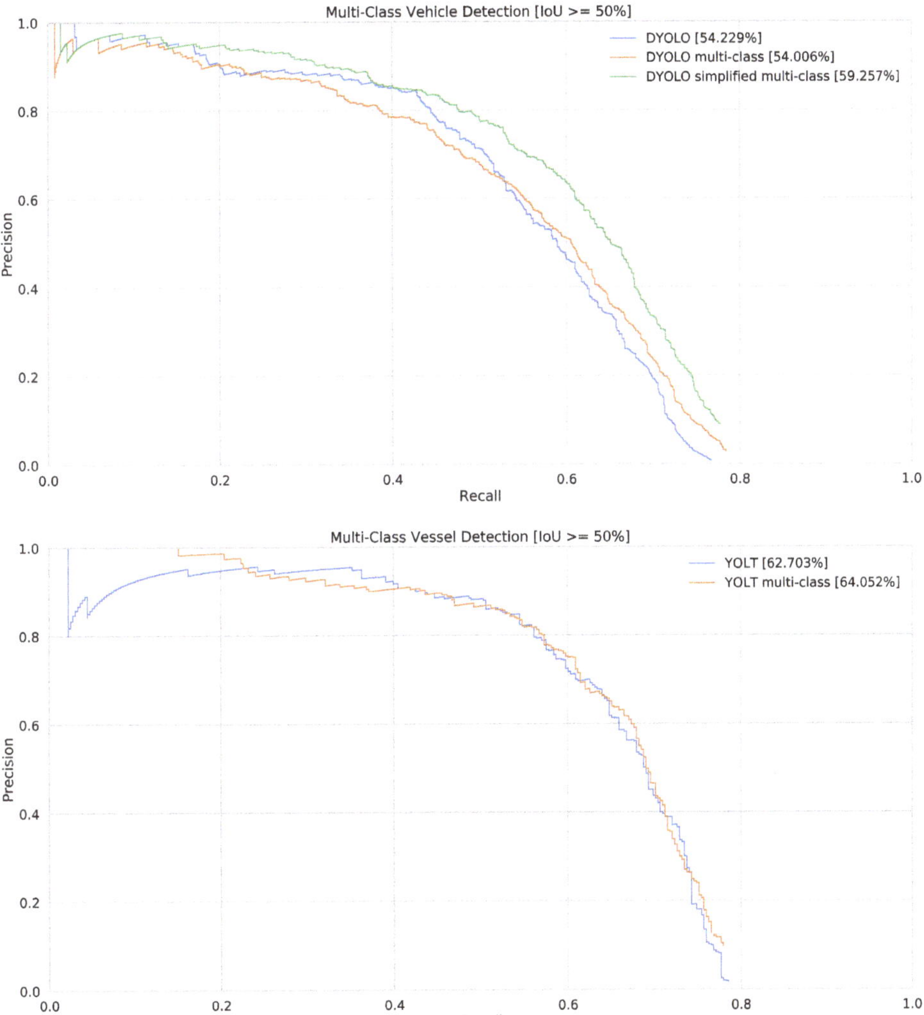

Figure 11. PR-curves of our multi-class detectors. AP values are written in the legend. Please note that for these curves, we disregard the labels, to only measure the detection accuracy.

Looking at the top-1 classification accuracy (see Table 5), the results seem promising and useful at a first glance. However, when we analyze the confusion matrices more closely (see Table 6), we can see that for the case of vehicle detection our classification pipeline just labels most of the objects as light vehicles. Our annotation statistics (see Table 3) show that there is a huge class imbalance between the different categories, with light vehicles being by far the biggest category. The detector thus learned to classify most of the objects as light vehicles, as this indeed gives the best results overall. This can also be seen when plotting the individual PR-curves of the different classes (see Figure 12).

Table 5. Top-1 accuracy of our multi-class detectors.

	Vehicle	Vessel
Top-1 (%)	58.86	68.72

Table 6. Confusion matrix of the classification results of our multi-class detectors.

Annotation \ Detection	Light Vehicle	Machinery	Bus	Truck	False Negative
Light vehicle	452	2	8	77	147
Machinery	4	4	2	3	13
Bus	3	0	4	8	1
Truck	47	8	11	68	35
False positive	14,653	3020	1428	6395	

Annotation \ Detection	Small Vessel	Medium Vessel	Large Vessel	False Negative
Small vessel	169	12	0	65
Medium vessel	17	64	0	12
Large vessel	0	4	13	2
False positive	1535	703	352	

The results for multi-class vessel detection are more promising, with most of the misclassification errors made among the small and medium vessels. As the classes were arbitrarily defined at 20 m in length, it is possible that the detector faces difficulties in classifying vessels that are around that length. The individual PR-curves of the different vessel classes are also closer to one another, demonstrating that a more balanced dataset is key to achieving good overall results.

Please note that our confusion matrices contain an extra 'False positive' row and 'False negative' row. This is because we are working with detectors and thus, they indicate erroneous and missing detections, respectively. The confidence threshold of the detectors was set to 0.1%, resulting in a lot of false positives, but we believe this is the best for showing the classification potential. Indeed, setting the threshold higher could result in an object that was correctly classified, to be discarded because its confidence value is not high enough.

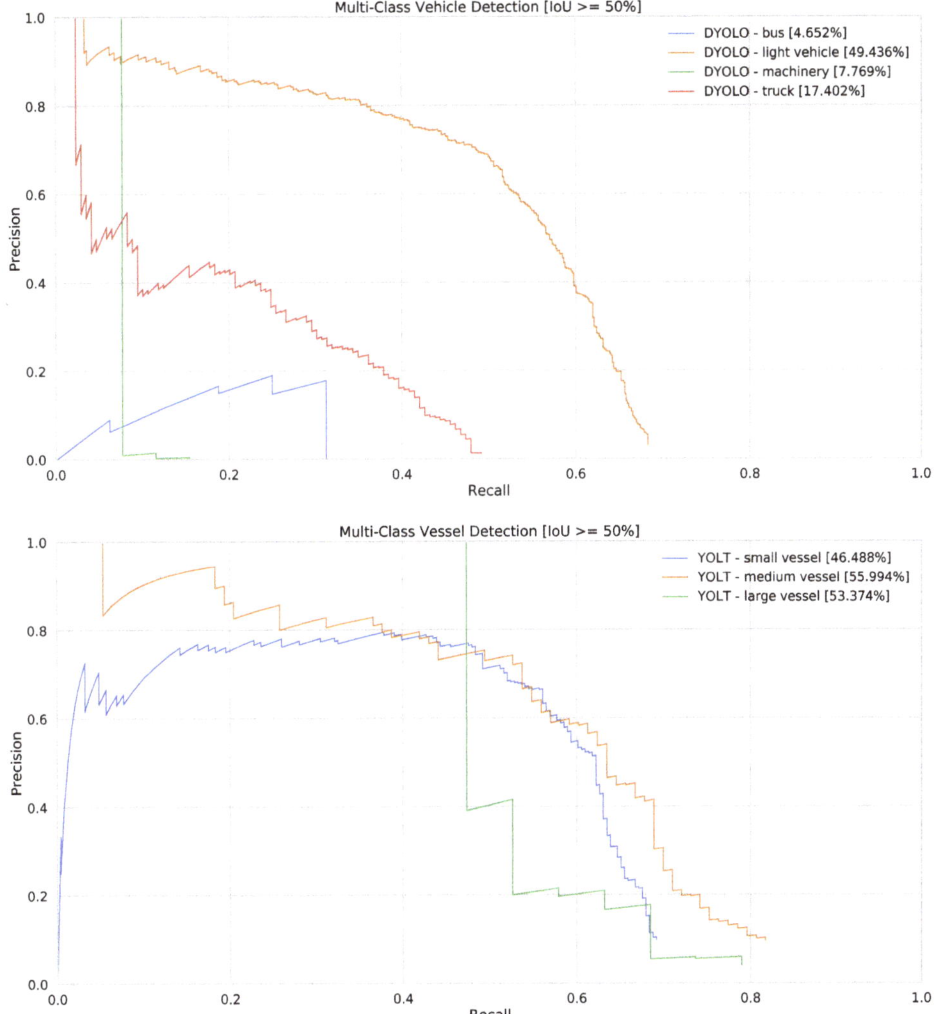

Figure 12. PR-curves per label of our multi-class detectors. AP values are written in the legend.

3.6. Speed

Besides performance, speed is another important factor which impacts the operational use of the detectors. As the current trend of satellite on-board processing is developing [26], it is important that the AI algorithms are adapted to use few resources and run in a timely fashion on the constrained hardware. Speed is also important when running the detectors in e.g., a datacenter, solely because of the sheer amount of data that needs to be processed.

To measure the speed of a detector, we need to choose a working point at which it will operate and thus we have to select a threshold to filter which bounding boxes get accepted and which do not. This has a significant influence on the post-processing time and as such it is important to set this up correctly. Because we have no specific precision or recall constraints for this study, we use the threshold with the highest F1-score of our precision and recall (see Figure 13).

Figure 14 shows the inference time for both the network and post-processing on a 416 × 416 image patch and was measured by averaging the time of running through the entire test set. The first things to note are that the tests give similar results for both vehicles and vessels and that the post-processing time can be neglected compared to the runtime of the network. When comparing the 4 different networks, we can clearly see that D-YOLO and YOLOV2 are the fastest, with an inference time of about 4 ms per patch. YOLT is slightly slower with 6 ms per patch and finally YOLOV3 is the slowest, taking twice as long to process a patch with 8 ms per patch.

The average timings of a single patch might appear to be insignificant. However, one needs to take into account that typical satellite images as the ones used in this study consist of 3500 patches on average. As a result, the processing time of one image for D-YOLO and YOLOV3 can be around 14 and 28 s respectively, which is significant and can be a determining factor for choosing one model over the other, seeing as the AP is similar for the different models (see Figure 15).

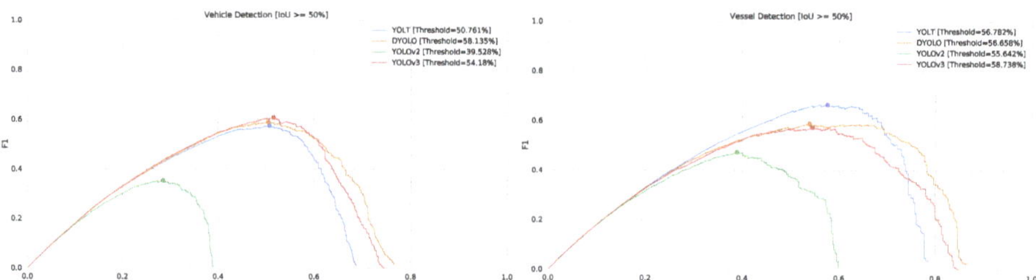

Figure 13. F1-curves of our detectors.

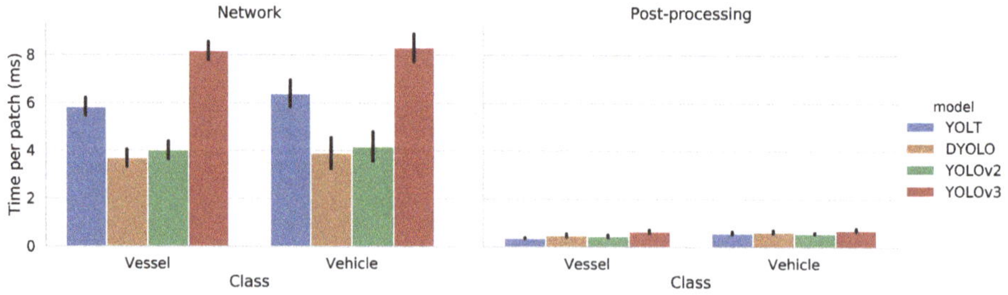

Figure 14. Inference timing results of our models with their threshold set at the best F1 value, averaged for a 416 × 416 pixel patch. This test was measured on the entire test dataset with an Nvidia GTX 1080 Ti GPU.

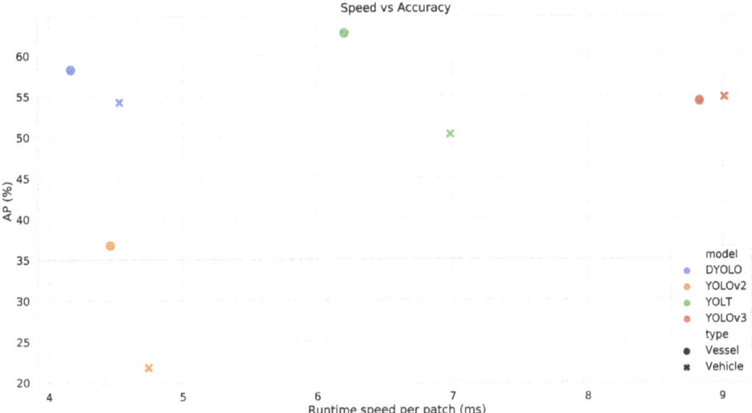

Figure 15. Speed vs. accuracy on the test set of our data. Accuracy is measured from the models with pretrained weights from ImageNet.

4. Discussion

In this study, we performed a practical implementation and experimental comparison on detection and classification of vehicles and vessels using optical satellite imagery with spatial resolutions of 0.3–0.5 m. A series of experiments was performed to understand the advantages and weaknesses of each one of the tested models.

In general, terms, YOLOV2 appears to be the less accurate to deal with the detection of small objects, while the three other networks seem to reach more similar performance (see Table 7). One remarkable finding is that for the specific case of vessel detection, YOLT outperforms both YOLOV3 and D-YOLO by a significant margin; however for vehicle detection it is the exact opposite. When looking at the accuracy of the models with pretrained DOTA weights, we can see that this difference has disappeared. This can be explained by looking at the different network architectures. The YOLT network is smaller and thus needs less data to properly train, while D-YOLO is deeper. The results from DOTA do show that with more data, D-YOLO seems to be a more capable network, reaching significantly better results than YOLT for both cases of vehicle and vessel detection. The dataset ablation study seems to support this hypothesis, as more data generally leads to better results in that experiment as well. In the future it might be interesting to also train YOLOV3 on the DOTA dataset and use those weights, as YOLOV3 is an even deeper network which might benefit from more data as well. Another alternative is to simply annotate more satellite images, creating a bigger dataset, thus eliminating the need for pretrained DOTA weights, which cannot be used outside of academic research.

Table 7. Overview of the speed and accuracy of the different models, measured on the entire test set. Speed is measured per 416×416-pixel patch.

	Vehicles				Vessels			
	YOLT	D-YOLO	YOLOV2	YOLOV3	YOLT	D-YOLO	YOLOV2	YOLOV3
$AP_{ImageNet}$ (%)	50.29	54.23	21.80	54.80	62.70	58.27	36.73	54.33
AP_{DOTA} (%)	52.22	59.52	22.41	-	62.02	66.02	46.57	-
Speed per patch (ms)	6.98	4.53	4.74	9.01	6.20	4.16	4.46	8.82

When we observe that both YOLOV3 and D-YOLO achieve a similar performance, the deciding factor might be inference time. When comparing runtime speeds, D-YOLO outperforms YOLOV3 by a factor of two and is thus a better fit for the task of satellite-based vehicle and vessel detection. This counters the general belief that deeper networks are better, as D-YOLO has significantly less convolutional layers and still reaches the same or a better accuracy in our experiments.

One might wonder whether we trained the networks properly and although we did fine-tune the hyperparameters to the best of our ability, the fact that we tuned those hyperparameters on the GE subset of our data, might not give the best results overall. We do believe this decision to be justified as it introduces less bias in the experiments, but investigating the influence of tweaking on a subset of data is something that needs more research in the future. Performing a full exhaustive search on the hyperparameters—while taking a long time—might also prove to be beneficial to reach even better accuracy, but care must be taken not to overfit the test set, which might result in a lower real accuracy.

For the case of multi-label detection, we conclude that this task is rather challenging for the detectors examined in this study. It is noted, however, that the classification scheme used here is very demanding, as regards the size of the minimum detectable object with respect to the spatial resolution of the dataset. In addition, there is a class imbalance, which makes training on specific classes very difficult, therefore a bigger dataset might be necessary to create useful results. Another issue to consider is annotator bias. This dataset was created by a single person and is thus possibly biased towards that person's background (*i.c.* computer vision engineer). While this problem is not so bad for the detection part, as vehicles and vessels are usually clearly distinguishable in this data, correctly classifying these objects has proven to be really challenging. Besides increasing the size of this dataset, it might be necessary to have these objects labeled by multiple experienced image analysts and to average the different results together. One final note about this is that while deep learning can achieve impressive results and even outperform humans at specific tasks, it cannot detect things that are completely undetectable by humans in any way. Indeed, a lot of satellite analysts use external information, like port locations or even GPS data of vessels, to label them. While this might be nice to speed up the annotation process and generate objectively better labels, there must still be some kind of visual clue the detector can exploit, to correctly classify these objects. Another approach might be to perform some kind of data fusion, combining e.g., GPS localization data with our image data, in order to create a better classification model.

5. Conclusions

The main contributions of this work are that we demonstrated the feasibility of vehicle and vessel detection in optical satellite imagery with a variable spatial resolution between 0.3 and 0.5 m, in which these objects are down to a few pixels in size. We selected four single-shot detection network architectures for their fast execution time and compared these, in the meantime optimizing the many hyperparameters for such a case of small object detection in large images. We empirically showed that there is a need to tune the hyperparameters differently for satellite object detection, to reach a good accuracy. A good understanding of the different hyperparameters is primordial for such a task, and thus we explained some of the hyperparameters specific to the loss function of single-shot detection networks. The implementation of these four models can be found in our open-source Lightnet library [22].

From our results, we can conclude that D-YOLO seems to be the most optimal detector, reaching the highest accuracy ($AP_{vehicle}$: 60%, AP_{vessel}: 66%) and fastest runtime speeds (± 4 ms per 416×416 patch). While our best results might not be good enough for fully automated detection pipelines, they can already be deployed as a tool for helping data analysts, speeding up their workflow tremendously. It is in this setting that the speed of our networks is primordial, as it promotes a convenient and fast workflow for data analysts and allows them to keep on working with the tool without long waiting times.

The problem of automatic satellite object detection can certainly not be considered solved. There is a dire need for bigger datasets with lots of variability, and even more so for the step of fine-grained classification, where we also need a more balanced dataset, which has multiple examples of each class of vehicle and vessel. The aforementioned technique of using already existing object detection models, such as the ones presented in this paper, but keeping a human in the loop as supervisor could prove to be a valuable tool in order to more easily scale up datasets, after which we might be able to create stronger models, capable of fully automatic small object detection in satellite imagery.

Author Contributions: Conceptualization, S.P., V.K. and J.-P.R.; Data curation, T.O., S.P., V.K. and J.-P.R.; Formal analysis, T.O.; Funding acquisition, S.P. and T.G.; Investigation, T.O.; Methodology, T.O. and S.P.; Project administration, S.P., V.K. and T.G.; Software, T.O.; Supervision, S.P. and T.G.; Validation, T.O., S.P., V.K. and J.-P.R.; Visualization, T.O.; Writing—original draft, T.O.; Writing—review and editing, S.P., V.K. and T.G. All authors have read and agreed to the published version of the manuscript.

Funding: This work was partially funded by the EU (Tender Lot 6 SatCen RP-01-2018) and FWO (SBO Project Omnidrone).

Acknowledgments: We thank the SADL research group of KU Leuven for their project managing efforts during this project and help with annotating.

Conflicts of Interest: The authors declare no conflict of interest. The funders provided the raw image data, outlined their needs and current workflow to assess the necessary research and, finally, helped in the writing of this manuscript.

Disclaimer: The views and opinions expressed in this article are those of the authors solely and do not reflect any official policy or position of their employers.

Abbreviations

The following abbreviations are used in this manuscript:

AI	Artificial Intelligence
AIS	Automatic Identification System
AP	Average Precision
CNN	Convolutional Neural Network
COTS	Commercial Off-The-Shelf components
D-YOLO	Deconvolutional YOLO; A variation of the YOLOV2 detector
DOTA	Dataset for Object Detection in Aerial images
GPS	Global Positioning System
GPU	Graphics Processing Unit
IMINT	Image Intelligence
IoU	Intersection over Union
PR	Precision-Recall curve
RGB	Red–Green–Blue; channels of an image
R-CNN	Regions with CNN features; An example of a two-staged detector
YOLO	You Only Look Once; An example of a single-shot detector
YOLT	You Only Look Twice; A variation of the YOLOV2 detector

References

1. Denis, G.; Claverie, A.; Pasco, X.; Darnis, J.P.; de Maupeou, B.; Lafaye, M.; Morel, E. Towards disruptions in Earth observation? New Earth Observation systems and markets evolution: Possible scenarios and impacts. *Acta Astronaut.* **2017**, *137*, 415–433. [CrossRef]
2. Van Tricht, K.; Gobin, A.; Gilliams, S.; Piccard, I. Synergistic Use of Radar Sentinel-1 and Optical Sentinel-2 Imagery for Crop Mapping: A Case Study for Belgium. *Remote Sens.* **2018**, *10*, 1642. [CrossRef]
3. Van Etten, A. You only look twice: Rapid multi-scale object detection in satellite imagery. *arXiv* **2018**, arXiv:1805.09512.

4. Acatay, O.; Sommer, L.; Schumann, A.; Beyerer, J. Comprehensive Evaluation of Deep Learning based Detection Methods for Vehicle Detection in Aerial Imagery. In Proceedings of the 15th IEEE International Conference on Advanced Video and Signal Based Surveillance (AVSS), Auckland, New Zealand, 27–30 November, 2018; pp. 1–6.

5. Peng, D.; Zhang, Y.; Guan, H. End-to-End Change Detection for High Resolution Satellite Images Using Improved UNet++. *Remote Sens.* **2019**, *11*, 1382. [CrossRef]

6. Xia, G.S.; Bai, X.; Ding, J.; Zhu, Z.; Belongie, S.; Luo, J.; Datcu, M.; Pelillo, M.; Zhang, L. DOTA: A large-scale dataset for object detection in aerial images. In Proceedings of the IEEE Conference on Computer Vision and Pattern Recognition, Salt Lake City, UT, USA, 18–22 June 2018; pp. 3974–3983.

7. Mundhenk, T.N.; Konjevod, G.; Sakla, W.A.; Boakye, K. A large contextual dataset for classification, detection and counting of cars with deep learning. In Proceedings of the 14th European Conference on Computer Vision, Amsterdam, The Netherlands, 8–16 October 2016; pp. 785–800.

8. Alganci, U.; Soydas, M.; Sertel, E. Comparative research on deep learning approaches for airplane detection from very high-resolution satellite images. *Remote Sens.* **2020**, *12*, 458. [CrossRef]

9. Weir, N.; Lindenbaum, D.; Bastidas, A.; Van Etten, A.; McPherson, S.; Shermeyer, J.; Kumar, V.; Tang, H. SpaceNet MVOI: A Multi-View Overhead Imagery Dataset. *arXiv* **2019**, arXiv:1903.12239.

10. Felzenszwalb, P.; McAllester, D.; Ramanan, D. A discriminatively trained, multiscale, deformable part model. In Proceedings of the IEEE Conference on Computer Vision and Pattern Recognition, Anchorage, AK, USA, 24–26 June 2008; pp. 1–8.

11. Dollár, P.; Tu, Z.; Perona, P.; Belongie, S. Integral Channel Features. Available online: https://authors.library.caltech.edu/60048/1/dollarBMVC09ChnFtrs.pdf (accessed on 8 April 2020).

12. Dollár, P.; Appel, R.; Belongie, S.; Perona, P. Fast feature pyramids for object detection. *IEEE TPAMI* **2014**, *36*, 1532–1545. [CrossRef]

13. Girshick, R.; Donahue, J.; Darrell, T.; Malik, J. Rich feature hierarchies for accurate object detection and semantic segmentation. In Proceedings of the IEEE Conference on Computer Vision and Pattern Recognition, Columbus, OH, USA, 23–28 June 2014; pp. 580–587.

14. Girshick, R. Fast R-CNN. In Proceedings of the 2015 IEEE International Conference on Computer Vision, Santiago, Chile, 7–13 December 2015; pp. 1440–1448. [CrossRef]

15. Ren, S.; He, K.; Girshick, R.; Sun, J. Faster R-CNN: Towards real-time object detection with region proposal networks. In Proceedings of the 28th International Conference on Neural Information Processing Systems, Montreal, QC, Canada, 7–12 December 2015; pp. 91–99.

16. Redmon, J.; Divvala, S.; Girshick, R.; Farhadi, A. You only look once: Unified, real-time object detection. In Proceedings of the 2016 IEEE Conference on Computer Vision and Pattern Recognition, Las Vegas, NV, USA, 27–30 June 2016; pp. 779–788.

17. Redmon, J.; Farhadi, A. YOLO9000: Better, Faster, Stronger. In Proceedings of the 2017 IEEE Conference on Computer Vision and Pattern Recognition, Honolulu, HI, USA, 21–26 July 2017; pp. 6517–6525. [CrossRef]

18. Redmon, J.; Farhadi, A. Yolov3: An Incremental Improvement. Technical Report, 2018. Available online: https://arxiv.org/pdf/1804.02767.pdf (accessed on 8 April 2020).

19. Pekel, J.F.; Cottam, A.; Gorelick, N.; Belward, A.S. High-resolution mapping of global surface water and its long-term changes. *Nature* **2016**, *540*, 418. [CrossRef] [PubMed]

20. Kanjir, U.; Greidanus, H.; Oštir, K. Vessel detection and classification from spaceborne optical images: A literature survey. *Remote Sens. Environ.* **2018**, *207*, 1–26. [CrossRef] [PubMed]

21. Paszke, A.; Gross, S.; Massa, F.; Lerer, A.; Bradbury, J.; Chanan, G.; Killeen, T.; Lin, Z.; Gimelshein, N.; Antiga, L.; et al. PyTorch: An Imperative Style, High-Performance Deep Learning Library. In *Advances in Neural Information Processing Systems 32*; Wallach, H., Larochelle, H., Beygelzimer, A., d'Alché Buc, F., Fox, E., Garnett, R., Eds.; Curran Associates, Inc.: New York, NY, USA, 2019; pp. 8024–8035.

22. Ophoff, T. Lightnet: Building Blocks to Recreate Darknet Networks in Pytorch. Available online: https://gitlab.com/EAVISE/lightnet (accessed on 8 April 2020).

23. Lin, T.Y.; Dollár, P.; Girshick, R.; He, K.; Hariharan, B.; Belongie, S. Feature pyramid networks for object detection. In Proceedings of the 2017 IEEE Conference on Computer Vision and Pattern Recognition, Honolulu, HI, USA, 21–26 July 2017; pp. 2117–2125.

24. Deng, J.; Dong, W.; Socher, R.; Li, L.J.; Li, K.; Fei-Fei, L. ImageNet: A Large-Scale Hierarchical Image Database. In Proceedings of the 2009 IEEE Conference on Computer Vision and Pattern Recognition, Miami, Florida, USA, 20–25 June 2009.
25. Huh, M.; Agrawal, P.; Efros, A.A. What makes ImageNet good for transfer learning? *arXiv* **2016**, arXiv:1608.08614.
26. Lansing, F.; Lemmerman, L.; Walton, A.; Bothwell, G.; Bhasin, K.; Prescott, G. Needs for communications and onboard processing in the vision era. In Proceedings of the IEEE International Geoscience and Remote Sensing Symposium, Toronto, ON, Canada, 24–28 June 2002; Volume 1, pp. 375–377.

 *remote sensing*

Article

Semantic Labeling in Remote Sensing Corpora Using Feature Fusion-Based Enhanced Global Convolutional Network with High-Resolution Representations and Depthwise Atrous Convolution

Teerapong Panboonyuen [1], Kulsawasd Jitkajornwanich [2], Siam Lawawirojwong [3], Panu Srestasathiern [3] and Peerapon Vateekul [1,*]

1   Chulalongkorn University Big Data Analytics and IoT Center (CUBIC), Department of Computer Engineering, Faculty of Engineering, Chulalongkorn University, Phayathai Rd, Pathumwan, Bangkok 10330, Thailand; teerapong.panboonyuen@gmail.com
2   Data Science and Computational Intelligence (DSCI) Laboratory, Department of Computer Science, Faculty of Science, King Mongkut's Institute of Technology Ladkrabang, Chalongkrung Rd, Ladkrabang, Bangkok 10520, Thailand; kulsawasd.ji@kmitl.ac.th
3   Geo-Informatics and Space Technology Development Agency (Public Organization), 120, The Government Complex, Chaeng Wattana Rd, Lak Si, Bangkok 10210, Thailand; siam@gistda.or.th (S.L.); panu@gistda.or.th (P.S.)
*   Correspondence: peerapon.v@chula.ac.th

Received: 5 March 2020; Accepted: 9 April 2020; Published: 12 April 2020

Abstract: One of the fundamental tasks in remote sensing is the semantic segmentation on the aerial and satellite images. It plays a vital role in applications, such as agriculture planning, map updates, route optimization, and navigation. The state-of-the-art model is the Enhanced Global Convolutional Network (GCN152-TL-A) from our previous work. It composes two main components: (*i*) the backbone network to extract features and (*ii*) the segmentation network to annotate labels. However, the accuracy can be further improved, since the deep learning network is not designed for recovering low-level features (e.g., river, low vegetation). In this paper, we aim to improve the semantic segmentation network in three aspects, designed explicitly for the remotely sensed domain. First, we propose to employ a modern backbone network called "High-Resolution Representation (HR)" to extract features with higher quality. It repeatedly fuses the representations generated by the high-to-low subnetworks with the restoration of the low-resolution representations to the same depth and level. Second, "Feature Fusion (FF)" is added to our network to capture low-level features (e.g., lines, dots, or gradient orientation). It fuses between the features from the backbone and the segmentation models, which helps to prevent the loss of these low-level features. Finally, "Depthwise Atrous Convolution (DA)" is introduced to refine the extracted features by using four multi-resolution layers in collaboration with a dilated convolution strategy. The experiment was conducted on three data sets: two private corpora from Landsat-8 satellite and one public benchmark from the "ISPRS Vaihingen" challenge. There are two baseline models: the Deep Encoder-Decoder Network (DCED) and our previous model. The results show that the proposed model significantly outperforms all baselines. It is the winner in all data sets and exceeds more than 90% of $F1$: 0.9114, 0.9362, and 0.9111 in two Landsat-8 and ISPRS Vaihingen data sets, respectively. Furthermore, it achieves an accuracy beyond 90% on almost all classes.

Keywords: deep learning; convolutional neural network; global convolution network; feature fusion; depthwise atrous convolution; high-resolution representations; ISPRS vaihingen; Landsat-8

1. Introduction

Semantic segmentation in a medium resolution (MR) image, e.g., a Landsat-8 (LS-8) image, and very high resolution (VHR) images, e.g., aerial images, is a long-standing issue and problem in the domains of remote sensing-based information. Natural objects such as roads, water, forests, urban, and agriculture fields regions are operated in various tasks such as route optimization to create imperative remotely sensed applications.

Deep learning, especially the Deep Convolutional Neural Network (CNN), is an acclaimed approach for automatic feature learning. In previous research, CNN-based segmentation approaches are proposed to perform semantic labeling [1–5]. To achieve such a challenging task, features from various levels are fused together [5–7]. Specifically, a lot of approaches fuse low-level and high-level features together [5–9]. In remote sensing corpora, ambiguous human-made objects need high-level features for a more well-defined recognition (e.g., roads, building roofs, and bicycle runways), while fine-structured objects (e.g., low vegetations, cars, and trees) could benefit from comprehensive low-level features [10]. Consequently, the performance will be affected by the different numbers of layers and/or different fusion techniques of the deep learning model.

In recent years, the Global Convolutional Network (GCN) [11], the modern CNN, has been introduced, in which the valid receptive field and large filter enable dense connections between pixel-based classifiers and activation maps, which enhances the capability to cope with different transformations. The GCN is aimed at addressing both the localization and segmentation problems for image labeling and presents Boundary Refinement (BR) to refine the object boundaries further as well. Our previous work [12] extended the GCN by enhancing three approaches as illustrated in Figures 1 and 2. First, "Transfer Learning" [13–15] was employed to relieve the shortage problem. Next, we varied the backbone network using ResNet152, ResNet101, and ResNet50. Last, "Channel Attention Block" [16,17] was applied to allocate CNN parameters for the output of each layer in the front-end of the deep learning architecture.

Nevertheless, our previous work still disregards the local context, such as low-level features in each stage. Moreover, most feature fusion methods are just a summation of the features from adjacent stages and they do not consider the representations of diversity (critical for the performance of the CNN). This leads to unpredictable results that suffer from measuring the performance such as the $F1$ score. This, in fact, is the inspiration for this work.

In summary, although the current enhanced Global Convolutional Network (GCN152-TL-A) method [12] has achieved significant breakthroughs in semantic segmentation on remote sensing corpora, it is still laborious to manually label the MR images in river and pineapple areas and the VHR images in low vegetation and car areas. The two reasons are as follows: (i) previous approaches are less efficient to recover low-level features for accurate labeling, and (ii) they ignore the low-level features learned by the backbone network's shallow layers with long-span connections, which is caused by semantic gaps in different-level contexts and features.

In this paper, motivated by the above observation, we propose a novel Global Convolutional Network ("HR-GCN-FF-DA") for segmenting multi-objects from satellite and aerial images, as illustrated in Figure 3. This paper aims to further improve the state-of-the-art on semantic segmentation in MR and VHR images. In this paper, there are three contributions, as follows:

- Applying a new backbone called "High-Resolution Representation (HR)" to GCN for the restoration of the low-resolution representations of the same depth and similar level.
- Proposing the "Feature Fusion (FF)" block into our network to fuse each level feature from the backbone model and the global model of GCN to enrich local and global features.
- Proposing "Depthwise Atrous Convolution (DA)" to bridge the semantic gap and implement durable multi-level feature aggregation to extract complementary information from very shallow features.

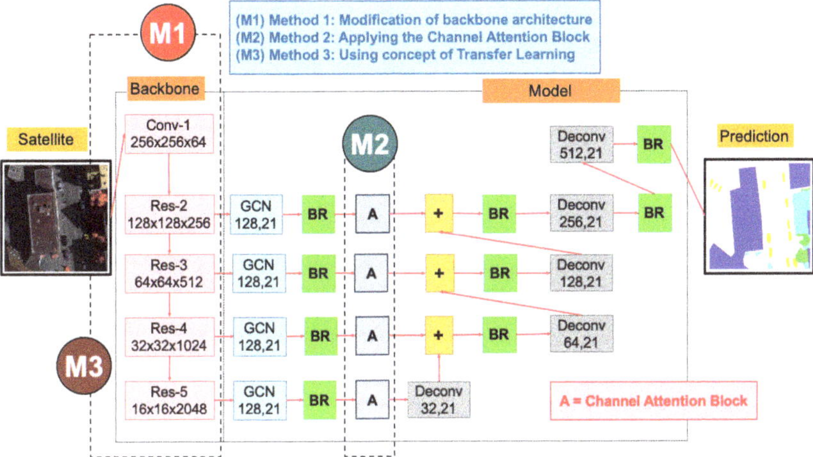

Figure 1. An overview of enhanced GCN (Global Convolution Network [11]) with transfer learning and attention mechanism (GCN152-TL-A) [12].

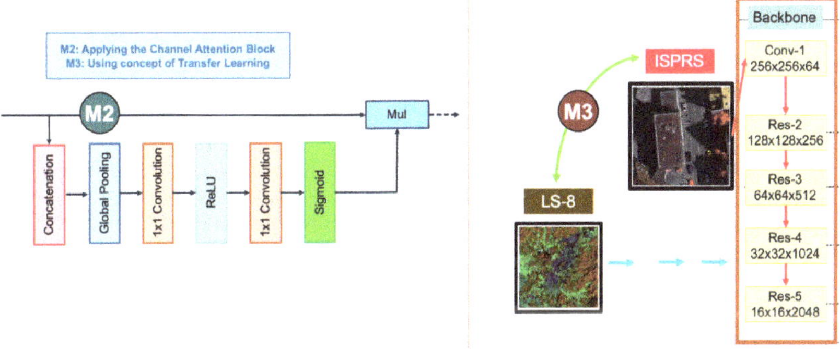

Figure 2. An Attention Mechanism (A) block (left) and the Transfer Learning (TL) approach (right) transfer knowledge (from pre-trained weights) across two corpora—medium and very high resolution images from [12].

The experiments were conducted using the widespread aerial imagery, ISPRS (Stuttgart) Vaihingen [18] data set and GISTDA (Geo-Informatics and Space Technology Development Agency (Public Organization)), organized by the government in our country, data sets (captured by the Landsat-8 satellite). The results revealed that our proposed method surpasses the two baselines: Deep Convolutional Encoder-Decoder Network (DCED) [19–21] and the enhanced Global Convolutional Network (GCN152-TL-A) method [12] in terms of $F1$ score.

The remainder of this paper is organized as follows: Section 2 discusses related work. Our proposed methods are detailed in Section 3. Next, Section 4 provides the details on remote sensing corpora. Section 5 presents our performance evaluation. Then, Section 6 reports the experimental results, and Section 7 is the discussion. Last, we close with the conclusions in Section 8.

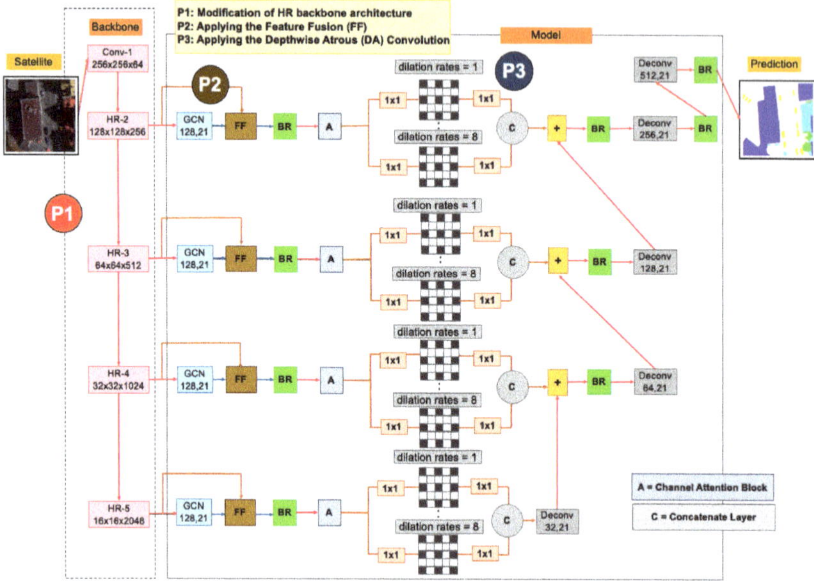

Figure 3. The HR-GCN-FF-DA: an enhanced GCN architecture with feature fusion and depthwise atrous convolution.

2. Related Work

The CNN has been outstandingly utilized for the data analysis of remote sensing domains, in particular, land cover classification or segmentation of agriculture or forest districts [10,12,22–26]. It has rapidly become a successful method for accelerating the process of computer vision tasks, e.g., image classification, object detection, or semantic segmentation with high precision results [4,27–33] and is a fast-growing area.

It is separated into two subsections: (*i*) we demonstrate modern CNN architectures for semantic labeling on both traditional computer vision and remote sensing tasks and (*ii*) the novel techniques of deep learning, especially playing with images, are discussed.

2.1. Modern CNN Architecture for Semantic Labeling

In early research, several DCED-based approaches have obtained a high performance in the various baseline corpora [16,19–21,26,34–36]. Nevertheless, most of them also struggle with issues with performance accuracy. Consequently, much research on novel CNN architectures has been introduced, such as a high-resolution representation [37,38] network that supports high-resolution representations in all processes by connecting high-to-low and low-to-high-resolution convolutions to keep high and low-resolution representations. CSRNet [8] proposed an atrous (dilated) CNN to comprehend highly congested scenes through crowd counting and generating high-quality density maps. They deployed the first ten layers from VGG-16 as the backbone convolutional models and dilated convolution layers as the backend to enlarge receptive fields and extract deeper features without losing resolutions. SeENet [6] enhanced shallow features to alleviate the semantic gap between deep features and shallow features and presented feature attention, which involves discovering complementary information from low-level features to enhance high-level features for precise segmentation. It also was constructed with the parallel pyramid to implement precise semantic segmentation. ExFuse [7] proposed to boost the feature fusion by bridging the semantic and resolution gap between low-level and high-level feature maps. They proposed more semantic information into low-level features with three aspects:

(*i*) semantic supervision, (*ii*) semantic embedding branch, and (*iii*) layer rearrangement. They also embedded spatial information into high-level features. In the remote sensing corpus, ResUNet [25] proposed a trustworthy structure for performance effects for the job of image labeling of aerial images. They used a VGG16 network as a backbone, combined with the pyramid scene parsing pooling and dilated deep neural network. They also proposed a new generalized dice loss for semantic segmentation. TreeUNet (also known as adaptive tree convolutional neural networks) [24] proposed a tree-cutting algorithm and an adequate deep neural network with inadequate binary links to increase the classification percentage at the pixel level for subdecimeter aerial imagery segmentation, by sending kernel maps within concatenating connections and fusing multi-scale features. From the ISPRS Vaihingen Challenge and Landsat-8 corpus, the enhanced Global Convolutional Network (also known as "GCN152-TL-A"), illustrated in Figure 1, Panboonyuen et al. (2019) [12] presented an enhanced GCN for semantic labeling with three main contributions. First, "Domain-Specific Transfer Learning" (TL) [13–15], illustrated in Figure 2 (right), aims to restate the weights obtained from distinct fields' inputs. It is currently prevalent in various tasks, such as Natural Language Processing (NLP), and has also become popular in Computer Vision (CV) in the past few years. It allows you to reach a deep learning model with comparatively inadequate data. They prefaced to relieve the lack of issue on the training set by appropriating other remote sensing data sets with various satellites with an essentially pre-trained weight. Next, "Channel Attention", shown in Figure 2 (left), proposed with their network to select the most discriminative kernels (feature maps). Finally, they enhanced the GCN network by improving its backbone by using "ResNet152". "GCN152-TL-A" has surpassed state-of-the-art (SOTA) approaches and become the new SOTA. Hence, "GCN152-TL-A" is selected as our baseline in this work.

2.2. Modern Technique of Deep Learning

A novel technique of deep learning is an essential agent for improving the precision of deep learning, especially the CNN. While the most prevalent contemporary designs tick all the boxes for image labeling responsibilities, e.g., the atrous convolution (also known as dilated convolution), channel attention mechanism, refinement residual block, and feature fusion, and have been utilized to boost the performance of the deep learning model.

Atrous convolution [5,6,9,39,40], also known as multi-scale context aggregation, is proposed to regularly aggregate multi-scale contextual information devoid of losing resolution. In this paper, we use the technique of "Depthwise Atrous Convolution (DA)" [6] to extract complementary information from very shallow features and enhance the deep features for improving feature fusion from our feature fusion step.

The channel attention mechanism [16,17] generates a one-dimensional tensor for allowed feature maps, which is activated by the softmax function. It focuses on global features found in some feature maps and has attracted broad interest in extracting rich features in the computer vision domain and offers great potential in improving the performance of the CNN. In previous work, GCN152-TL-A [12], the self attention and utilize channel attention modules are applied to pick the features similar to [16].

Refinement residual block [16] is part of the enhanced CNN model with ResNet-backbones, e.g., ResNet101 or ResNet152. This block is used after the GCN module and during the deconvolution layer. It is used to refine the object boundaries further. In our previous work, GCN152-TL-A [12], we employed the boundary refinement block (BR) that is based on the "Refinement Residual Block" from [11].

Feature fusion [7,41–44] is regularly manipulated in semantic labeling for different purposes and concepts. It presents a concept that combines multiplied, added, or concatenate CNN layers for improving a process of dimensionality reduction to recover and/or prevent the loss of some important features such as low-level features (e.g., lines, dots, or gradient orientation with the content of an image scene). In another way, it can also recover high-level features by using the technique of "high-to-low and low-to-high" [37,38] to produce high-resolution representations.

3. Proposed Method

Our proposed deep learning architecture, "HR-GCN-FF-DA", is demonstrated in an overview architecture in Figure 3. The network, based on GCN152-TL-A [12], consists primarily of three parts: (*i*) changing the backbone architecture (the P1 block in Figure 3), (*ii*) implementing the "Feature Fusion" (the P2 block in Figure 3), and (*iii*) using the concept of "Depthwise Atrous Convolution" (the P3 block in Figure 3).

3.1. Data Preprocessing and Augmentation

In this work, three benchmarks were used with the experiments, these were the (*i*) Landsat-8w3c, (*ii*) Landsat-8w5c, and (*iii*) ISPRS Vaihingen (Stuttgart) Challenge data sets. Before a discussion about the model, it is important to deploy a data preprocessing, e.g., pixel standardization, scale pixel values (to have unit variance), and a zero mean into the data sets. In the image domain, the mean subtraction, calculated by the per-channel mean from the training set, is executed in order to improve the model convergence.

Furthermore, a data augmentation (also known as the "ImageDataGenerator" function in TensorFlow/Keras library) is employed, since it can help the model to avoid an overfitting issue and somewhat enlarge the training data—a strategy used to increase the amount of data. To augment the data, each image is width and height-shifted and flipped horizontally and vertically. Then, unwanted outer areas are removed into 512×512 pixels with a resolution of 81 cm^2/pixel in the ISPRS and 900 m^2/pixel in the Landsat-8 data set.

3.2. The GCN with High-Resolution Representations (HR) Front-End

The GCN152-TL-A [12], as shown in Figure 1, is our prior attempt that surpasses a traditional semantic segmentation model, e.g., deep convolutional encoder-decoder (DCED) networks [19–21]. By using GCN as our core model, our previous work was improved in three aspects. First, its backbone was revised by varying ResNet-50, ResNet-101, and ResNet-152 networks, as shown in M1 in Figure 1. Second, the "Channel Attention Mechanism" was employed (shown in M2 in Figure 1). Third, the "Domain-Specific Transfer Learning" (TL) was employed to reuse the pre-trained weights obtained from training on other data sets in the remote sensing domain. This strategy is important in the deep learning domain to overcome the limited amount of training data. In our work, there are two main data sets: Landsat-8 and ISPRS. To train the Landsat-8 model, the pre-trained network is obtained by utilizing the ISPRS data. This can also be explained conversely—the pre-trained network can be obtained by Landsat-8 data.

Although the GCN152-TL-A network has determined an encouraging forecast performance, it can still be possible to improve it further through changing the frontend using high-resolution representation (HR) [37,38] instead of ResNet-152 [25,45]. HR has surpassed all existing deep learning methods on semantic segmentation, multi-person pose estimation, object detection, and pose estimation tasks in the COCO, which is large-scale object detection, segmentation, and captioning corpora. It is a parallel structure to enable the deep learning model to link multi-resolution subnetworks in an effective and modern way. HR connects high-to-low subnetworks in parallel. It maintains high-resolution representations through the whole process for a spatially precise heatmap estimation. It creates reliable high-resolution representations through repeatedly fusing the representations generated by the high-to-low subnetworks. It introduces "exchange units" which shuttle across different subnetworks, enabling each one to receive information from other parallel subnetworks. Representations of HR can be obtained by repeating this process. There are four stages as the 2nd, 3rd, 4th, and 5th stages are formed by repeating modularized multi-resolution blocks. A multi-resolution block consists of a multi-resolution group convolution and a multi-resolution convolution, which is illustrated as P1 in Figure 3 (backbone model) and this proposed method is named the "HR-GCN" method.

3.3. Feature Fusion

Inspired by the idea of feature fusion [41–44] that integrates multiplication, additional, or concatenate layers. Convolution with 1×1 filters is used to transform features with different dimensions into the shape, which can be fused. The fusion method contains an addition process. Each layer of the backbone network such as VGG, Inception, ResNet, or HR creates the feature map for specific. We proposed to combine output with low-level features (front-end network) with the deep model and refine the feature information.

As shown in Figure 4, the kernel maps after fusing will be calculated as Equation (1) :

$$Z_{add} = X_1 \oplus X_2 \oplus X_3 \cdots \oplus X_i \cdots \oplus X_j \tag{1}$$

where j adverts to the index of the layer, X_k is a set of output activation maps of one layer and $\oplus$ adverts to element-wise addition.

Hence, the nature of the addition process encourages essential information to build classifiers to comprehend the feature details. It denotes all bands of Z_{add} to hold more feature information.

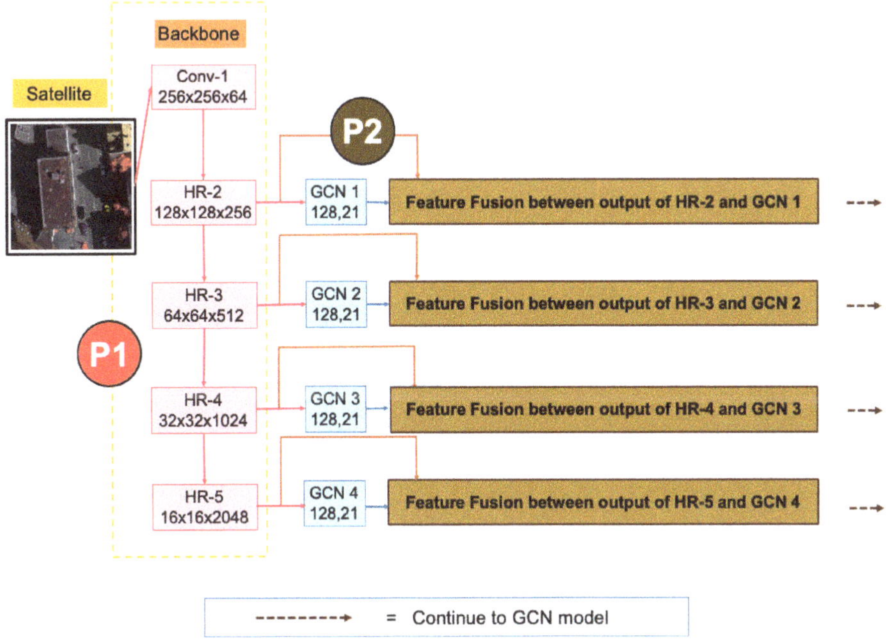

Figure 4. The framework of our feature fusion strategy.

Equation (2) shows the relationship between input and output. Thus, we take the fusion activation map into the model again, it can be performed as Equation (4):

$$\bar{y}^i = ReLU(w^T x^i + b) \tag{2}$$

where x is the input and output of layer of the convolution recorded as y^i; b and w refer to bias and weight. The cost function in this work is demonstrated via Equation (3).

$$J(w,b) = -\frac{1}{m} \times [(1 - y^{(i)})log(1 - \bar{y}^{(i)} + (y^{(i)}log(\bar{y}^{(i)})] \tag{3}$$

where y refers to segmentation target of input (each image) and J, w, and b are the loss, weight, and bias value, respectively.

$$Y_{add} = f(W_k Z_{add} + B_k) \tag{4}$$

The feature fusion procedure always transforms into the same thing when using additional procedures. In this work, we use addition fusion elements, as shown in Figure 4.

3.4. Depthwise Atrous Convolution (DA)

Depthwise Atrous Convolution (DA) [6,9,39] is presented to settle the contradictory requirements between the larger region of the input space that affects a particular unit of the deep network (receptive fields) and activation map resolution.

DA is a robust operation to reduce the number of parameters (weights) in the layer of the CNN while maintaining a similar performance that includes the computation cost and tunes the kernel's field-of-view in order to capture a generalized standard convolution operation and multi-scale information. An atrous filter can be a dilated kernel in varied rates, e.g., rate = 1, 2, 4, 8, by inserting zeros into appropriate positions in the kernel mask.

Basically, the DA module uses atrous convolutions to aggregate multi-scale contextual information without dissipating resolution orderly in each layer. It generalizes "Kronecker-factored" convolutional kernels, and it allows for broad receptive fields, while only expanding the number of weights logarithmically. In other words, DA can apply the same kernel at distinct scales using various atrous factors.

Compared to the ordinary convolution operator, atrous (dilated) convolution is able to achieve a larger receptive field size without increasing the numbers of kernel parameters.

Our motivation is to apply DA to solve challenging scale variations and to trade off precision in aerial and satellite images, as shown in Figure 5.

In a one-dimensional (1D) case, let $x[i]$ denote input signal, and $y[i]$ denote output signal. The dilated convolution is formulated as Equation (5):

$$y[i] = \sum_{j=1}^{J} x[i + a \cdot k] \cdot w[j] \tag{5}$$

where a is the atrous (dilated) rate, $w[j]$ denotes the j-th parameter of the kernel, and J is the filter size. This equation reduces to a standard convolution when d = 1, 2, 4, and 8, respectively.

In the cascading mode from DeepLabV3 [46,47] and Atrous Spatial Pyramid Pooling (ASPP) [9], multi-scale contextual information can be encoded by probing the incoming features with dilated convolution to capture sharper object boundaries by continuously recovering the spatial characteristic. DA has been applied to increase the computational ability and achieve the performance by factorizing a traditional convolution into a depth-wise convolution followed by a point-wise convolution, such as 1×1 convolution (it is often applied on the low-level attributes to decrease the whole of the bands (kernel maps)).

To simplify notations, $H_{J,a}(x)$ is term of a dilated convolution, and ASPP can be performed as Equation (6).

$$y = H_{3,1}(x) + H_{3,2}(x) + H_{3,4}(x) + H_{3,8}(x) \tag{6}$$

To improve the semantics of shallow features, we apply the idea of multiple dilated convolution with different sampling rates to the input kernel map before continuing with the decoder network and adjusting the dilation rates (1, 2, 4, and 8) to configure the whole process of our proposed method called "HR-GCN-FF-DA", shown in P3 in Figures 3 and 5.

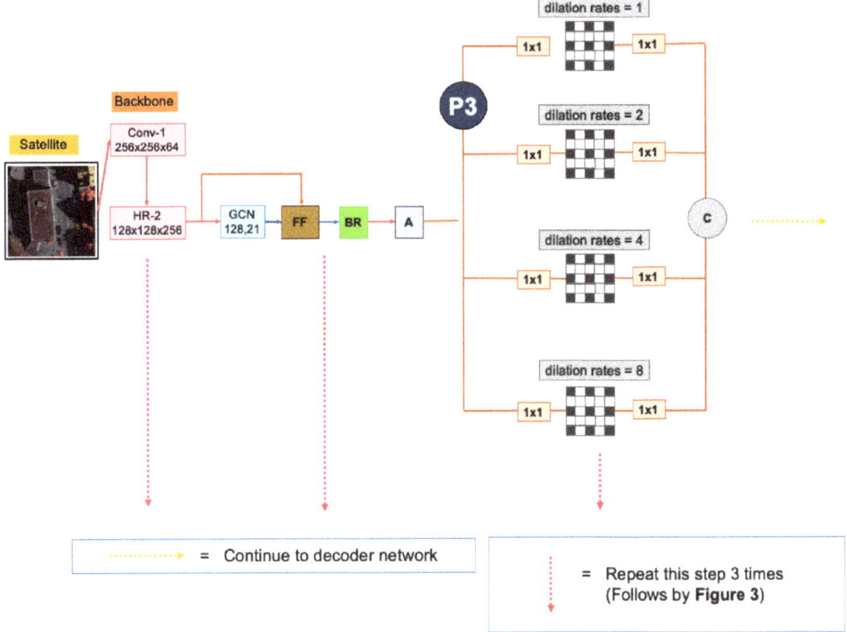

Figure 5. The Depthwise Atrous Convolution (DA) module in the proposed parallel pyramid method for improving feature fusion.

4. Remote Sensing Corpora

In our experiments, there are two main sources of data: public and private corpora. The private corpora is the medium resolution imagery received from the satellite "Landsat-8" used by the government organization in Thailand called GISTDA. Since there are two variations of annotations, the Landsat-8 data is considered as two data sets: one with three classes and the other with five classes, as shown in Table 1. The public corpora is very high-resolution imagery from the standard benchmark called "ISPRS Vaihingen (Stuttgart)". Evaluations based on classification/segmentation metrics, e.g., *F1 Score*, *Precision*, *Recall* and *Average Accuracy* are deployed with all experiments.

Table 1. Abbreviations on our Landsat-8 corpora.

Abbreviation	Description
Landsat-8w3c corpus	Landsat-8 corpus with 3 classes
Landsat-8w5c corpus	Landsat-8 corpus with 5 classes

4.1. Landsat-8w3c Corpus

For this corpus, there is a new benchmark that differs from our previous work. All images are taken in the area of the northern provinces (Changwat) of Thailand. The data set is made from the Landsat-8 satellite consisting of 1420 satellite images, some samples are shown in Figure 6. This data set contains a massive collection of medium resolution imagery of (20,921 × 17,472) pixels. There are three classes: para rubber (red), pineapple (green), and corn (yellow). From a total of 1390 images, the images are separated into 1000 training and 230 validation images, as well as 190 test images to compare with other baseline methods.

4.2. Landsat-8w5c Corpus

This data set is the same corpus from Landsat-8, but it is annotated with five class labels: agriculture, forest, miscellaneous (misc), urban, and water as shown in Figure 7. There are 1012 medium resolution satellite images of 17,200 × 16,300 pixels. From the total 1039 images, the images are separated into 700 training and 239 validation images, as well as 100 test images to comparison to other baseline methods.

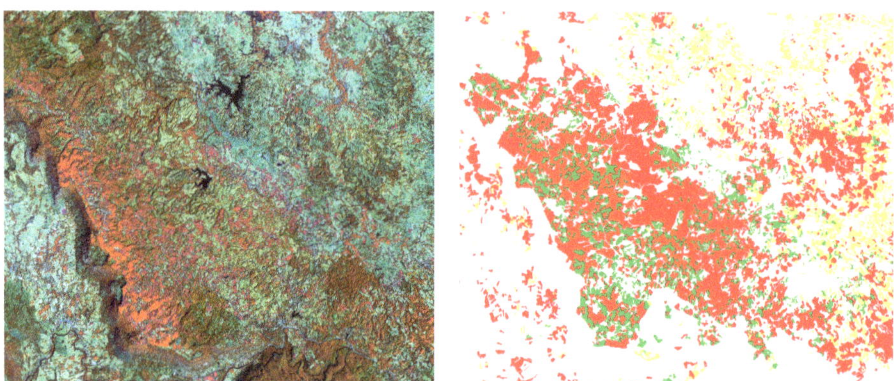

Figure 6. The example of satellite images from the Landsat-8w3c corpus, northern province (left) and target image (right). The ground-truth of the medium resolution data set includes three classes: para rubber (red), pineapple (green), and corn (Yellow).

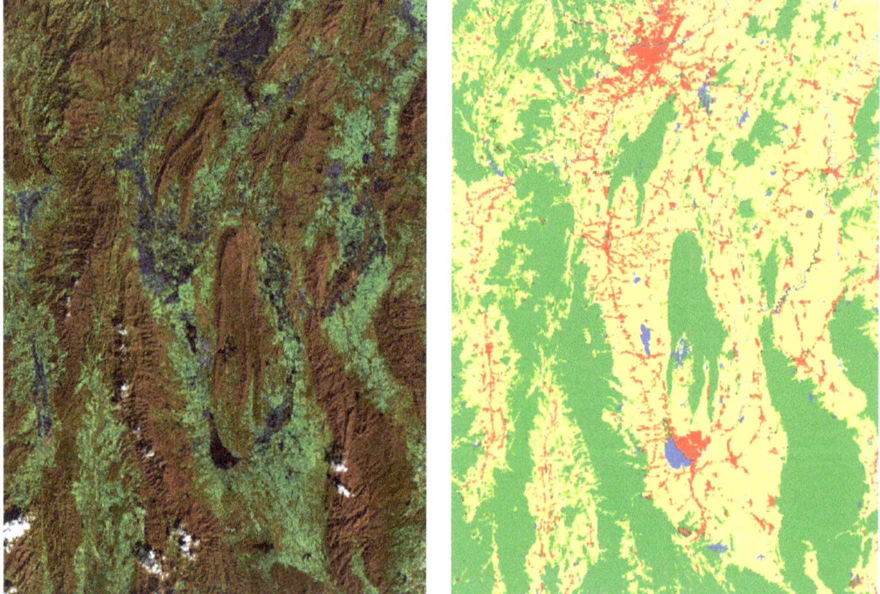

Figure 7. The example of satellite images from Landsat-8w5c corpus, northern province (left) and target image (right). The ground-truth of medium resolution data set includes five classes: urban (red), forest (green), water (blue), agriculture or harvested area (yellow), and miscellaneous or misc (brown).

4.3. ISPRS Vaihingen Corpus

The challenge of ISPRS semantic segmentation at Vaihingen (Stuttgart) [18] (Figures 8 and 9) is used to be our standard corpus. They were captured over Vaihingen in Germany. The data set is a subset of the data used for the test of digital aerial cameras carried out by the German Association of Photogrammetry and Remote Sensing (DGPF).

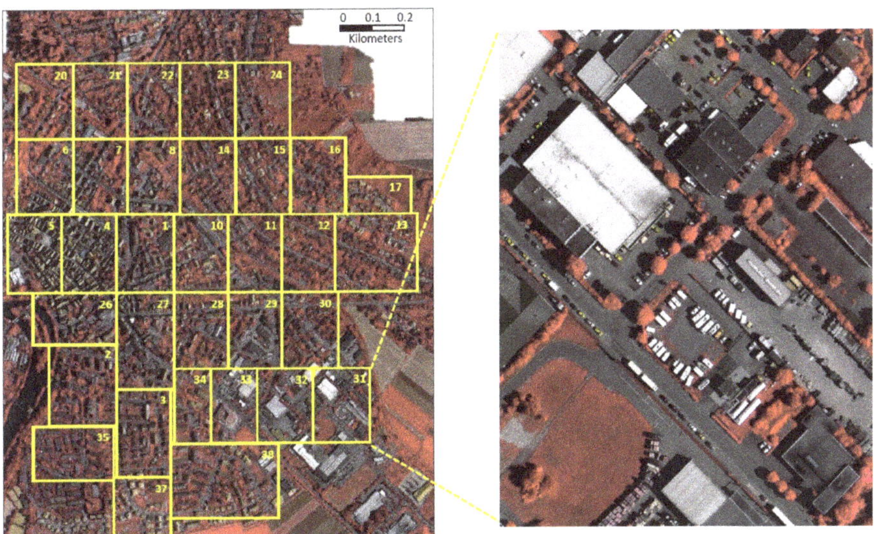

Figure 8. ISPRS 2D Vaihingen segmentation corpus (33 scenes).

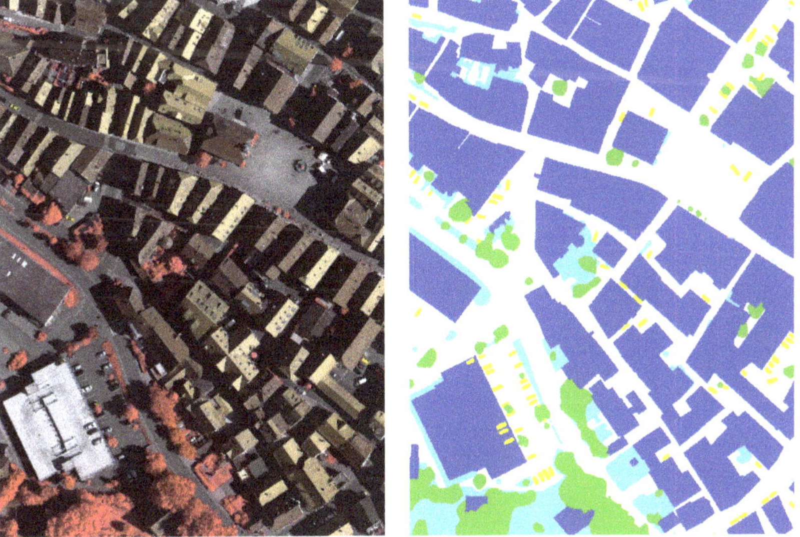

Figure 9. Sample of input scene from Figure 8 (left) and target image (right). The annotated Vaihingen corpus has five categories: tree (green), building (blue), clutter/background (red), low vegetation or LV (greenish-blue), and impervious surface or imp surf (white)

It consists of three spectral bands such as NDSM, DSM, near-infrared bands, red, and green data. For our work, NDSM and DSM data are not used in this corpus. They provide 33 images of about 2500 × 2000 pixels of about 9 cm of resolution. Following other methods, four scenes such as scene 5, 7, 23, and 30 are removed from the training set as a validation set. All experimental results are announced on the validation set if not specified.

5. Performance Evaluation

The performance of "HR-GCN-FF-DA" is evaluated in all corpora for $F1$ and *Average Accuracy*. To assess class-specific performance, the $F1$, *precision, recall*, and *Average Accuracy* metric are used. It is computed as the symphonious average between recall and precision. We carry *precision, recall*, and $F1$ as fundamental metrics and also incorporate the *Average Accuracy*, which calculates the number of correctly classified positions and divides it by the total number of the reference positions. The *Average Accuracy* and $F1$ metrics can be assessed using Equations (7)–(10).

The confusion matrix for pixel-level classification [18] and the false positive (denoted as FP) are computed from the summation of the column. In contrast, the false negative (denoted as FN) is the summation of the horizontal axis, excluding the principal diagonal factor. Next, the true positive (denoted as TP) is the value of the identical oblique elements, and the true negative (denote as TN) contrasts TP.

$$Precision = \frac{TP}{TP + FP} \tag{7}$$

$$Recall = \frac{TP}{TP + FN} \tag{8}$$

$$F1 = \frac{2 \times Precision \times Recall}{Precison + Recall} \tag{9}$$

$$Average Accuracy = \frac{TP + TN}{TP + +FP + FN + TN} \tag{10}$$

6. Experimental Results

For a python deep learning framework, we use "Tensorflow (TF)" [48], an end-to-end open source platform for deep learning. The whole experiment was implemented on servers with Intel® 2066 core I9-10900X, 128 GB of memory, and the NVIDIA RTX™ 2080Ti (11 GB) x 4 cards.

For the training phrase, the adaptive learning rate optimization algorithm (extension to the stochastic gradient descent (SGD)) [49] and batch normalization [50], a technique for improving the performance, stability, and speed of deep learning, were applied and standardized to ease the training in every experiment.

For the learning rate schedules tasks, [9,16,32], we selected the polylearning rate policy. As shown in Equation (11), the learning rate is scheduled by multiplying a decaying factor to the initial learning rate (4×10^{-3}).

$$learning\ rate = init\_learning\ rate \times (1 - \frac{epoch}{MaxEpoch})^{0.9} \tag{11}$$

All deep CNN models are trained for 30 epochs on the Landsat-8w3c corpus and ISPRS Vaihingen data sets. It is increased to be 50 epochs for the Landsat-8w5c data set. Each image is resized to 521 × 521 pixels along with augmented data using a randomly cropping strategy. Weights are updated using the mini-batch of 4.

This section explains the elements of our experiments. The proposed CNN architecture is based on the victor from our previous work called "GCN152-TL-A" [12]. In our work, there are three proposed improvements: (*i*) adjusting backbones using high-resolution representations, (*ii*) the feature fusion module, and (*iii*) depthwise atrous convolution. From all proposed policies, there are four acronyms of procedures, as shown in Table 2.

Table 2. Acronyms of our proposed deep learning approaches.

Acronym	Representation
A	Channel Attention Block
DA	Depthwise Atrous Convolution
FF	Feature Fusion
HR	High-Resolution Representations

There are three subsections to discuss the experimental results of each data set: (*i*) Landsat-8w3c, (*ii*) Landsat-8w5c, and (*iii*) ISPRS Vaihingen data sets.

There are two baseline models of the semantic labeling task in the domains of remote sensing-based information on the computer vision. The first baseline is DCED, which is commonly used in much segmentation work [19–21]. The second baseline is the winner of our previous work called "GCN152-A-TL" [12]. Note that "GCN-A-TL" is abbreviated using just "GCN", since we always employ the attention and transfer-learning strategies into our proposed models.

Each proposed tactic can elevate the completion of the baseline approach shown via the whole experiment. First, the effect of our new backbone (HRNET) is investigated by using HRNET on the GCN framework called "HR-GCN". Second, the effect of our feature fusion is shown by adding it into our model, called "HR-GCN-FF". Third, the effect of the depthwise atrous convolution is explained by using it on top of a traditional convolution mechanism, called "HR-GCN-FF-DA".

6.1. The Results of Landsat-8w3c Data Set

The Landsat-8w3c corpus was used in all experiments. We distinguished between the alterations of the proposed approaches and CNN baselines. "HR-GCN-FF-DA", the full proposed method, is the winner with $F1$ of 0.9114. Furthermore, it is also the winner of all classes. More detailed results are given in the next subsection. Presented in Tables 3 and 4 are the results of this corpus, Landsat-8w3c.

6.1.1. HR-GCN Model: Effect of Heightened GCN with High-Resolution Representations on Landsat-8w3c

The previous enhanced GCN network is improved to increase the $F1$ score by using the High-Resolution Representations (HR) backbone instead of the ResNet-152 backbone (best frontend network from our previous work). $F1$ of HR-GCN (0.8763) outperforms that of the baseline methods. DCED (0.8114) and GCN152-TL-A (0.8727) refer to Tables 3 and 4. The result returns a higher $F1$ at 6.50% and 0.36%, respectively. Hence, it means the features extracted from HRNET are better than those from ResNet-152.

For the analysis of each class, HR-GCN achieved an average accuracy on para rubber, pineapple, and corn for 0.8371, 0.8147, and 0.8621, consecutively. Compared to DCED, it won in two classes. para rubber and corn. However, it won against our previous work (GCN152-TL-A) only in the pineapple class.

6.1.2. HR-GCN-FF Model: Effect of Using "Feature Fusion" on Landsat-8w3c

Next, we apply "Feature Fusion" to capture low-level features to decorate the feature information of CNN. HR-GCN-FF (0.8852) is higher than that of HR-GCN (0.8763), GCN152-TL-A (0.8727), and DCED (0.8113), shown in Tables 3 and 4. It gives a higher $F1$ score at 0.89%, 1.26%, and 7.39%, consecutively.

It is interesting that the FF module can really improve the performance in all classes, especially in the para rubber and pineapple classes. It outperforms both HR-GCN and all baselines in all classes. To further investigate the results, Figures 10e and 11e show that the model with FF can capture pineapple (green area) surrounded in para rubber (red area).

Table 3. Effects on the testing set of the Landsat-8w3c data set.

	Pretrained	Frontend	Model	*Precision*	*Recall*	*F1*
Baseline	-	VGG16	DCED [19–21]	0.8546	0.7723	0.8114
	TL	Res152	GCN-A [12]	0.8732	0.8722	0.8727
Proposed Method	TL	HRNET	GCN-A	0.8693	0.8836	0.8764
	TL	HRNET	GCN-A-FF	0.8797	0.8910	0.8853
	TL	HRNET	GCN-A-FF-DA	**0.8999**	**0.9233**	**0.9114**

Table 4. Effects on the testing set of the Landsat-8w3c data set among each class with our proposed procedures in terms of *Average Accuracy*.

	Model	Para Rubber	Pineapple	Corn
Baseline	DCED [19–21]	0.8218	0.8618	0.8084
	GCN152-TL-A [12]	0.9127	0.7778	0.8878
Proposed Method	HR-GCN	0.8371	0.8147	0.8621
	HR-GCN-FF	0.9179	0.8689	0.8989
	HR-GCN-FF-DA	**0.9386**	**0.8881**	**0.9184**

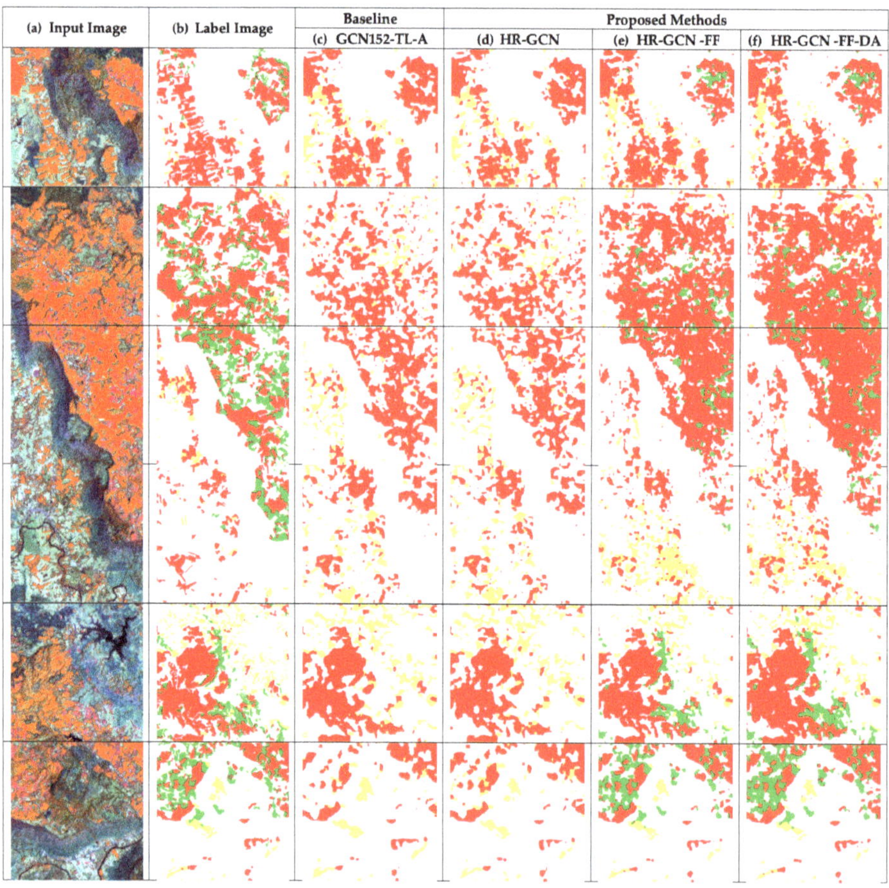

Figure 10. Comparisons between "HR-GCN-FF-DA" and other published methods of the Landsat-8w3c corpus testing set.

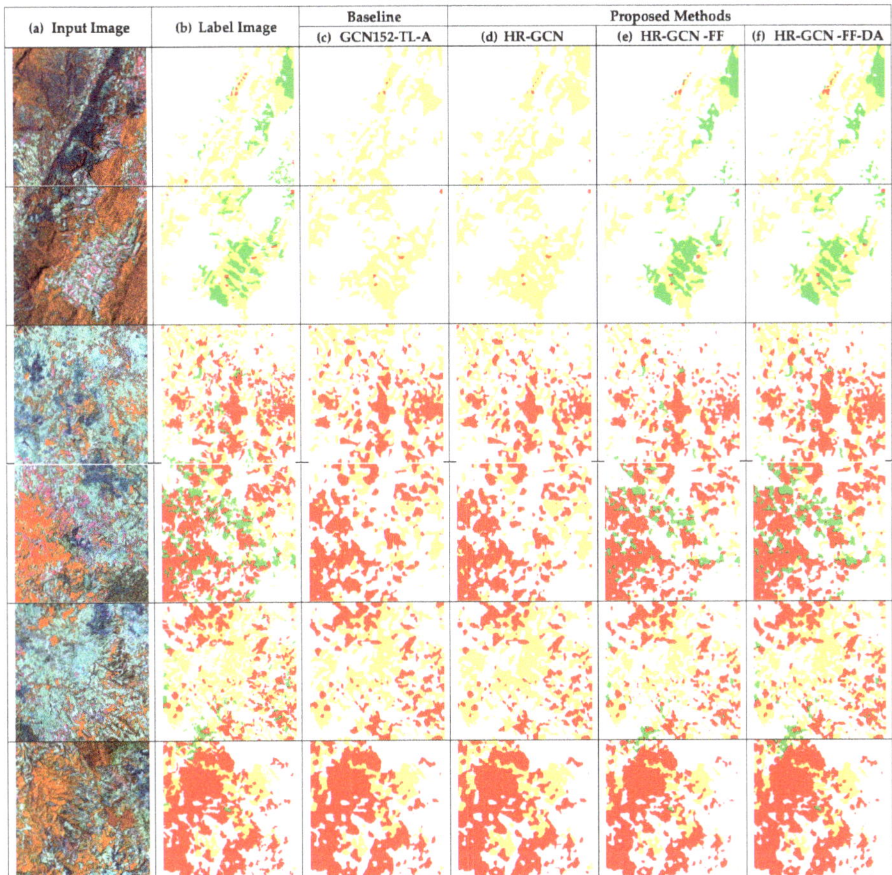

Figure 11. Comparisons between "HR-GCN-FF-DA" and beyond baseline methods of the Landsat-8w3c corpus testing set.

6.1.3. HR-GCN-FF-DA Model: Effect of Using "Depthwise Atrous Convolution" on Landsat-8w3c

The last strategy aims to use an approach of "Depthwise Atrous Convolution" (details in Section 3.4) by extracting complementary information from very shallow features and enhancing the deep features for improving feature fusion of the Landsat-8w3c corpus. The "HR-GCN-FF-DA" method is the victor. $F1$ is obviously more distinguished than DCED at 10.00% and GCN152-TL-A (the best benchmark) at 3.87%, as shown in Tables 3 and 4.

For an analysis of each class, our model is clearly the winner in all classes with an accuracy beyond 90% in two classes: para rubber and corn. Figures 10 and 11 show twelve sample outputs from our proposed methods (column ($d\ to\ f$)) compared to the baseline (column (c)) to expose improvements in its results. From our investigation, we found that the dilated convolutional concept can make our model have better overview information, so it can capture larger areas of data.

There is a lower discrepancy (peak) in the validation data of "HR-GCN-FF-DA", Figure 12a, than that in the baseline, Figure 13a. Moreover, Figures 13b and 12b show three learning graphs such as precision, recall, and F1 lines. The loss graph of the "HR-GCN-FF-DA" model seems flatter (very smooth) than the baseline in Figure 13a. The epoch at number 27 was selected to be a pre-trained model for testing and transfer learning procedures.

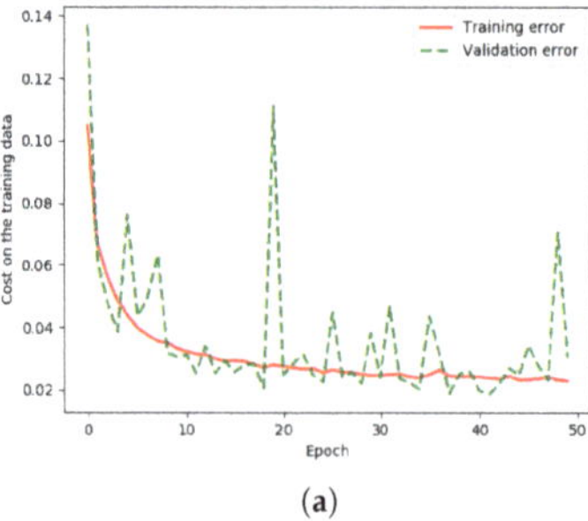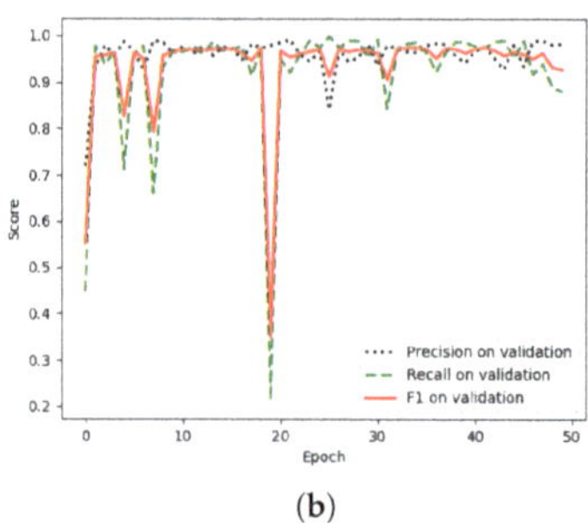

(**a**) (**b**)

Figure 12. Graph (learning curves) of the Landsat-8w3c data set of the proposed approach, "HR-GCN-FF-DA"; x refers to epochs, and y refers to different measures (**a**) Plot of model loss (cross-entropy) on training and validation corpora; (**b**) performance plot on the validation corpus.

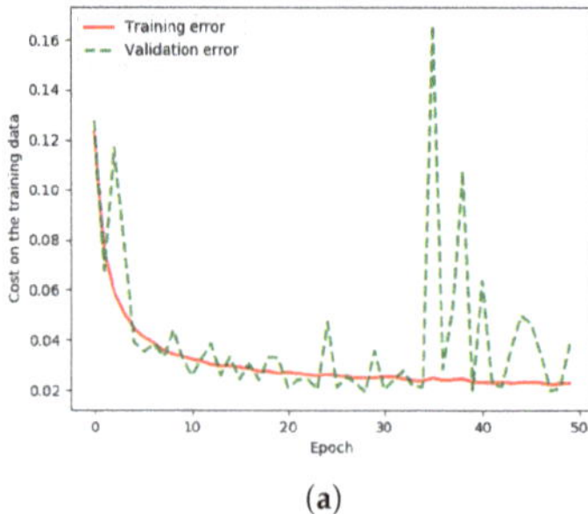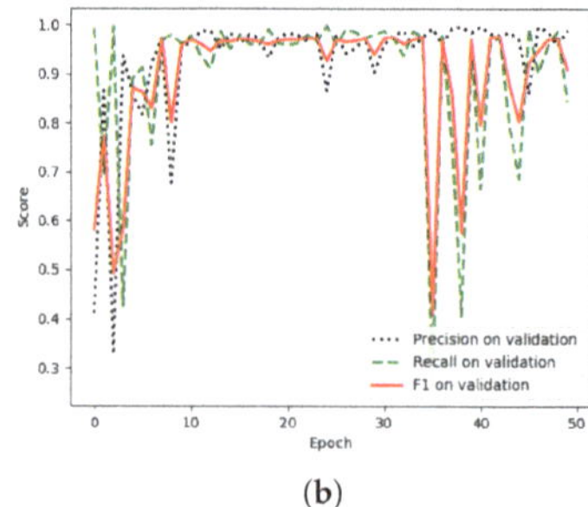

(**a**) (**b**)

Figure 13. Graph (learning curves) of the Landsat-8w3c data set of the baseline approach, GCN152-TL-A; x refers to epochs, and y refers to different measures (**a**) Plot of model loss (cross-entropy) on training and validation corpora; (**b**) performance plot on the validation corpus. [12].

6.2. The Results on Landsat-8w5c Data Set

In this subsection, the Landsat-8w5c corpus was conducted on all experiments. We compare "HR-GCN-FF-DA" network (column (f)) to CNN baselines via Tables 5 and 6. "HR-GCN-FF-DA" is the winner with a $F1$ of 0.9111. Furthermore, it is also the winner in all classes especially water and urban class that are composed with low-level features. More detailed results are described in the next subsection and are presented in Tables 5 and 6 for the results of this data set, Landsat-8w5c.

6.2.1. HR-GCN Model: Effect of Heightened GCN with High-Resolution Representations on Landsat-8w5c

The $F1$ score of HR-GCN (0.8897) outperforms that of baseline methods: DCED (0.8505) and GCN152-TL-A (0.8791); $F1$ at 3.92% and 1.07% respectively. The main reason is due to both higher recall and precision. This can imply that features extracted from HRNET are also better than those from ResNet-152 on Landsat-8 images as well, shown in Tables 5 and 6.

Table 5. Effects on the testing set of Landsat-8w5c data set.

	Pre-trained	Frontend	Model	*Precision*	*Recall*	*F1*
Baseline	-	VGG16	DCED [19–21]	0.8571	0.8441	0.8506
	TL	Res152	GCN-A [12]	0.8616	0.8973	0.8791
Proposed Method	TL	HRNET	GCN-A	0.8918	0.8877	0.8898
	TL	HRNET	GCN-A-FF	0.9209	0.9181	0.9195
	TL	HRNET	GCN-A-FF-DA	**0.9338**	**0.9385**	**0.9362**

Table 6. Effects on the testing set of Landsat-8w5c data set among each class with our proposed procedures in terms of *Average Accuracy*.

	Model	Agriculture	Forest	Misc	Urban	Water
Baseline	DCED [19–21]	0.9819	**0.9619**	0.7628	0.8538	0.7250
	GCN152-TL-A [12]	0.9757	0.9294	0.6847	0.9288	0.7846
Proposed Method	HR-GCN	0.9755	0.9501	0.8231	0.9133	0.7972
	HR-GCN-FF	0.9741	0.9526	0.8641	0.9335	0.8282
	HR-GCN-FF-DA	**0.9856**	0.9531	**0.9176**	**0.9561**	**0.8437**

For the analysis on each class, HR-GCN achieved an averaging accuracy in agriculture, forest, miscellaneous, urban, and water for 0.9755, 0.9501, 0.8231, 0.9133, and 0.7972, consecutively. Compared to DCED, it won in three classes: forest, miscellaneous and water. However, it won against our previous work (GCN152-TL-A) in the pineapple class, and it showed about the same performance in the agriculture class.

6.2.2. HR-GCN-FF Model: Effect of Using "Feature Fusion" on Landsat-8w5c

The second mechanism focuses on utilizing 'Feature Fusion" to fuse each level feature for enriching the feature information. From Tables 5 and 6, the *F1* of HR-GCN-FF (0.9195) is greater than that of HR-GCN (0.8897), GCN152-TL-A (0.8791), and DCED (0.8505). It produces a more precise *F1* score at 2.97%, 4.04%, and 6.89%.

To further analyze the results, Figures 14e and 15e show that the FF module can better capture low-level details. Especially in the water class, it can recover the missing water area, resulting in an improvement of accuracy from 0.7972 to 0.8282 (3.1%).

6.2.3. HR-GCN-FF-DA Model: Effect of Using "Depthwise Atrous Convolution" on Landsat-8w5c

The last policy points to the performance of the method of "Depthwise Atrous Convolution" by enhancing the features of CNN for improving the previous step. The *F1* score of the "HR-GCN-FF-DA" approach is the conqueror. It is more eminent than DCED and GCN152-TL at 8.56% and 5.71%, consecutively, shown in Tables 5 and 6.

In the dilated convolution, filters are boarder, which can capture better overview details resulting in (*i*) larger coverage areas and (*ii*) connected small areas together.

For an analysis of each class, our final model is clearly the winner in all classes with an accuracy beyond 95% in two classes: agriculture and urban classes. Figures 14 and 15 show twelve sample outputs from our proposed methods (column (*d to f*)) compared to the baseline (column (*c*)) to expose improvements in its results and that founds that Figures 14f and 15f are likewise to the ground images. From our investigation, we found that the dilated convolutional concept can make our model have better overview information, so it can capture larger areas of data.

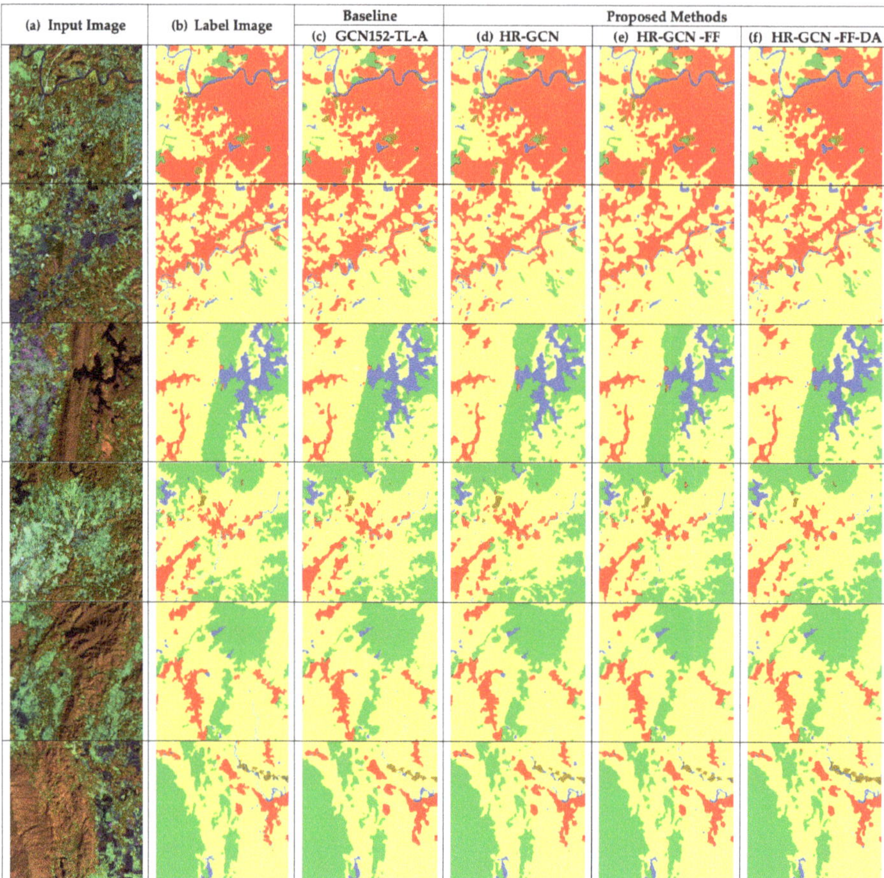

Figure 14. Comparisons between "HR-GCN-FF-DA" and beyond baseline methods on the Landsat-8w5c corpus testing set.

Considering the loss graphs, our model in Figure 16a can learn smoother than the baseline (our previous work) in Figure 17a, since the discrepancy (peak) in the validation error (green line) is lower in our model. There is a lower discrepancy (peak) in the validation data of "HR-GCN-FF-DA", Figure 16a, than that in the baseline, Figure 17a. Moreover, Figures 17b and 16b show three learning graphs such as precision, recall, and F1 lines. The loss graph of the "HR-GCN-FF-DA" model seems flatter (very smooth) than the baseline in Figure 17a and the epoch at number 40 out of 50 was selected to be a pre-trained model for testing and transfer learning procedures.

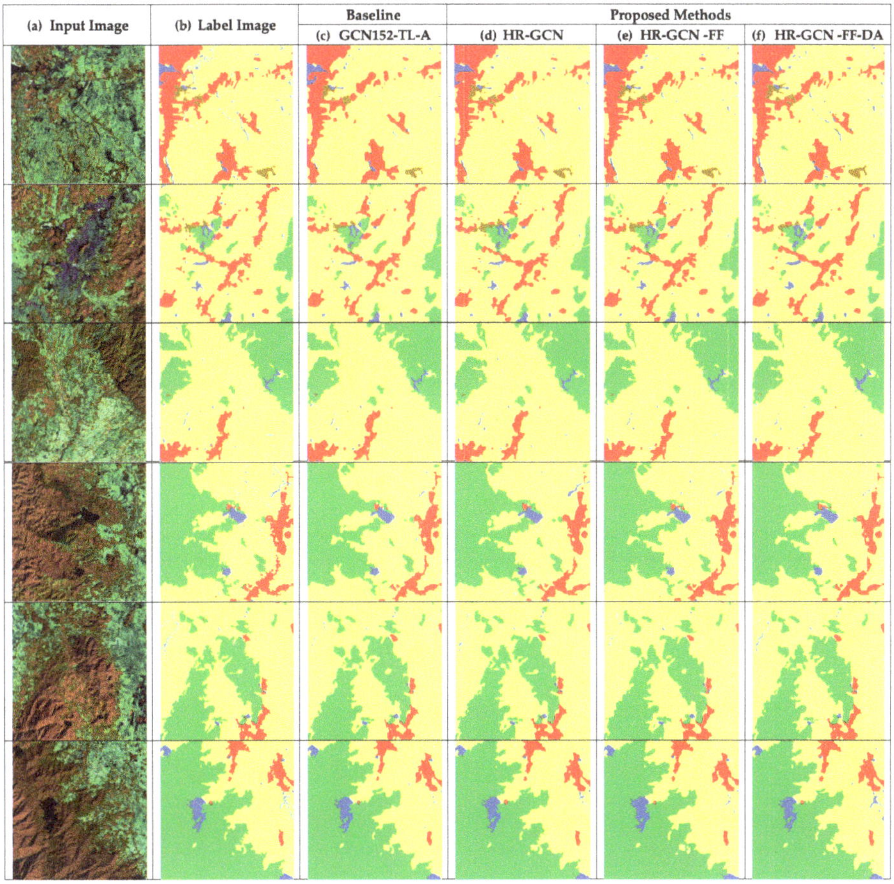

(a) Input Image	(b) Label Image	Baseline	Proposed Methods		
		(c) GCN152-TL-A	(d) HR-GCN	(e) HR-GCN -FF	(f) HR-GCN -FF-DA

Figure 15. Comparisons between "HR-GCN-FF-DA" and beyond baseline methods on the Landsat-8w5c corpus testing set.

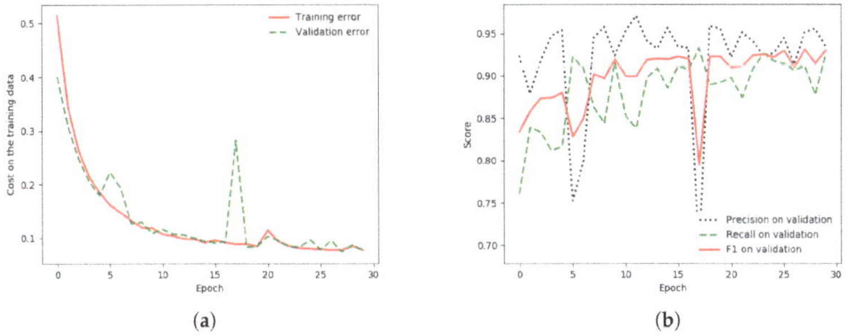

Figure 16. Graph (learning curves) on Landsat-8w5c data set of the proposed approach, "HR-GCN-FF-DA"; x refers to epochs, and y refers to different measures (**a**) Plot of model loss (cross-entropy) on training and validation corpora; (**b**) performance plot on the validation corpus.

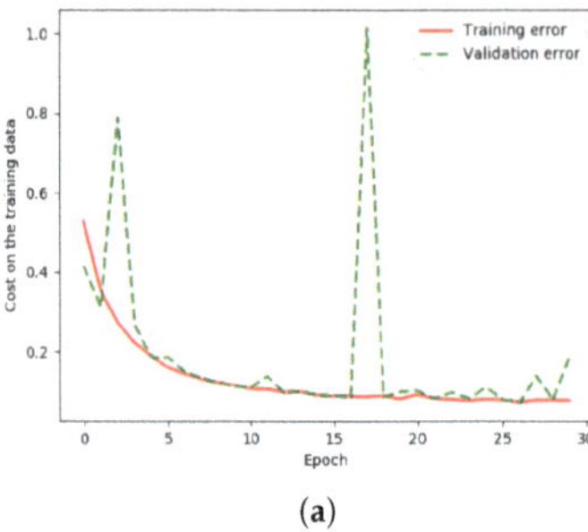
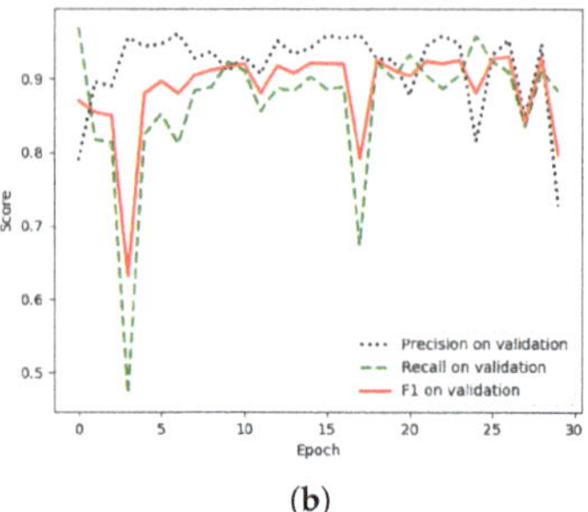

(a)   (b)

Figure 17. Graph (learning curves) on Landsat-8w5c data set of the baseline approach, GCN152-TL-A [12]; x refers to epochs, and y refers to different measures (**a**) Plot of model loss (cross-entropy) on training and validation corpora; (**b**) performance plot on the validation corpus.

6.3. The Results in ISPRS Vaihingen Challenge Data Set

In this subsection, the ISPRS Vaihingen (Stuttgart) Challenge corpus was used in all experiments. The "HR-GCN-FF-DA" is the winner with $F1$ of 0.9111. Furthermore, it is also the winner of all classes. More detailed results will be provided in the next subsection, and the consequences of our proposed method with CNN baselines for this data set are shown in Tables 7 and 8.

6.3.1. HR-GCN Model: Effect of Heightened GCN with High-Resolution Representations in ISPRS Vaihingen

The $F1$ score of HR-GCN (0.8701) exceeds that of the baseline methods: DCED (0.8580) and GCN152-TL-A (0.8620). It complies a higher $F1$ at 1.21% and 0.81%, respectively. This shows that the enhanced GCN with HR backbone is also more significantly streamlined than the GCN152-TL-A style, shown in Tables 7 and 8.

The goal of the HR module is to help prevent the loss of some important features, such as low-level features, so it can significantly improve the accuracy of the car class from 0.8034 to 0.8202 (1.68%) and the building class from 0.8725 to 0.9282 (5.57%).

Table 7. Effects on the testing set of ISPRS Vaihingen (Stuttgart) challenge data set.

	Pre-trained	Frontend	Model	*Precision*	*Recall*	*F1*
Baseline	-	VGG16	DCED [19–21]	0.8672	0.8490	0.8580
	TL	Res152	GCN-A [12]	0.8724	0.8520	0.8620
Proposed Method	TL	HRNET	GCN-A	0.8717	0.8686	0.8701
	TL	HRNET	GCN-A-FF	0.8981	0.8812	0.8896
	TL	HRNET	GCN-A-FF-DA	**0.9228**	**0.8997**	**0.9111**

Table 8. Effects on the testing set of ISPRS Vaihingen (Stuttgart) challenge data set among each class with our proposed procedures in terms of *Average Accuracy*.

	Model	IS	Buildings	LV	Tree	Car
Baseline	DCED [19–21]	0.8721	0.8932	0.8410	0.9144	0.8153
	GCN152-TL-A [12]	0.8758	0.8725	0.8567	**0.9534**	0.8034
Proposed Method	HR-GCN	0.8864	0.9282	0.8114	0.8945	0.8202
	HR-GCN-FF	0.8279	0.9458	0.9264	0.9475	0.8502
	HR-GCN-FF-DA	**0.9075**	**0.9589**	**0.9266**	0.9299	**0.8710**

6.3.2. HR-GCN-FF Model: Effect of Using "Feature Fusion" on ISPRS Vaihingen

Next, we propose "Feature Fusion" to fuse each level feature for enriching the feature information. From Tables 7 and 8, the *F*1 of HR-GCN-FF (0.8895) is greater than that of HR-GCN (0.8701), GCN152-TL-A (0.8620), and DCED (0.8580). It also returns a higher *F*1 score at 1.95%, 2.76%, and 3.16%, respectively.

The goal of the FF module is to capture low-level features, so it can significantly improve the accuracy of the low vegetation class (LV) from 0.8114 to 0.9264 (11.5%), the accuracy of the tree class from 0.8945 to 0.9475 (5.3%), and the accuracy of the car class from 0.8202 to 0.8502 (3%). This finding is shown in Figures 18e and 19e.

6.3.3. HR-GCN-FF-DA Model: Effect of Using "Depthwise Atrous Convolution" on ISPRS Vaihingen

Finally, our last approach is to apply "Depthwise Atrous Convolution" to intensify the deep features from the previous step. From Tables 7 and 8 we see that the *F*1 of the "HR-GCN-FF-DA" method is also the conqueror in this data set. The *F*1 score of "HR-GCN-FF-DA" is also more precise than the DCED and GCN152-TL-A at 5.31% and 4.96%, consecutively.

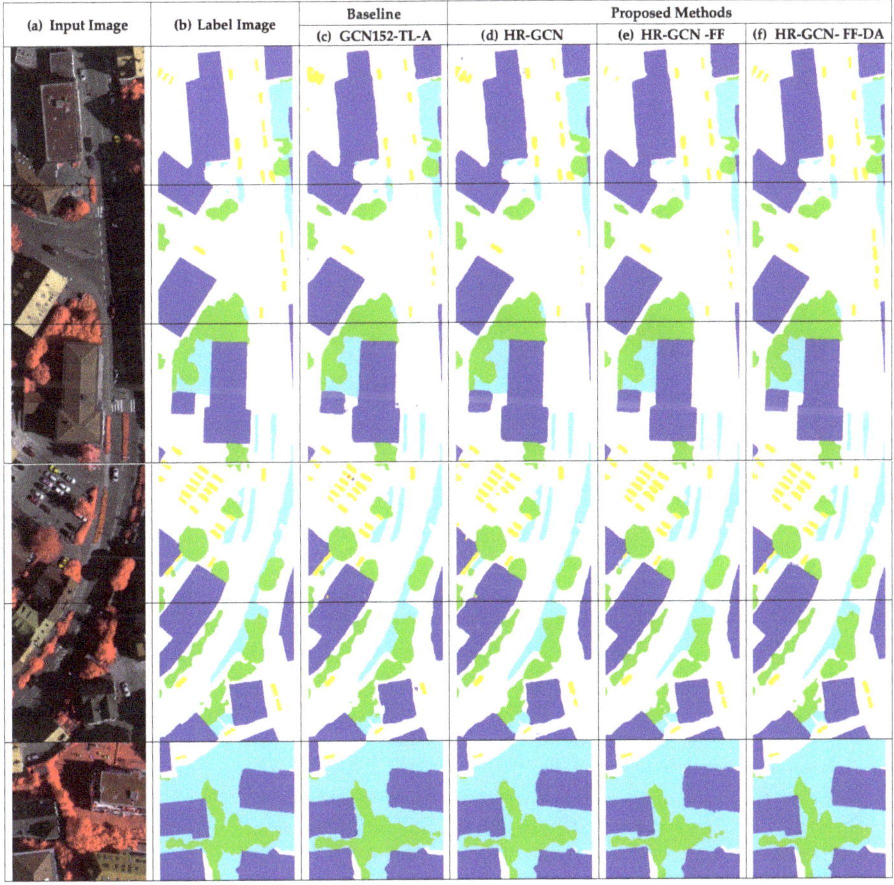

Figure 18. Comparisons between "HR-GCN-FF-DA" and beyond baseline methods on the ISPRS Vaihingen (Stuttgart) challenge corpus testing set.

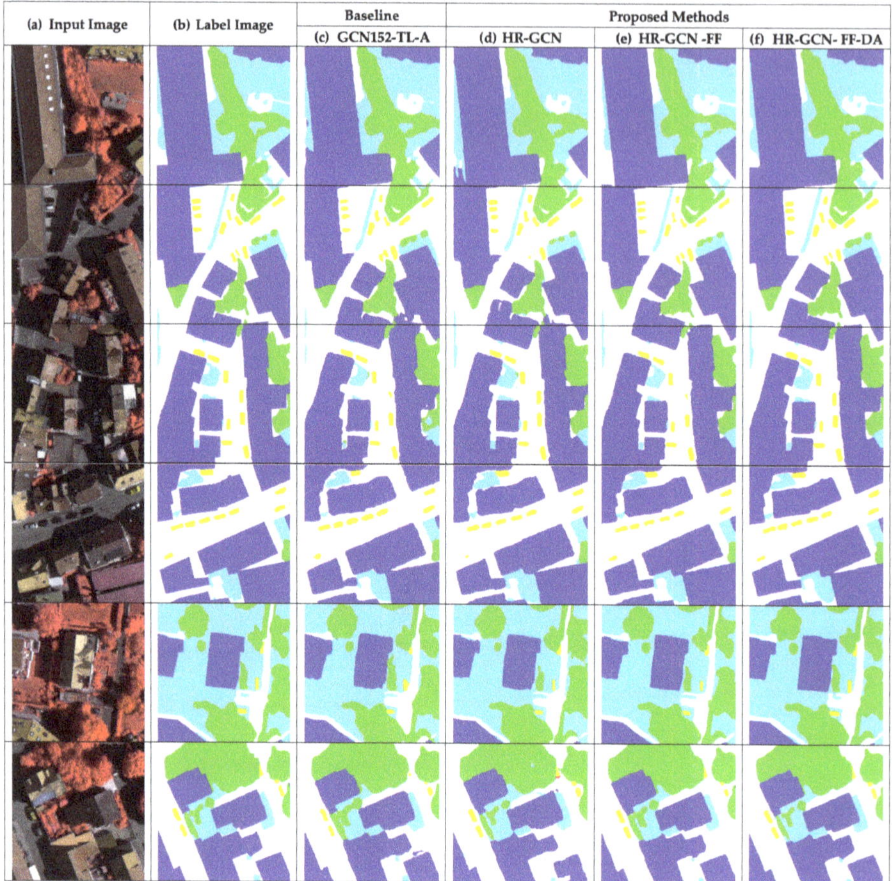

Figure 19. Comparisons between "HR-GCN-FF-DA" and beyond baseline methods on the ISPRS Vaihingen (Stuttgart) challenge corpus testing set.

It is very impressive that our model with all its strategies can improve the accuracy in almost all classes to be greater than 90%. Although the accuracy of car is 0.8710, it improves on the baseline (0.8034) by 6.66%.

7. Discussion

In the Landsat-8w3c corpus, for an analysis of each class, our model is clearly the winner in all classes with an accuracy beyond 90% in two classes: para rubber and corn. Figures 10 and 11 show twelve sample outputs from our proposed methods (column (d to f)) compared to the baseline (column (c)) to expose improvements in its results and shows that Figures 10f and 11f are similar to the target images. From our investigation, we found that the dilated convolutional concept can make our model have better overview information, so it can capture larger areas of data.

There is a lower discrepancy (peak) in the validation data of "HR-GCN-FF-DA" Figure 12a than that in the baseline Figure 13a. Moreover, Figures 13b and 12b show three learning graphs such as precision, recall, and F1 lines. The loss graph of the "HR-GCN-FF-DA" model seems flatter (very smooth) than the baseline in Figure 13a. The epoch at number 27 was selected to be a pre-trained model for testing and transfer learning procedures.

In the Landsat-8w5c corpus, for an analysis of each class, our final model is clearly the winner in all classes with an accuracy beyond 95% in two classes: agriculture and urban classes. Figures 14 and 15 show twelve sample outputs from our proposed methods (column ($d\ to f$)) compared to the baseline (column (c)) to expose improvements in its results and shows that Figures 14f and 15f are similar to the ground images. From our investigation, we found that the dilated convolutional concept can make our model have better overview information, so it can capture larger areas of data.

Considering the loss graphs, our model in Figure 16a can learn smoother than the baseline (our previous work) in Figure 17a, since the discrepancy (peak) in the validation error (green line) is lower in our model. There is a lower discrepancy (peak) in the validation data of "HR-GCN-FF-DA", Figure 16a, than that in the baseline Figure 17a. Moreover, Figures 17b and 16b show three learning graphs such as precision, recall, and F1 lines. The loss graph of the "HR-GCN-FF-DA" model seems flatter (very smooth) than the baseline in Figure 17a. The epoch at number 40 out of 50 was selected to be the pre-trained model for testing and transfer-learning procedures.

In the ISPRS Vaihingen corpus, for an analysis of each class, our model is clearly the winner in all classes with an accuracy beyond 90% in four classes: impervious surface, building, low vegetation, and trees. Figure 18 shows twelve sample outputs from our proposed methods (column ($d\ to f$)) compared to the baseline (column (c)) to expose improvements in its results and shows that Figures 18f and 19f are similar to the target images. From our investigation, we found that the dilated (atrous) convolutional idea can make our deep CNN model have better overview learning, so that it can capture more ubiquitous areas of data.

For the loss graph, it is similar to the results in our previous experiments. There is a lower discrepancy (peak) in the validation data of our model (Figure 20a) than that in the baseline (Figure 21a). Moreover, Figures 21b and 20b explicate a trend that represents a high-grade model performance. Lastly, the epoch at number 26 (out of 30) was selected to be a pre-trained model for testing and transfer learning procedures.

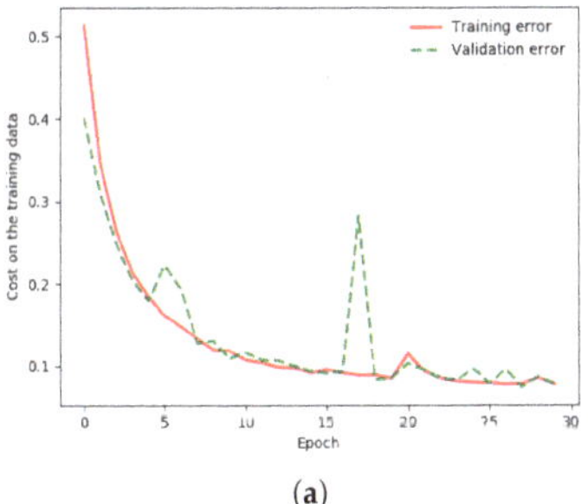

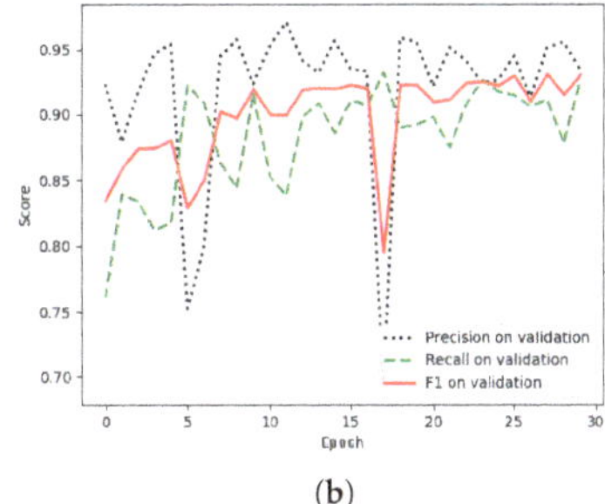

(**a**) (**b**)

Figure 20. Graph (learning curves) on ISPRS Vaihingen data set of the proposed approach, "HR-GCN-FF-DA"; x refers to epochs, and y refers to different measures (**a**) Plot of model loss (cross-entropy) on training and validation corpora; (**b**) performance plot on the validation corpus.

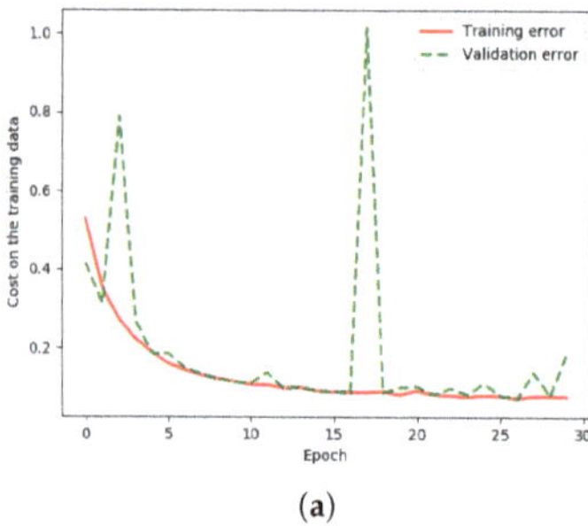

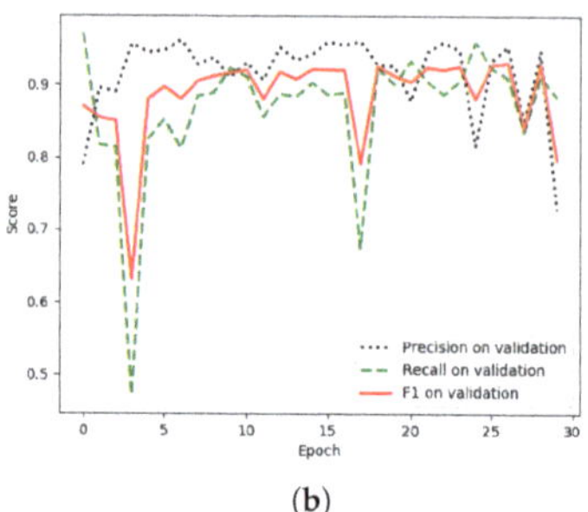

(a)   (b)

Figure 21. Graph (learning curves) in ISPRS data set of the baseline approach, GCN152-TL-A [12]; x refers to epochs, and y refers to different measures (**a**) Plot of model loss (cross-entropy) on training and validation corpora; (**b**) performance plot on the validation corpus.

8. Conclusions

We propose a novel CNN architecture to achieve image labeling on remote-sensed images. Our best-proposed method, "HR-GCN-FF-DA", delivers an excellent performance in regards to three aspects: (i) modifying the backbone architecture with "High-Resolution Representations (HR)", (ii) applying the "Feature Fusion (FF)", and (iii) using the concept of "Depthwise Atrous Convolution (DA)". Each proposed strategy can really improve $F1$-results by 4.82%, 4.08%, and 2.14% by adding HR, FF, and DA modules, consecutively. The FF module can really capture low-level features, resulting in a higher accuracy of river and low-vegetation classes. The DA module can refine the features and provide more coverage areas, resulting in a higher accuracy of pineapple and miscellaneous classes. The results demonstrate that the "HR-GCN-FF-DA" model significantly exceeds all baselines. It is the victor in all data sets and exceeds more than 90% of $F1$: 0.9114, 0.9362, and 0.9111 of the Landsat-8w3c, Landsat-8w5c, and ISPRS Vaihingen corpora, respectively. Moreover, it reaches an accuracy surpassing 90% in almost all classes.

Author Contributions: Conceptualization, T.P.; Data curation, K.J., S.L. and P.S.; Formal analysis, T.P.; Investigation, T.P.; Methodology, T.P.; Project administration, T.P.; Resources, T.P.; Software, T.P.; Supervision, T.P. and P.V.; Validation, T.P.; Visualization, T.P.; Writing–original draft, T.P.; Writing–review and editing, T.P. and P.V. All authors have read and agreed to the published version of the manuscript.

Acknowledgments: Teerapong Panboonyuen, also known as Kao Panboonyuen appreciates and thanks to the scholarship from the 100th Anniversary Chulalongkorn University Fund for the Doctoral Scholarship and the 90th Anniversary Chulalongkorn University Fund (Ratchadaphiseksomphot Endowment Fund. Teerapong Panboonyuen greatly acknowledges the Geo-Informatics and Space Technology Development Agency (GISTDA), Thailand, and Kao thanks to the staff from the GISTDA (Thanwarat Anan, Bussakon Satta, and Suwalak Nakya) for providing the remote sensing corpora used in this study.

Conflicts of Interest: The authors declare no conflict of interest.

Abbreviations

The following acronyms are used in this article:

A	Channel Attention
BR	Boundary Refinement
DA	Depthwise Atrous Convolution
DSM	Digital Surface Model
FF	Feature Fusion
HR	High-Resolution Representations
IS	Impervious Surfaces
Misc	Miscellaneous
NDSM	Normalized Digital Surface Mode
LV	Low Vegetation
TL	Transfer Learning

References

1. He, K.; Gkioxari, G.; Dollár, P.; Girshick, R. Mask r-cnn. In Proceedings of the 2017 IEEE International Conference on Computer Vision (ICCV), Venice, Italy, 22–29 October 2017; pp. 2980–2988.
2. Yu, C.; Wang, J.; Peng, C.; Gao, C.; Yu, G.; Sang, N. Bisenet: Bilateral segmentation network for real-time semantic segmentation. In Proceedings of the European Conference on Computer Vision (ECCV), Munich, Germany, 8–14 September 2018; pp. 325–341.
3. Zhang, H.; Dana, K.; Shi, J.; Zhang, Z.; Wang, X.; Tyagi, A.; Agrawal, A. Context encoding for semantic segmentation. In Proceedings of the IEEE Conference on Computer Vision and Pattern Recognition (CVPR), Salt Lake City, UT, USA, 18–22 June 2018.
4. Li, Y.; Qi, H.; Dai, J.; Ji, X.; Wei, Y. Fully convolutional instance-aware semantic segmentation. In Proceedings of the IEEE Conference on Computer Vision and Pattern Recognition, Honolulu, HI, USA, 21–26 July 2017; pp. 2359–2367.
5. Zhao, H.; Shi, J.; Qi, X.; Wang, X.; Jia, J. Pyramid scene parsing network. In Proceedings of the IEEE Conference on Computer Vision and Pattern Recognition (CVPR), Honolulu, HI, USA, 21–26 July 2017; pp. 2881–2890.
6. Pang, Y.; Li, Y.; Shen, J.; Shao, L. Towards bridging semantic gap to improve semantic segmentation. In Proceedings of the IEEE International Conference on Computer Vision, Seoul, Korea, 27 October–2 November 2019; pp. 4230–4239.
7. Zhang, Z.; Zhang, X.; Peng, C.; Xue, X.; Sun, J. Exfuse: Enhancing feature fusion for semantic segmentation. In Proceedings of the European Conference on Computer Vision (ECCV), Munich, Germany, 8–14 September 2018; pp. 269–284.
8. Li, Y.; Zhang, X.; Chen, D. Csrnet: Dilated convolutional neural networks for understanding the highly congested scenes. In Proceedings of the IEEE Conference on Computer Vision and Pattern Recognition, Salt Lake City, UT, USA, 18–22 June 2018; pp. 1091–1100.
9. Yang, M.; Yu, K.; Zhang, C.; Li, Z.; Yang, K. DenseASPP for Semantic Segmentation in Street Scenes. In Proceedings of the IEEE Conference on Computer Vision and Pattern Recognition, Salt Lake City, UT, USA, 18–22 June 2018, pp. 3684–3692.
10. Liu, Y.; Fan, B.; Wang, L.; Bai, J.; Xiang, S.; Pan, C. Semantic labeling in very high resolution images via a self-cascaded convolutional neural network. *ISPRS J. Photogramm. Remote. Sens.* **2018**, *145*, 78–95. [CrossRef]
11. Peng, C.; Zhang, X.; Yu, G.; Luo, G.; Sun, J. Large kernel matters—Improve semantic segmentation by global convolutional network. In Proceedings of the 2017 IEEE Conference on Computer Vision and Pattern Recognition (CVPR), Honolulu, HI, USA, 21–26 July 2017; pp. 1743–1751.
12. Panboonyuen, T.; Jitkajornwanich, K.; Lawawirojwong, S.; Srestasathiern, P.; Vateekul, P. Semantic segmentation on remotely sensed images using an enhanced global convolutional network with channel attention and domain specific transfer learning. *Remote. Sens.* **2019**, *11*, 83. [CrossRef]
13. Xie, M.; Jean, N.; Burke, M.; Lobell, D.; Ermon, S. Transfer learning from deep features for remote sensing and poverty mapping. In Proceedings of the Thirtieth AAAI Conference on Artificial Intelligence, Phoenix, AZ, USA, 12–17 February 2016.

14. Yosinski, J.; Clune, J.; Bengio, Y.; Lipson, H. How transferable are features in deep neural networks? In Proceedings of the Advances in Neural Information Processing Systems, Montreal, QC, Canada, 8–13 December 2014; pp. 3320–3328.

15. Liu, J.; Wang, Y.; Qiao, Y. Sparse Deep Transfer Learning for Convolutional Neural Network. In Proceedings of the AAAI, San Francisco, CA, USA, 4–9 February 2017; pp. 2245–2251.

16. Yu, C.; Wang, J.; Peng, C.; Gao, C.; Yu, G.; Sang, N. Learning a discriminative feature network for semantic segmentation. In Proceedings of the IEEE Conference on Computer Vision and Pattern Recognition, Salt Lake City, UT, USA, 18–22 June 2018; pp. 1857–1866.

17. Hu, J.; Shen, L.; Sun, G. Squeeze-and-excitation networks. In Proceedings of the IEEE Conference on Computer Vision and Pattern Recognition, Salt Lake City, UT, USA, 18–22 June 2018; pp. 7132–7141.

18. International Society for Photogrammetry and Remote Sensing. 2D Semantic Labeling Challenge. Available online: http://www2.isprs.org/commissions/comm3/wg4/semantic-labeling.html (accessed on 9 September 2018).

19. Badrinarayanan, V.; Handa, A.; Cipolla, R. Segnet: A deep convolutional encoder-decoder architecture for robust semantic pixel-wise labelling. *arXiv* **2015**, arXiv:1505.07293.

20. Badrinarayanan, V.; Kendall, A.; Cipolla, R. Segnet: A deep convolutional encoder-decoder architecture for image segmentation. *IEEE Trans. Pattern Anal. Mach. Intell.* **2017**, *39*, 2481–2495. [CrossRef] [PubMed]

21. Kendall, A.; Badrinarayanan, V.; Cipolla, R. Bayesian segnet: Model uncertainty in deep convolutional encoder-decoder architectures for scene understanding. *arXiv* **2015**, arXiv:1511.02680.

22. Wang, H.; Wang, Y.; Zhang, Q.; Xiang, S.; Pan, C. Gated convolutional neural network for semantic segmentation in high-resolution images. *Remote Sens.* **2017**, *9*, 446. [CrossRef]

23. Zhu, X.X.; Tuia, D.; Mou, L.; Xia, G.S.; Zhang, L.; Xu, F.; Fraundorfer, F. Deep learning in remote sensing: A comprehensive review and list of resources. *IEEE Geosci. Remote. Sens. Mag.* **2017**, *5*, 8–36. [CrossRef]

24. Yue, K.; Yang, L.; Li, R.; Hu, W.; Zhang, F.; Li, W. TreeUNet: Adaptive Tree convolutional neural networks for subdecimeter aerial image segmentation. *ISPRS J. Photogramm. Remote. Sens.* **2019**, *156*, 1–13. [CrossRef]

25. Diakogiannis, F.I.; Waldner, F.; Caccetta, P.; Wu, C. Resunet-a: A deep learning framework for semantic segmentation of remotely sensed data. *ISPRS J. Photogramm. Remote. Sens.* **2020**, *162*, 94–114. [CrossRef]

26. Sun, T.; Chen, Z.; Yang, W.; Wang, Y. Stacked U-Nets With Multi-Output for Road Extraction. In Proceedings of the CVPR Workshops, Salt Lake City, UT, USA, 18–22 June 2018; pp. 202–206.

27. Lateef, F.; Ruichek, Y. Survey on semantic segmentation using deep learning techniques. *Neurocomputing* **2019**, *338*, 321–348. [CrossRef]

28. Wang, P.; Chen, P.; Yuan, Y.; Liu, D.; Huang, Z.; Hou, X.; Cottrell, G. Understanding convolution for semantic segmentation. In Proceedings of the 2018 IEEE Winter Conference on Applications of Computer Vision (WACV), Lake Tahoe, NV, USA, 12–15 March 2018; pp. 1451–1460.

29. Ghosh, S.; Das, N.; Das, I.; Maulik, U. Understanding deep learning techniques for image segmentation. *ACM Comput. Surv. (CSUR)* **2019**, *52*, 1–35. [CrossRef]

30. Zhao, H.; Qi, X.; Shen, X.; Shi, J.; Jia, J. Icnet for real-time semantic segmentation on high-resolution images. In Proceedings of the European Conference on Computer Vision (ECCV), Munich, Germany, 8–14 September 2018; pp. 405–420.

31. Dai, J.; He, K.; Sun, J. Instance-aware semantic segmentation via multi-task network cascades. In Proceedings of the IEEE Conference on Computer Vision and Pattern Recognition, Las Vegas, NV, USA, 27–30 June 2016; pp. 3150–3158.

32. Krizhevsky, A.; Sutskever, I.; Hinton, G.E. Imagenet classification with deep convolutional neural networks. In Proceedings of the Advances in Neural Information Processing Systems, Lake Tahoe, NV, USA, 3–6 December 2012; pp. 1097–1105.

33. Li, Y.; Zhang, H.; Xue, X.; Jiang, Y.; Shen, Q. Deep learning for remote sensing image classification: A survey. *Wiley Interdiscip. Rev. Data Min. Knowl. Discov.* **2018**, *8*, e1264. [CrossRef]

34. Long, J.; Shelhamer, E.; Darrell, T. Fully convolutional networks for semantic segmentation. In Proceedings of the IEEE Conference on Computer Vision and Pattern Recognition, Boston, MA, USA, 7–12 June 2015; pp. 3431–3440.

35. Noh, H.; Hong, S.; Han, B. Learning deconvolution network for semantic segmentation. In Proceedings of the IEEE International Conference on Computer Vision, Santiago, Chile, 7–13 December 2015; pp. 1520–1528.

36. Ronneberger, O.; Fischer, P.; Brox, T. U-net: Convolutional networks for biomedical image segmentation. In Proceedings of the International Conference on Medical Image Computing and Computer-Assisted Intervention, Munich, Germany, 5–9 October2015; Springer: Berlin, Germany, 2015; pp. 234–241.

37. Sun, K.; Zhao, Y.; Jiang, B.; Cheng, T.; Xiao, B.; Liu, D.; Mu, Y.; Wang, X.; Liu, W.; Wang, J. High-resolution representations for labeling pixels and regions. *arXiv* **2019**, arXiv:1904.04514.

38. Sun, K.; Xiao, B.; Liu, D.; Wang, J. Deep high-resolution representation learning for human pose estimation. In Proceedings of the IEEE Conference on Computer Vision and Pattern Recognition, Long Beach, CA, USA, 16–20 June 2019; pp. 5693–5703.

39. Chen, L.C.; Zhu, Y.; Papandreou, G.; Schroff, F.; Adam, H. Encoder-decoder with atrous separable convolution for semantic image segmentation. In Proceedings of the European Conference on Computer Vision (ECCV), Munich, Germany, 8–14 September 2018; pp. 801–818.

40. Kampffmeyer, M.; Jenssen, R.; Salberg, A.B. Dense Dilated Convolutions Merging Network for Semantic Mapping of Remote Sensing Images. In Proceedings of the 2019 Joint Urban Remote Sensing Event (JURSE), Vannes, France, 22–24 May 2019; pp. 1–4.

41. Yang, W.; Wang, W.; Zhang, X.; Sun, S.; Liao, Q. Lightweight feature fusion network for single image super-resolution. *IEEE Signal Process. Lett.* **2019**, *26*, 538–542. [CrossRef]

42. Ma, C.; Mu, X.; Sha, D. Multi-Layers Feature Fusion of Convolutional Neural Network for Scene Classification of Remote Sensing. *IEEE Access* **2019**, *7*, 121685–121694. [CrossRef]

43. Du, Y.; Song, W.; He, Q.; Huang, D.; Liotta, A.; Su, C. Deep learning with multi-scale feature fusion in remote sensing for automatic oceanic eddy detection. *Inf. Fusion* **2019**, *49*, 89–99. [CrossRef]

44. Duarte, D.; Nex, F.; Kerle, N.; Vosselman, G. Multi-resolution feature fusion for image classification of building damages with convolutional neural networks. *Remote Sens.* **2018**, *10*, 1636. [CrossRef]

45. He, K.; Zhang, X.; Ren, S.; Sun, J. Deep residual learning for image recognition. In Proceedings of the IEEE Conference on Computer Vision and Pattern Recognition, Las Vegas, NV, USA, 27–30 June 2016; pp. 770–778.

46. Chen, L.C.; Papandreou, G.; Schroff, F.; Adam, H. Rethinking atrous convolution for semantic image segmentation. *arXiv* **2017**, arXiv:1706.05587.

47. Chen, L.C.; Papandreou, G.; Kokkinos, I.; Murphy, K.; Yuille, A.L. Deeplab: Semantic image segmentation with deep convolutional nets, atrous convolution, and fully connected crfs. *IEEE Trans. Pattern Anal. Mach. Intell.* **2017**, *40*, 834–848. [CrossRef]

48. Abadi, M.; Barham, P.; Chen, J.; Chen, Z.; Davis, A.; Dean, J.; Devin, M.; Ghemawat, S.; Irving, G.; Isard, M.; et al. Tensorflow: A system for large-scale machine learning. In Proceedings of the OSDI, Savannah, GA, USA, 2–4 November 2016; Volume 16, pp. 265–283.

49. Kingma, D.P.; Ba, J. Adam: A method for stochastic optimization. *arXiv* **2014**, arXiv:1412.6980.

50. Ioffe, S.; Szegedy, C. Batch normalization: Accelerating deep network training by reducing internal covariate shift. *arXiv* **2015**, arXiv:1502.03167.

Article

Small-Object Detection in Remote Sensing Images with End-to-End Edge-Enhanced GAN and Object Detector Network

Jakaria Rabbi [1],*, Nilanjan Ray [1], Matthias Schubert [2] and Subir Chowdhury [3] and Dennis Chao [3]

[1] Department of Computing Science, 2-32 Athabasca Hall, University of Alberta, Edmonton, AB T6G 2E8, Canada; nray1@ualberta.ca
[2] Institute for Informatic, Ludwig-Maximilians-Universität München, Oettingenstraße 67, D-80333 Munich, Germany; schubert@dbs.ifi.lmu.de
[3] Alberta Geological Survey, Alberta Energy Regulator, Edmonton, AB T6B 2X3, Canada; subir.chowdhury@aer.ca (S.C.); dennis.chao@aer.ca (D.C.)
* Correspondence: jakaria@ualberta.ca

Received: 19 March 2020; Accepted: 28 April 2020; Published: 1 May 2020

Abstract: The detection performance of small objects in remote sensing images has not been satisfactory compared to large objects, especially in low-resolution and noisy images. A generative adversarial network (GAN)-based model called enhanced super-resolution GAN (ESRGAN) showed remarkable image enhancement performance, but reconstructed images usually miss high-frequency edge information. Therefore, object detection performance showed degradation for small objects on recovered noisy and low-resolution remote sensing images. Inspired by the success of edge enhanced GAN (EEGAN) and ESRGAN, we applied a new edge-enhanced super-resolution GAN (EESRGAN) to improve the quality of remote sensing images and used different detector networks in an end-to-end manner where detector loss was backpropagated into the EESRGAN to improve the detection performance. We proposed an architecture with three components: ESRGAN, EEN, and Detection network. We used residual-in-residual dense blocks (RRDB) for both the ESRGAN and EEN, and for the detector network, we used a faster region-based convolutional network (FRCNN) (two-stage detector) and a single-shot multibox detector (SSD) (one stage detector). Extensive experiments on a public (car overhead with context) dataset and another self-assembled (oil and gas storage tank) satellite dataset showed superior performance of our method compared to the standalone state-of-the-art object detectors.

Keywords: object detection; faster region-based convolutional neural network (FRCNN); single-shot multibox detector (SSD); super-resolution; remote sensing imagery; edge enhancement; satellites

1. Introduction

1.1. Problem Description and Motivation

Object detection on remote sensing imagery has numerous prospects in various fields, such as environmental regulation, surveillance, military [1,2], national security, traffic, forestry [3], oil and gas activity monitoring. There are many methods for detecting and locating objects from images, which are captured using satellites or drones. However, detection performance is not satisfactory for noisy and low-resolution (LR) images, especially when the objects are small [4]. Even on high-resolution (HR) images, the detection performance for small objects is lower than that for large objects [5].

Current state-of-the-art detectors have excellent accuracy on benchmark datasets, such as ImageNet [6] and Microsoft common objects in context (MSCOCO) [7]. These datasets consist of everyday natural images with distinguishable features and comparatively large objects.

On the other hand, there are various objects in satellite images like vehicles, small houses, small oil and gas storage tanks etc., only covering a small area [4]. The state-of-the-art detectors [8–11] show a significant performance gap between LR images and their HR counterparts due to a lack of input features for small objects [12]. In addition to the general object detectors, researchers have proposed specialized methods, algorithms, and network architectures to detect particular types of objects from satellite images such as vehicles [13,14], buildings [15], and storage tanks [16]. These methods are object-specific and use fixed resolution for feature extraction and detection.

To improve detection accuracy on remote sensing images, researchers have used deep convolutional neural network (CNN)-based super-resolution (SR) techniques to generate artificial images and then detect objects [5,12]. Deep CNN-based SR techniques such as single image super-resolution convolutional networks (SRCNN) [17] and accurate image super-resolution using very deep convolutional networks (VDSR) [18] showed excellent results on generating realistic HR imagery from LR input data. Generative Adversarial Network (GAN)-based [19] methods such as super-resolution GAN (SRGAN) [20] and enhanced super-resolution GAN (ESRGAN) [21] showed remarkable performance in enhancing LR images with and without noise. These models have two subnetworks: a generator and a discriminator. Both subnetworks consist of deep CNNs. Datasets containing HR and LR image pairs are used for training and testing the models. The generator generates HR images from LR input images, and the discriminator predicts whether generated image is a real HR image or an upscaled LR image. After sufficient training, the generator generates HR images that are similar to the ground truth HR images, and the discriminator cannot correctly discriminate between real and fake images anymore.

Although the resulting images look realistic, the compensated high-frequency details such as image edges may cause inconsistency with the HR ground truth images [22]. Some works showed that this issue negatively impacts land cover classification results [23,24]. Edge information is an important feature for object detection [25], and therefore, this information needs to be preserved in the enhanced images for acceptable detection accuracy.

In order to obtain clear and distinguishable edge information, researchers proposed several methods using separate deep CNN edge extractors [26,27]. The results of these methods are sufficient for natural images, but the performance degrades on LR and noisy remote sensing images [22]. A recent method [22] used the GAN-based edge-enhancement network (EEGAN) to generate a visually pleasing result with sufficient edge information. EEGAN employs two subnetworks for the generator. One network generates intermediate HR images, and the other network generates sharp and noise-free edges from the intermediate images. The method uses a Laplacian operator [28] to extract edge information and in addition, it uses a mask branch to obtain noise-free edges. This approach preserves sufficient edge information, but sometimes the final output images are blurry compared to a current state-of-the-art GAN-based SR method [21] due to the noises introduced in the enhanced edges that might hurt object detection performance.

Another important issue with small-object detection is the huge cost of HR imagery for large areas. Many organizations are using very high-resolution satellite imagery to fulfill their purposes. When it comes to continuous monitoring of a large area for regulation or traffic purposes, it is costly to buy HR imagery frequently. Publicly available satellite imagery such as Landsat-8 [29] (30 m/pixel) and Sentinel-2 [30] (10 m/pixel) are not suitable for detecting small objects due to the high ground sampling distance (GSD). Detection of small objects (e.g., oil and gas storage tanks and buildings) is possible from commercial satellite imagery such as 1.5-m GSD SPOT-6 imagery but the detection accuracy is low compared to HR imagery, e.g., 30-cm GSD DigitalGlobe imagery in Bing map.

We have identified two main problems to detect small-objects from satellite imagery. First, the accuracy of small-object detection is lower compared to large objects, even in HR imagery due to sensor noise, atmospheric effects, and geometric distortion. Secondly, we need to have access to HR imagery, which is very costly for a vast region with frequent updates. Therefore, we need a solution to increase the accuracy of the detection of smaller objects from LR imagery. To the best of our knowledge, no work employed both SR network with edge-enhancement and object detector network in an end-to-end manner, i.e., using joint optimization to detect small remote sensing objects.

In this paper, we propose an end-to-end architecture where object detection and super-resolution is performed simultaneously. Figure 1 shows the significance of our method. State-of-the-art detectors miss objects when trained on the LR images; in comparison, our method can detect those objects. The detection performance improves when we use SR images for the detection of objects from two different datasets. Average precision (AP) versus different intersection over union (IoU) values (for both LR and SR) are plotted to visualize overall performance on test datasets. From Figure 1, we observe that for both the datasets, our proposed end-to-end method yields significantly better IoU values for the same AP. In Section 4.2, we discuss AP and IoU in more detail and these results are discussed in Section 4.

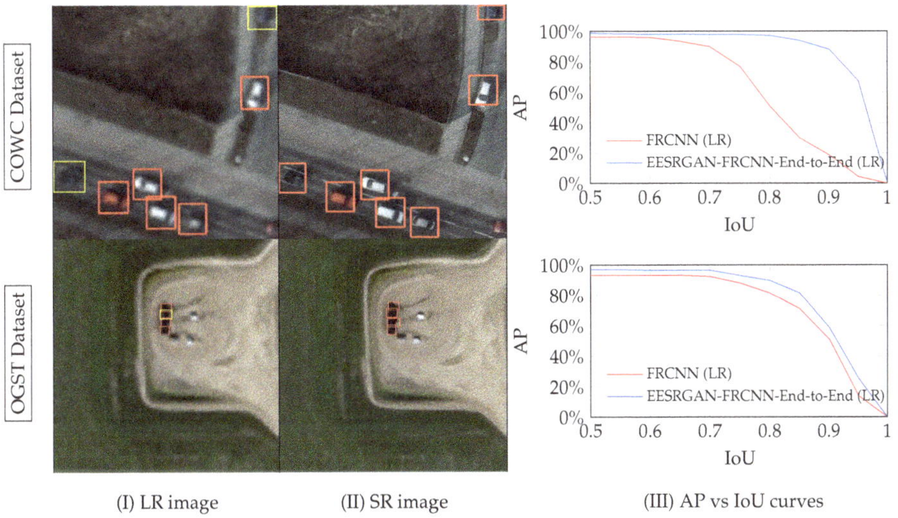

Figure 1. Detection on LR (low-resolution) images (60 cm/pixel) is shown in (I); in (II), we show the detection on generated SR (super-resolution) images (15 cm/pixel). The first row of this figure represents the COWC (car overhead with context) dataset [31], and the second row represents the OGST (oil and gas storage tank) dataset [32]. AP (average precision) values versus different IoU (intersection over union) values for the LR test set and generated SR images from the LR images are shown in (III) for both the datasets. We use FRCNN (faster region-based CNN) detector on LR images for detection. Then instead of using LR images directly, we use our proposed end-to-end EESRGAN (edge-enhanced SRGAN) and FRCNN architecture (EESRGAN-FRCNN) to generate SR images and simultaneously detect objects from the SR images. The red bounding boxes represent true positives, and yellow bounding boxes represent false negatives. IoU = 0.75 is used for detection.

1.2. Contributions of Our Method

Our proposed architecture consists of two parts: EESRGAN network and a detector network. Our approach is inspired by EEGAN and ESRGAN networks and showed a remarkable improvement over EEGAN to generate visually pleasing SR satellite images with enough edge information. We employed a generator subnetwork, a discriminator subnetwork, and an edge-enhancement subnetwork [22] for the SR network. For the generator and edge-enhancement network, we used residual-in-residual dense blocks (RRDB) [21]. These blocks contain multi-level residual networks with dense connections that showed good performance on image enhancement.

We used a relativistic discriminator [33] instead of a normal discriminator. Besides GAN loss and discriminator loss, we employed Charbonnier loss [34] for the edge-enhancement network. Finally, we used different detectors [8,10] to detect small objects from the SR images. The detectors acted like the discriminator as we backpropagated the detection loss into the SR network and, therefore, it improved the quality of the SR images.

We created the oil and gas storage tank (OGST) dataset [32] from satellite imagery (Bing map), which has 30 cm and 1.2 m GSD. The dataset contains labeled oil and gas storage tanks from the Canadian province of Alberta, and we detected the tanks on SR images. Detection and counting of the tanks are essential for the Alberta Energy Regulator (AER) [35] to ensure safe, efficient, orderly, and environmentally responsible development of energy resources. Therefore, there is a potential use of our method for detecting small objects from LR satellite imagery. The OGST dataset is available on Mendeley [32].

In addition to the OGST dataset, we applied our method on the publicly available car overhead with context (COWC) [31] dataset to compare the performance of detection for varying use-cases. During training, we used HR and LR image pairs but only required LR images for testing. Our method outperformed standalone state-of-the-art detectors for both datasets.

The remainder of this paper is structured as follows. We discuss related work in Section 2. In Section 3, we introduce our proposed method and describe every part of the method. The description of datasets and experimental results are shown in Section 4, final discussion is stated in Section 5 and Section 6 concludes our paper with a summary.

2. Related Works

Our work consists of an end-to-end edge enhanced image SR network with an object detector network. In this section, we discuss existing methods related to our work.

2.1. Image Super-Resolution

Many methods were proposed on SR using deep CNNs. Dong et al. proposed super-resolution CNN (SRCNN) [17] to enhance LR images in an end-to-end training outperforming previous SR techniques. The deep CNNs for SR evolved rapidly, and researchers introduced residual blocks [20], densely connected networks [36], and residual dense block [37] for improving SR results. He et al. [38] and Lim et al. [39] used deep CNNs without the batch normalization (BN) layer and observed significant performance improvement and stable training with a deeper network. These works were done on natural images.

Liebel et al. [40] proposed deep CNN-based SR network for multi-spectral remote sensing imagery. Jiang et al. [22] proposed a new SR architecture for satellite imagery that was based on GAN. They introduced an edge-enhancement subnetwork to acquire smooth edge details in the final SR images.

2.2. Object Detection

Deep learning-based object detectors can be categorized into two subgroups, region-based CNN (R-CNN) models that employ two-stage detection and uniform models using single stage detection [41]. Two-stage detectors comprise of R-CNN [42], Fast R-CNN [43], Faster R-CNN [8] and the most used single stage detectors are SSD [10], You only look once (YOLO) [11] and RetinaNet [9]. In the first stage of a two-stage detector, regions of interest are determined by selective search or a region proposal network. Then, in the second stage, the selected regions are checked for particular types of objects and minimal bounding boxes for the detected objects are predicted. In contrast, single-stage detectors omit the region proposal network and run detection on a dense sampling of all possible locations. Therefore, single-stage detectors are faster but, usually less accurate. RetinaNet [9] uses a focal loss function to deal with the data imbalance problem caused by many background objects and often showed similar performance as the two-stage approaches.

Many deep CNN-based object detectors were proposed on remote sensing imagery to detect and count small objects, such as vehicles [13,44,45]. Tayara et al. [13] introduced a convolutional regression neural network to detect vehicles from satellite imagery. Furthermore, a deep CNN-based detector was proposed [44] to detect multi oriented vehicles from remote sensing imagery. A method combining a deep CNN for feature extraction and a support vector machine (SVM) for object classification was proposed [45]. Ren et al. [46] modified the faster R-CNN detector to detect small objects in remote sensing images. They changed the region proposal network and incorporated context information into the detector. Another modified faster R-CNN detector was proposed by Tang et al. [47]. They used a hyper region proposal network to improve recall and used a cascade boosted classifier to verify candidate regions. This classifier can reduce false detection by mining hard negative examples.

An SSD-based end-to-end airplane detector with transfer learning was proposed, where, the authors used a limited number of airplane images for training [48]. They also proposed a method to solve the input size restrictions by dividing a large image into smaller tiles. Then they detected objects on smaller tiles and finally, mapped each image tile to the original image. They showed that their method performed better than the SSD model. In [49], the authors showed that finding a suitable parameter setting helped to boost the object detection performance of convolutional neural networks on remote sensing imagery. They used YOLO [11] as object detector to optimize the parameters and infer the results.

In [3], the authors detected conifer seedlings along recovering seismic lines from drone imagery. They used a dataset from different seasons and used faster R-CNN to infer the detection accuracy. There is another work [50] related to plant detection, where authors detected palm trees from satellite imagery using sliding window techniques and an optimized convolutional neural network.

Some works produced excellent results in detecting small objects. Lin et al. [51] proposed feature pyramid networks, which is a top-down architecture with lateral connections. The architecture could build high-level semantic feature maps at all scales. These feature maps boosted the object detection performance, especially for small object detection, when used as a feature extractor for faster R-CNN. Inspired by the receptive fields in human visual systems, Liu et al. [52] proposed a receptive field block (RFB) module that used the relationship between the size and eccentricity of receptive fields to enhance the feature discrimination and robustness. Hence, the module increased the detection performance of objects with various sizes when used as the replacement of the top convolutional layers of SSD.

A one-stage detector called single-shot refinement neural network (RefineDet) [53] was proposed to increase the detection accuracy and also enhance the inference speed. The detector worked well for small object detection. RefineDet used two modules in its architecture: an anchor refinement module to remove negative anchors and an object detection module that took refined anchors as the input. The refinement helped to detect small objects more efficiently than previous methods. In [54], feature fusion SSD (FSSD) was proposed where features from different layers with different scales were concatenated together, and then some downsampling blocks were used to generate new feature pyramids. Finally, the features were fed to multibox detector for prediction. The feature fusion in FSSD increased the detection performance for both large and small objects. Zhu et al. [55] trained single-shot object detectors from scratch and obtained state-of-the-art performance on various benchmark datasets. They removed the first downsampling layer of SSD and introduced root block (with modified convolutional filters) to exploit more local information from an image. Therefore, the detector was able to extract powerful features for small object detection.

All of the aforementioned works were proposed for natural images. A method related to small object detection on remote sensing imagery was proposed by Yang et al. [56]. They used modified faster R-CNN to detect both large and small objects. They proposed rotation dense feature pyramid networks (R-DFPN), and the use of this network helped to improve the detection performance of small objects.

There is an excellent review paper by Zhao et al. [57], where the authors showed a thorough review of object detectors and also showed the advantages and disadvantages of different object detectors. The effect of object size was also discussed in the paper. Another survey paper about object detection in remote sensing images by Li et al. [58] showed review and comparison of different methods.

2.3. Super-Resolution Along with Object Detection

The positive effects of SR on object detection tasks was discussed in [5] where the authors used remote sensing datasets for their experiments. Simultaneous CNN-based image enhancement with object detection using single-shot multibox detector (SSD) [10] was done in [59]. Haris et al. [60] proposed a deep CNN-based generator to generate a HR image from a LR image and then used a multi-task network as a discriminator and also for localization and classification of objects. These works were done on natural images, and LR and HR image pairs were required. In another work [12], a method using simultaneous super-resolution with object detection on satellite imagery was proposed. The SR network in this approach was inspired by the cycle-consistent adversarial network [61]. A modified faster R-CNN architecture was used to detect vehicles from enhanced images produced by the SR network.

3. Method

In this paper, we aim to improve the detection performance of small objects on remote sensing imagery. Towards this goal, we propose an end-to-end network architecture that consists of two modules: A GAN-based SR network and a detector network. The whole network is trained in an end-to-end manner and HR and LR image pairs are needed for training.

The SR network has three components: generator (G), discriminator (D_{Ra}), and edge-enhancement network (EEN). Our method uses end-to-end training as the gradient of the detection loss from the dectector is backpropagated into the generator. Therefore, the detector also works like a discriminator and encourages the generator G to generate realistic images similar to the ground truth. Our entire network structure can also be divided into two parts: A generator consisting of the EEN and a discriminator, which includes the D_{Ra} and the detector network. In Figure 2, we show the role of the detector as a discriminator.

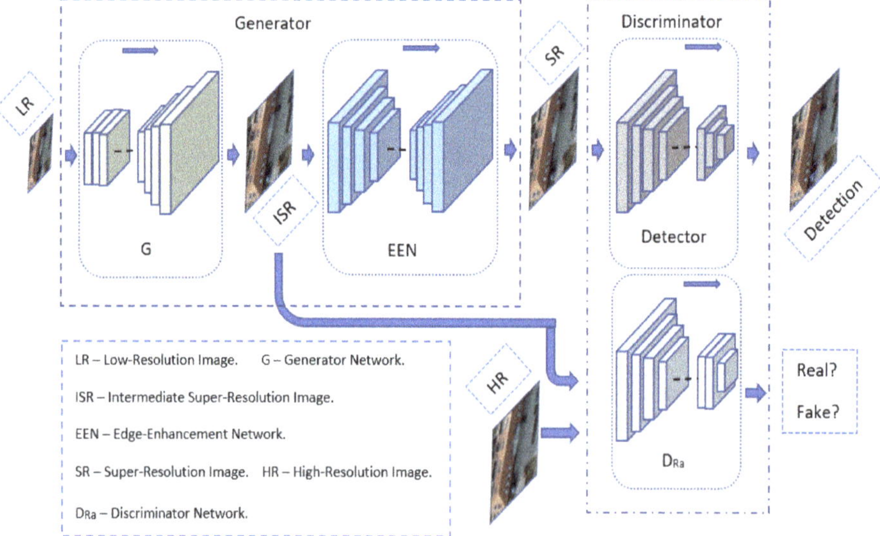

Figure 2. Overall network architecture with a generator and a discriminator module.

The generator G generates intermediate super-resolution (ISR) images, and then final SR images are generated after applying the EEN network. The discriminator (D_{Ra}) discriminates between ground truth (GT) HR images and ISR. The inverted gradients of D_{Ra} are backpropagated into the generator G in order to create SR images allowing for accurate object detection. Edge information is extracted from ISR, and the EEN network enhances these edges. Afterwards, the enhanced edges are again added to the ISR after subtracting the original edges extracted by the Laplacian operator and we get the output SR images with enhanced edges. Finally, we detect objects from the SR images using the detector network.

We use two different loss functions for EEN: one compares the difference between SR and ground truth images, and the other compares the difference between the extracted edge from ISR and ground truth. We also use the VGG19 [62] network for feature extraction that is used for perceptual loss [21]. Hence, it generates more realistic images with more accurate edge information. We divide the whole pipeline as a generator, and a discriminator, and these two components are elaborated in the following.

3.1. Generator

Our generator consists of a generator network G and an edge-enhancement network EEN. In this section, we describe the architectures of both networks and the corresponding loss function.

3.1.1. Generator Network G

We use the generator architecture from ESRGAN [21], where all batch normalization (BN) layers are removed, and RRDB is used. The overall architecture of generator G is shown in Figure 3, and the RRDB is depicted in Figure 4.

Inspired by the architecture of ESRGAN, we remove BN layers to increase the performance of the generator G and to reduce the computational complexity. The authors of ESRGAN also state that the BN layers tend to introduce unpleasant artifacts and limit the generalization ability of the generator when the statistics of training and testing datasets differ significantly.

We use RRDB as the basic blocks of the generator network G that uses a multi-level residual network with dense connections. Those dense connections increase network capacity, and we also use residual scaling to prevent unstable conditions during the training phase [21]. We use the parametric rectified linear unit (PReLU) [38] for the dense blocks to learn the parameter with the other neural network parameters. As discriminator (D_{Ra}), we employ a relativistic average discriminator similar to the work represented in [21].

In Equations (1) and (2), the relativistic average discriminator is formulated for our architecture. Our generator G depends on the discriminator D_{Ra}, and hence we briefly discuss the discriminator D_{Ra} here and then, describe all details in Section 3.2. The discriminator predicts the probability that a real image (I_{HR}) is relatively more realistic than a generated intermediate image (I_{ISR}).

$$D_{Ra}(I_{HR}, I_{ISR}) = \sigma(C(I_{HR}) - \mathbb{E}_{I_{ISR}}[C(I_{ISR})]) \rightarrow 1 \quad \text{More Realistic than fake data?} \tag{1}$$

$$D_{Ra}(I_{ISR}, I_{HR}) = \sigma(C(I_{ISR}) - \mathbb{E}_{I_{HR}}[C(I_{HR})]) \rightarrow 0 \quad \text{Less realistic than real data?} \tag{2}$$

In Equations (1) and (2), σ, $C(\cdot)$ and $\mathbb{E}_{I_{ISR}}$ represents the sigmoid function, discriminator output and operation of calculating mean for all generated intermediate images in a mini-batch. The generated intermediate images are created by the generator where $I_{ISR} = G(I_{LR})$. It is evident from Equation (3) that the adversarial loss of the generator contains both I_{HR} and I_{ISR} and hence, it benefits from the gradients of generated and ground truth images during the training process. The discriminator loss is depicted in Equation (4).

$$L_G^{Ra} = -\mathbb{E}_{I_{HR}}[\log(1 - D_{Ra}(I_{HR}, I_{ISR}))] - \mathbb{E}_{I_{ISR}}[\log(D_{Ra}(I_{ISR}, I_{HR}))] \tag{3}$$

$$L_D^{Ra} = -\mathbb{E}_{I_{HR}}[\log(D_{Ra}(I_{HR}, I_{ISR}))] - \mathbb{E}_{I_{ISR}}[\log(1 - D_{Ra}(I_{ISR}, I_{HR}))] \tag{4}$$

We use two more losses for generator G: one is perceptual loss (L_{percep}), and another is content loss (L_1) [21]. The perceptual loss is calculated using the feature map ($vgg_{fea}(\cdot)$) before the activation layers of a fine-tuned VGG19 [62] network, and the content loss calculates the 1-norm distance between I_{ISR} and I_{HR}. Perceptual loss and content loss is shown in Equations (5) and (6).

$$L_{percep} = \mathbb{E}_{I_{LR}} ||vgg_{fea}(G(I_{LR}) - vgg_{fea}(I_{HR})||_1 \tag{5}$$

$$L_1 = \mathbb{E}_{I_{LR}} ||G(I_{LR}) - I_{HR}||_1 \tag{6}$$

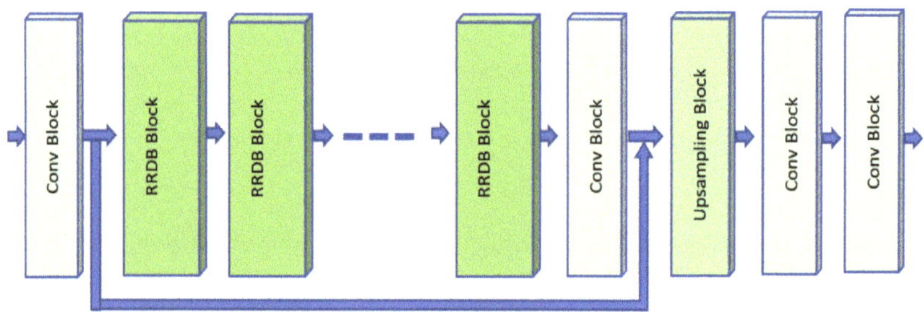

Figure 3. Generator G with RRDB (residual-in-residual dense blocks), convolutional and upsampling blocks.

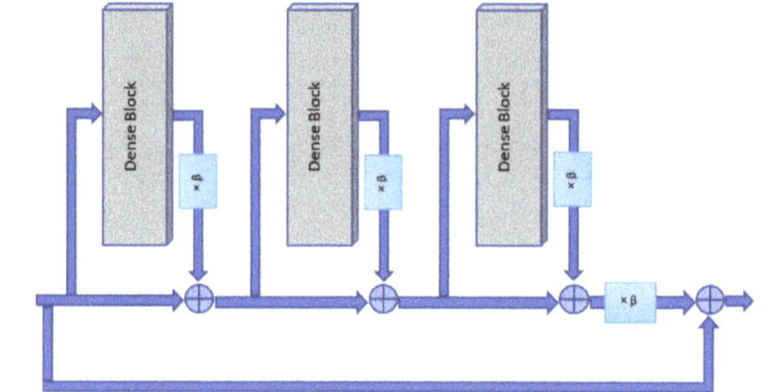

(**a**) RRDB from generator.

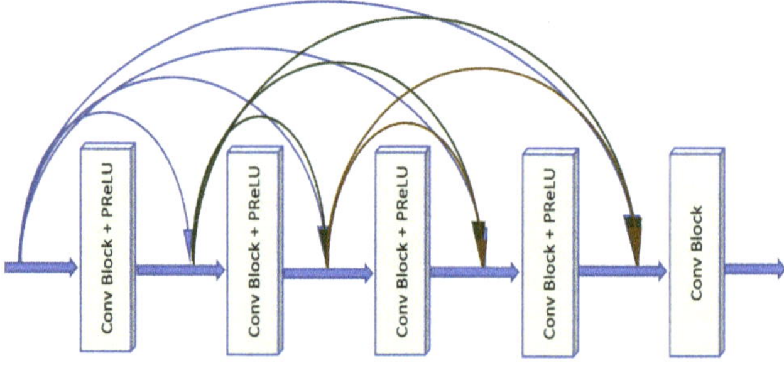

(**b**) Dense block from RRDB.

Figure 4. Internal diagram of RRDB (residual-in-residual dense blocks).

3.1.2. Edge-Enhancement Network EEN

The EEN network removes noise and enhances the extracted edges from an image. An overview of the network is depicted in Figure 5. In the beginning, Laplacian operator [28] is used to extract edges from the input image. After the edge information is extracted, it is passed through convolutional, RRDB, and upsampling blocks. There is a mask branch with sigmoid activation to remove edge noise as described in [22]. Finally, the enhanced edges are added to the input images where the edges extracted by the Laplacian operator were subtracted.

The EEN network is similar to the edge-enhancement subnetwork proposed in [22] with two improvements. First, we replace the dense blocks with RRDB. The RRDB shows improved performance according to ESRGAN [21]. Hence, we replace the dense block for improved performance of the EEN network. Secondly, we introduce a new loss term to improve the reconstruction of the edge information.

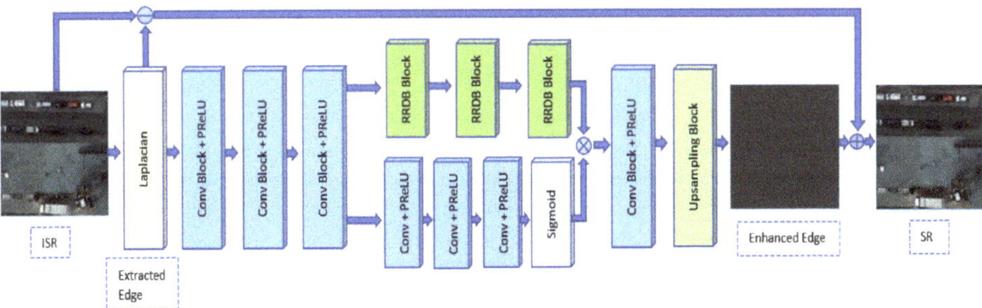

Figure 5. Edge-enhancement network where input is an ISR (intermediate super-resolution) image and output is a SR (super-resolution) image.

In [22], authors extracted the edge information from I_{ISR} and enhanced the edges using an edge-enhancement subnetwork which is afterwards added to the edge-subtracted I_{ISR}. To train the network, [22] proposed to use Charbonnier loss [34] between the I_{ISR} and I_{HR}. This function is called consistency loss for images (L_{img_cst}) and helps to get visually pleasant outputs with good edge information. However, sometimes the edges of some objects are distorted and produce some noises and consequently, do not give good edge information. Therefore, we introduce a consistency loss for the edges (L_{edge_cst}) as well. To compute L_{edge_cst} we evaluate the Charbonnier loss between the extracted edges (I_{edge_SR}) from I_{SR} and the extracted edges (I_{edge_HR}) from I_{HR}. The two consistency losses are depicted in Equations (7) and (8) where $\rho(\cdot)$ is the Charbonnier penalty function [63]. The total consistency loss is finally calculated for both images and edges by summing up the individual loss. The loss of our EEN is shown in Equation (9).

$$L_{img\_cst} = \mathbb{E}_{I_{SR}}[\rho(I_{HR} - I_{SR})] \tag{7}$$

$$L_{edge\_cst} = \mathbb{E}_{I_{edge\_SR}}[\rho(I_{edge\_HR} - I_{edge\_SR})] \tag{8}$$

$$L_{een} = L_{img\_cst} + L_{edge\_cst} \tag{9}$$

Finally, we get the overall loss for the generator module by adding the losses of the generator G and the EEN network. The overall loss for the generator module is shown in Equation (10) where λ_1, λ_2, λ_3, and λ_4 are the weight parameters to balance different loss components. We empirically set the values as $\lambda_1 = 1$, $\lambda_2 = 0.001$, $\lambda_3 = 0.01$, and $\lambda_4 = 5$.

$$L_{G\_een} = \lambda_1 L_{percep} + \lambda_2 L_G^{Ra} + \lambda_3 L_1 + \lambda_4 L_{een} \tag{10}$$

3.2. Discriminator

As described in the previous section, we use the relativistic discriminator D_{Ra} for training the generator G. The architecture of the discriminator is taken from ESRGAN [21] which employs the VGG-19 [62] architecture. We use Faster R-CNN [8] and SSD [10] for our detector networks. The discriminator (D_{Ra}) and the detector network jointly act as discriminator for the generator module. We briefly describe these two detectors in the next two sections.

3.2.1. Faster R-CNN

The Faster R-CNN [8] is a two-stage object detector and contains two networks: a region proposal network (RPN) to generate region proposals from an image and another network to detect objects from these proposals. In addition, the second network also tries to fit the bounding boxes around the detected objects.

The task of the RPN is to return image regions that have a high probability of containing an object. The RPN network uses a backbone network such as VGG [62], ResNet, or ResNet with feature pyramid network [51]. These networks are used as feature extractors, and different types of feature extractors can be chosen based on their performance on public datasets. We use ResNet-50-FPN [51] as a backbone network for our faster R-CNN. We use this network because it displayed a higher precision than VGG-19 and ResNet-50 without FPN (especially for small object detection) [51]. Even though the use of a larger network might lead to a further performance improvement, we chose ResNet-50-FPN due to its comparably moderate hardware requirements and more efficient convergence times.

After the RPN, there are two branches for detection: a classifier and a regressor. The classification branch is responsible for classifying a proposal to a specific object, and the regression branch finds the accurate bounding box of the object. In our case, both datasets contain objects with only one class, and therefore, our classifier infers only two classes: the background class and the object class.

3.2.2. SSD

The SSD [10] is a single-shot multibox detector that detects objects in a single stage. Here, single-stage means that classification and localization are done in a single forward pass through the network. Like Faster R-CNN, SSD also has a feature extractor network, and different types of networks can be used. To serve the primary purpose of SSD, which is speed, we use VGG-16 [62] as a feature extractor network. After this network, SSD has several convolutional feature layers of decreasing sizes. This representation can seem like a pyramid representation of images at different scales. Therefore, the detection of objects happens in every layer, and finally, we get the object detection output as class values and coordinates of bounding boxes.

3.2.3. Loss of the Discriminator

The relativistic discriminator loss (L_D^{Ra}) is already described in the previous section and depicted in Equation (4). This loss is added to the detector loss to get the final discriminator loss.

Both Faster R-CNN and SSD have similar regression/localization losses but different classification losses. For regression/localization, both use smooth L_1 [8] loss between detected and ground truth bounding box coordinates (t_*). Classification (L_{cls_frcnn}) and regression loss (L_{reg_frcnn}) and overall loss (L_{det_frcnn}) of Faster R-CNN are given in the following:

$$L_{cls\_frcnn} = \mathbb{E}_{I_{LR}}[-\log(Det_{cls\_frcnn}(G_{G\_een}(I_{LR})))] \tag{11}$$

$$L_{reg\_frcnn} = \mathbb{E}_{I_{LR}}[smooth_{L1}(Det_{reg\_frcnn}(G_{G\_een}(I_{LR})), t_*)] \tag{12}$$

$$L_{det\_frcnn} = L_{cls\_frcnn} + \lambda L_{reg\_frcnn} \tag{13}$$

Here, λ is used to balance the losses, and it is set to 1 empirically. Det_{cls_frcnn} and Det_{reg_frcnn} are the classifier and regressor for the Faster R-CNN. Classification (L_{cls_ssd}), regression loss (L_{reg_ssd}) and overall loss (L_{det_ssd}) of SSD are as following:

$$L_{cls\_ssd} = \mathbb{E}_{I_{LR}}[-\log(softmax(Det_{cls\_ssd}(G_{G\_een}(I_{LR}))))] \tag{14}$$

$$L_{reg\_ssd} = \mathbb{E}_{I_{LR}}[smooth_{L1}(Det_{reg\_ssd}(G_{G\_een}(I_{LR})), t_*)] \tag{15}$$

$$L_{det\_ssd} = L_{cls\_ssd} + \alpha L_{reg\_ssd} \tag{16}$$

Here, α is used to balance the losses, and it is set to 1 empirically. Det_{cls_ssd} and Det_{reg_ssd} are the classifier and regressor for the SSD.

3.3. Training

Our architecture can be trained in separate steps or jointly in an end-to-end way. We discuss the details of these two types of training in the next two sections.

3.3.1. Separate Training

In separate training, we train the SR network (generator module and discriminator D_{Ra}) and the detector separately. Detector loss is not backpropagated to the generator module. Therefore, the generator is not aware of the detector and thus, it only gets feedback from the discriminator D_{Ra}. For example, in Equation (11), no error is backpropagated to the G_{G_een} network (the network is detached during the calculation of the detector loss) while calculating the loss L_{cls_frcnn}.

3.3.2. End-to-End Training

In end-to-end training, we train the whole architecture end-to-end that means the detector loss is backpropagated to the generator module. Therefore, the generator module revceives gradients from both detector and discriminator D_{Ra}. We get the final discriminator loss (L_{D_det}) as following:

$$L_{D\_det} = L_D^{Ra} + \eta L_{det} \tag{17}$$

Here, η is the parameter to balance the contribution of the detector loss and we empirically set it to 1. Detection loss from SSD or Faster R-CNN is denoted by L_{det}. Finally, we get an overall loss ($L_{overall}$) for our architecture as follows.

$$L_{overall} = L_{G\_een} + L_{D\_det} \tag{18}$$

4. Experiments

As mentioned above, we trained our architecture separately and in an end-to-end manner. For separate training, we first trained the SR network until convergence and then trained the detector networks based on the SR images. For end-to-end training, we also employed separate training as pre-training step for weight initialization. Afterwards SR and object detection networks were jointly trained, i.e., the gradients from the the object detector were propagated into the generator network.

In the training process, the learning rate was set to 0.0001 and halved after every 50 K iterations. The batch size was set to 5. We used Adam [64] as optimizer with $\beta_1 = 0.9$, $\beta_2 = 0.999$ and updated the whole architecture weights until convergence. We used 23 RRDB blocks for the generator G and five RRDB blocks for the EEN network. We implemented our architecture with the PyTorch framework [65] and trained/tested using two NVIDIA Titan X GPUs. The end-to-end training with COWC took 96 h for 200 epochs. The average inference speed using faster R-CNN was approximately four images/second and seven images/second for SSD. Our implementation can be found in GitHub [66].

4.1. Datasets

4.1.1. Cars Overhead with Context Dataset

Cars overhead with context (COWC) dataset [31] contains 15 cm (one pixel cover 15 cm distance at ground level) satellite images from six different regions. The dataset contains a large number of unique cars and covers regions from Toronto in Canada, Selwyn in New Zealand, Potsdam and Vaihingen in Germany, Columbus and Utah in the United States. Out of these six regions, we used the dataset from Toronto and Potsdam. Therefore, when we refer to the COWC dataset, we refer to the dataset from these two regions. There are 12,651 cars in our selected dataset. The dataset contains only RGB images, and we used these images for training and testing.

We used 256-by-256 image tiles, and every image tile contains at least one car. The average length of a car was between 24 and 48 pixels, and the width was between 10 and 20 pixels. Therefore, the area of a car was between 240 and 960 pixels, which can be considered as a small object relative to the other large satellite objects. We used bi-cubic downsampling to generate LR images from the COWC dataset. The downscale factor was $4\times$, and therefore, we had 64 pixels to 64 pixels size for LR images. We had a text file associated with each image tile containing the coordinates of the bounding box for each car.

Our experiments considered the dataset having only one class, car, and did not consider any other type of object. Figure 6 shows examples from the COWC dataset. We experimented with a total of 3340 tiles for training and testing. Our train/test split was 80%/20%, and the training set was further divided into a training and a validation set by an 80% to 20% ratio. We trained our end-to-end architecture with an augmented training dataset with random horizontal flips and ninety-degree rotations.

(**a**) LR image   (**b**) HR image   (**c**) GT image

Figure 6. COWC (car overhead with context) dataset: LR-HR (low-resolution and high-resolution) image pairs are shown in (**a**,**b**) and GT (ground truth) images with bounding boxes for cars are in (**c**).

4.1.2. Oil and Gas Storage Tank Dataset

The oil and gas storage tank (OGST) dataset has been complied in Alberta Geological Survey (AGS) [67], a branch of the Alberta Energy Regulatory (AER) [35]. AGS provides geoscience information and support to AER's regulatory functions on energy developments to be carried out in a manner to ensure public and environmental safety. To assist AER with sustainable land management and compliance assurance [68], AGS is utilizing remote sensing imagery for identifying the number of oil and gas storage tanks inside well pad footprints in Alberta.

While the SPOT-6 satellite imagery at 1.5 m pixel resolution provided by the AGS has sufficient quality and details for many regulatory functions, it is difficult to detect small objects within well pads, e.g., oil and gas storage tanks with ordinary object detection methods. The diameter of a typical storage tank is about 3 m and their placements are usually vertical and side-by-side with less than 2 m. To train our architecture for this use-case, we needed a dataset for providing pairs of low and high-resolution images. Therefore, we have created the OGST dataset using free imagery from the Bing map [69].

The OGST dataset contains 30 cm resolution remote sensing images (RGB) from the Cold Lake Oil Sands region of Alberta, Canada where there is a high level of oil and gas activities and concentration of well pad footprints. The dataset contains 1671 oil and gas storage tanks from this area.

We used 512-by-512 image tiles, and there was no image without any oil and gas storage tank in our experiment. The average area covered by an individual tank was between 800 and 1600 pixels. Some industrial tanks were large, but most of the tanks covered small regions on the imagery. We downscaled the HR images using bi-cubic downsampling with the factor of $4\times$, and therefore, we got a LR tile of size 128-by-128 pixels. Every image tile was associated with a text file containing the coordinates of the bounding boxes for the tanks on a tile. We have showed examples from the OGST dataset in Figure 7.

(a) LR image   (b) HR image   (c) GT image

Figure 7. OGST (oil and gas storage tank) dataset: LR-HR (low-resolution and high-resolution) image pairs are shown in (**a**,**b**) and GT (ground truth) images with bounding boxes for oil and gas storage tanks are in (**c**).

As with the COWC dataset, our experiments considered one unique class here, tank, and we had a total of 760 tiles for training and testing. We used a 90%/10% split for our train/test data. The training data was further divided by 90%/10% for the train/validation split. The percentage of training data was higher here compared to the previous dataset to increase the training data because of the smaller size of the dataset. The dataset is available at [66].

4.2. Evaluation Metrics for Detection

We obtained our detection output as bounding boxes with associated classes. To evaluate our results, we used average precision (AP), and calculated intersection over union (IoU), precision, and recall for obtaining AP.

We denote the set of correctly detected objects as true positives (TP) and the set of falsely detected objects of false positives (FP). The precision is now the ratio between the number of TPs relative to all predicted objects:

$$Precision = \frac{|TP|}{|TP| + |FP|} \tag{19}$$

We denote the set of objects which are not detected by the detector as false negatives (FN). Then, the recall is defined as the ratio of detected objects (TP) relative to the number of all objects in the data set:

$$Recall = \frac{|TP|}{|TP| + |FN|} \tag{20}$$

To measure the localization error of predicted bounding boxes, IoU computes the overlap between two bounding boxes: the detected and the ground truth box. If we take all the boxes that have an IoU $\geq \tau$ as TP and consider all other detections as FP, then we get the precision at τ IoU. If we now vary τ from 0.5 to 0.95 IoU with a step size of 0.05, we receive ten different precision values which can be combined into the average precision (AP) at IoU = 0.5:0.95 [8]. Let us note that in the case of multi-class classification, we would need to compute the AP for object each class separately. To receive a single performance measure for object detection, the mean AP (mAP) is computed which is the most common performance measure for object detection quality.

In this paper, both of our datasets only contain single class, and hence, we used AP as our evaluation metric. We mainly showed the results of AP at IoU = 0.5:0.95 as our method performed increasingly better compared to other models when we increased the IoU values for AP calculation. We show this trend in Section 4.3.4.

4.3. Results

4.3.1. Detection without Super-Resolution

We ran the two detectors to document the object detection performance on both LR and HR images. We used SSD with vgg16 [62] network and Faster R-CNN (FRCNN) with ResNet-50-FPN [51] detector. We trained the two models with both HR and 4×-downscaled LR images. Testing was also done with both HR and LR images.

In Table 1, we show the results of the detection performance of the detectors with different train/test combinations. When we only used LR images for both training and testing, we observed 64% AP for Faster R-CNN. When training on HR images and testing with LR images, the accuracy dropped for both detectors. We also added detection results (using LR images for training/testing) for both the datasets using SSD with RFB modules (SSD-RFB) [52], where accuracy slightly increased from the base SSD.

The last two rows in Table 1 depict the accuracy of both detectors when training and testing on HR images. We have achieved up to 98% AP with the Faster R-CNN detector. This, shows the large impact of the resolution to the object detection quality and sets a natural upper bound on how close a SR-based method can get when working on LR images. In the next sections, we demonstrate that our approaches considerably improve the detection rate on LR imagery and get astonishingly close to the performance of directly working on HR imagery.

Table 1. Detection on LR (low-resolution) and HR (high-resolution) images without using super-resolution. Detectors are trained with both LR and HR images and AP (average precision) values are calculated using 10 different IoUs (intersection over union).

Model	Training Image Resolution-Test Image Resolution	COWC Dataset (Test Results) (AP at IoU = 0.5:0.95) (Single Class-15 cm)	OGST Dataset (Test Results) (AP at IoU = 0.5:0.95) (Single Class-30 cm)
SSD	LR-LR	61.9%	76.5%
SSD	HR-LR	58%	75.3%
FRCNN	LR-LR	64%	77.3%
FRCNN	HR-LR	59.7%	75%
SSD-RFB	LR-LR	63.1%	76.7%
SSD	HR-HR	94.1%	82.5%
FRCNN	HR-HR	98%	84.9%

4.3.2. Separate Training with Super-Resolution

In this experiment, we created $4\times$ upsampled images from the LR input images using bicubic upsampling and different SR methods. Let us note that no training was needed for applying bicubic upsampling since it is a parameter free function. We used the SR images as test data for two types of detectors. We compared three GAN architectures for generating SR images, our new EESRGAN architecture, ESRGAN [21] and EEGAN [22]. Each network was trained separately on the training set before the object detector was trained. For the evaluation, we again compared detectors being trained on the SR images from the particular architecture and detectors being directly trained on the HR images.

In Table 2, the detection output of the different combinations of SR methods and detectors is shown with the different combinations of train/test pairs. As can be seen, our new EESRGAN architecture displayed the best results already getting close to the detection rates which could be observed when working with HR images only. However, after training EESRGAN can be directly applied to LR imagery where no HR data is available and still achieved very good results. Furthermore, we could observe that other SR methods EEGAN and ESRGAN have already improved the AP considerably when used for preprocessing of LR images. However, for both data sets, EESRGAN have outperformed the other two methods.

Table 2. Detection on SR (super-resolution) images with separately trained SR network. Detectors are trained with both SR and HR (high-resolution) images and AP (average precision) values are calculated using 10 different IoUs (intersection over union).

Model	Training Image Resolution-Test Image Resolution	COWC Dataset (Test Results) (AP at IoU = 0.5:0.95) (Single Class-15 cm)	OGST Dataset (Test Results) (AP at IoU = 0.5:0.95) (Single Class-30 cm)
Bicubic + SSD	SR-SR	72.1%	77.6%
Bicubic + SSD	HR-SR	58.3%	76%
Bicubic + FRCNN	SR-SR	76.8%	78.5%
Bicubic + FRCNN	HR-SR	61.5%	77.1%

Table 2. *Cont.*

Model	Training Image Resolution-Test Image Resolution	COWC Dataset (Test Results) (AP at IoU = 0.5:0.95) (Single Class-15 cm)	OGST Dataset (Test Results) (AP at IoU = 0.5:0.95) (Single Class-30 cm)
EESRGAN + SSD	SR-SR	86%	80.2%
	HR-SR	83.1%	79.4%
EESRGAN + FRCNN	SR-SR	**93.6%**	**81.4%**
	HR-SR	92.9%	80.6%
ESRGAN + SSD	SR-SR	85.8%	80.2%
	HR-SR	82.5%	78.9%
ESRGAN + FRCNN	SR-SR	92.5%	81.1%
	HR-SR	91.8%	79.3%
EEGAN + SSD	SR-SR	86.1%	79.1%
	HR-SR	83.3%	77.5%
EEGAN + FRCNN	SR-SR	92%	79.9%
	HR-SR	91.1%	77.9%

4.3.3. End-to-End Training with Super-Resolution

We trained our EESRGAN network and detectors end-to-end for this experiment. The discriminator (D_{Ra}), and the detectors jointly acted as a discriminator for the entire architecture. Detector loss was backpropagated to the SR network, and therefore, the loss contributed to the enhancement of LR images. At training time, LR-HR image pairs were used to train the EEGAN part, and then the generated SR images were sent to the detector for training. At test time, only the LR images were fed to the network. Our architecture first generated a SR image of the LR input before object detection was performed.

We also compared our results with different architectures. We used ESRGAN [21] and EEGAN [22] with the detectors for comparison. Table 3 clearly shows that our method delivers superior results compared to others.

Table 3. Detection with end-to-end SR (super-resolution) network. Detectors are trained with SR images and AP (average precision) values are calculated using 10 different IoUs (intersection over union).

Model	Training Image Resolution-Test Image Resolution	COWC Dataset (Test Results) (AP at IoU = 0.5:0.95) (Single Class-15 cm)	OGST Dataset (Test Results) (AP at IoU = 0.5:0.95) (Single Class-30 cm)
EESRGAN + SSD	SR-SR	89.3%	81.8%
EESRGAN + FRCNN	SR-SR	**95.5%**	**83.2%**
ESRGAN + SSD	SR-SR	88.5%	81.1%
ESRGAN + FRCNN	SR-SR	93.6%	82%
EEGAN + SSD	SR-SR	88.1%	80.8%
EEGAN + FRCNN	SR-SR	93.1%	81.3%

4.3.4. AP Versus IoU Curve

We have calculated the AP values on different IoUs. In Figure 8, we plot the AP versus IoU curves for our datasets. The performance of EESRGAN-FRCNN, end-to-end EESRGAN-FRCNN, and FRCNN is shown in the figure. The end-to-end EESRGAN-FRCNN network has performed better than the separately trained network. The difference is most evident for the higher IoUs on the COWC dataset.

Our results indicate excellent performance compared to the highest possible AP values obtained from standalone FRCNN (trained and tested on HR images).

The OGST dataset has displayed less performance variation compared to the COWC dataset. The object size of the OGST dataset is larger than that of the COWC dataset. Therefore, the performance difference was not similar to the COWC dataset when we compared between standalone FRCNN and our method on the OGST dataset. To conclude, training our new architecture in an end-to-end manner has displayed an improvement for both the datasets.

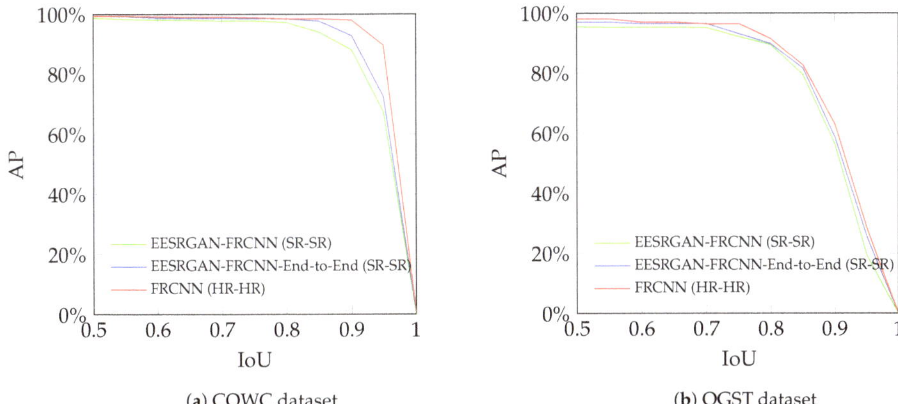

(a) COWC dataset   (b) OGST dataset

Figure 8. AP-IoU (average precision-intersection over union) curves for the datasets. Plotted results show the detection performance of standalone faster R-CNN on HR (high-resolution) images and our proposed method (with and without end-to-end training) on SR (super-resolution) images.

4.3.5. Precision Versus Recall

In Figure 9, precision-recall curves are shown for both of our datasets. The precision-recall curve for COWC dataset is depicted in Figure 9a,b represents the curve for OGST dataset. For each dataset, we plot the curves for standalone faster R-CNN with LR training/testing images, and our method with/without end-to-end training. We used IoU = 0.5 to calculate precision and recall.

The precision-recall curves for both datasets show that our method has higher precision values in higher recall values compared to the standalone faster R-CNN models. Our models with end-to-end training performed better than our models without the end-to-end training. In particular, the end-to-end models have detected more than 99% of the cars with 96% AP in the COWC dataset. For the OGST dataset, our end-to-end models have detected more than 81% of the cars with 97% AP.

4.3.6. Effects of Dataset Size

We trained our architecture with different training set sizes and tested with a fixed test set. In Figure 10, we plot the AP values (IoU = 0.5:0.95) against different numbers of labeled objects for both of our datasets (training data). We used five different dataset sizes: $\{500, 1000, 3000, 6000, 10,000 \ (cars)\}$ and $\{100, 200, 400, 750, 1491 \ (tanks)\}$ to train our model with and without the end-to-end setting.

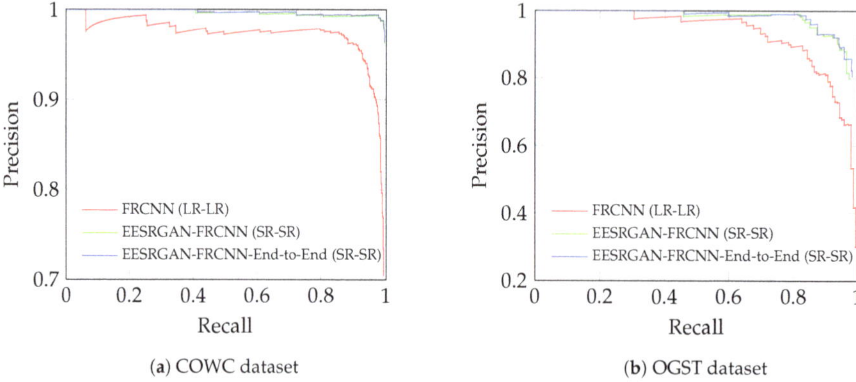

(**a**) COWC dataset　　　　　　　　(**b**) OGST dataset

Figure 9. Precision-recall curve for the datasets. Plotted results show the detection performance of standalone faster R-CNN on LR (low-resolution) images and our proposed method (with and without end-to-end training) on SR (super-resolution) images.

We got the highest AP value of 95.5% with our full COWC training dataset (10,000 cars), and we used the same test dataset (1000 cars) for all combinations of the training dataset (with end-to-end setting). We also used another set of 1000 labeled cars for validation. Using 6000 cars, we got an AP value near to the highest AP, as shown with the plot of AP versus dataset size (COWC). The AP value decreased significantly when we used only 3000 labeled cars as training data. We got the lowest AP using only 500 labeled cars, and the trend of AP was further decreasing as depicted in Figure 10a. Therefore, we can infer that we needed around 6000 labeled cars to get precision higher than 90% for the COWC dataset. We observed slightly lower AP values for all sizes of COWC datasets when we did not use the end-to-end setting, and we observed higher differences between the two settings (with and without end-to-end) when we used less than 6000 labeled cars.

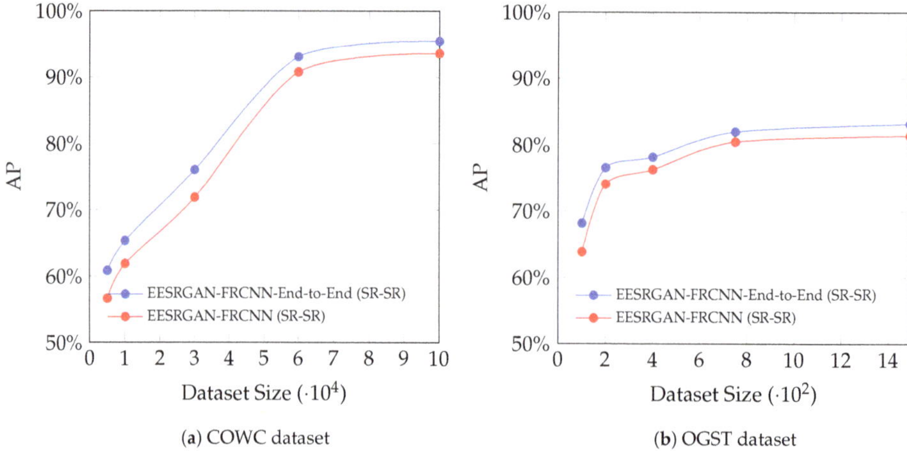

(**a**) COWC dataset　　　　　　　　(**b**) OGST dataset

Figure 10. AP (average precision) with varying number of training sets from the datasets. Plotted results show the detection performance of our proposed method (with and without end-to-end training) on SR (super-resolution) images.

The OGST dataset gave 83.2% AP (with end-to-end setting) using the full training dataset (1491 tanks), and we used 100 labeled tanks as test and same amount as validation data for all combinations of the training dataset. We got high AP values with 50% of our full training dataset as depicted in Figure 10b. AP values dropped below 80% when we further decreased the training data. Similar to the COWC datasets, we also got comparatively lower AP values for all sizes of OGST datasets. We observed slightly higher differences between the two settings (with and without end-to-end) when the dataset consisted of less than 400 labeled tanks, as shown in the plot of AP versus dataset size (OGST dataset).

We used 90% of the OGST dataset for training while we used the 80% of the COWC dataset for the same purpose. The accuracy of the testing data (OGST) slightly increased when we added more training data, as depicted in Figure 10b. Therefore, we used a larger percentage of training data for the OGST dataset than for the COWC dataset, and it slightly helped to improve the relatively low accuracy of the OGST test data.

4.3.7. Enhancement and Detection

In Figure 11, we have shown our input LR images, corresponding generated SR image, enhanced edge information and final detection. The image enhancement has helped the detectors to get high AP values and also makes the images visually good enough to identify the objects easily. It is evident from the figure that the visual quality of the generated SR images is quite good compared to the corresponding LR images, and the FRCNN detector has detected most of the objects correctly.

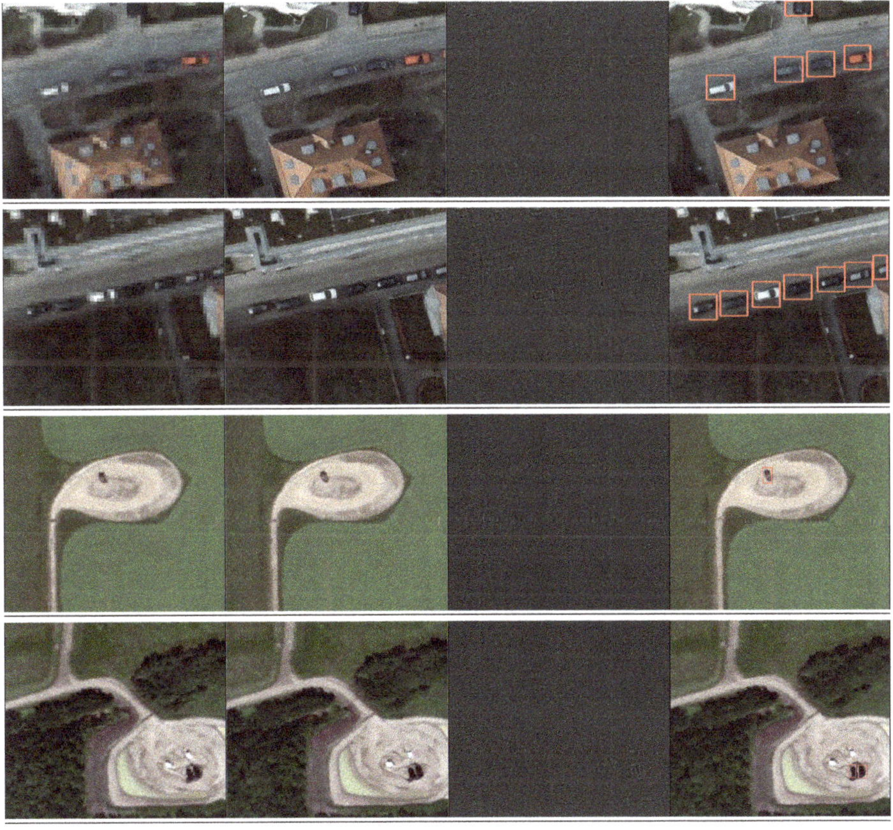

(**a**) Input LR image   (**b**) Generated SR image   (**c**) Enhanced edge   (**d**) Detection

Figure 11. Examples of SR (super-resolution) images that are generated from input LR (low-resolution) images are shown in (**a**,**b**). The enhanced edges and detection results are shown in (**c**,**d**).

4.3.8. Effects of Edge Consistency Loss (L_{edge_cst})

In EEGAN [22], only image consistency loss (L_{img_cst}) was used for enhancing the edge information. This loss generated edge information with noise, and as a result, the final SR images became blurry. The blurry output with noisy edge using only L_{img_cst} loss is shown in Figure 12a. The blurry final images gave lower detection accuracy compared to sharp outputs.

Therefore, we have introduced edge consistency loss (L_{edge_cst}) in addition to L_{img_cst} loss that gives noise-free enhanced edge information similar to the edge extracted from ground truth images and the effects of the L_{edge_cst} loss is shown in Figure 12b. The ground truth HR image with extracted edge is depicted in Figure 12c.

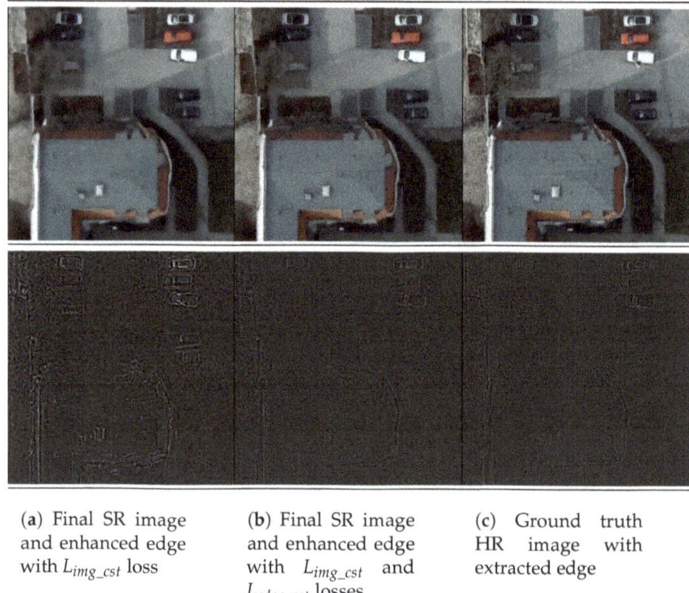

(**a**) Final SR image and enhanced edge with L_{img_cst} loss

(**b**) Final SR image and enhanced edge with L_{img_cst} and L_{edge_cst} losses

(**c**) Ground truth HR image with extracted edge

Figure 12. Effects of edge consistency loss (L_{edge_cst}) on final SR (super-resolution) images and enhanced edges compared to the extracted edges from HR (high-resolution) images.

5. Discussion

The detection results of our method presented in the previous section have indicated that our end-to-end SR-detector network improved detection accuracy compared to several other methods. Our method outperformed the standalone state-of-the-art methods such as SSD or faster R-CNN when implemented in low-resolution remote sensing imagery. We used EESRGAN, EEGAN, and ESRGAN as the SR network with the detectors. We showed that our EESRGAN with the detectors performed better than the other methods and the edge-enhancement helped to improve the detection accuracy. The AP improvement was higher in high IoUs and not so much in the lower IoUs. We have also showed that the precision increased with higher resolution. The improvement of AP values for the OGST dataset was lower than that for the COWC dataset because the area covered by a tank was slightly bigger than that of a car, and tanks sizes and colors were less diverse than the cars.

Our experimental results indicated that AP values of the output could be improved slightly with the increase of training data. The results also demonstrated that we could use less training data for both the datasets to get a similar level of accuracy that we obtained from our total training data.

The faster R-CNN detector gave us the best result, but it took a longer time than an SSD detector. If we need detection results from a vast area, then SSD would be the right choice sacrificing some amount of accuracy.

We had large numbers of cars from different regions in the COWC dataset, and we obtained high AP values using different IoUs. On the other hand, the OGST dataset needed more data to get a general detection result because we used data from a specific area and for a specific season and this was one of the limitations of our experiment. Most likely, more data from different regions and seasons would make our method more robust for the use-case of oil and gas storage tank detection. Another limitation of our experiment was that we showed the performance of the datasets that contain only one class with less variation. We would be looking forward to exploring the performance of our method on a broader range of object types and landscapes from different satellite datasets.

We have used LR-HR image pairs to train our architecture, and the LR images were generated artificially from the HR counterparts. To our knowledge, there is no suitable public satellite dataset that contains both real HR and real LR image pairs and ground truth bounding boxes for detecting small objects. Therefore, we have created the LR images which do not precisely correspond to true LR images. However, improvement of resolution through deep learning always improved object detection performance on remote sensing images (for both artificial and real low-resolution images), as discussed in the introduction and related works section of this paper [5]. Impressive works [61,70] exist in literature to create realistic LR images from HR images. For future work, we are looking forward to exploring the works to create more accurate LR images for training.

6. Conclusions

In this paper, we propose an end-to-end architecture that takes LR satellite imagery as input and gives object detection results as outputs. Our architecture contains a SR network and a detector network. We have used a different combination of SR systems and detectors to compare the AP values for detection using two different datasets. Our experimental results show that the proposed SR network with faster R-CNN has yielded the best results for small objects on satellite imagery. However, we need to add more diverse training data in the OGST dataset to make our model robust in detecting oil and gas storage tanks. We also need to explore diverse datasets and the techniques to create more realistic LR images. In conclusion, our method has combined different strategies to provide a better solution to the task of small-object detection on LR imagery.

Author Contributions: Conceptualization, J.R., N.R. and M.S.; methodology, J.R., N.R. and M.S.; software, J.R.; validation, J.R.; formal analysis, J.R.; investigation, J.R.; resources, N.R.; data curation, J.R., S.C. and D.C.; writing–original draft preparation, J.R.; writing–review and editing, J.R., N.R., M.S., S.C. and D.C.; visualization, J.R.; supervision, N.R. and M.S.; project administration, N.R.; funding acquisition, N.R., S.C. and D.C. All authors have read and agreed to the published version of the manuscript.

Funding: This research was partially supported by Alberta Geological Survey (AGS) and NSERC discovery grant.

Acknowledgments: The first and the second authors acknowledge support from the Department of Computing Science, University of Alberta and Compute Canada.

Conflicts of Interest: The authors declare no conflict of interest.

Abbreviations

The following acronyms are used in this paper:

SRCNN	Single image Super-Resolution Convolutional Neural Network
VDSR	Very Deep Convolutional Networks
GAN	Generative Adversarial Network
SRGAN	Super-Resolution Generative Adversarial Network
ESRGAN	Enhanced Super-Resolution Generative Adversarial Network
EEGAN	Edge-Enhanced Generative Adversarial Network
EESRGAN	Edge-Enhanced Super-Resolution Generative Adversarial Network

RRDB   Residual-in-Residual Dense Blocks
EEN   Edge-Enhancement Network
SSD   Single-Shot MultiBox Detector
YOLO   You Only Look Once
CNN   Convolutional Neural Network
R-CNN   Region-based Convolutional Neural Network
FRCNN   Faster Region-based Convolutional Neural Network
VGG   Visual Geometry Group
BN   Batch Normalization
MSCOCO   Microsoft Common Objects in Context
OGST   Oil and Gas Storage Tank
COWC   Car Overhead With Context
GSD   Ground Sampling Distance
G   Generator
D   Discriminator
ISR   Intermediate Super-Resolution
SR   Super-Resolution
HR   High-Resolution
LR   Low-Resoluton
GT   Ground Truth
FPN   Feature Pyramid Network
RPN   Region Proposal Network
AER   Alberta Energy Regulator
AGS   Alberta Geological Survey
AP   Average Precision
IoU   Intersection over Union
TP   True Positive
FP   False Positive
FN   False Negative

References

1. Colomina, I.; Molina, P. Unmanned aerial systems for photogrammetry and remote sensing: A review. *ISPRS J. Photogramm. Remote Sens.* **2014**, *92*, 79–97. [CrossRef]
2. Zhang, F.; Du, B.; Zhang, L.; Xu, M. Weakly supervised learning based on coupled convolutional neural networks for aircraft detection. *IEEE Trans. Geosci. Remote Sens.* **2016**, *54*, 5553–5563. [CrossRef]
3. Fromm, M.; Schubert, M.; Castilla, G.; Linke, J.; McDermid, G. Automated Detection of Conifer Seedlings in Drone Imagery Using Convolutional Neural Networks. *Remote Sens.* **2019**, *11*, 2585. [CrossRef]
4. Pang, J.; Li, C.; Shi, J.; Xu, Z.; Feng, H. $\mathcal{R}^2$ -CNN: Fast Tiny Object Detection in Large-Scale Remote Sensing Images. *IEEE Trans. Geosci. Remote Sens.* **2019**, *57*, 5512–5524. [CrossRef]
5. Shermeyer, J.; Van Etten, A. The effects of super-resolution on object detection performance in satellite imagery. In Proceedings of the IEEE Conference on Computer Vision and Pattern Recognition Workshops (CVPR 2019), Long Beach, CA, USA, 16–20 June 2019; pp. 1–10.
6. Russakovsky, O.; Deng, J.; Su, H.; Krause, J.; Satheesh, S.; Ma, S.; Huang, Z.; Karpathy, A.; Khosla, A.; Bernstein, M.; et al. ImageNet Large Scale Visual Recognition Challenge; *Int. J. Comput. Vis.* **2015**. *115*, 211–252. [CrossRef]
7. Lin, T.Y.; Maire, M.; Belongie, S.; Hays, J.; Perona, P.; Ramanan, D.; Dollár, P.; Zitnick, C.L. Microsoft coco: Common objects in context. In *European Conference on Computer Vision*; Springer: Berlin/Heidelberg, Germany, 2014; pp. 740–755.
8. Ren, S.; He, K.; Girshick, R.; Sun, J. Faster R-CNN: Towards Real-Time Object Detection with Region Proposal Networks. *IEEE Trans. Pattern Anal. Mach. Intell.* **2017**, *39*, 1137–1149. [CrossRef] [PubMed]
9. Lin, T.Y.; Goyal, P.; Girshick, R.; He, K.; Dollar, P. Focal Loss for Dense Object Detection. In Proceedings of the 2017 IEEE International Conference on Computer Vision (ICCV), Venice, Italy, 22–29 October 2017.

10. Liu, W.; Anguelov, D.; Erhan, D.; Szegedy, C.; Reed, S.; Fu, C.Y.; Berg, A.C. SSD: Single Shot MultiBox Detector. In *European Conference on Computer Vision*; Lecture Notes in Computer Science; Springer: Cham, Switzerland, 2016; pp. 21–37._2. [CrossRef]
11. Redmon, J.; Divvala, S.; Girshick, R.; Farhadi, A. You Only Look Once: Unified, Real-Time Object Detection. In Proceedings of the 2016 IEEE Conference on Computer Vision and Pattern Recognition (CVPR), Las Vegas, NV, USA, 26 June–1 July 2016. [CrossRef]
12. Ji, H.; Gao, Z.; Mei, T.; Ramesh, B. Vehicle Detection in Remote Sensing Images Leveraging on Simultaneous Super-Resolution. In *IEEE Geoscience and Remote Sensing Letters*; IEEE-INST ELECTRICAL ELECTRONICS ENGINEERS INC.: New York, NY, USA, 2019; pp. 1–5. [CrossRef]
13. Tayara, H.; Soo, K.G.; Chong, K.T. Vehicle detection and counting in high-resolution aerial images using convolutional regression neural network. *IEEE Access* **2017**, *6*, 2220–2230. [CrossRef]
14. Yu, X.; Shi, Z. Vehicle detection in remote sensing imagery based on salient information and local shape feature. *Opt.-Int. J. Light Electron. Opt.* **2015**, *126*, 2485–2490. [CrossRef]
15. Stankov, K.; He, D.C. Detection of buildings in multispectral very high spatial resolution images using the percentage occupancy hit-or-miss transform. *IEEE J. Sel. Top. Appl. Earth Obs. Remote Sens.* **2014**, *7*, 4069–4080. [CrossRef]
16. Ok, A.O.; Başeski, E. Circular oil tank detection from panchromatic satellite images: A new automated approach. *IEEE Geosci. Remote Sens. Lett.* **2015**, *12*, 1347–1351. [CrossRef]
17. Dong, C.; Loy, C.C.; He, K.; Tang, X. Image Super-Resolution Using Deep Convolutional Networks. *IEEE Trans. Pattern Anal. Mach. Intell.* **2016**, *38*, 295–307. [CrossRef] [PubMed]
18. Kim, J.; Lee, J.K.; Lee, K.M. Accurate Image Super-Resolution Using Very Deep Convolutional Networks. In Proceedings of the 2016 IEEE Conference on Computer Vision and Pattern Recognition (CVPR), Las Vegas, NV, USA, 26 June–1 July 2016. [CrossRef]
19. Goodfellow, I.; Pouget-Abadie, J.; Mirza, M.; Xu, B.; Warde-Farley, D.; Ozair, S.; Courville, A.; Bengio, Y. Generative adversarial nets. In Proceedings of the Advances in Neural Information Processing Systems, Montreal, QC, Canada, 8–13 December 2014; pp. 2672–2680.
20. Ledig, C.; Theis, L.; Huszar, F.; Caballero, J.; Cunningham, A.; Acosta, A.; Aitken, A.; Tejani, A.; Totz, J.; Wang, Z.; et al. Photo-Realistic Single Image Super-Resolution Using a Generative Adversarial Network. In Proceedings of the 2017 IEEE Conference on Computer Vision and Pattern Recognition (CVPR), Honolulu, HI, USA, 21–26 July 2017. [CrossRef]
21. Wang, X.; Yu, K.; Wu, S.; Gu, J.; Liu, Y.; Dong, C.; Qiao, Y.; Loy, C.C. ESRGAN: Enhanced Super-Resolution Generative Adversarial Networks. In Proceedings of the Computer Vision—ECCV 2018 Workshops, Munich, Germany, 8–14 September 2018; pp. 63–79._5. [CrossRef]
22. Jiang, K.; Wang, Z.; Yi, P.; Wang, G.; Lu, T.; Jiang, J. Edge-Enhanced GAN for Remote Sensing Image Superresolution. *IEEE Trans. Geosci. Remote Sens.* **2019**, *57*, 5799–5812. [CrossRef]
23. Jiang, J.; Ma, J.; Wang, Z.; Chen, C.; Liu, X. Hyperspectral Image Classification in the Presence of Noisy Labels. *IEEE Trans. Geosci. Remote Sens.* **2019**, *57*, 851–865. [CrossRef]
24. Tong, F.; Tong, H.; Jiang, J.; Zhang, Y. Multiscale union regions adaptive sparse representation for hyperspectral image classification. *Remote Sens.* **2017**, *9*, 872. [CrossRef]
25. Zhan, C.; Duan, X.; Xu, S.; Song, Z.; Luo, M. An improved moving object detection algorithm based on frame difference and edge detection. In Proceedings of the Fourth International Conference on Image and Graphics (ICIG 2007), Sichuan, China, 22–24 August 2007; pp. 519–523.
26. Mao, Q.; Wang, S.; Wang, S.; Zhang, X.; Ma, S. Enhanced image decoding via edge-preserving generative adversarial networks. In Proceedings of the 2018 IEEE International Conference on Multimedia and Expo (ICME), San Diego, CA, USA, 23–27 July 2018; pp. 1–6.
27. Yang, W.; Feng, J.; Yang, J.; Zhao, F.; Liu, J.; Guo, Z.; Yan, S. Deep Edge Guided Recurrent Residual Learning for Image Super-Resolution. *IEEE Trans. Image Process.* **2017**, *26*, 5895–5907. [CrossRef] [PubMed]
28. Kamgar-Parsi, B.; Kamgar-Parsi, B.; Rosenfeld, A. Optimally isotropic Laplacian operator. *IEEE Trans. Image Process.* **1999**, *8*, 1467–1472. [CrossRef]
29. Landsat 8. Available online: https://www.usgs.gov/land-resources/nli/landsat/landsat-8 (accessed on 11 February 2020).
30. Sentinel-2. Available online: http://www.esa.int/Applications/Observing_the_Earth/Copernicus/Sentinel-2 (accessed on 11 February 2020).

31. Mundhenk, T.N.; Konjevod, G.; Sakla, W.A.; Boakye, K. A large contextual dataset for classification, detection and counting of cars with deep learning. In *European Conference on Computer Vision*; Springer: Cham, Switzerland, 2016; pp. 785–800.

32. Rabbi, J.; Chowdhury, S.; Chao, D. Oil and Gas Tank Dataset. In *Mendeley Data, V3*; 2020. Available online: https://data.mendeley.com/datasets/bkxj8z84m9/3 (accessed on 30 April 2020). [CrossRef]

33. Jolicoeur-Martineau, A. The Relativistic Discriminator: A Key Element Missing from Standard GAN. *arXiv* **2018**, arXiv:1807.00734.

34. Charbonnier, P.; Blanc-Féraud, L.; Aubert, G.; Barlaud, M. Two deterministic half-quadratic regularization algorithms for computed imaging. *Proc. Int. Conf. Image Process.* **1994**, *2*, 168–172.

35. Alberta Energy Regulator. Available online: https://www.aer.ca (accessed on 5 February 2020).

36. Tai, Y.; Yang, J.; Liu, X.; Xu, C. MemNet: A Persistent Memory Network for Image Restoration. In Proceedings of the 2017 IEEE International Conference on Computer Vision (ICCV), Venice, Italy, 22–29 October 2017. [CrossRef]

37. Zhang, Y.; Tian, Y.; Kong, Y.; Zhong, B.; Fu, Y. Residual Dense Network for Image Super-Resolution. In Proceedings of the 2018 IEEE/CVF Conference on Computer Vision and Pattern Recognition, Salt Lake City, UT, USA, 18–23 June 2018. [CrossRef]

38. He, K.; Zhang, X.; Ren, S.; Sun, J. Delving Deep into Rectifiers: Surpassing Human-Level Performance on ImageNet Classification. In Proceedings of the 2015 IEEE International Conference on Computer Vision (ICCV), Santiago, Chile, 7–13 December 2015. [CrossRef]

39. Lim, B.; Son, S.; Kim, H.; Nah, S.; Lee, K.M. Enhanced Deep Residual Networks for Single Image Super-Resolution. In Proceedings of the 2017 IEEE Conference on Computer Vision and Pattern Recognition Workshops (CVPRW), Honolulu, HI, USA, 21–26 July 2017. [CrossRef]

40. Liebel, L.; Körner, M. Single-image super resolution for multispectral remote sensing data using convolutional neural networks. *ISPRS Int. Arch. Photogramm. Remote Sens. Spat. Inf. Sci.* **2016**, *41*, 883–890. [CrossRef]

41. Tayara, H.; Chong, K. Object detection in very high-resolution aerial images using one-stage densely connected feature pyramid network. *Sensors* **2018**, *18*, 3341. [CrossRef]

42. Girshick, R.; Donahue, J.; Darrell, T.; Malik, J. Rich Feature Hierarchies for Accurate Object Detection and Semantic Segmentation. In Proceedings of the 2014 IEEE Conference on Computer Vision and Pattern Recognition, Columbus, OH, USA, 24–27 June 2014. [CrossRef]

43. Girshick, R. Fast R-CNN. In Proceedings of the 2015 IEEE International Conference on Computer Vision (ICCV), Santiago, Chile, 7–13 December 2015. [CrossRef]

44. Li, Q.; Mou, L.; Xu, Q.; Zhang, Y.; Zhu, X.X. R3-Net: A Deep Network for Multioriented Vehicle Detection in Aerial Images and Videos. *IEEE Trans. Geosci. Remote Sens.* **2019**, *57*, 5028–5042. [CrossRef]

45. Ammour, N.; Alhichri, H.; Bazi, Y.; Benjdira, B.; Alajlan, N.; Zuair, M. Deep learning approach for car detection in UAV imagery. *Remote Sens.* **2017**, *9*, 312. [CrossRef]

46. Ren, Y.; Zhu, C.; Xiao, S. Small object detection in optical remote sensing images via modified faster R-CNN. *Appl. Sci.* **2018**, *8*, 813. [CrossRef]

47. Tang, T.; Zhou, S.; Deng, Z.; Zou, H.; Lei, L. Vehicle detection in aerial images based on region convolutional neural networks and hard negative example mining. *Sensors* **2017**, *17*, 336. [CrossRef] [PubMed]

48. Chen, Z.; Zhang, T.; Ouyang, C. End-to-end airplane detection using transfer learning in remote sensing images. *Remote Sens.* **2018**, *10*, 139. [CrossRef]

49. Radovic, M.; Adarkwa, O.; Wang, Q. Object recognition in aerial images using convolutional neural networks. *J. Imaging* **2017**, *3*, 21. [CrossRef]

50. Li, W.; Fu, H.; Yu, L.; Cracknell, A. Deep learning based oil palm tree detection and counting for high-resolution remote sensing images. *Remote Sens.* **2017**, *9*, 22. [CrossRef]

51. Lin, T.Y.; Dollar, P.; Girshick, R.; He, K.; Hariharan, B.; Belongie, S. Feature Pyramid Networks for Object Detection. In Proceedings of the 2017 IEEE Conference on Computer Vision and Pattern Recognition (CVPR), Honolulu, HI, USA, 21–26 July 2017. [CrossRef]

52. Liu, S.; Huang, D. Receptive field block net for accurate and fast object detection. In Proceedings of the European Conference on Computer Vision (ECCV), Munich, Germany, 8–14 September 2018; pp. 385–400.

53. Zhang, S.; Wen, L.; Bian, X.; Lei, Z.; Li, S.Z. Single-shot refinement neural network for object detection. In Proceedings of the IEEE conference on Computer Vision and Pattern Recognition, Salt Lake City, UT, USA, 18–22 June 2018; pp. 4203–4212.

54. Li, Z.; Zhou, F. FSSD: Feature fusion single shot multibox detector. *arXiv* **2017**, arXiv:1712.00960.

55. Zhu, R.; Zhang, S.; Wang, X.; Wen, L.; Shi, H.; Bo, L.; Mei, T. ScratchDet: Training single-shot object detectors from scratch. In Proceedings of the IEEE Conference on Computer Vision and Pattern Recognition, Long Beach, CA, USA, 15–20 June 2019; pp. 2268–2277.

56. Yang, X.; Sun, H.; Fu, K.; Yang, J.; Sun, X.; Yan, M.; Guo, Z. Automatic ship detection in remote sensing images from google earth of complex scenes based on multiscale rotation dense feature pyramid networks. *Remote Sens.* **2018**, *10*, 132. [CrossRef]

57. Zhao, Z.Q.; Zheng, P.; Xu, S.T.; Wu, X. Object detection with deep learning: A review. *IEEE Trans. Neural Netw. Learn. Syst.* **2019**, *30*, 3212–3232. [CrossRef]

58. Li, K.; Wan, G.; Cheng, G.; Meng, L.; Han, J. Object detection in optical remote sensing images: A survey and a new benchmark. *ISPRS J. Photogramm. Remote Sens.* **2020**, *159*, 296–307. [CrossRef]

59. Bai, Y.; Zhang, Y.; Ding, M.; Ghanem, B. Sod-mtgan: Small object detection via multi-task generative adversarial network. In Proceedings of the European Conference on Computer Vision (ECCV), Munich, Germany, 8–14 September 2018; pp. 206–221.

60. Haris, M.; Shakhnarovich, G.; Ukita, N. Task-driven super resolution: Object detection in low-resolution images. *arXiv* **2018**, arXiv:1803.11316.

61. Zhu, J.Y.; Park, T.; Isola, P.; Efros, A.A. Unpaired image-to-image translation using cycle-consistent adversarial networks. In Proceedings of the IEEE International Conference on Computer Vision, Venice, Italy, 22–29 October 2017; pp. 2223–2232.

62. Simonyan, K.; Zisserman, A. Very Deep Convolutional Networks for Large-Scale Image Recognition. *arXiv* **2014**, arXiv:1409.1556.

63. Lai, W.S.; Huang, J.B.; Ahuja, N.; Yang, M.H. Deep Laplacian Pyramid Networks for Fast and Accurate Super-Resolution. In Proceedings of the 2017 IEEE Conference on Computer Vision and Pattern Recognition (CVPR), Honolulu, HI, USA, 21–26 July 2017. [CrossRef]

64. Kingma, D.P.; Ba, J. Adam: A method for stochastic optimization. *arXiv* **2014**, arXiv:1412.6980.

65. Paszke, A.; Gross, S.; Massa, F.; Lerer, A.; Bradbury, J.; Chanan, G.; Killeen, T.; Lin, Z.; Gimelshein, N.; Antiga, L.; et al. PyTorch: An Imperative Style, High-Performance Deep Learning Library. In *Advances in Neural Information Processing Systems 32*; Wallach, H., Larochelle, H., Beygelzimer, A., d'Alché Buc, F., Fox, E., Garnett, R., Eds.; Curran Associates, Inc.: New York, NY, USA, 2019; pp. 8024–8035.

66. Rabbi, J. Edge Enhanced GAN with Faster RCNN for End-to-End Object Detection from Remote Sensing Imagery. 2020. Available online: https://github.com/Jakaria08/Filter_Enhance_Detect (accessed on 28 April 2020).

67. Alberta Geological Survey. Available online: https://ags.aer.ca (accessed on 5 February 2020).

68. Chowdhury, S.; Chao, D.K.; Shipman, T.C.; Wulder, M.A. Utilization of Landsat data to quantify land-use and land-cover changes related to oil and gas activities in West-Central Alberta from 2005 to 2013. *GISci. Remote Sens.* **2017**, *54*, 700–720. [CrossRef]

69. Bing Map. Available online: https://www.bing.com/maps (accessed on 5 February 2020).

70. Bulat, A.; Yang, J.; Tzimiropoulos, G. To learn image super-resolution, use a gan to learn how to do image degradation first. In Proceedings of the European Conference on Computer Vision (ECCV), Munich, Germany, 8–14 September 2018; pp. 185–200.

 *remote sensing*

Article

Deep Open-Set Domain Adaptation for Cross-Scene Classification based on Adversarial Learning and Pareto Ranking

Reham Adayel, Yakoub Bazi *, Haikel Alhichri and Naif Alajlan

Computer Engineering Department, College of Computer and Information Sciences, King Saud University, Riyadh 11543, Saudi Arabia; 437204127@student.ksu.edu.sa (R.A.); hhichri@ksu.edu.sa (H.A.); najlan@ksu.edu.sa (N.A.)
* Correspondence: ybazi@ksu.edu.sa; Tel.: +966-114696297

Received: 28 April 2020; Accepted: 25 May 2020; Published: 27 May 2020

Abstract: Most of the existing domain adaptation (DA) methods proposed in the context of remote sensing imagery assume the presence of the same land-cover classes in the source and target domains. Yet, this assumption is not always realistic in practice as the target domain may contain additional classes unknown to the source leading to the so-called open set DA. Under this challenging setting, the problem turns to reducing the distribution discrepancy between the shared classes in both domains besides the detection of the unknown class samples in the target domain. To deal with the openset problem, we propose an approach based on adversarial learning and pareto-based ranking. In particular, the method leverages the distribution discrepancy between the source and target domains using min-max entropy optimization. During the alignment process, it identifies candidate samples of the unknown class from the target domain through a pareto-based ranking scheme that uses ambiguity criteria based on entropy and the distance to source class prototype. Promising results using two cross-domain datasets that consist of very high resolution and extremely high resolution images, show the effectiveness of the proposed method.

Keywords: scene classification; open-set domain adaptation; adversarial learning; min-max entropy; pareto ranking

---

1. Introduction

Scene classification is the process of automatically assigning an image to a class label that describes the image correctly. In the field of remote sensing, scene classification gained a lot of attention and several methods were introduced in this field such as bag of word model [1], compressive sensing [2], sparse representation [3], and lately deep learning [4]. To classify a scene correctly, effective features are extracted from a given image, then classified by a classifier to the correct label. Early studies of remote sensing scene classification were based on handcrafted features [1,5,6]. In this context, deep learning techniques showed to be very efficient in terms compared to standard solutions based on handcrafted features. Convolutional neural networks (CNN) are considered the most common deep learning techniques for learning visual features and they are widely used to classify remote sensing images [7–9]. Several approaches were built around these methods to boost the classification results such as integrating local and global features [10–12], recurrent neural networks (RNNs) [13,14], and generative adversarial networks (GANs) [15].

The number of remote sensing images has been steadily increasing over the years. These images are collected using different sensors mounted on satellites or airborne platforms. The type of the sensor results in different image resolutions: spatial, spectral, and temporal resolution. This leads to huge number of images with different spatial and spectral resolutions [16]. In many real-world applications, the training data used to learn a model may have different distributions from the data used for testing, when images are acquired over different locations and with different sensors. For such purpose, it becomes necessary to reduce the distribution gap between the source and target domains to obtain acceptable classification results [17]. The main goal of domain adaptation is to learn a classification model from a labeled source domain and apply this model to classify an unlabeled target domain. Several methods have been introduced related to domain adaptation in the field of remote sensing [18–21].

The above methods assume that the sample belongs to one of a fixed number of known classes. This is called a closed set domain adaptation, where the source and target domains have shared classes only. This assumption is violated in many cases. In fact, many real applications follow an open set environment, where some test samples belong to classes that are unknown during training [22]. These samples are supposed to be classified as unknown, instead of being classified to one of the shared classes. Classifying the unknown image to one of the shared classes leads to negative transfer. Figure 1 shows the difference between open and closed set classification. This problem is known in the community of machine learning as open set domain adaptation. Thus, in an open set domain adaptation one has to learn robust feature representations for the source labeled data, reduce the data-shift problem between the source and target distributions, and detect the presence of new classes in the target domain.

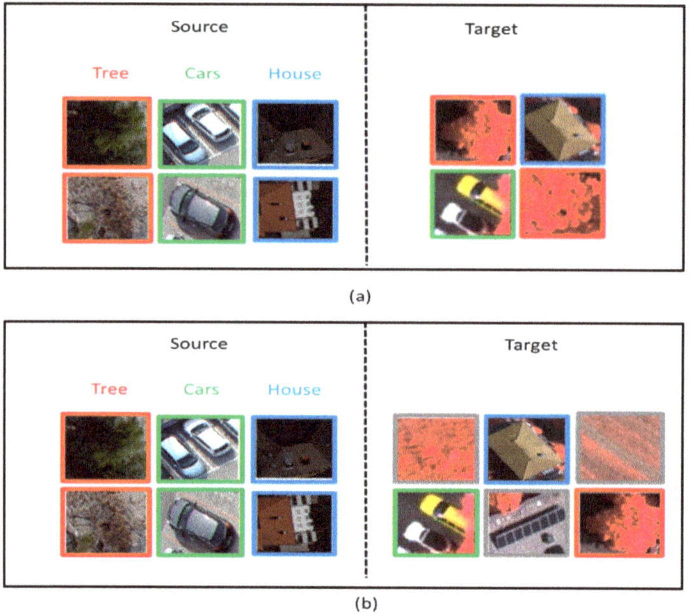

Figure 1. (**a**) Closed set domain adaptation: source and target domains share the same classes. (**b**) Open set domain adaptation: target domain contains unknown classes (in the grey boxes).

Open set classification is a new research area in the remote sensing field. Few works have introduced the problem of open set. Anne Pavy and Edmund Zelnio [23] introduced a method to classify Synthetic Aperture Radar (SAR) images in the test samples that are in the training samples and reject those not in the training set as unknown. The method uses a CNN as a feature extractor and SVM for classification and rejection of unknowns. Wang et al. [24] addressed the open set problem in high range resolution profile (HRRP) recognition. The method is based on random forest (RF) and extreme value theory, the RF is used to extract the high-level features of the input image, then used as input to the open set module. The two approaches use a single domain for training and testing.

To the best of our knowledge, no previous works have addressed the open-set domain adaptation problem for remote sensing images where the source domain is different from the target domain. To address this issue, the method we propose is based on adversarial learning and pareto-based ranking. In particular, the method leverages the distribution discrepancy between the source and target domain using min-max entropy optimization. During the alignment process, it identifies candidate samples of the unknown class from the target domain through a pareto-based ranking scheme that uses ambiguity criteria based on entropy and the distance to source class porotypes.

2. Related Work on Open Set Classification

Open set classification is a more challenging and more realistic approach, thus it has gained a lot of attention by researchers lately and many works are done in this field. Early research on open set classification depended on traditional machine learning techniques, such as support vector machines (SVMs). Scheirer et al. [25] first proposed 1-vs-Set method which used a binary SVM classifier to detect unknown classes. The method introduced a new open set margin to decrease the region of the known class for each binary SVM. Jain et al. [26] invoked the extreme value theory (EVT) to present a multi-class open set classifier to reject the unknowns. The authors introduced the Pi-SVM algorithm to estimate the un-normalized posterior class inclusion likelihood. The probabilistic open set SVM (POS-SVM) classifier proposed by Scherreik et al. [27] empirically determines the unique reject threshold for each known class. Sparse representation techniques were used in open set classification, where the sparse representation based classifier (SRC) [28] looks for the sparest possible representation of the test sample to correctly classify the sample [29]. Bendale and Boult [30] presented the nearest non-outlier (NNO) method to actively detect and learn new classes, taking into account the open space risk and metric learning.

Deep neural networks (DNNs) were very interesting in several tasks lately, including open set classification. Bendale and Boult [31] first introduced OpenMax model, which is a DNN to perform open set classification. The OpenMax layer replaced the softmax layer in a CNN to check if a given sample belongs to an unknown class. Hassen and Chan [32] presented a method to solve the open set problem, by keeping instances belonging to the same class near each other, and instances that belong to different or unknown classes farther apart. Shu et al. [33] proposed deep open classification (DOC) for open set classification, that builds a multi-class classifier which used instead of the last softmax layer a 1-vs-rest layer of sigmoids to make the open space risk as small as possible. Later, Shu et al. [34] presented a model for discovering unknown classes that combines two neural networks: open classification network (OCN) for seen classification and unseen class rejection, and a pairwise classification network (PCN) which learns a binary classifier to predict if two samples come from the same class or different classes.

In the last years generative adversarial networks (GANs) [35] were introduced to the field of open set classification. Ge et al. [36] presented the Generative OpenMax (G-OpenMax) method. The algorithm adapts OpenMax to generative adversarial networks for open set classification. The GAN trained the network by generating unknown samples, then combined it with an OpenMax layer to reject samples belonging to the unknown class. Neal et al. [37] proposed another GAN-based algorithm to generate counterfactual images that do not belong to any class; instead are unknown, which are used to train a classifier to correctly classify unknown images. Yu et al. [38] also proposed a GAN that generated negative samples for known classes to train the classifier to distinguish between known and unknown samples.

Most of the previous studies mentioned in the literature of scene classification assume that one domain is used for both training and testing. This assumption is not always satisfied, due to the fact that some domains have images that are labeled, on the other hand many new domains have shortage in labeled images. It will be time-consuming and expensive to generate and collect large datasets of labeled images. One suggestion to solve this issue is to use labeled images from one domain as training data for different domains. Domain adaptation is one part of transfer learning where transfer of knowledge occurs between two domains, source and target. Domain adaptation approaches differ from each other in the percentage of labeled images in the target domain. Some works have been done in the field of open set domain adaptation. First, Busto et al. [22] introduced open set domain adaptation in their work, by allowing the target domain to have samples of classes not belonging to the source domain and vice versa. The classes not shared or uncommon are joined as a negative class called "unknown". The goal was to correctly classify target samples to the correct class if shared between source and target, and classify samples to unknown if not shared between domains. Saito et al. [39] proposed a method where unknown samples appear only in the target domain, which is more challenging. The proposed approach introduced adversarial learning where the generator can separate target samples to known and unknown classes. The generator can decide to reject or accept the target image. If accepted, it is classified to one of the classes in the source domain. If rejected it is classified as unknown.

Cao et al. [40] introduced a new partial domain adaptation method, they assumed that the target dataset contained classes that are a subset of the source dataset classes. This makes the domain adaptation problem more challenging due to the extra source classes, which could result in negative transfer problems. The authors used a multi-discriminator domain adversarial network, where each discriminator has the responsibility of matching the source and target domain data after filtering unknown source classes. Zhang et al. [41] also introduced the problem of transferring from big source to target domain with subset classes. The method requires only two domain classifiers instead of multiple classifiers one for each domain as shown by the previous method. Furthermore, Baktashmotlagh et al. [42] proposed an approach to factorize the data into shared and private sub-spaces. Source and target samples coming from the same, known classes can be represented by a shared subspace, while target samples from unknown classes were modeled with a private subspace.

Lian et al. [43] proposed Known-class Aware Self-Ensemble (KASE) that was able to reject unknown classes. The model consists of two modules to effectively identify known and unknown classes and perform domain adaptation based on the likeliness of target images belonging to known classes. Lui et al. [44] presented Separate to Adapt (STA), a method to separate known from unknown samples in an advanced way. The method works in two steps: first a classifier was trained to measure the similarity between target samples and every source class with source samples. Then, high and low values of similarity were selected to be known and unknown classes. These values were used to train a classifier to correctly classify target images. Tan et al. [45] proposed a weakly supervised method, where the source and target domains had some labeled images. The two domains learn from each other through the few labeled images to correctly classify the unlabeled images in both domains. The method aligns the source and target domains in a collaborative way and then maximizes the margin for the shared and unshared classes.

In the contest of remote sensing, open set domain adaptation is a new research field and no previous work was achieved.

3. Description of the Proposed Method

Assume a labeled source domain $D_s = \left\{ \left(X_i^{(s)}, y_i^{(s)} \right) \right\}_{i=1}^{n_s}$ composed of $X_i^{(s)}$ images and their corresponding class labels $y_i^{(s)} \in \{1, 2, \ldots, K\}$, where n_s is the number of images and K is the number of classes. Additionally, we assume an unlabeled target domain $D_t = \left\{ \left(X_j^{(t)} \right) \right\}_{j=1}^{n_t}$ with n_t unlabeled images. In an open set setting, the target domain contains $K + 1$ classes, where K classes are shared with the source domain, and an addition unknown class (can be many unknown classes but grouped in one class). The objective of this work is twofold: (1) reduce the distribution discrepancy between the source and target domains, and (2) detect the presence of the unknown class in the target domain. Figure 2 shows the overall description of the proposed adversarial learning method, which relies on the idea of min-max entropy for carrying out the domain adaptation and uses an unknown class detector based on pareto ranking.

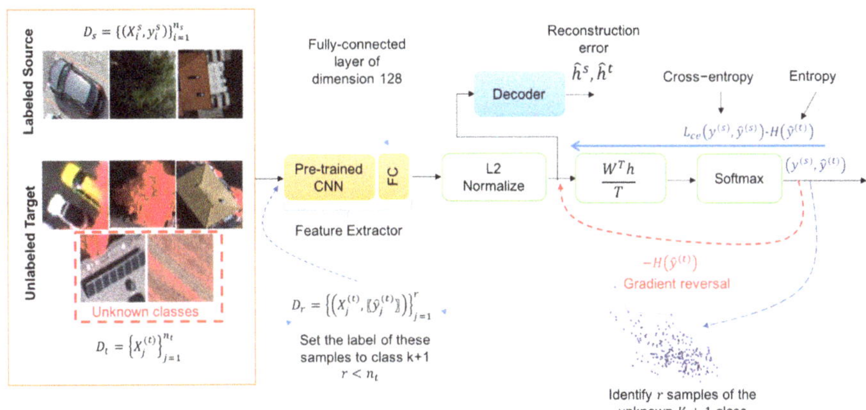

Figure 2. Proposed open-set domain adaptation method.

3.1. Network Architecture

Our model uses EfficientNet-B3 network [46] from Google as a feature extractor although other networks could be used as well since the method is independent of the pre-trained CNN. The choice of this network is motivated by its ability to generate high classification accuracies but with reduced parameters compared to other architectures. EfficientNets are based on the concept of scaling up CNNs by means of a compound coefficient, which jointly integrates the width, depth, and resolution. Basically, each dimension is scaled in a balanced way using a set of scaling coefficients. We truncate this network by removing its original ImageNet-based softmax classification layer. For computation convenience, we set $\{h_i^{(s)}\}_{i=1}^{n_s}$ and $\{h_j^{(t)}\}_{j=1}^{n_t}$ as the feature representations for both source and target data obtained at the output of this trimmed CNN (each feature is a vector of dimension 1536). These features are further subject to dimensionality reduction via a fully-connected layer F acting as a feature extractor yielding new feature representations $\{z_i^{(s)}\}_{i=1}^{n_s}$ and $\{z_j^{(t)}\}_{j=1}^{n_t}$ each of dimension 128. The output of F is further normalized using l_2 normalization and fed as input to a decoder D and similarity-based classifier C.

The decoder D has the task to constrain the mapping spaces of F with reconstruction ability to the original features provided by the pre-trained CNN in order to reduce the overlap between classes during adaptation. On the other side, the similarity classifier C aims to assign images to the corresponding classes including the unknown one (identified using ranking criteria) by computing the similarity measure of their related representations to its weight $W = [w_1, w_2, \ldots, w_K, w_{K+1}]$. These weights are viewed as estimated porotypes for the K-source classes and the unknown class with index $K+1$.

3.2. Adversarial Learning with Reconstruction Ability

We reduce the distribution discrepancy using an adversarial learning approach with reconstruction ability based on min-max entropy optimization [47]. To learn the weights of F, C, and D we use both labeled sources and unlabeled target samples. We learn the network to discriminate between the labeled classes in the source domain and the unknown class samples identified iteratively in the target domain (using the proposed pareto ranking scheme), while clustering the remaining target samples to the most suitable class prototypes. For such purpose, we jointly minimize the following loss functions:

$$
\begin{cases}
L_F = L_s\left(y^{(s)}, \hat{y}^{(s)}\right) + L_{K+1}\left(y^{(t)}, \hat{y}^{(t)}\right) + \lambda H\left(\hat{y}^{(t)}\right) \\
L_C = L_s\left(y^{(s)}, \hat{y}^{(s)}\right) + L_{K+1}\left(y^{(t)}, \hat{y}^{(t)}\right) - \lambda H\left(\hat{y}^{(t)}\right) \\
L_D = \frac{1}{n_s}\sum_{i=1}^{n_s} \|h_i^{(s)} - \hat{h}_i^{(s)}\|^2 + \frac{1}{n_t}\sum_{j=1}^{n_t} \|h_i^{(t)} - \hat{h}_i^{(t)}\|^2
\end{cases}
\tag{1}
$$

where λ is a regulalrization parameter which controls the contribution of the entropy to the total loss. L_s is the categorical cross-entropy loss computed for the source domain:

$$
L_s\left(y^{(s_k)}, \hat{y}^{(s_k)}\right) = -\frac{1}{n_{s_k}} \sum_{i=1}^{n_{s_k}} \sum_{k=1}^{K} \mathbb{1}\left(k = y_{ik}^{(s_k)}\right) \log\left(\hat{y}_{ik}^{(s_k)}\right)
\tag{2}
$$

L_{K+1} is the cross-entropy loss computed for the samples iteratively identified as unknown class:

$$
L_{K+1}\left(y^{(t_{K+1})}, \hat{y}^{(t_{K+1})}\right) = -\frac{1}{n_{t_{k+1}}} \sum_{j=1}^{n_{t_{K+1}}} y_j^{(t_{K+1})} \log\left(\hat{y}_j^{(t_{K+1})}\right)
\tag{3}
$$

$H\left(\hat{y}^{(t)}\right)$ is the entropy computed for the samples of the target domain:

$$
H\left(\hat{y}^{(t)}\right) = -\frac{1}{n_t} \sum_{i=1}^{n_t} \sum_{k=1}^{K+1} \mathbb{1}\left(k = \hat{y}_{ik}^{(t)}\right) \log\left(\hat{y}_{ik}^{(t)}\right)
\tag{4}
$$

and L_D is the reconstruction loss.

From Equation (1), we observe that both C and F are used to learn discriminative features for the labeled samples. The classifier C makes the target samples closer to the estimated prototypes by increasing the entropy. On the other side, the feature extractor F tries to decrease it by assigning the target samples to the most suitable class prototype. On the other side, the decoder D constrains the projection to the reconstruction space to control the overlap between the samples of the different classes. In the experiments, we will show that this learning mechanism allows to boost the classification accuracy of the target samples. In practice, we use a gradient reversal layer to flip the gradients of $H\left(\hat{y}^{(t_k)}\right)$ between C and F. To this end, we use a gradient reverse layer [48] between C and F to flip the sign of gradient to simplify the training process. The gradient reverse layer aims to flip the sign of the input by multiplying it with a negative scalar in the backpropagation, while leaving it as it is in the forward propagation.

Pareto Ranking for Unknown Class Sample Selection

During the alignment of the source and target distributions, we strive for detecting the most $r < n_t$ ambiguous samples and assign a soft label to them (unknown class K+1). Indeed, the adversarial learning will push the target samples to the most suitable class prototypes in the source domain, while the most ambiguous ones will potentially indicate the presence of a new class. In this work, we use the entropy measure and the distance from the class prototypes as a possible solution for identifying these samples. In particular, we propose to rank the target samples using both measures.

An important aspect of pareto ranking is the concept of dominance widely applied in multi-objective optimization, which involves finding a set of pareto optimal solutions rather than a single one. This set of pareto optimal solutions contains solutions that cannot be improved on one of the objective functions without affecting the other functions. In our case, we formulate the problem as finding a sub-set P from the unlabeled samples that maximizes the two objective functions f_1 and f_2, where

$$
f_1^j = cosine\left(z_j^{(t)}, \bar{z}_k^{(s)}\right), \; j = 1,\ldots,n_t
\tag{5}
$$

$$f_2^j = -\frac{1}{n_t} \sum_{i=1}^{n_t} \sum_{k=1}^{K+1} 1\!\left(k = \hat{y}_{ik}^{(t)}\right) log\!\left(\hat{y}_{ik}^{(t)}\right) \tag{6}$$

where f_1^j is the cosine distance of the representation $z_j^{(t)}$ of the target sample $X_j^{(t)}$ with respect to the class porotypes $\bar{z}_k^{(s)} = \frac{1}{n_{sk}} \sum_{i=1}^{n_{sk}} z_{ik}$ of each source class, and f_2^j is the cross-entropy loss computed for the samples of the target domain.

Many samples in Figure 3 are considered undesirable choices due to having low values of entropy and distance which should be dominated by other points. The samples of the pareto-set P should dominate all other samples in the target domain. Thus, the samples in this set are said to be non-dominated and forms the so-called Pareto front of optimal solutions.

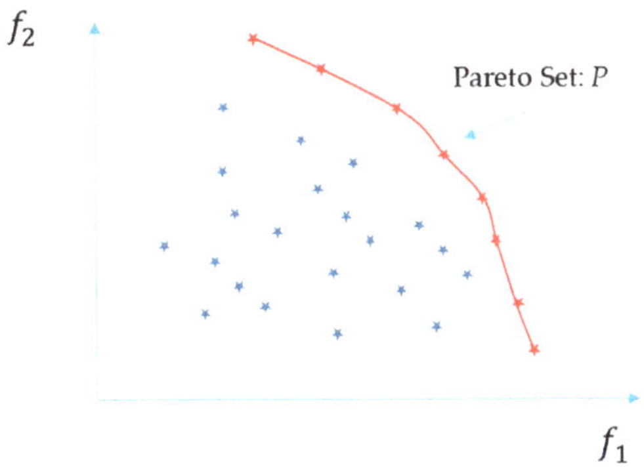

Figure 3. Pareto-front samples potentially indicating the presence of the unknown class.

The following Algorithm 1 provides the main steps for training the open-set DA with its nominal parameters:

Algorithm 1: Open-Set DA

Input: Source domain: $D_s = \left\{\left(X_i^{(s)}, y_i^{(s)}\right)\right\}_{i=1}^{n_s}$, and target domain: $D_t = \left\{X_j^{(t)}\right\}_{j=1}^{n_t}$

Output: Target labels

1: Network parameters:

- Number of iterations $num\_iter = 100$
- Mini-batch size: $b = 100$
- Adam optimizer: learning rate: 0.0001, exponential decay rate for the first and second moments $\beta_1 = 0.9$, $\beta_2 = 0.999$ and epsilon$=1e^{-8}$

2: Get feature representations from EfficientNet-B3: $h_i^{(s)} = Net\!\left(X^{(s)}\right)$ and $h_i^{(t)} = Net\!\left(X^{(t)}\right)$

3: Train the extra network on the source domain only by optimizing the loss $L_s\!\left(y^{(s_k)}, \hat{y}^{(s_k)}\right)$ for num_iter

4: Classify the target domain and get r samples forming the pareto set P and assign them to class K+1

5: for $epoch = 1 : num\_iter$

    5.1: Shuffle the labeled samples and organize them into $r_b = \frac{n_s + r}{b}$ groups each of size b

    5.2: for $k = 1 : r_b$

- Pick minibatch k from these labeled samples referred as $h_k^{(s)}$
- Pick randomly another minibatch of size b from the target domain $h_k^{(t)}$
- Update the parameters of the network by optimizing the loss in (1)

    5.3: Feed the target domain samples to the network and form a new pareto set P

6: Classify the target domain data.

---

4. Experimental Results

4.1. Dataset Description

To test the performance of the proposed architecture, we used two benchmark datasets. The first dataset consists of very high resolution (VHR) images customized from three well-known remote sensing datasets, which is the Merced dataset [1] consisting of 21 category classes each with 100 images. This dataset contains images with size of 256×256 pixels and with 0.3-m resolution. The AID dataset contains a large number of images more than 10,000 images of size 600×600 pixels with a pixel resolution varying from 8 m to about 0.5 m per pixel [49]. The images are classified to different 30 classes. The NWPU dataset contains images of size of 256×256 pixels with spatial resolutions varying from resolution 30 to 0.2 m per pixel [50]. These images correspond to 45 category classes with 700 images for each. From these three heterogonous datasets, we build cross-domain datasets, by extracting 12 common classes (see Figure 4), where each class contains 100 images.

Figure 4. Example of samples from cross-scene dataset 1 composed of very high resolution (VHR) images.

The second dataset consists of extremely high resolution (EHR) images collected by two different Aerial Vehicle platforms. The Vaihingen dataset was captured using a Leica ALS50 system at an altitude of 500 m over Vaihingen city in Germany. Every image in this dataset is represented by three channels: near infrared (NIR), red (R), and green (G) channels. The Trento dataset contains unmanned aerial vehicles (UAV) images acquired over Trento city in Italy. These images were captured using a Canon EOS 550D camera with 2 cm resolution. Both datasets contain seven classes as shown in Figure 5 with 120 images per class.

Figure 5. Example of samples from cross-scene dataset 2 composed of extremely high resolution (EHR) images.

4.2. Experiment Setup

For training the proposed architecture, we used the Adam optimization method with a fixed learning rate of 0.001. We fixed the mini-batch size to 100 samples and we set the regularization parameter of the reconstruction error and entropy terms to 1 and 0.1, respectively.

We evaluated our approach using three proposed ranking criteria for detecting the samples of the unknown class including entropy, cosine distance, and the combination of both measures using pareto-based ranking.

We present the results in terms of (1) closed set (CS) accuracy related to the shared classes between the source and target domains, which is the number of correctly classified samples divided by the total number of tested samples of the shared classes only; (2) the open set accuracy (OS) including known and unknown classes; (3) the accuracy of the unknown class itself termed as (Unk); which is the number of correctly classified unknown samples divided by the total number of tested samples of the unknown class only, and (4) the F-measure, which is the harmonic mean of Precision and Recall:

$$F = 2 \times \frac{Precision \times Recall}{Precision + Recall} \tag{7}$$

where Recall is calculated as

$$Recall = \frac{TP}{TP + FN} \tag{8}$$

and Precision is calculated as

$$\text{Precision} = \frac{\text{TP}}{\text{TP} + \text{FP}} \tag{9}$$

where TP, FN, and FP are for true positive, false negative, and false positive, respectively. F-measure gives a value between 0 and 1. High F-measure values result in better performance for the image classification system.

For the openness measure, which is the percentage of classes that appear in the target domain and are not known in the source domain, we define it as

$$openness = 1 - \frac{C_S}{C_S + C_u} \tag{10}$$

where C_S is the number of classes in the source domain shared with the target domain and C_u is the number of unknown classes in the target domain. Thus, when removing three classes from the source domain this leads to nine classes in the source ($C_s = 9$). The number of unknown classes C_u is 3, which leads to an openness of $1 - \frac{9}{9+3} = 0.25$. Increasing the value of the openness leads to increasing the number of unknown classes in the target domain that are not shared by the source domain. Setting the openness to 0 is similar to the closed set architecture where all the classes are shared between source and target domains with no unknown classes in the target domain.

4.3. Results

As we are dealing with open set domain adaptation, we propose in this first set of experiments to remove three classes from each source dataset corresponding to an openness of 25%. This means that the source dataset contains nine classes while the target dataset contains 12 classes (three are unknown). Figure 6 shows the selection of the pareto-samples from the target set for the scenario AID→Merced and AID→NWPU during the adaptation process for the first and last iterations. Here we recall that the number of selected samples is automatically determined by the ranking process.

As can be seen from Tables 1–3, the proposed approach exhibits promising results in leveraging the shift between the source and target distributions and detecting the presence of samples belonging to the unknown class. Table 1 shows the results when the Merced dataset is the target and the AID and NWPU are the sources, respectively. The results show that applying the domain adaptation always increases the accuracy for all scenarios, for example in the AID → Merced, the closed set accuracy (CS) 79.11% is lower than the results when applying the domain adaptation in all the three approaches, distance 97.77%, entropy 94.55%, and the Pareto approach 96.66%. The open set accuracy (OS) also achieves better results when the domain adaptation is applied with a minimum of 28.88% increase in the accuracy.

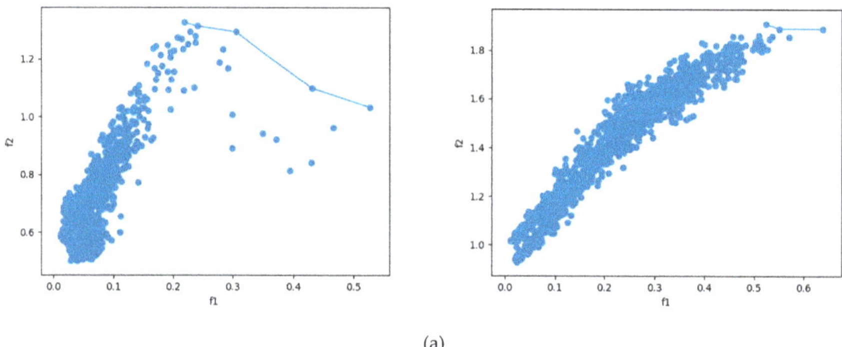

(a)

Figure 6. *Cont.*

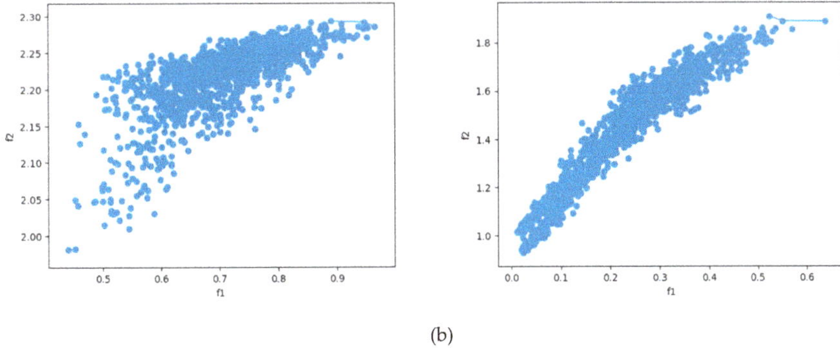

(b)

Figure 6. Pareto set selection from the target domain: (**a**) Merced→AID, and (**b**) AID-NWPU.

The unknown accuracy is always 0 for all the scenarios without domain adaptation due to the negative transfer problem. The F-measure value for the AID →Merced scenario shows a degrade when no domain adaptation is applied with a minimum percentage of 34.97% from all other approaches. For the first scenario AID →Merced, the highest closed set accuracy (CS) 97.77% is achieved by the distance approach, which also gives the better open set accuracy (OS) 90.75% and F-measure value 88.31%. For the same scenario, the entropy approach results the highest unknown accuracy (Unk) 71.66%. Among the proposed selection criteria, the Pareto-based ranking achieves highest accuracies for all four metrics CS, OS, the unknown class, and the F-measure compared to other approaches in the scenario NWPU →Merced. The accuracy of all classes including the unknown class (OS) for this scenario is 85.08%.

Table 2 shows two scenarios where the AID dataset is the target and the Merced and NWPU are the sources, respectively. The Pareto approach gives an 88.33%, 93.44%, and 85.22% for the OS, CS, and the F-measure value, respectively, in the Merced→AID scenario. For the same scenario, the highest unknown accuracy (Unk) 86% is achieved by the entropy approach. The Pareto approach achieves highest accuracies in the NWPU → AID scenario for the OS, Unk, and the F-measure, while the best CS accuracy is resulted from the distance method. Table 3 shows the results of the two scenarios Merced→ NWPU and AID→ NWPU. The Pareto method can achieve higher results for the CS and OS 72.77% and 67.75%, respectively, for the Merced→ NWPU scenario, while the best unknown accuracy 65.66% is achieved by the entropy method. The AID→ NWPU scenario shows different results with different values of the metrics for the methods. The Pareto approach results the best unknown accuracy 68.33%, while the highest CS 89.44% is achieved by the distance approach. For the same scenario, the entropy method gives the better results for the OS and F-measure, with the values 79.41% and 72.83%, respectively. Compared to the base non-adaptation method, the Pareto approach achieves better results in all for metrics for both scenarios.

Table 1. Classification results obtained for the scenarios: AID→ Merced and NWPU→ Merced for an openness = 25%.

	Target: Merced							
	Source: AID				Source: NWPU			
	CS	OS	Unk	F	CS	OS	Unk	F
No adapt.	79.11	59.33	0	51.81	80.33	60.25	0	53.57
Distance	97.77	90.75	69.66	88.31	97.55	73.16	0	76.25
Entropy	94.55	88.83	71.66	86.88	97.66	77.41	16.66	80.81
Pareto	96.66	88.21	62.83	86.78	97.77	85.08	47	84.48

Table 2. Classification results obtained for the scenarios: Merced→ AID and NWPU→ AID for an openness = 25%.

| | Target: AID | | | | | | | |
| | Source: Merced | | | | Source: NWPU | | | |
	CS	OS	Unk	F	CS	OS	Unk	F
No adapt.	71.55	53.66	0	46.76	89.77	67.33	0	63.9
distance	87.44	83.25	70.66	74.89	98.66	83.91	39.66	82.44
Entropy	78.77	80.58	86	73.58	92.33	81.08	47.33	78.81
Pareto	93.44	88.33	73	85.22	94.66	87.33	67.33	85.07

Table 3. Classification results obtained for the scenarios: Merced→ NWPU and AID→ NWPU for an openness = 25%.

| | Target: NWPU | | | | | | | |
| | Source: Merced | | | | Source: AID | | | |
	CS	OS	Unk	F	CS	OS	Unk	F
No adapt.	70.08	53.16	0	45.52	85	63.75	0	56.91
distance	69.88	61.75	37.33	55.03	89.44	74.25	28.66	68.14
Entropy	62.55	63.33	65.66	54.78	85.79	79.41	60.66	72.83
Pareto	72.77	67.75	52.66	57.6	82.77	79.16	68.33	72.22

The Pareto method shows better results in most of the scenarios. Table 4 gives the results of the average accuracy (AA) for all six scenarios in Tables 1–3. The highest average accuracy for the OS is 82.64% given by the Pareto method, while for the CS is 90.12% given by the distance method which is near the 89.68% accuracy resulted by the Pareto method. The Pareto approach achieves 61.86% in the average score of unknown class. This is 3.87% higher than other methods. The Pareto approach also achieves the highest F-measure value among all other methods with an accuracy of 78.56%. The average results in Table 4 shows the effectiveness of the proposed method compared to the non-adaptation method, where the values of all four metrics in the non-adaptation method are increased by at least 10.37% in the proposed Pareto method.

Table 4. Average performances obtained for the VHR dataset.

	CS	OS	Unk	F
No adapt.	79.31	59.58	0	53.08
Distance	90.12	77.85	40.99	74.18
Entropy	85.28	78.44	57.99	74.62
Pareto	89.68	82.64	61.86	78.56

For the EHR dataset, we tested two datasets, Vaihingen and Trento. Tables 5 and 6 show the results of the scenarios Trento→Vaihingen and Vaihingen→Trento, respectively. For the first scenario, the highest open set accuracy (OS) is 82.02% achieved by the Pareto approach, which also results in the highest closed set accuracy (CS) and F-measure values of 98.66% and 82.22%, respectively. The highest unknown accuracy (Unk) 51.66% is achieved by the distance approach for this scenario. The proposed Pareto method achieves better results in all four metrics compared to the base non-adaptation method. The Pareto approach achieves the highest results in all four metrics in the second scenario as shown in Table 6. The average accuracy (AA) for the two scenarios in Table 7 show that the Pareto approach achieves the best accuracies among all other approaches with an average OS accuracy 80.27%. The same method also results in the better percentage in all other three metrics used for evaluation. The average results for the two scenarios show that the proposed approach achieves a 40.52% higher open set accuracy (OS) compared to the approach where no domain adaptation is applied.

Table 5. Classification results obtained for the scenario Trento→ Vaihingen.

	CS	OS	Unk	F
No adapt.	55.16	39.4	0	29.17
Distance	65.5	61.54	51.66	54.96
Entropy	97.5	71.19	5.41	72.16
Pareto	98.66	82.02	40.41	82.22

Table 6. Classification results obtained for the scenario Vaihingen→ Trento.

	CS	OS	Unk	F
No adapt.	56.16	40.11	0	33.89
Distance	77.83	60	15.41	52.29
Entropy	68.83	67.38	63.75	65.29
Pareto	81.33	78.52	71.66	67.65

Table 7. Average performances obtained for the EHR dataset.

	CS	OS	Unk	F
No adapt.	55.66	39.75	0	31.53
Distance	71.66	60.77	33.53	53.62
Entropy	83.16	69.28	34.58	68.72
Pareto	89.99	80.27	56.04	74.93

5. Discussion

5.1. Effect of the Openness

In this section, we compare the robustness of the proposed Pareto approach over several openness values. The performance of the method was measured with different numbers of classes between the source and domain. As the value of openness increased, the number of unknown samples also increased which was more difficult for the classifier, compared to classifying only shared classes in the closed set classification. Table 8 shows the results obtained using different openness values for the VHR dataset. In the first scenario, we removed three classes from each source dataset corresponding to an openness of 25%. This means that the source dataset contained nine classes while the target dataset contained 12 classes (three were unknown). The highest OS accuracy was 88.33% achieved by the Merced→ AID scenario, while the lowest accuracy 67.75% achieved by the Merced→ NWPU scenario, which was 14.59% higher than the non-adaptation approach for the same scenario. The second scenario we removed four classes from the source dataset, which led to eight classes in the source dataset and 12 classes in the target dataset (four were unknown). In this scenario, the value of the accuracy degraded, resulting in 80.58% as the highest from the Merced→ AID scenario and 69.91% as the lowest accuracy from the AID→ NWPU scenario. The third and fourth scenario, we removed five and six classes, respectively. The results showed that although the accuracy decreased in both scenarios, the results were still better than the non-adaptation approach. When computing the average accuracy for all six scenarios, the Pareto method achieved higher accuracy than the approach with no adaptation in all values of openness, even with an openness of 50% the Pareto method achieved an average of 63.80% accuracy with a 24.65% increase than the non-adaptation method for the same openness.

Table 8. Sensitivity analysis with respect to the openness for the VHR dataset. Results are expressed in terms of open set accuracy (OS) (%) and average accuracy (AA) (%).

Datasets	Openness (Number of Classes Removed)							
	25% (3)		33.3% (4)		41.6% (5)		50% (6)	
	No Adapt.	Pareto	No Adapt.	Pareto	No Adapt.	Pareto	No Adapt.	Pareto
AID→Merced	59.33	88.21	50.83	79.91	47.08	61.75	38.16	61.91
NWPU → Merced	60.25	85.08	50.25	79.75	44.41	78.58	39.25	66.83
Merced→AID	53.66	88.33	47.75	80.58	44.5	77.58	33.0	55.33
NWPU→AID	67.33	87.33	56.16	76.58	46.33	73.91	44.75	76.08
Merced→NWPU	53.16	67.75	49.25	71.41	46.41	71.58	36.16	50.16
AID→NWPU	63.75	79.16	55.75	69.91	46.5	69.75	43.58	72.5
AA (%)	59.58	82.64	51.66	76.35	45.87	72.19	39.15	63.80

The results of the EHR dataset, shown in Table 9 were achieved using different openness values. In the first scenario, we removed three classes from the source domain while the target domain contained seven classes leading to an openness of 42.85%. The scenario Trento→Vaihingen resulted in 82.02% accuracy higher than the Vaihingen→Trento which resulted in 78.52% for the openness of 42.85%. For this scenario, the Pareto approach was 40.52% higher in accuracy than the non-adaptation method which resulted in an average 39.75% accuracy for the 42.85% openness. For the second scenario we removed four classes from the source dataset. The accuracy degraded in this scenario to the values of 60.11% and 48.33% for the two scenarios, respectively. The third scenario, we removed five classes from the source dataset, resulting in an openness of 71.14%. The results of the accuracy decreased to 51.66% achieved by the Trento→Vaihingen scenario. As a conclusion, we found that the proposed approach outperforms the accuracy of the non-adaptation method for different values of openness with at least 20.18%.

Table 9. Sensitivity analysis with respect to the openness for the EHR dataset. Results are expressed in terms of OS (%) and AA (%).

Datasets	Openness (Number of Classes Removed)					
	42.85% (3)		57.14% (4)		71.42% (5)	
	No Adapt.	Pareto	No Adapt.	Pareto	No Adapt.	Pareto
Trento →Vaihingen	39.4	82.02	36.19	60.11	20.71	51.66
Vaihingen →Trento	40.11	78.52	31.90	48.33	20.35	31.90
AA (%)	39.75	80.27	34.04	54.22	20.53	41.78

5.2. Effect of the Reconstruction Loss

Table 10 shows the results of the proposed method with setting the regularization parameter λ to different values in the range [0,1] for the VHR dataset. We made three scenarios with regularization parameter values of 0, 0.5, and 1. For the first scenario, the λ was set to 0, which corresponds to the removal of the decoder part. The average accuracy dropped to 78.94%, which indicated the importance of the decoder part in the proposed method. As we can see from Table 10, setting the regularization parameter to 1 resulted in the best accuracy percentage for all scenarios except the NWPU→ AID which gave the highest accuracy 89.5%, when the regularization parameter was set to 0.5. The average accuracy (AA) results in Table 10 suggested that setting the regularization parameter to 1 gives better accuracy results.

Table 10. Sensitivity analysis with respect to the regularization parameter for the VHR dataset. Results are expressed in terms of OS (%) and AA (%).

Datasets	Regularization Parameter λ		
	0	0.5	1
AID→Merced	83.91	84.08	88.21
NWPU →Merced	83.25	84.16	85.08
Merced→AID	79.25	80.36	88.33
NWPU→AID	84.58	89.5	87.33
Merced→NWPU	66.5	66.9	67.75
AID→NWPU	76.16	69.75	79.16
AA (%)	78.94	79.12	82.64

The results in Table 11 show the effect of setting the regularization parameter λ to different values in the range [0,1] for the EHR dataset. We made three scenarios with regularization parameter values of 0, 0.5, and 1. The first scenario, we removed the decoder part by setting the regularization parameter λ to 0. The second and third scenario, the regularization parameter was set to 0.5 and 1, respectively. For the Trento→Vaihingen scenario, removing the decoder resulted in an accuracy of 66.54%, which was a noticeable decrease from the highest accuracy 82.02% achieved when the regularization parameter was set to 1. The second scenario Vaihingen→Trento also resulted in the better accuracy with the regularization parameter 1, while setting the regularization to 0.5 resulted in the worst accuracy of 51.30%. From the results shown in Table 11, setting the regularization parameter to 1 gave better accuracy results.

Table 11. Sensitivity analysis with respect to the regularization parameter for the EHR dataset. Results are expressed in terms of OS (%) and AA (%).

Datasets	Regularization Parameter λ		
	0	0.5	1
Trento→Vaihingen	66.54	76.90	82.02
Vaihingen→Trento	59.88	51.30	78.52
AA (%)	63.21	64.1	80.27

6. Conclusions

In this paper, we addressed the problem of open-set domain adaptation in remote sensing imagery. Different to the widely known closed set domain adaptation, open set domain adaptation shares a subset of classes between the source and target domains, whereas some of the target domain samples are unknown to the source domain. Our proposed method aims to leverage the domain discrepancy between source and target domains using adversarial learning, while detecting the samples of the unknown class using a pareto-based raking scheme, which relies on the two metrics based on distance and entropy. Experiment results obtained on several remote sensing datasets showed promising performance of our model, the proposed method resulted an 82.64% openset accuracy for the VHR dataset, outperforming the method with no-adaptation by 23.06%. In the EHR dataset, the pareto approach resulted an 80.27% accuracy for the openset accuracy. For future developments, we plan to investigate other criteria for identifying the unknown samples to improve further the performance of the model. In addition, we plan to extend this method to more general domain adaptation problems such as the universal domain adaptation.

Author Contributions: R.A., Y.B. designed and implemented the method, and wrote the paper. H.A., N.A. contributed to the analysis of the experimental results and paper writing. All authors have read and agreed to the published version of the manuscript.

Funding: Deanship of Scientific Research at King Saud University.

Remote Sens. **2020**, *12*, 1716

Acknowledgments: The authors extend their appreciation to the Deanship of Scientific Research at King Saud University for funding this work through research group no (RG-1441-055).

Conflicts of Interest: The authors declare no conflict of interest.

References

1. Yang, Y.; Newsam, S. Bag-of-visual-words and spatial extensions for land-use classification. In *Proceedings of the 18th SIGSPATIAL International Conference on Advances in Geographic Information Systems—GIS '10*; ACM Press: San Jose, CA, USA, 2010; p. 270. [CrossRef]
2. Mekhalfi, M.L.; Melgani, F.; Bazi, Y.; Alajlan, N. Land-Use Classification with Compressive Sensing Multifeature Fusion. *IEEE Geosci. Remote Sens. Lett.* **2015**, *12*, 2155–2159. [CrossRef]
3. Cheriyadat, A.M. Unsupervised Feature Learning for Aerial Scene Classification. *IEEE Trans. Geosci. Remote Sens.* **2014**, *52*, 439–451. [CrossRef]
4. Othman, E.; Bazi, Y.; Alajlan, N.; Alhichri, H.; Melgani, F. Using convolutional features and a sparse autoencoder for land-use scene classification. *Int. J. Remote Sens.* **2016**, *37*, 2149–2167. [CrossRef]
5. Huang, L.; Chen, C.; Li, W.; Du, Q. Remote Sensing Image Scene Classification Using Multi-Scale Completed Local Binary Patterns and Fisher Vectors. *Remote Sens.* **2016**, *8*, 483. [CrossRef]
6. Lazebnik, S.; Schmid, C.; Ponce, J. Beyond Bags of Features: Spatial Pyramid Matching for Recognizing Natural Scene Categories. In Proceedings of the 2006 IEEE Computer Society Conference on Computer Vision and Pattern Recognition—Volume 2 (CVPR'06), IEEE, New York, NY, USA, 17–22 June 2006; Volume 2, pp. 2169–2178. [CrossRef]
7. Nogueira, K.; Miranda, W.O.; Dos Santos, J.A. Improving Spatial Feature Representation from Aerial Scenes by Using Convolutional Networks. In Proceedings of the 2015 28th SIBGRAPI Conference on Graphics, Patterns and Images, Salvador, Brazil, 26–29 August 2015; pp. 289–296. [CrossRef]
8. Marmanis, D.; Datcu, M.; Esch, T.; Stilla, U. Deep Learning Earth Observation Classification Using ImageNet Pretrained Networks. *IEEE Geosci. Remote Sens. Lett.* **2016**, *13*, 105–109. [CrossRef]
9. Maggiori, E.; Tarabalka, Y.; Charpiat, G.; Alliez, P. Convolutional Neural Networks for Large-Scale Remote-Sensing Image Classification. *IEEE Tran. Geosci. Remote Sens.* **2017**, *55*, 645–657. [CrossRef]
10. Zeng, D.; Chen, S.; Chen, B.; Li, S. Improving Remote Sensing Scene Classification by Integrating Global-Context and Local-Object Features. *Remote Sens.* **2018**, *10*, 734. [CrossRef]
11. Zhu, Q.; Zhong, Y.; Liu, Y.; Zhang, L.; Li, D. A Deep-Local-Global Feature Fusion Framework for High Spatial Resolution Imagery Scene Classification. *Remote Sens.* **2018**, *10*, 568. [CrossRef]
12. Liu, B.-D.; Xie, W.-Y.; Meng, J.; Li, Y.; Wang, Y. Hybrid Collaborative Representation for Remote-Sensing Image Scene Classification. *Remote Sens.* **2018**, *10*, 1934. [CrossRef]
13. Lakhal, M.I.; Çevikalp, H.; Escalera, S.; Ofli, F. Recurrent neural networks for remote sensing image classification. *IET Comput. Vis.* **2018**, *12*, 1040–1045. [CrossRef]
14. Wang, Q.; Liu, S.; Chanussot, J.; Li, X. Scene Classification with Recurrent Attention of VHR Remote Sensing Images. *IEEE Trans. Geosci. Remote Sens.* **2019**, *57*, 1155–1167. [CrossRef]
15. Xu, S.; Mu, X.; Chai, D.; Zhang, X. Remote sensing image scene classification based on generative adversarial networks. *Remote Sens. Lett.* **2018**, *9*, 617–626. [CrossRef]
16. Zhang, L.; Zhang, L.; Du, B. Deep Learning for Remote Sensing Data: A Technical Tutorial on the State of the Art. *IEEE Geosci. Remote Sens. Mag.* **2016**, *4*, 22–40. [CrossRef]
17. Ye, M.; Qian, Y.; Zhou, J.; Tang, Y.Y. Dictionary Learning-Based Feature-Level Domain Adaptation for Cross-Scene Hyperspectral Image Classification. *IEEE Trans. Geosci. Remote Sens.* **2017**, *55*, 1544–1562. [CrossRef]
18. Othman, E.; Bazi, Y.; Melgani, F.; Alhichri, H.; Alajlan, N.; Zuair, M. Domain Adaptation Network for Cross-Scene Classification. *IEEE Trans. Geosci. Remote Sens.* **2017**, *55*, 4441–4456. [CrossRef]
19. Ammour, N.; Bashmal, L.; Bazi, Y.; Rahhal, M.M.A.; Zuair, M. Asymmetric Adaptation of Deep Features for Cross-Domain Classification in Remote Sensing Imagery. *IEEE Geosci. Remote Sens Lett.* **2018**, *15*, 597–601. [CrossRef]
20. Wang, Z.; Du, B.; Shi, Q.; Tu, W. Domain Adaptation with Discriminative Distribution and Manifold Embedding for Hyperspectral Image Classification. *IEEE Geosci. Remote Sens Lett.* **2019**, *1155-1159*. [CrossRef]

21. Bashmal, L.; Bazi, Y.; AlHichri, H.; AlRahhal, M.; Ammour, N.; Alajlan, N. Siamese-GAN: Learning Invariant Representations for Aerial Vehicle Image Categorization. *Remote Sens.* **2018**, *10*, 351. [CrossRef]
22. Busto, P.P.; Gall, J. Open Set Domain Adaptation. In Proceedings of the 2017 IEEE International Conference on Computer Vision (ICCV) IEEE, Venice, Italy, 22–29 October 2017; pp. 754–763. [CrossRef]
23. Zelnio, E.; Pavy, A. Open set SAR target classification. In *Proceedings of the Algorithms for Synthetic Aperture Radar Imagery XXVI*; International Society for Optics and Photonics: Bellingham, WA, USA, 2019; Volume 10987, p. 109870J. [CrossRef]
24. Wang, Y.; Chen, W.; Song, J.; Li, Y.; Yang, X. Open Set Radar HRRP Recognition Based on Random Forest and Extreme Value Theory. In Proceedings of the 2018 International Conference on Radar (RADAR), IEEE, Brisbane, QLD, Australia, 27–30 August 2018; pp. 1–4. [CrossRef]
25. Scheirer, W.; Rocha, A.; Sapkota, A.; Boult, T. Toward Open Set Recognition. *IEEE Trans. Pattern Anal. Mach. Intell.* **2013**, *35*, 1757–1772. [CrossRef]
26. Jain, L.P.; Scheirer, W.J.; Boult, T.E. Multi-class Open Set Recognition Using Probability of Inclusion. In *Computer Vision—ECCV 2014*; Fleet, D., Pajdla, T., Schiele, B., Tuytelaars, T., Eds.; Springer International Publishing: Cham, Switzerland, 2014; Volume 8691, pp. 393–409, ISBN 978-3-319-10577-2. [CrossRef]
27. Scherreik, M.D.; Rigling, B.D. Open set recognition for automatic target classification with rejection. *IEEE Trans. Aerosp. Electron. Syst.* **2016**, *52*, 632–642. [CrossRef]
28. Wright, J.; Yang, A.Y.; Ganesh, A.; Sastry, S.S.; Ma, Y. Robust Face Recognition via Sparse Representation. *IEEE Trans. Pattern Anal. Mach. Intell.* **2009**, *31*, 210–227. [CrossRef]
29. Zhang, H.; Patel, V.M. Sparse Representation-Based Open Set Recognition. *IEEE Trans. Pattern Anal. Mach. Intell.* **2017**, *39*, 1690–1696. [CrossRef] [PubMed]
30. Bendale, A.; Boult, T. Towards Open World Recognition. In *Proceedings of the 2015 IEEE Conference on Computer Vision and Pattern Recognition (CVPR)*; IEEE: Boston, MA, USA, 2015; pp. 1893–1902. [CrossRef]
31. Bendale, A.; Boult, T.E. Towards Open Set Deep Networks. In *Proceedings of the 2016 IEEE Conference on Computer Vision and Pattern Recognition (CVPR)*; IEEE: Las Vegas, NV, USA, 2016; pp. 1563–1572. [CrossRef]
32. Hassen, M.; Chan, P.K. Learning a Neural-network-based Representation for Open Set Recognition. In Proceedings of the 2020 SIAM International Conference on Data Mining; Society for Industrial and Applied Mathematics: Philadelphia, PA, USA, 2020; pp. 154–162.
33. Shu, L.; Xu, H.; Liu, B. DOC: Deep Open Classification of Text Documents. In Proceedings of the 2017 Conference on Empirical Methods in Natural Language Processing; Association for Computational Linguistics: Copenhagen, Denmark, 2017; pp. 2911–2916. [CrossRef]
34. Shu, L.; Xu, H.; Liu, B. Unseen Class Discovery in Open-world Classification. *arXiv* **2018**, arXiv:1801.05609. Available online: https://arxiv.org/abs/1801.05609 (accessed on 26 December 2019).
35. Goodfellow, I.; Pouget-Abadie, J.; Mirza, M.; Xu, B.; Warde-Farley, D.; Ozair, S.; Courville, A.; Bengio, Y. Generative Adversarial Nets. In *Advances in Neural Information Processing Systems 27*; Ghahramani, Z., Welling, M., Cortes, C., Lawrence, N.D., Weinberger, K.Q., Eds.; Curran Associates, Inc.: Red Hook, NY, USA, 2014; pp. 2672–2680.
36. Ge, Z.; Demyanov, S.; Chen, Z.; Garnavi, R. Generative OpenMax for multi-class open set classification. In *Proceedings of the British Machine Vision Conference Proceedings 2017*; British Machine Vision Association and Society for Pattern Recognition: Durham, UK, 2017. [CrossRef]
37. Neal, L.; Olson, M.; Fern, X.; Wong, W.-K.; Li, F. Open Set Learning with Counterfactual Images. In *Computer Vision—ECCV 2018*; Ferrari, V., Hebert, M., Sminchisescu, C., Weiss, Y., Eds.; Springer International Publishing: Cham, Switzerland, 2018; Volume 11210, pp. 620–635, ISBN 978-3-030-01230-4. [CrossRef]
38. Yu, Y.; Qu, W.-Y.; Li, N.; Guo, Z. Open-category classification by adversarial sample generation. In Proceedings of the 26th International Joint Conference on Artificial Intelligence; AAAI Press: Melbourne, Australia, 2017; pp. 3357–3363.
39. Saito, K.; Yamamoto, S.; Ushiku, Y.; Harada, T. Open Set Domain Adaptation by Backpropagation. In *Computer Vision—ECCV 2018*; Ferrari, V., Hebert, M., Sminchisescu, C., Weiss, Y., Eds.; Springer International Publishing: Cham, Switzerland, 2018; Volume 11209, pp. 156–171, ISBN 978-3-030-01227-4. [CrossRef]
40. Cao, Z.; Long, M.; Wang, J.; Jordan, M.I. Partial Transfer Learning with Selective Adversarial Networks. In Proceedings of the 2018 IEEE/CVF Conference on Computer Vision and Pattern Recognition, Salt Lake City, UT, USA, 18–22 June 2018; pp. 2724–2732. [CrossRef]

41. Zhang, J.; Ding, Z.; Li, W.; Ogunbona, P. Importance Weighted Adversarial Nets for Partial Domain Adaptation. In Proceedings of the 2018 IEEE/CVF Conference on Computer Vision and Pattern Recognition, Salt Lake City, UT, USA, 18–22 June 2018; pp. 8156–8164. [CrossRef]

42. Baktashmotlagh, M.; Faraki, M.; Drummond, T.; Salzmann, M. Learning factorized representations for open-set domain adaptation. In Proceedings of the International Conference on Learning Representations (ICLR), New Orleans, LA, USA, 6–9 May 2019.

43. Lian, Q.; Li, W.; Chen, L.; Duan, L. Known-class Aware Self-ensemble for Open Set Domain Adaptation. *arXiv* **2019**, arXiv:1905.01068. Available online: https://arxiv.org/abs/1905.01068 (accessed on 26 December 2019).

44. Liu, H.; Cao, Z.; Long, M.; Wang, J.; Yang, Q. Separate to Adapt: Open Set Domain Adaptation via Progressive Separation. In Proceedings of the 2019 IEEE/CVF Conference on Computer Vision and Pattern Recognition (CVPR), IEEE, Long Beach, CA, USA, 16–20 June 2019; pp. 2922–2931. [CrossRef]

45. Tan, S.; Jiao, J.; Zheng, W.-S. Weakly Supervised Open-Set Domain Adaptation by Dual-Domain Collaboration. In Proceedings of the 2019 IEEE/CVF Conference on Computer Vision and Pattern Recognition (CVPR), IEEE, Long Beach, CA, USA, 16–20 June 2019; pp. 5389–5398. [CrossRef]

46. Tan, M.; Le, Q. EfficientNet: Rethinking Model Scaling for Convolutional Neural Networks. In Proceedings of the International Conference on Machine Learning, Long Beach, CA, USA, 9–15 June 2019; pp. 6105–6114.

47. Saito, K.; Kim, D.; Sclaroff, S.; Darrell, T.; Saenko, K. Semi-Supervised Domain Adaptation via Minimax Entropy. In Proceedings of the 2019 IEEE/CVF International Conference on Computer Vision (ICCV), IEEE, Seoul, Korea, 27 October–2 November 2019; pp. 8049–8057.

48. Ganin, Y.; Lempitsky, V. Unsupervised domain adaptation by backpropagation. In *Proceedings of the 32nd International Conference on Machine Learning—Volume 37; Lille, France, 7–9 July 2015*; JMLR.org: Lille, France, 2015; pp. 1180–1189.

49. Xia, G.-S.; Hu, J.; Hu, F.; Shi, B.; Bai, X.; Zhong, Y.; Zhang, L. AID: A Benchmark Dataset for Performance Evaluation of Aerial Scene Classification. *IEEE Trans. Geosci. Remote Sens.* **2017**, *55*, 3965–3981. [CrossRef]

50. Cheng, G.; Han, J.; Lu, X. Remote Sensing Image Scene Classification: Benchmark and State of the Art. *Proc. IEEE* **2017**, *105*, 1865–1883. [CrossRef]

Article

Deep Learning with Open Data for Desert Road Mapping

Christopher Stewart [1,*], Michele Lazzarini [2], Adrian Luna [2] and Sergio Albani [2]

[1] European Space Agency (ESA), Earth Observation Programmes, Future Systems Department, 00044 Frascati, Italy
[2] European Union Satellite Centre (SatCen), 28850 Madrid, Spain; Michele.Lazzarini@satcen.europa.eu (M.L.); Adrian.Luna@satcen.europa.eu (A.L.); Sergio.Albani@satcen.europa.eu (S.A.)
* Correspondence: chris.stewart@esa.int

Received: 10 June 2020; Accepted: 10 July 2020; Published: 15 July 2020

Abstract: The availability of free and open data from Earth observation programmes such as Copernicus, and from collaborative projects such as Open Street Map (OSM), enables low cost artificial intelligence (AI) based monitoring applications. This creates opportunities, particularly in developing countries with scarce economic resources, for large–scale monitoring in remote regions. A significant portion of Earth's surface comprises desert dune fields, where shifting sand affects infrastructure and hinders movement. A robust, cost–effective and scalable methodology is proposed for road detection and monitoring in regions covered by desert sand. The technique uses Copernicus Sentinel–1 synthetic aperture radar (SAR) satellite data as an input to a deep learning model based on the U–Net architecture for image segmentation. OSM data is used for model training. The method comprises two steps: The first involves processing time series of Sentinel–1 SAR interferometric wide swath (IW) acquisitions in the same geometry to produce multitemporal backscatter and coherence averages. These are divided into patches and matched with masks of OSM roads to form the training data, the quantity of which is increased through data augmentation. The second step includes the U–Net deep learning workflow. The methodology has been applied to three different dune fields in Africa and Asia. A performance evaluation through the calculation of the Jaccard similarity coefficient was carried out for each area, and ranges from 84% to 89% for the best available input. The rank distance, calculated from the completeness and correctness percentages, was also calculated and ranged from 75% to 80%. Over all areas there are more missed detections than false positives. In some cases, this was due to mixed infrastructure in the same resolution cell of the input SAR data. Drift sand and dune migration covering infrastructure is a concern in many desert regions, and broken segments in the resulting road detections are sometimes due to sand burial. The results also show that, in most cases, the Sentinel–1 vertical transmit–vertical receive (VV) backscatter averages alone constitute the best input to the U–Net model. The detection and monitoring of roads in desert areas are key concerns, particularly given a growing population increasingly on the move.

Keywords: synthetic aperture radar; SAR; Sentinel–1; Open Street Map; deep learning; U–Net; desert; road; infrastructure; mapping; monitoring

---

1. Introduction

The mapping and monitoring of roads in desert regions are key concerns. Population growth and an increase in the development of urban centres have led to a corresponding expansion of transportation networks [1,2]. These networks are constantly evolving [1,3]. An awareness of the location and state of road systems is important to help monitor human activity and to identify any maintenance that may be required for the infrastructure. In many desert regions, roads and tracks are used for illicit activities, such as smuggling [4]. Sand drift and dune migration can rapidly bury roads, thus necessitating intervention [5–7].

Ground techniques used for surveying and monitoring road networks are expensive and time consuming [2]. This is especially true for desert regions, given the extensive areas involved, the often inhospitable landscapes, and, in some cases, the political instability [8,9]. Remote sensing techniques have the ability to acquire information over large areas simultaneously, at frequent intervals, and at a low cost [10,11]. The application of emerging technologies, such as big data, cloud computing, interoperable platforms and artificial intelligence (AI), have opened new scenarios in different geospatial domains [12], such as the monitoring of critical infrastructure, i.e., roads.

Previously developed algorithms to automatically extract road features using techniques such as classification, segmentation, edge and line detection and mathematical morphology are summarised in a number of review papers, such as [13–16]. Since deep convolutional neural networks proved their effectiveness in the 2012 ImageNet Large Scale Visual Recognition Challenge (ILSVRC), deep learning has significantly gathered pace. Among the first to apply deep learning for road extraction were Mnih and Hinton [17]. Saito and his colleagues later achieved even better results with convolutional neural networks (CNNs) [18]. Techniques using CNNs are now considered to be standard for image segmentation [19] with many studies proposing different CNN architectures for road detection and monitoring, e.g., [1,3,20–24]. This is a fast evolving domain, and new research is regularly published on architectures and methods to address some of the limitations of CNNs. These include, for example, the significant computing and memory requirements [25], the fact that much training data is often needed, and the difficulty in adapting models to varying conditions [26]. A particularly effective CNN model for semantic segmentation is the U–Net architecture. Devised by Ronneberger and his colleagues for medical image segmentation [27], U–Net has become a standard technique for semantic segmentation in many applications since it won the IEEE International Symposium on Biomedical Imaging (ISBI) cell tracking challenge in 2015 by a large margin. The popularity of this architecture, which consists of a contracting path to capture the context and a symmetric expanding path that enables precise localisation, is due partly to its speed, and its ability to be trained end–to–end with very few images [27]. Many have applied variations of U–Net for road detection, e.g., [2,3,22–24], the majority basing their models on dedicated benchmark datasets of optical images for road identification, such as the Massachusetts roads data, created by Mihn and Hinton [17].

Most remote sensing based techniques for road detection and monitoring have relied on very high resolution (VHR) optical data [13]. However, in desert regions the spectral signatures of roads are often similar to the surrounding landscape, making them difficult to distinguish. Synthetic aperture radar (SAR) data has characteristics which make it efficient in the retrieval of roads in desert regions [9,28]. These include the sensitivity of the radar to surface roughness and the relative permittivity of targets, and the fact that SAR is a coherent system [29]. Dry sand usually has a very low relative permittivity and is therefore not a high reflector of microwave radiation. Sand covered areas are thus usually characterised by a very low SAR backscatter. Roads on the other hand may display a very different type of backscatter, which can contrast highly with the surrounding sand, even if the roads are significantly narrower than the SAR resolution cell [9]. These characteristics can be exploited to retrieve roads from SAR amplitude data. SAR coherence can also help to detect roads in desert regions. The low relative permittivity of dry sand causes the transmission of the microwave SAR signal into the sand volume [30,31]. Coherence is rapidly lost in such areas due to volume decorrelation [32]. This low coherence may contrast with the higher coherence of roads, often made from materials with a higher relative permittivity, such as asphalt, tarmac, or gravel, which therefore are not affected by volume decorrelation.

Some studies nonetheless have demonstrated methodologies for road detection and monitoring using SAR data. A good review of many of these is provided by [14]. More recently, a few studies have successfully applied deep learning techniques for SAR based road detection, e.g., [1,2,21], but these have mainly focused on relatively small, local areas, in developed landscapes, where good ground truth and training data have been available. Some have also used SAR for detecting roads and tracks in desert regions, e.g., Abdelfattah and his colleagues proposed a semi–automatic technique for SAR based road detection over a local area in the Tunisian–Libyan border [4], but again, this was applied to a specific area, and was not fully automatic.

Robust methodologies are required for operational road detection and monitoring in desert regions over large areas without the need to acquire expensive reference data. Many desert areas are situated in developing countries, such as in North Africa, where accurate and abundant training data are not available, and budgets for infrastructure surveying are low.

The work presented in this paper aims to demonstrate a methodology for road detection and monitoring in desert regions, using free input and reference data that can be scaled to desert regions globally. This approach takes input SAR data from the free and open Copernicus Sentinel–1 satellite constellation over the area to be surveyed. The input data comprises both the amplitude and coherence averages from a time series of around two and a half months acquired in the same geometry (around seven scenes). The time series average contributes to removing image speckle and improves the model performance. The reference data on the other hand includes freely available Open Street Map (OSM) data. The combined use of OSM and Earth observation (EO) data in semantic segmentation has been much discussed, e.g., [33–35], but in most cases it has been used either with very high resolution (VHR) EO data, or for general classes with much less class imbalance than the road, no–road distinction. Roads are then extracted using a version of U–Net. With its architecture consisting of a contracting path to capture the context and a symmetric expanding path that enables precise localisation, U–Net has the well–known advantages that it can be trained end–to–end with very few images, and is fast [27]. This makes it suitable for cases where abundant, high quality reference data may not be available. One of the many versions of this architecture adapted to Earth observation data includes one proposed by Jerin Paul that was previously applied successfully to VHR optical data [36]. This was the version adopted in this methodology. Despite the fact that it was developed for use with optical data, it performed well on SAR based inputs with similar class imbalance. This U–Net model is trained with SAR amplitude and coherence averages, with OSM reference masks, for each desert region. The model is then applied to detect roads in each of the desert areas for which is was trained.

The method proposed here for SAR based deep learning segmentation, trained on OSM data, has been applied to a number of test areas in various deserts in Africa and Asia. The high accuracy of the results suggests that a robust methodology involving the use of freely available input and reference data could potentially be used for operational road network mapping and monitoring.

This study has been carried out in the framework of a joint collaboration between the European Space Agency (ESA) Φ–Lab, and the European Union Satellite Centre (SatCen) Research, Technology Development and Innovation (RTDI) Unit. The ESA Φ–Lab carries out research in transformative technologies that may impact the future of EO. The SatCen RTDI Unit provides new solutions supporting the operational needs of the SatCen and its stakeholders by looking at the whole EO data lifecycle.

2. Materials and Methods

This section presents the methodology for road detection and monitoring using free and open data. The process can be divided into two steps: The first is a SAR pre–processing step, to obtain temporal averages of the calibrated backscatter amplitude and consecutive coherence for each time series, over each area. The second is the deep learning workflow. In this second step, the input SAR layers are divided into adjacent, non–overlapping patches of 256 × 256 pixels. Each patch is matched with a corresponding mask of the same resolution and dimension showing the location of any OSM roads. In these masks, pixels coinciding with roads have a value of 1, while all other pixels have a value of 0. All SAR patches, which included OSM roads in their corresponding masks, were used to train the U–Net model, initiated with random weights, using the OSM data as a reference. Subsequently, the model was applied to all patches in each area of interest (AOI) to extract the roads not included in the OSM dataset. The AOIs included three areas in three different desert environments in Africa and Asia, each the size of one Sentinel–1 IW scene (250 × 170 km).

While the OSM was used as the reference for model training, a more precise dataset was needed for an accuracy assessment. The reference masks only recorded roads present in the OSM dataset. The possibility existed that roads were present in the coverage of any given reference mask patch, but not included in the OSM. Moreover, due to the varying quality of the OSM and the varying width of roads, precise overlap between the model detected roads and OSM reference masks was difficult to achieve. To maintain automation and ensure the scalability of the method, there was no manual editing of these patches. Nonetheless, for the purpose of model training, the procedure to use the OSM as the reference worked well. For a reliable accuracy assessment however, a more rigorous technique was adopted: a subset area was randomly selected in each desert region in which all roads were manually digitised. These data were then used as the reference for a more precise accuracy assessment.

2.1. Areas of Interest (AOIs)

Three AOIs in different types of sand covered deserts were chosen to apply the method. These include most of the North Sinai Desert of Egypt, a large part of the Grand Erg Oriental in the Algerian and Tunisian Sahara, and the central part of the Taklimakan Desert of China (see Figure 1). The size of each of these three areas corresponds to the extent of one Sentinel–1 interferometric wide swath (IW) scene: 250 × 170 km, covering an area of 47,500 km^2 in each desert region. They were chosen for their geographic and morphological variety, each having very different sand dune forms and local conditions.

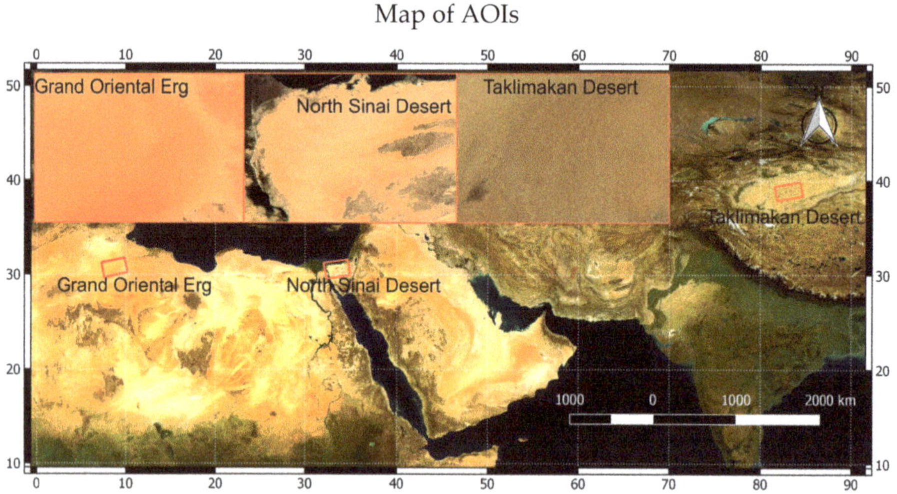

Figure 1. Map showing the location of areas of interest (AOIs) on an ENVISAT MERIS true colour mosaic, in a geographic latitude, longitude map system, World Geodetic System 1984 (WGS84) datum. Insets show a close–up of the AOIs. Each AOI and inset has the dimension of one Sentinel–1 IW footprint (250 km East–West, 170 km North–South). Credits: CHELYS srl for the world map and the European Space Agency (ESA) GlobCover for insets.

The North Sinai Desert, in the north of the Sinai Peninsula, is composed mainly of aeolian sand dune fields and interdune areas. The sand dunes include barchan, seif or longitudinal linear dunes trending east–west, transverse and star dunes [37]. Linear dunes are the main aeolian form in North Sinai [5]. The climate of the study area is arid. The average annual rainfall is about 140 mm at El Arish [38], but drops in the south, where it does not exceed 28 mm per year [5].

The Grand Erg Oriental is a sand dune field in the Sahara desert, mainly in Algeria, but with its north–eastern edge in Tunisia. It is characterised by four large–scale dune pattern types with gradual transitions between them. These include large, branching linear dunes; small and widely spaced star and dome dunes; a network type created mostly from crescentic dunes; and large, closely spaced star dunes [39]. The average annual rainfall does not exceed 70 mm [40].

The Taklimakan Desert is the world's second–largest shifting sand desert, located in China, in the rain shadow of the Tibetan Plateau [41]. Three types of sand dunes exist in the Taklimakan Desert: compound, complex crescent dunes and crescent chains; compound dome dunes; and compound, complex linear dunes [42]. The mean annual precipitation varies between 22 and 70 mm [43].

2.2. SAR and OSM Data

To achieve the objective of demonstrating a robust and cost–effective methodology that can be applied globally, it was decided to exploit the Copernicus Sentinel–1 archive. The Sentinel–1 data are acquired at regular intervals worldwide and are available under a free and open access policy [44]. Over each of the AOIs, a time series was obtained of 7 images acquired every 12 days over an approximately two–and–a–half–month period (June/July to August/September 2019). The images were all interferometric wide swath (IW), all in ascending geometry and dual polarisation: vertical transmit–vertical receive, and vertical transmit–horizontal receive (VV and VH, respectively). In order to explore the use of both amplitude and coherence in road detection, the time series over each area was obtained in both ground range detected (GRD) and single look complex (SLC) formats. The spatial resolution of the Sentinel–1 IW data is approximately 20 × 20 metres for the GRD and 5 × 20 metres for the SLC. The pixel spacing of the GRD data is 10 × 10 metres. Table 1 shows the details of the Sentinel–1 data used in each AOI. All GPT graphs and bash scripts are available on Github (A Github repository has been created which contains all scripts that were used in this research, including the Bash files and GPT graphs for the Sentinel–1 data processing, and the Python code for the deep learning workflow available in a Jupyter Notebook. In this repository are also results in a shapefile format of the road detections over each of the AOIs. Supplementary data—Available online: https://github.com/ESA-PhiLab/infrastructure) [45].

Table 1. Details of the Sentinel–1 time series used in each of the AOIs.

AOI	Orbit	Polarisation	Time Series Length
North Sinai Desert	Ascending	VV/VH	7 scenes acquired from 4 July to 14 September 2019
Grand Erg Oriental	Ascending	VV/VH	7 scenes acquired from 4 June to 15 August 2019
Taklimakan Desert	Ascending	VV/VH	7 scenes acquired from 11 June to 22 August 2019

OSM data, including all roads, was downloaded at continental scale. From the original XML formatted osm files, they were converted to vector shapefiles; subsequently, the OSM data were subset for each AOI and attribute fields were reduced to the sole roads identification, in order to limit the file size.

2.3. SAR Pre–Processing

Given that roads in desert areas can be distinguished in both SAR amplitude and coherence, it was decided to include both as inputs to the U–Net model. A virtual machine (VM), with Ubuntu as the operating system, was used for the Sentinel–1 pre–processing. This VM was connected to the CreoDIAS cloud environment, containing archive Sentinel–1 data. Processing was carried out automatically on the cloud using the command line graph processing tool (GPT) of the open source ESA Sentinel application platform (SNAP) software. Two GPT graphs, including all steps of each of the SLC and GRD processing chains, were applied in batch to the time series of data over each area using Linux bash scripts.

2.3.1. Amplitude Processing

For the amplitude processing, each Sentinel–1 scene, in a GRD format, was calibrated to σ^0 backscatter. The calibrated data was then terrain corrected to the European Petroleum Survey Group (EPSG) 4326 map system, i.e., geographic latitude and longitude with the World Geodetic System 1984 (WGS84) datum. The topographic distortion was corrected with the aid of the shuttle radar topography mission (SRTM) 3 s global digital elevation model (DEM). The output pixel spacing was 10 m. The stack of calibrated and terrain corrected scenes was then co–registered using cross correlation. The co–registered stack was averaged into one scene to reduce speckle. This average was finally converted from the linear backscatter scale to logarithmic decibel, to improve visualisation and facilitate further pre–processing during the deep learning workflow.

Some very good multitemporal speckle filters exist that preserve the spatial resolution while also keeping the temporal backscatter differences, such as the De Grandi speckle filter [46]. This allows for the monitoring of temporal intervals of less than the length of the time series. However, the emphasis of the study was to demonstrate a robust methodology that uses open data and tools. The most effective way to sufficiently reduce speckle while completely preserving the spatial resolution using the tools available was to average the data. Figure 2 shows the steps of the processing chain applied automatically to the time series of the Sentinel–1 GRD data.

Sentinel–1 intensity processing chain

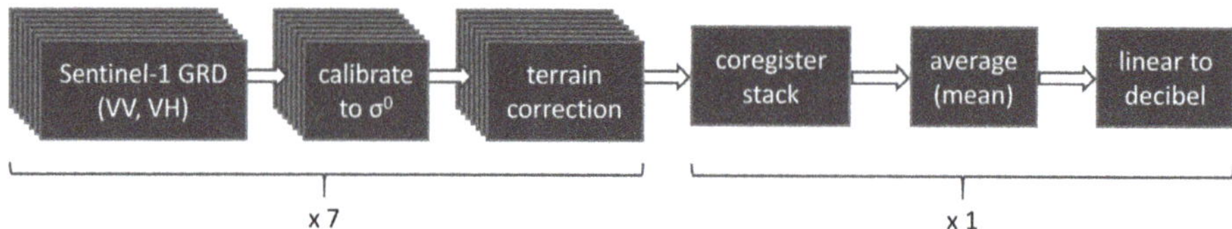

Figure 2. The intensity processing chain applied automatically to the Sentinel–1 data in the CreoDIAS cloud environment. The numbers below show how the time series of seven images are reduced to one layer for each area in which the model is applied.

2.3.2. Coherence Processing

For the coherence processing, the interferometric coherence was calculated for each consecutively acquired Sentinel–1 image pair in SLC format. For a time series of seven images therefore, six coherence images were obtained. These were then averaged to reduce clutter and better distinguish roads from the surrounding sand.

The steps in the coherence generation workflow began with the calculation of precise orbits. The three subswaths of each pair were then split to enable back–geocoding, coherence generation and TOPSAR–debursting to be carried out per subswath. These were then merged, before the coherence for each full scene pair was terrain corrected, to the same map system as used for the amplitude data processing. All terrain corrected coherences were co–registered, using a cross correlation, and averaged by taking the mean coherence for each pixel. The coherence average was finally resampled to the same pixel spacing as the amplitude average for each area, to enable the simultaneous (amplitude and coherence) input to the U–Net model. Figure 3 shows the main steps of the processing chain applied automatically to the time series of Sentinel–1 SLC data.

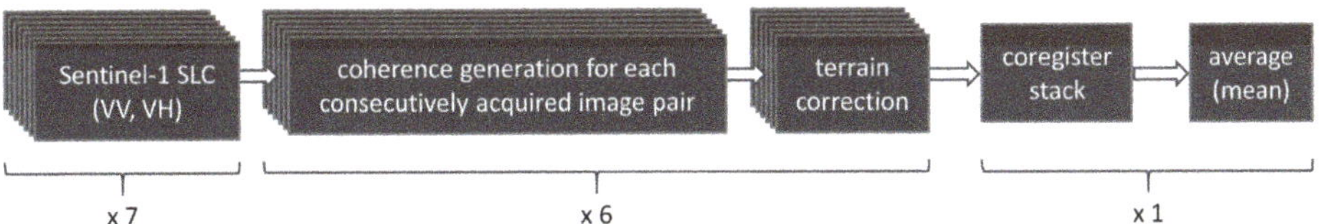

Figure 3. The main steps of the coherence processing chain applied automatically to the Sentinel–1 data in the CreoDIAS cloud environment. The numbers below show how the time series of seven images are reduced to one layer for each area in which the model is applied.

2.4. Deep Learning Workflow

The output of the Sentinel–1 pre–processing included three separate backscatter and coherence averages for each of the three desert areas, each covering the extent of one Sentinel–1 IW footprint (250 × 170 km). These, together with the OSM ancillary data, comprised the input to the second part of the methodology, which included the deep learning workflow. This second part was developed in a sandbox environment for AI projects, called Sandy, belonging to the ESA Advanced Concepts Team. It includes 10 NVIDIA GTX1080 Ti graphics processing units (GPUs), suitable for training deep neural networks, although only one GPU was necessary for the model training. The deep learning workflow was implemented in Python 3, with Keras and Tensorflow. This Jupyter Notebook is available on Github (Supplementary data—Available online: https://github.com/ESA-PhiLab/infrastructure) [45].

2.4.1. OSM and SAR Data Preparation for Deep Learning

For each backscatter and coherence scene average, a corresponding mask was produced of the same extent and spatial resolution, in which pixels overlapping with OSM roads have a value of 1, and all other pixels have a value of 0. These masks were created by converting the OSM road vectors to raster.

Each SAR derived scene average and corresponding OSM mask were split into 256 × 256 adjacent, non–overlapping patches, and the SAR patches were normalised to contain pixel values from 0 to 1.

2.4.2. Data Augmentation

Those SAR and mask patches in which OSM roads were present, comprised the samples to train the U–Net model. The number of such patches per AOI varied from around 400 to 800. Data augmentation was therefore used to increase the number of training samples. The data augmentation included a random rotation through a 360–degree range. It also included a horizontal and vertical flip. These random transformations were chosen as they preserve exactly the width and backscatter characteristics of roads and surrounding features, while changing only their orientation. This was considered the best choice given that roads may feasibly have any orientation, while it is uncertain as to how much their width and backscatter properties may vary. To fill gaps in the image patches following transformation, a reflect mode was selected, i.e., any blank areas, e.g., corners of patches following rotation, are converted to the mirror image of the same number of non–blank pixels on the other side of the dividing line. This was the best possible fill mode given that the lines are not broken, but continue as would be expected with the roads. Figure 4 shows an example of random data augmentation for an input SAR patch and its corresponding OSM derived reference mask.

The data augmentation was implemented through an instance of the image data generator class of the Keras library for Python 3.7.

Data augmentation.

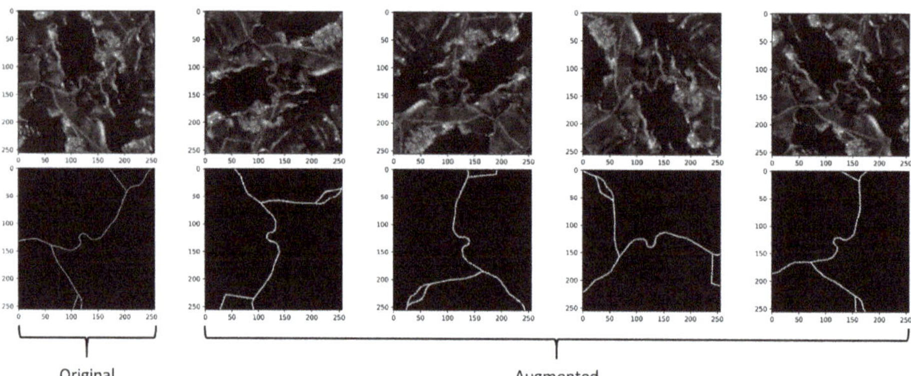

Original   Augmented

Figure 4. An example of data augmentation for an input synthetic aperture radar (SAR) patch (upper leftmost) and its corresponding reference mask (lower leftmost). Note how the augmented versions continue segments without breaking lines, using the "reflect" mode to fill gaps following rotation. Contains modified Copernicus Sentinel–1 data 2020.

2.4.3. U–Net Model

The deep learning model for image segmentation that was chosen is the modified U–Net architecture proposed by Jerin Paul [36]. This architecture has 58 layers, divided into a downsampling encoder and upsampling decoder parts, which are connected via skip connections. The convolution layers are all 3×3, with exponential linear units as the activation and He normal initialiser. The only exception to this is the last output layer, which is a 1×1 convolution layer with sigmoid activation. In between the convolution layers, batch normalisation, max pooling and data dropout layers were included. The data dropout layers applied a dropout rate varying from 0.1 to 0.3. The total number of parameters were 1,946,993. The trainable parameters were 1,944,049. All models were initiated with random weights.

The input to the network included up to three layers of average backscatter intensity in VV and VH, and average coherence. The models returned segmented images for each input patch, with pixels ranging in value from 0 to 1. Values closer to 0 have a high prediction probability of belonging to the class of non–roads, while values closer to 1 have a high probability of being a road.

2.4.4. Loss Function and Performance Metric

Road detection in desert regions is an unbalanced segmentation task, since in any given scene there are many more pixels falling into the non–road category than into the road category. The loss function applied during model training, and the accuracy metric to assess the performance of the model, need to take into account this class imbalance. The loss function that was applied in this case is the soft Dice loss. Based on the Dice coefficient, the soft Dice loss normalises for class imbalance. For the accuracy metric, it was decided to use the Jaccard index, also known as the Jaccard similarity coefficient, or intersection over union (IoU), which likewise considers class imbalance [47].

The formulae for soft Dice loss and the Jaccard index for a model predicted class (A) and a known class (B) are the following:

$$\text{Soft Dice Loss} = 1 - \frac{2|A \cap B|}{|A| + |B|} \tag{1}$$

$$\text{Jaccard index} = \frac{|A \cap B|}{|A \cup B|} \tag{2}$$

2.4.5. Hyperparameters and Model Training

Different approaches were attempted for the model training. One approach was to include the available training data from all areas with the aim of training one model applicable to every sand covered desert landscape with characteristics similar to those of the test areas. It was soon apparent that this was not feasible, due mainly to the greatly varying sand dune forms between different desert environments. This led to systematic false detections and almost no positive road detections over any area. Another approach was to choose one model, which would be trained for each specific desert region, with the available training data from that area. With this approach, even with much less training data, the model performed much better.

In addition to experimenting with the geographic coverage, different types of Sentinel–1 input were tested. Various combinations of the VV and VH backscatter and coherence were included as inputs to the model, from individual bands, to combinations of two, or all three.

The model hyperparameters are listed in Table 2. After testing different values for each parameter, these provided the best results for all the regions in which the method was applied, and with all the options for the SAR inputs. The only area specific parameter to be adjusted is the number of steps per epoch, which depends on the amount of training samples over a given region. Apart from the steps per epoch, these are the same hyperparameters as is in the model of Jerin Paul [36].

Table 2. Model hyperparameters.

Model Hyperparameters	
Epochs	100
Steps per epoch	Training samples/Batch size, (minimum of 50)
Learning rate	0.0001
Batch size	16

After randomly shuffling all samples, 10 percent were set aside for validation to assess the performance of the model during training, and another 10 percent for testing. After a review of this, a second round of training was carried out using all available data. Given the incompleteness of the OSM data in any given area, and the poor overlap between the OSM and detected roads discussed above, a more reliable accuracy assessment was carried out with the test data comprising manually digitised roads over subset areas. This is described in Section 2.5 below.

2.4.6. Post–Processing and Final Map Generation

Over each area, having trained the model with the image patches containing the available OSM data, the model was applied to predict the presence of roads in all patches. The pixels in the resulting segmented patches ranged from 0 to 1. Values closer to 0 represented a high probability of belonging to the no–road class, while those closer to 1 were considered likely to be roads. To create binary masks, all pixels with a value of less than 0.5 were converted to 0, and those greater than or equal to 0.5 were converted to 1. The patches were then put together to reconstruct the original image scene. Finally, the resulting single raster mask was converted to a vector layer containing all the predicted roads as polygon vectors in one shapefile.

2.5. Performance Evaluation

A performance evaluation of the methodology was carried out by manually digitising all the roads in subset areas within each AOI, and comparing these with the model detections through the calculation of the Jaccard similarity coefficient and the rank distance. The rank distance in turn is a measure which combines the completeness (percentage of reference data covered by model detections), and correctness (percentage of model detections covered by reference data) [48]. A performance evaluation with manually digitised reference data was necessary given the following limitations of using the OSM data as a reference.

1. The quality of the OSM data varied, in some cases road meanders were represented by straight lines (see Figure 5). This caused a misregistration between actual and OSM roads. In these cases, the U–Net model could still associate the same roads in the SAR and OSM masks by downscaling through layers of convolution filters, but for the automatically calculated IoU, the misregistration could mean no overlap (unless large buffers are used), and hence the SAR and OSM roads would not match.
2. In addition to the misregistration between the modelled and reference data, another challenge was the missing roads in the OSM data. The intention of the methodology was to detect roads unrecorded in the OSM dataset, but in the same geographic area. The chances of roads being missed in many of the reference OSM mask patches was therefore high. In the interest of demonstrating a robust and scalable methodology, manual editing to improve the mask patches was avoided.

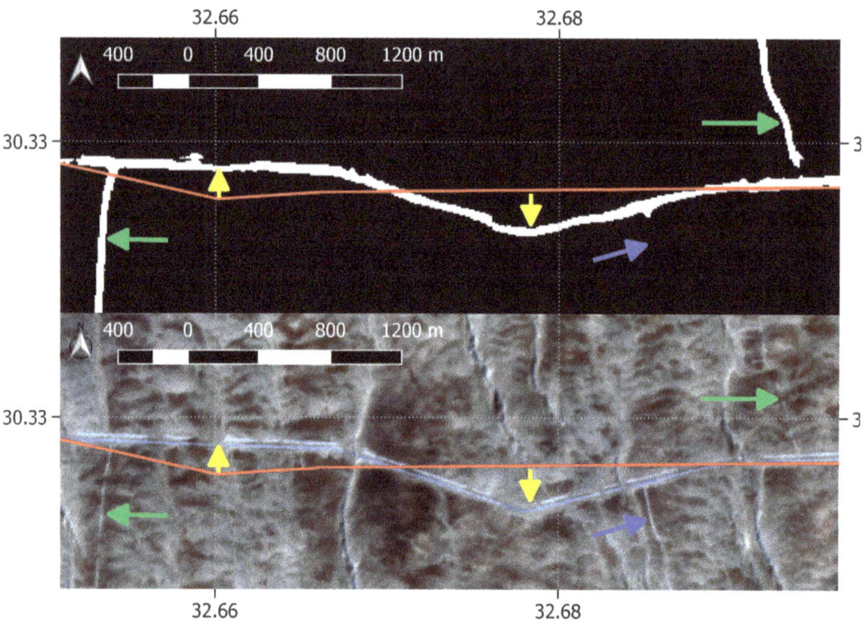

Figure 5. Above: mask of detected roads. Below: Sentinel–2 image. Red line shows the Open Street Map (OSM) road data overlaid. The yellow arrows highlight the misregistration between the OSM road and both detected (mask) and actual (Sentinel–2) roads. Green arrows show roads which are not in the OSM dataset. Blue arrows point to a road which was neither in the OSM data, nor detected by the model. Contains modified Copernicus Sentinel–2 data 2020.

Figure 5 demonstrates the success in using, in some cases, low quality OSM data to train the U–Net model (accurate detections despite misregistration of the OSM with actual roads), while also highlighting the various problems with using OSM data as a reference for performance evaluation: the varying width of roads, missing roads in the OSM data, misregistration of the OSM. The top part of this figure shows the mask of detected roads over a part of the North Sinai AOI, while the lower part is a true colour Sentinel–2 image acquired on 2 August 2019 (roughly in the middle of the Sentinel–1 time series used as an input to the model). White lines in the mask correspond with road detections. In this case the model detected the correct location of the road despite the misregistration of the OSM, which was used to train the model. Roads branching off the main road segment are not in the OSM dataset, but have been detected by the model (apart from one branch which was not detected).

These challenges resulted in the automatically calculated Jaccard index rarely exceeding 0.5 during model training, and the loss function seldom dropping below 0.3. As a relative assessment of performance, this was sufficient for model training and validation. For a more accurate quantitative assessment of results however, this was not sufficient, and a more thorough technique was adopted.

For a more robust accuracy assessment, the following was carried out: For each area, a 0.2×0.2 degree image subset was randomly selected. To avoid subsets with sparse detections, a threshold was applied to enable only those with at least 7000 road pixels to be considered. In the selected subsets, all roads were manually digitised as vectors, using the Sentinel–1 data and Sentinel–2 data from the same date range as the references. The model detected roads for the same area were similarly digitised. The resulting vector layers were visually inspected and the model detected vector components were assigned labels for true or false positives. All vector layers were converted to raster (one pixel width). A confusion matrix was created by quantifying the true and false positives and negatives. Based on this confusion matrix, the Jaccard index was calculated. Any OSM roads present in the selected subsets were discarded from the analysis since these had been used for training. While this method was suitable for quantifying true or false positives and negatives, another metric was required to assess the positional accuracy of the detections. For this, buffers of a two pixel width (40 m) were first created around both the reference and model detections. The percentage of reference data covered by model detections (completeness) and the percentage of model detections covered by the reference data (correctness) were calculated [15]. From these, the rank distance was derived, using the formula [47]:

$$\text{Rank Distance} \equiv \sqrt{\frac{\%\ \text{Complete}^2 + \%\ \text{Correct}^2}{2}} \tag{3}$$

The two pixel buffer was necessary to account for the varying width of roads and errors in manual digitising, but is a reasonable value when compared to other studies, e.g., [4].

The manually digitised reference subsets in each AOI could have been used to assess the accuracy of the OSM data. However, each validation subset only had a small quantity of OSM roads—in the Taklimakan Desert subset there were none at all (the minimum 7000 road pixel threshold applied only to model detected roads). An assessment of the accuracy of these would not have been representative. Moreover, there have been several dedicated studies on the accuracy of OSM data, e.g., [49–51].

3. Results

The model performed well in all areas, despite the diverse landscape forms and road conditions encountered in each. Especially in VV polarisation, many sand dunes produced a high backscatter. This is typical when the incidence angle of the SAR system equals the angle of repose of sand dunes [52]. The model proved nonetheless capable of distinguishing roads from sand dunes, rock formations and other natural features with similar backscatter characteristics as roads.

Of the various SAR input data types, the VV backscatter average alone proved the most effective for both the North Sinai Desert and Grand Erg Oriental. Only for the Taklimakan Desert site did all three layers of coherence, VV backscatter and VH backscatter yield the best results. Table 3 lists the IoU accuracies for each of the single SAR input layers, and for all three input layers for each of the AOIs. The fact that the use of all the input layers improved the results in only one area shows that more information provided to a model is not necessarily better. The decreased accuracy caused by the inclusion of the VH backscatter and coherence is perhaps due to the increase in speckle in these layers. This may have hindered the models in distinguishing particularly challenging roads, such as those that may be narrow, unpaved, or partially buried. The VV backscatter return over this type of landscape is generally much stronger than the VH backscatter, and enables a clearer distinction of roads. The exception of the Taklimakan desert is perhaps due to the predominant type of road surface and surrounding context of sand dunes. The results for each area are discussed in more detail in the subsections below.

Table 3. Intersection over union (IoU) accuracy of road detection models with different SAR input types. The best results for each area are highlighted in bold.

SAR Input to U–Net Model	IoU Accuracy (in %)		
	North Sinai	Grand Erg Oriental	Taklimakan Desert
VV	**89**	**84**	68
VH	65	57	77
Coherence	64	66	71
VV + VH + Coherence	74	66	**89**

3.1. North Sinai Desert

Figure 6 shows roads detected by the model over a part of the North Sinai AOI. The area includes the location of the randomly selected subset (0.2 × 0.2 degree area) in which a more accurate performance evaluation was carried out. This subset is shown in more detail in Figure 7. Figure 8 shows the corresponding area of the SAR layer used as a model input, which was the Sentinel–1 multitemporal backscatter average of the VV polarisation only. Figure 9 shows a Sentinel–2 true colour image of the subset with the available OSM data for this area overlaid. The Sentinel–2 image was acquired on 2 August 2019, which is approximately in the middle of the Sentinel–1 time series (see Table 1).

The confusion matrix for the accuracy assessment is shown in Table 4. Table 5 shows the values of various accuracy indices. The average Jaccard similarity coefficient is 89% and the rank distance is 80%. There were few false positives, i.e., non–roads classified as roads, despite the many natural features of high backscatter that could have been misinterpreted by the model as roads, such as sand dune ridges. However, there were many more false negatives, i.e., undetected road segments. In some cases, the broken segments shown in the mask were correct in that the actual road was partially buried in segments (see Figure 10 and arrows in Figures 7–9).

The VH backscatter over the entire area was much lower, and road features much less clearly defined, particularly those that were unpaved and partially sand covered. This may be the reason why the VH backscatter input degraded the results. The coherence layer highlighted very clearly the roads, but overall had much more speckle, which is possibly why this also reduced the quality of the detection.

Table 4. Confusion matrix for true and detected roads, with only the VV backscatter as input, calculated for the same area as in Figures 7–9.

North Sinai Desert Confusion Matrix	Predicted Roads	Predicted Non–Roads
True roads	7671	2060
True non–roads	218	4,881,581

Table 5. Accuracy indices calculated for the North Sinai results.

IoU Accuracy	Rank Distance	Completeness	Correctness
89%	80%	71%	88%

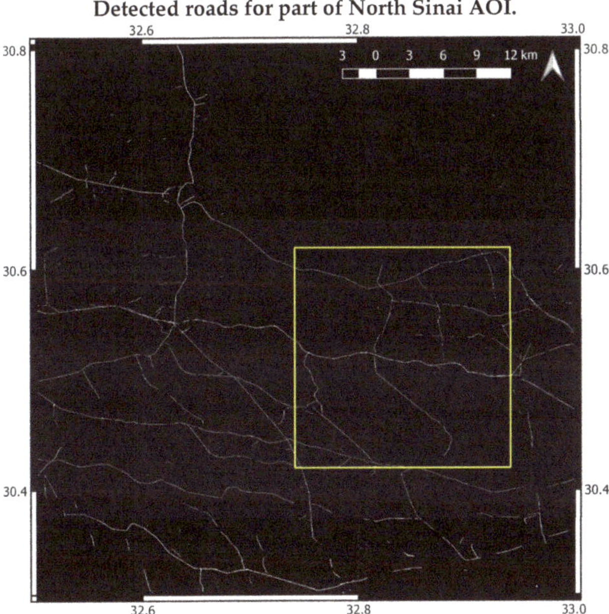

Figure 6. Detected roads for part of the North Sinai AOI. The yellow rectangle shows a 0.2 × 0.2 degree subset over which roads were manually digitised and a performance evaluation carried out. This area is shown in more detail in Figure 7.

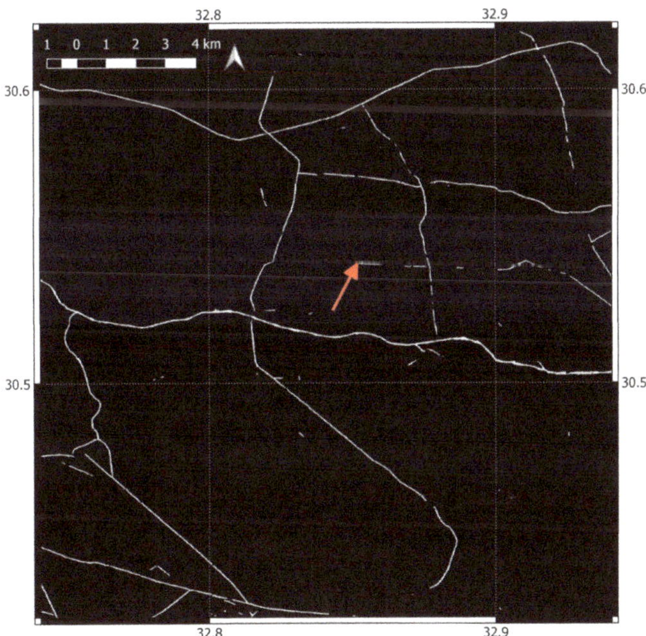

Figure 7. Detected roads for a randomly selected 0.2 × 0.2 degree subset over the North Sinai area. White lines correspond with detected roads. The red arrow points to an example of a buried road segment.

Sentinel–1 VV backscsatter average input to North Sinai model

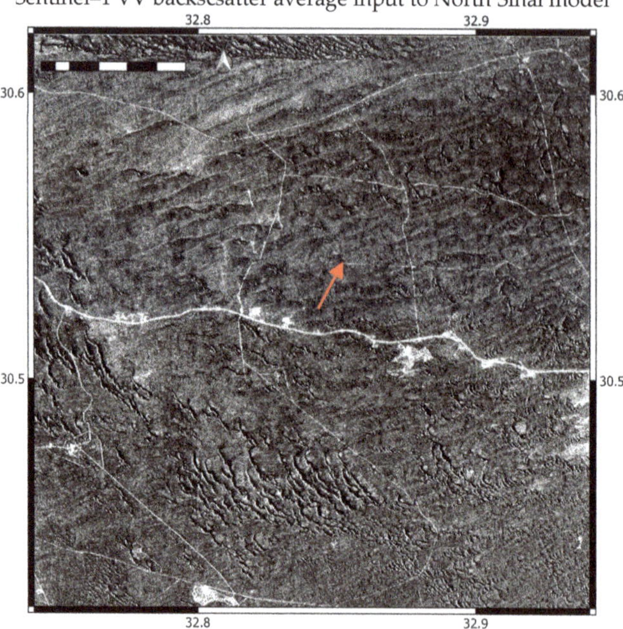

Figure 8. The Sentinel–1 average vertical transmit–vertical receive (VV) backscatter used as an input to the model for North Sinai. The area is the same as that of Figure 7. Contains modified Copernicus Sentinel–1 data 2020.

Sentinel–2 image of North Sinai AOI subset

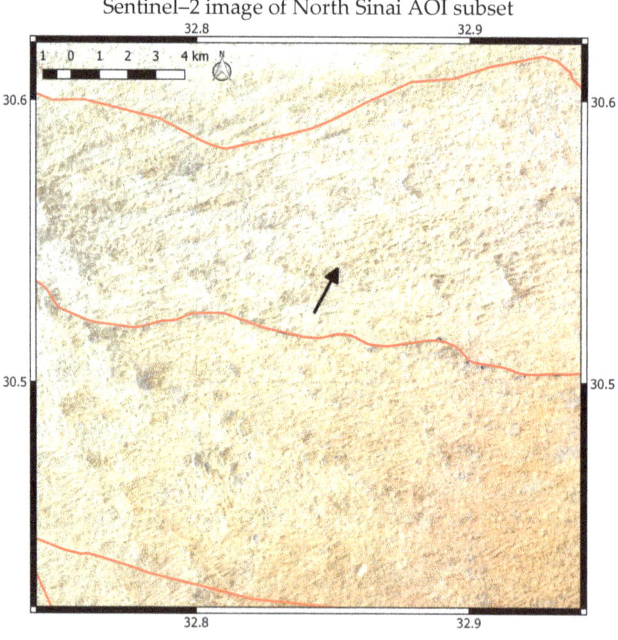

Figure 9. Sentinel–2 image of the same area as in Figure 7. The image was acquired on 2 August 2019 (roughly in the middle of the Sentinel–1 time series). It is displayed in true colour, bands 4,3,2 as red,

green, and blue, respectively. Overlaid in red are the available OSM roads for this area. Contains modified Copernicus Sentinel–2 data 2020.

Figure 10. Close–up of the buried road segment shown in the very high resolution (VHR) optical data available on Google Earth Pro. The area corresponds with that shown by the red arrow in Figures 7–9. The imagery date is reported to be 5 May 2010.

3.2. Grand Erg Oriental

Figure 11 shows the roads detected by the model over a part of the Grand Erg Oriental AOI. The area includes the location of the randomly selected subset (0.2 × 0.2 degree area) in which a more accurate performance evaluation was carried out. This subset is shown in more detail in Figure 12. Figure 13 displays the Sentinel–1 input (VV backscatter) of the same area, and Figure 14 the Sentinel–2 image with the location of OSM roads overlaid. The Sentinel–2 image was acquired on 23 July 2019, which is roughly in the middle of the Sentinel–1 time series (see Table 1).

The confusion matrix for the accuracy assessment is shown in Table 6. Table 7 shows the values of various accuracy indices. The average Jaccard similarity coefficient calculated is 84% and the rank distance is 76%. As with the North Sinai evaluation, there were a few false positives, but many more false negatives. However, the evaluation subset area contains infrastructure in addition to roads. A large segment of missed detections, for example, includes a road running parallel with a large infrastructure installation (see Figure 15). Given the width of the structure, in particular as it appears on the Sentinel–1 data (Figure 13), the model may have misinterpreted it as a natural feature.

As with the North Sinai area, the VH backscatter over the entire area was much lower, with road features less clearly defined, while the coherence layer had greater speckle. These may be the reasons why the results were better with the VV backscatter alone.

Table 6. Confusion matrix for true and detected roads calculated for same area as in Figures 12–14.

Grand Erg Oriental Confusion Matrix	Predicted Roads	Predicted Non–Roads
True roads	14,692	6430
True non–roads	282	4,929,956

Table 7. Accuracy indices calculated for the Grand Erg Oriental results.

IoU Accuracy	Rank Distance	Completeness	Correctness
84%	76%	64%	86%

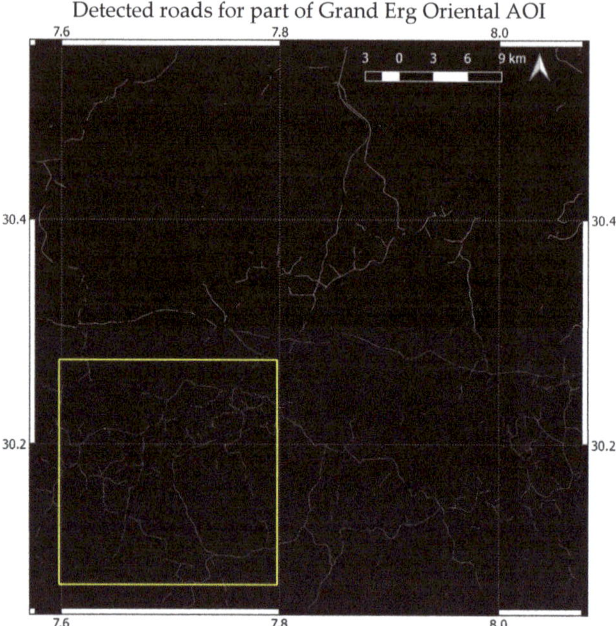

Figure 11. Detected roads for part of the Grand Erg Oriental AOI. The yellow rectangle shows a 0.2 × 0.2 degree subset over which roads were manually digitised and a performance evaluation carried out. This area is shown in more detail in Figure 12.

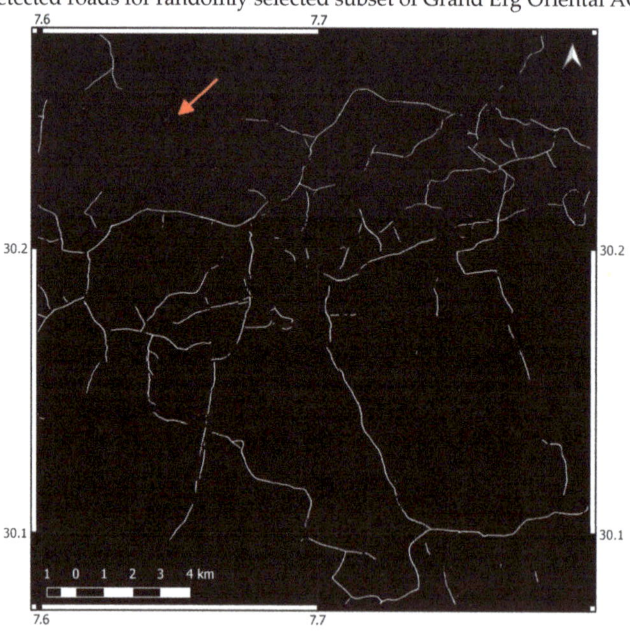

Figure 12. Detected roads for a randomly selected 0.2 × 0.2 degree subset over the Grand Erg Oriental AOI. White lines correspond with detected roads. The red arrow points to an example of a missed detection, perhaps due to mixed infrastructure.

Sentinel–1 VV backscsatter average input to Grand Erg Oriental model

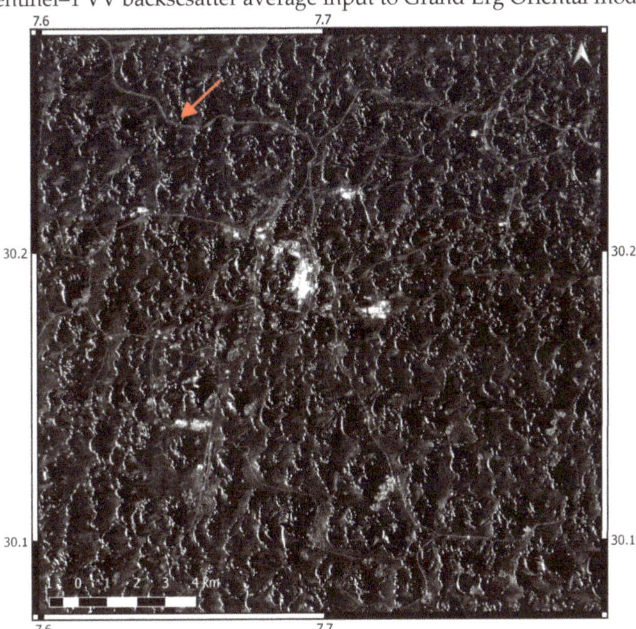

Figure 13. The Sentinel–1 average VV backscatter used as an input to the model for the Grand Erg Oriental. The area is the same as that of Figure 12. Contains modified Copernicus Sentinel–1 data 2020.

Sentinel–2 image of Grand Erg Oriental AOI subset

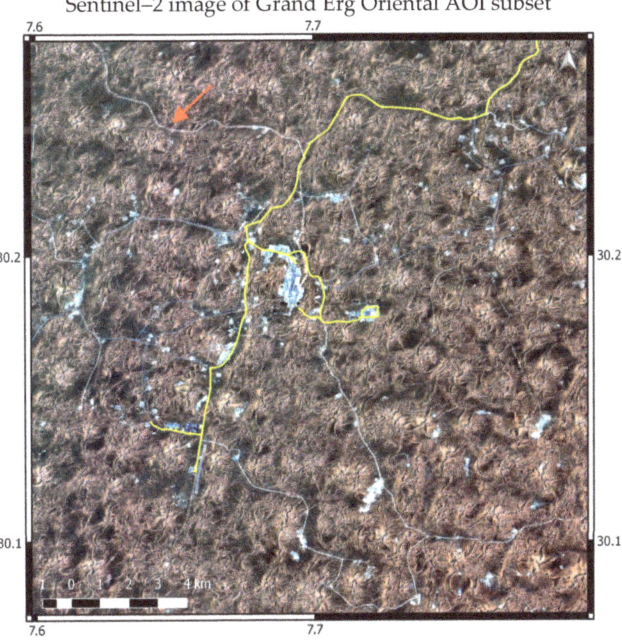

Figure 14. Sentinel–2 image of the same area as in Figure 12. The image was acquired on 23 July 2019 (roughly in the middle of the Sentinel–1 time series). It is displayed in true colour, bands 4,3,2 as red,

green, and blue, respectively. Overlaid in yellow are the available OSM roads for this area. Contains modified Copernicus Sentinel data 2020.

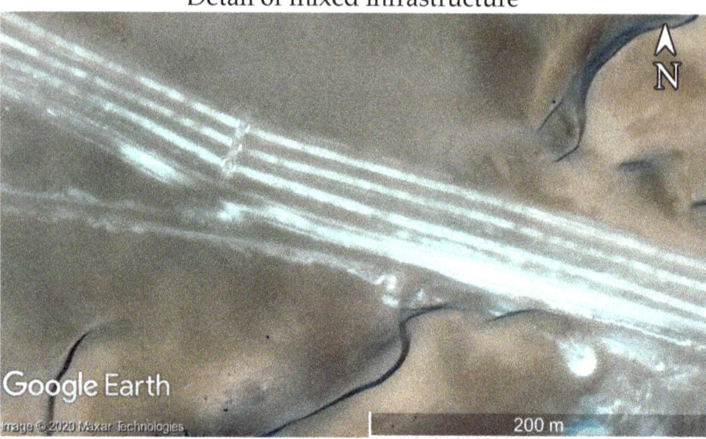

Figure 15. Close–up of a road segment in VHR optical data available on Google Earth Pro. The area corresponds with that shown by the red arrow in Figures 12–14. The road segment was not detected by the model. As evident in the figure, the road runs parallel with other infrastructure, which may have affected the ability of the model to correctly interpret the scene. The imagery date is reported to be 13 September 2013.

3.3. Taklimakan Desert

Figure 16 shows roads detected by the model over a part of the Taklimakan Desert AOI. The area includes the location of the randomly selected subset (0.2 × 0.2 degree area) in which a more accurate performance evaluation was carried out. This subset is shown in more detail in Figure 17. Figure 18 shows the Sentinel–1 input as a red, green, and blue combination of VH and VV backscatter and coherence averages, respectively. Figure 19 shows a Sentinel–2 image of the same area. In this subset there were no OSM roads. The Sentinel–2 image was acquired on 31 July 2019, approximately in the middle of the Sentinel–1 time series (see Table 1).

The confusion matrix for the accuracy assessment is shown in Table 8. Table 9 shows the values of various accuracy indices. The average Jaccard similarity coefficient calculated is 89% and the rank distance is 75%. As with the other areas, there were more false negatives than false positives. Compared to the other areas, there appear to be more paved roads, with straighter paths. The sand dunes are larger and do not have the characteristic lines of high backscatter apparent in the other two areas, although some sparse misclassifications still arise over natural features.

This was the only area where the best results were obtained with all three SAR input layers of coherence, VV backscatter and VH backscatter. The backscatter over sand dunes is much lower in VH than VV, while the road features are still clearly defined, perhaps due to the high relative permittivity of the paved roads. This may be the added value of the VH layer. The coherence layer still displayed much speckle over the sand dunes, but the roads were very clearly defined, perhaps again due to the material of their construction.

Sand drift encroachment on roads in the Taklimakan desert is a serious problem and many efforts have been made to mitigate the issue [7,53,54]. Figure 20 shows a road segment of the subset in VHR optical data available on Google Earth Pro, the date of which is reported to be 26 October 2014. The road is partially buried in this image, but the model output shows a continuous, unbroken line. It would seem that maintenance had been carried out on this road in between the date of the VHR optical image acquisition and the date range of the Sentinel–1 time series used as an input to the model.

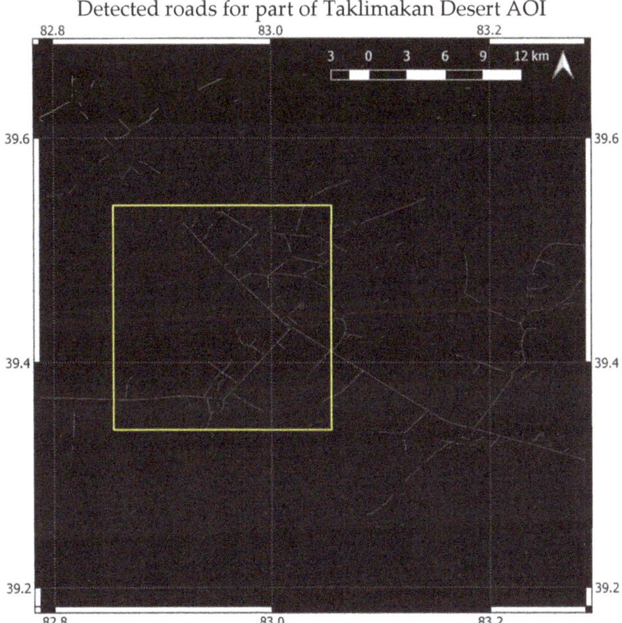

Figure 16. Detected roads for part of the Taklimakan Desert AOI. The yellow rectangle shows a 0.2 × 0.2 degree subset over which roads were manually digitised and a performance evaluation carried out. This area is shown in more detail in Figure 17.

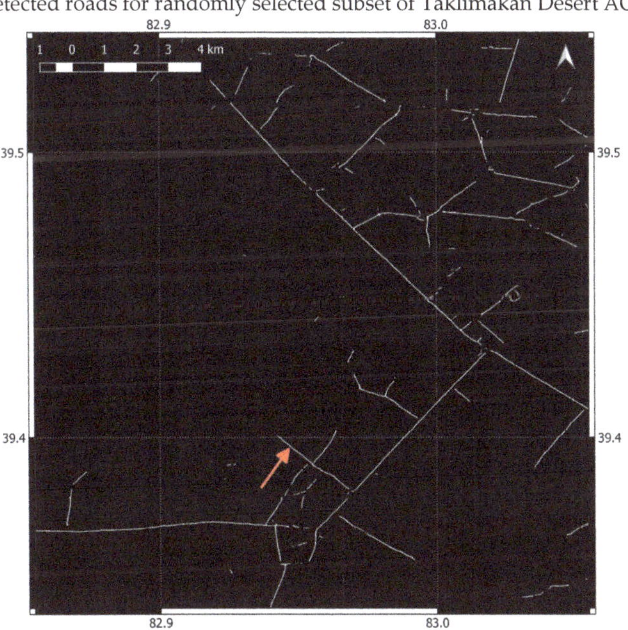

Figure 17. Detected roads for a randomly selected 0.2 × 0.2 degree subset over the Taklimakan Desert AOI. White lines correspond with detected roads. The red arrow points to an example of a formerly buried road segment.

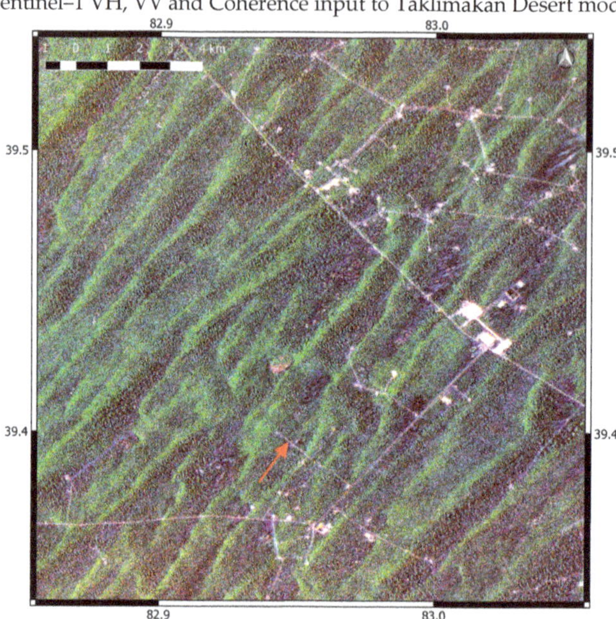

Figure 18. Sentinel–1 colour composite of time series averages of the VH backscatter in red, VV backscatter in green, and coherence in blue. This comprised the input to the model for the Taklimakan Desert area. The extent is the same as that for Figure 17. Contains modified Copernicus Sentinel data 2020.

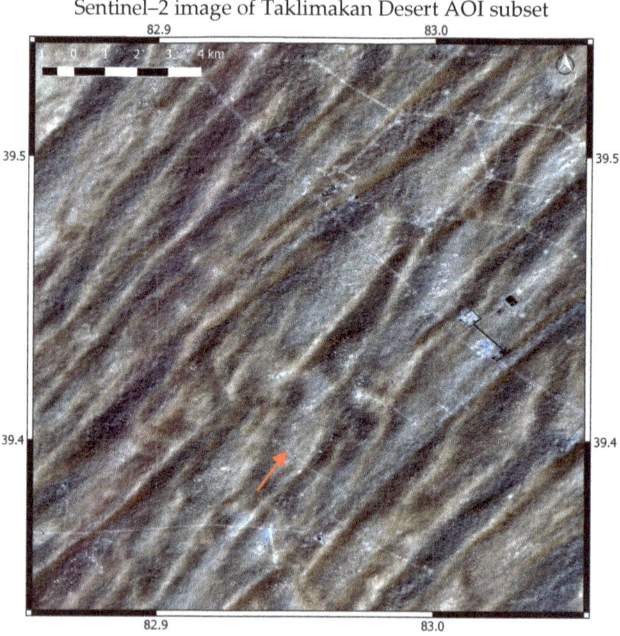

Figure 19. Sentinel–2 image of the same area as in Figure 17. The image was acquired on 31 July 2019 (roughly in the middle of the Sentinel–1 time series). It is displayed in true colour, bands 4,3,2 as red, green, and blue, respectively. Contains modified Copernicus Sentinel data 2020.

Table 8. Confusion matrix for true and detected roads calculated for the same area as in Figures 17–19.

Taklimakan Desert Confusion Matrix	Predicted Roads	Predicted Non–Roads
True roads	11,077	2622
True non–roads	428	4,940,949

Table 9. Accuracy metrics calculated for the Taklimakan Desert results.

IoU Accuracy	Rank Distance	Completeness	Correctness
89%	75%	69%	81%

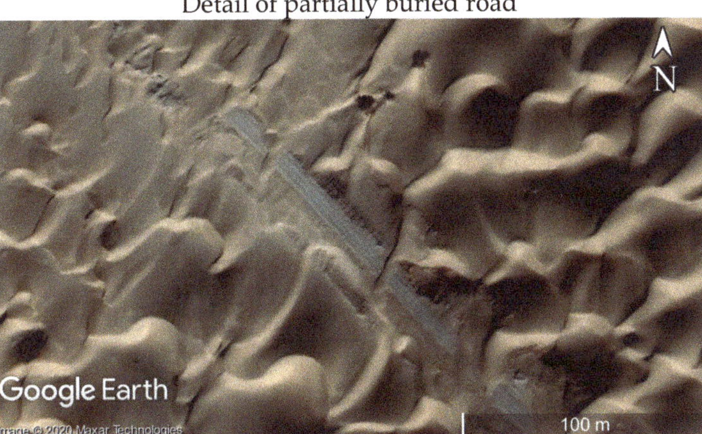

Figure 20. Close–up of a road segment in VHR optical data available on Google Earth Pro. The area corresponds with that shown by the red arrow in Figures 17–19. The road appears to be partially buried, while the output of the model shows a continuous line. It would appear the road was cleared some time between the acquisition of this image and the date range of the Sentinel–1 time series used as the model input. The imagery date is reported to be 26 October 2014.

4. Discussion

The road detection methodology proposed here aims above all to demonstrate the potential for low cost, but reliable mapping and monitoring of road networks, that can easily be adapted and transferred to extensive regions. It has been applied to three desert areas, each covering around 47,500 km2, each corresponding to the footprint of one Sentinel–1 IW scene. The results over all three areas have achieved an IoU accuracy of over 80%. This accuracy metric takes into account the class imbalance, which is typically the case for road detection in satellite imagery, where in any given area the non–road pixels are expected to greatly outnumber the road pixels. The rank distance is over 75% in all the areas tested, which demonstrates the proximity (in completeness and correctness space) between the reference and model detected roads. The recent study of Abdelfattah and Chokmani [4], who also used Sentinel–1 for road detection over a similar area, could be considered a benchmark to assess the performance of the methodology described here. The correctness and completeness of detections reported in their study are at least 10 percentage points below those described in this paper, for each of the AOIs, and despite the fact that the buffers surrounding the reference objects and detections were larger (three pixels as opposed to two pixels in the study reported here). However, the authors did focus on smaller roads and tracks, while the research of this paper includes major roads in addition to smaller unpaved tracks, so this comparison needs to be treated with caution.

Each desert area in which the methodology was tested has different characteristics, such as diverse sand dune forms, varying predominant road types, and the presence of other infrastructure. Despite these differences, the algorithm performs well and therefore demonstrates robustness. Moreover, all the required input is free and open, including the satellite data from which roads are detected, and the OSM reference data. The methodology is hence cost–effective. To scale to other areas, the model does require training. Attempts to apply pre–trained models to the other AOIs resulted in a poor performance, probably due in large part to the greatly varying sand dune morphology in each area. However, the U–Net model architecture is particularly efficient, and training with one NVIDIA GTX1080 Ti GPU never took more than around 25 min. The capability of the algorithm to work well over multiple areas with the available OSM data, without manual intervention for reference dataset cleaning, demonstrates scalability.

Despite the success of the model there are some limitations. While there were few false positives, a significantly higher number of false negatives were encountered, where the algorithm failed to detect roads. This was consistent across all test areas. Many of these false positives were due to the complexity of the context in which the road was situated. Some roads, for example, ran alongside other infrastructure, which in the resolution cell of the SAR input data could be misinterpreted as natural or other non–road features. A possible solution to mitigate these missed detections could be to expand the training set to include more OSM training samples over a wider area, or to include additional classes with a mixed infrastructure.

Another limitation is related to the required input data. In order to reduce speckle while preserving the spatial resolution, speckle filtering was carried out in the temporal domain. This requires processing of a time series, which is computationally expensive, especially for the average coherence generation with SLC format data. However, in most cases it was demonstrated that the VV average backscatter alone produces the best results. Only in one case did the coherence improve the results of the detection, but not significantly. The added value of the coherence is perhaps in those cases where there is a high proportion of roads made from material characterised by a high coherence, i.e., those that have a stable surface, such as paved roads. These would contrast highly with the surrounding sand, where volume decorrelation causes low coherence. In cases where roads are unpaved, or partially sand covered, the coherence is perhaps too noisy, despite the multitemporal averaging, and may degrade the results. Particularly in less developed areas, where there may be fewer paved roads, the coherence processing could therefore be discarded in the interest of a more computationally efficient algorithm.

The VH backscatter over all sites was much weaker than the VV. As with the coherence, the VH backscatter input only improved the detection results over the Taklimakan Desert site. Here, the VH backscatter was less pronounced over sand dunes, while still high enough over roads to enable their distinction. This may have helped reduce the ambiguity between the high backscatter encountered in the VV polarisation at certain sand dune inclinations with roads. Again, the unique suitability of the VH channel in this area alone may be due to the high relative permittivity of the roads, causing a high enough backscatter even in the weak cross polarisation channel. Elsewhere, however, the low backscatter return of the VH over less reflective roads may have contributed to the degradation of the results.

In terms of the utility of the algorithm for operational road detection in desert areas, the low number of false positives are advantageous in any alert system. Committing resources to detect human activity in remote areas is expensive and time consuming, especially in developing countries such as in North Africa where the means for such activities may be limited. As a monitoring system, the chosen time series length constrains the maximum frequency of monitoring to at least two and a half months. It may be possible to reduce this and still achieve good results. However, any changes significant enough to be observed at the spatial scale of the model are unlikely to occur at temporal frequencies significantly higher than this.

The algorithm proposed here is only a prototype. Improvements could be made for example to reduce the number of missed detections in challenging areas, perhaps by expanding the training set, and including other classes. It could be interesting also to assess the extent to which the Sentinel–1 stripmap (SM) mode may improve detections. While the higher resolution SM mode (around 10×10 m in GRD format) is not as systematically available as the Sentinel–1 IW mode, it may nonetheless be useful to detect smaller roads and tracks that may not be resolved in IW mode. This is particularly relevant for security applications [4]. However, the methodology applied with Sentinel–1 IW data nonetheless demonstrates the potential for regular and large–scale mapping and monitoring of desert roads.

5. Conclusions

The methodology proposed here for road detection in desert areas, using Sentinel–1 SAR data as an input and OSM data for training, has the potential to provide a robust, cost–effective and scalable solution for the mapping and monitoring of road networks in desert areas. This methodology is still a prototype that has been tested in three areas, each the size of one Sentinel–1 IW scene. More work is required to test its performance over a wider area and over different desert landscape types. Possible improvements with the Sentinel–1 SM mode could be explored. While the accuracy assessments over the AOIs resulted in Jaccard similarity coefficients above 84% and rank distances of over 75%, more work still needs to be done to improve the accuracy, in particular to reduce the number of missed detections. Future improvements may include the addition of other infrastructure classes, or mixed classes, to account for roads in the proximity of other structures. The methodology may be further tested to quantify model improvement according to the quantity of training data. Additionally, more experimentation can be carried out with additional data augmentation techniques, such as those that modify the intensity of pixels, rather than their spatial position alone. More importantly, the utility of the system needs to be tested by real end users. Its success should be measured against the available systems already in place. Such pre–existing systems are likely to vary between different users and geographic regions. Any improvements should be tailored to meet specific user requirements. The objective of the work presented here is to assess the benefits of EO and open data in combination with deep learning for cost–effective and large–scale monitoring. The ambition is to ultimately improve operational road detection and monitoring to support decision–making. With an increasing global population, dynamic migration patterns, and with expanding and evolving road networks, the need for efficient monitoring systems is ever more critical.

Author Contributions: Conceptualization, C.S.; Data curation, C.S., M.L. and A.L.; Formal analysis, C.S.; Investigation, C.S.; Methodology, C.S.; Project administration, C.S. and S.A.; Software, C.S.; Supervision, M.L. and A.L.; Validation, C.S.; Writing—original draft, C.S.; Writing—review & editing, M.L. and A.L. All authors have read and agreed to the published version of the manuscript.

Funding: This research received no external funding.

Acknowledgments: The authors would like to thank the following for their contribution to this research: Jerin Paul for the modified U–Net model that was adopted in this methodology [36]; the Copernicus programme, which provides free and open access to Sentinel data; Open Street Map, for the free provision of vector data of roads, and Lucio Colaiacomo (SatCen) for the preliminary preparation of this OSM data; CreoDIAS for access to the cloud processing environment for Sentinel–1 data processing; and the Advanced Concepts Team of ESA, for access to Sandy, the GPU sandbox environment.

Conflicts of Interest: The authors declare no conflict of interest.

References

1. Henry, C.; Azimi, S.M.; Merkle, N. Road segmentation in SAR satellite images with deep fully convolutional neural networks. *IEEE Geosci. Remote Sens. Lett.* **2018**, *15*, 1867–1871. [CrossRef]
2. Zhang, Q.; Kong, Q.; Zhang, C.; You, S.; Wei, H.; Sun, R.; Li, L. A new road extraction method using Sentinel–1 SAR images based on the deep fully convolutional neural network. *Eur. J. Remote Sens.* **2019**, *52*, 572–582. [CrossRef]
3. Abderrahim, N.Y.Q.; Abderrahim, S.; Rida, A. Road Segmentation using U–Net architecture. In Proceedings of the 2020 IEEE International Conference of Moroccan Geomatics (Morgeo), Casablanca, Morocco, 11–13 May 2020.

4. Abdelfattah, R.; Chokmani, K. A semi automatic off–roads and trails extraction method from Sentinel–1 data. In Proceedings of the 2017 IEEE International Geoscience and Remote Sensing Symposium (IGARSS), Fort Worth, TX, USA, 23–28 July 2017.

5. Hermas, E.; Leprince, S.; El–Magd, I.A. Retrieving sand dune movements using sub–pixel correlation of multi–temporal optical remote sensing imagery, northwest Sinai Peninsula, Egypt. *Remote Sens. Environ.* **2012**, *121*, 51–60. [CrossRef]

6. Misak, R.; Draz, M. Sand drift control of selected coastal and desert dunes in Egypt: Case studies. *J. Arid Environ.* **1997**, *35*, 17–28. [CrossRef]

7. Han, Z.; Wang, T.; Sun, Q.; Dong, Z.; Wang, X. Sand harm in Taklimakan Desert highway and sand control. *J. Geogr. Sci.* **2003**, *13*, 45–53. [CrossRef]

8. Gold, Z. *Security in the Sinai: Present and Future*; ICCT Research Paper; International Centre for Counter–Terrorism: The Hague, The Netherlands, 2014.

9. Stewart, C.; Montanaro, R.; Sala, M.; Riccardi, P. Feature Extraction in the North Sinai Desert Using Spaceborne Synthetic Aperture Radar: Potential Archaeological Applications. *Remote Sen.* **2016**, *8*, 27. [CrossRef]

10. Berger, M.; Moreno, J.; Johannessen, J.A.; Levelt, P.F.; Hanssen, R.F. ESA's sentinel missions in support of Earth system science. *Remote Sens. Environ.* **2012**, *120*, 84–90. [CrossRef]

11. Stewart, C.; Oren, E.D.; Cohen–Sasson, E. Satellite remote sensing analysis of the Qasrawet archaeological site in North Sinai. *Remote Sens.* **2018**, *10*, 1090. [CrossRef]

12. Albani, S.; Saameño, P.; Lazzarini, M.; Popescu, A.; Luna, A. Facing the Geospatial Intelligence challenges in the Big EO Data scenario. In *Proc. of the 2019 Conference on Big Data from Space (BiDS'2019)*; EUR 29660 EN; Publications Office of the European Union: Luxembourg, 2019; ISBN 978-92-76-00034-1. [CrossRef]

13. Wang, W.; Yang, N.; Zhang, Y.; Wang, F.; Cao, T.; Eklund, P. A review of road extraction from remote sensing images. *J. Traffic Transp. Eng.* **2016**, *3*, 271–282. [CrossRef]

14. Sun, N.; Zhang, J.; Huang, G.; Zhao, Z.; Lu, L. Review of road extraction methods from SAR image. In *IOP Conference Series: Earth and Environmental Science 17, Proceedings of the 35th International Symposium on Remote Sensing of Environment (ISRSE35), Beijing, China, 22–26 April 2013*; IOP Publishing Ltd: Bristol, UK, 2014. [CrossRef]

15. Crommelinck, S.; Bennett, R.; Gerke, M.; Nex, F.; Yang, M.Y.; Vosselman, G. Review of automatic feature extraction from high–resolution optical sensor data for UAV–based cadastral mapping. *Remote Sens.* **2016**, *8*, 689. [CrossRef]

16. Quackenbush, L.J. A review of techniques for extracting linear features from imagery. *Photogramm. Eng. Remote Sens.* **2004**, *70*, 1383–1392. [CrossRef]

17. Mnih, V.; Hinton, G.E. Learning to detect roads in high–resolution aerial images. In *European Conference on Computer Vision*; Springer: Berlin/Heidelberg, Germany, 2010.

18. Saito, S.; Yamashita, T.; Aoki, Y. Multiple object extraction from aerial imagery with convolutional neural networks. *Electron. Imaging* **2016**, *2016*, 1–9. [CrossRef]

19. Garcia–Garcia, A.; Orts–Escolano, S.; Oprea, S.; Villena–Martinez, V.; Garcia–Rodriguez, J. A review on deep learning techniques applied to semantic segmentation. *arXiv Preprint* **2017**, arXiv:1704.06857.

20. Cheng, G.; Wang, Y.; Xu, S.; Wang, H.; Xiang, S.; Pan, C. Automatic road detection and centerline extraction via cascaded end–to–end convolutional neural network. *IEEE Trans. Geosci. Remote Sens.* **2017**, *55*, 3322–3337. [CrossRef]

21. Şen, N.; Olgun, O.; Ayhan, Ö. Road and railway detection in SAR images using deep learning. In *Image and Signal Processing for Remote Sensing XXV*; International Society for Optics and Photonics: 2019; SPIE Remote Sensing: Strasbourg, France, 2019. [CrossRef]

22. Zhang, Z.; Liu, Q.; Wang, Y. Road extraction by deep residual u–net. *IEEE Geosci. Remote Sens. Lett.* **2018**, *15*, 749–753. [CrossRef]

23. Yang, X.; Li, X.; Ye, Y.; Lau, R.Y.; Zhang, X.; Huang, X. Road detection and centerline extraction via deep recurrent convolutional neural network u–net. *IEEE Trans. Geosci. Remote Sens.* **2019**, *57*, 7209–7220. [CrossRef]

24. Yuan, M.; Liu, Z.; Wang, F. Using the wide–range attention U–Net for road segmentation. *Remote Sens. Lett.* **2019**, *10*, 506–515. [CrossRef]

25. Zhang, L.; Wang, P.; Shen, C.; Liu, L.; Wei, W.; Zhang, Y.; van den Hengel, A. Adaptive importance learning for improving lightweight image super–resolution network. *Int. J. Comput. Vis.* **2020**, *128*, 479–499. [CrossRef]

26. Wu, Z.; Wang, X.; Gonzalez, J.E.; Goldstein, T.; Davis, L.S. ACE: Adapting to changing environments for semantic segmentation. In Proceedings of the IEEE International Conference on Computer Vision, Seoul, Korea, 27 October–2 November 2019.

27. Ronneberger, O.; Fischer, P.; Brox, T. U–net: Convolutional networks for biomedical image segmentation. In *Proceedings of the International Conference on Medical Image Computing and Computer–Assisted Intervention*; Springer: Berlin/Heidelberg, Germany, 2015.

28. Stewart, C.; Lasaponara, R.; Schiavon, G. ALOS PALSAR analysis of the archaeological site of Pelusium. *Archaeol. Prospect.* **2013**, *20*, 109–116. [CrossRef]

29. Ulaby, T.; Moore, K.; Fung, K. Microwave remote sensing. In *Volume II: Radar Remote Sensing and Surface Scattering and Emission Theory*; Addison Wesley: New York, NY, USA, 1982; Volume 2.

30. Paillou, P.; Lopez, S.; Farr, T.; Rosenqvist, A. Mapping subsurface geology in Sahara using L–Band SAR: First results from the ALOS/PALSAR imaging radar. *IEEE J. Sel. Top. Appl. Earth Obs. Remote Sens.* **2010**, *3*, 632. [CrossRef]

31. Farr, T.G.; Elachi, C.; Hartl, P.; Chowdhury, K. *Microwave Penetration and Attenuation in Desert Soil–A field Experiment with the Shuttle Imaging Radar*; IEEE: Piscataway, NJ, USA, 1986.

32. Stewart, C.; Gusmano, S.; Fea, M.; Vittozzi, G.C.; Montanaro, R.; Sala, M. A Review of the Subsurface Mapping Capability of SAR in Desert Regions, Bir Safsaf Revisited with Sentinel–1A and ENVISAT ASAR. In *2° Convegno Internazionale di Archeologia Aerea: "Dagli Aerostati ai Droni: Le Immagini Aeree in Archeologia"*; Claudio Grenzi sas: Rome, Italy, 2017.

33. Audebert, N.; le Saux, B.; Lefèvre, S. Joint learning from earth observation and openstreetmap data to get faster better semantic maps. In Proceedings of the IEEE Conference on Computer Vision and Pattern Recognition Workshops, Honolulu, HI, USA, 21–26 July 2017.

34. Yao, W.; Marmanis, D.; Datcu, M. *Semantic Segmentation Using Deep Neural Networks for SAR and Optical Image Pairs*; German Aerospace Center (DLR): Koln, Germany, 2017.

35. Kaiser, P.; Wegner, J.D.; Lucchi, A.; Jaggi, M.; Hofmann, T.; Schindler, K. Learning aerial image segmentation from online maps. *IEEE Trans. Geosci. Remote Sens.* **2017**, *55*, 6054–6068. [CrossRef]

36. Paul, J. Semantic Segmentation of Roads in Aerial Imagery. 2020. Available online: https://github.com/Paulymorphous/Road--Segmentation (accessed on 20 May 2020).

37. Hassan, M.A. Morphological and pedological characterization of sand dunes in the northern part of Sinai Peninsula using remote sensing integrated with field investigations. In Proceedings of the 2nd International Conference of Natural Resources in Africa, Cape Town, South Africa, 11–13 December 2009.

38. Gad, M.; Zeid, S.; Khalaf, S.; Abdel–Hamid, A.; Seleem, E.; el Feki, A. Optimal Management of Groundwater Resources in Arid Areas Case Study: North Sinai, Egypt. *Int. J. Water Resour. Arid Environ.* **2015**, *4*, 59–77.

39. Telbisz, T.; Keszler, O. DEM–based morphometry of large–scale sand dune patterns in the Grand Erg Oriental (Northern Sahara Desert, Africa). *Arab. J. Geosci.* **2018**, *11*, 382. [CrossRef]

40. Koull, N.; Benzaoui, T.; Sebaa, A.; Kherraze, M.; Berroussi, S. Grain size characteristics of dune sands of the Grand Erg Oriental (Algeria). *J. Algérien des Régions Arides (JARA)* **2016**, *13*, 1.

41. Sun, J.; Zhang, Z.; Zhang, L. New evidence on the age of the Taklimakan Desert. *Geology* **2009**, *37*, 159–162. [CrossRef]

42. Wang, X.; Dong, Z.; Zhang, J.; Chen, G. Geomorphology of sand dunes in the Northeast Taklimakan Desert. *Geomorphology* **2002**, *42*, 183–195. [CrossRef]

43. Wang, X.; Dong, Z.; Liu, L.; Qu, J. Sand sea activity and interactions with climatic parameters in the Taklimakan Sand Sea, China. *J. Arid Environ.* **2004**, *57*, 225–238. [CrossRef]

44. Torres, R.; Snoeij, P.; Geudtner, D.; Bibby, D.; Davidson, M.; Attema, E.; Potin, P.; Rommen, B.; Floury, N.; Brown, M. GMES Sentinel–1 mission. *Remote Sens. Environ.* **2012**, *120*, 9–24. [CrossRef]

45. Stewart, C. Mapping and Monitoring of Infrastructure in Desert Regions with Sentinel–1. 2020. Available online: https://github.com/ESA-PhiLab/infrastructure (accessed on 26 May 2020).

46. De Grandi, G.; Leysen, M.; Lee, J.; Schuler, D. Radar reflectivity estimation using multiple SAR scenes of the same target: Technique and applications. In Proceedings of the Geoscience and Remote Sensing, 1997, IGARSS'97. Remote Sensing–A Scientific Vision for Sustainable Development, 1997 IEEE International, Singapore, 3–8 August 1997; IEEE: Piscataway, NJ, USA, 2002.

47. Rahman, M.A.; Wang, Y. Optimizing Intersection–Over–Union in Deep Neural Networks for Image Segmentation. In *International Symposium on Visual Computing*; Springer: Berlin/Heidelberg, Germany, 2016.

48. Harvey, W. Performance evaluation for road extraction. *Bulletin de la Société Française de Photogrammétrie et Télédétection* **1999**, *153*, 79–87.

49. Haklay, M. How good is volunteered geographical information? A comparative study of OpenStreetMap and Ordnance Survey datasets. *Environ. Plan. B Plan. Des.* **2010**, *37*, 682–703. [CrossRef]

50. Comber, A.; See, L.; Fritz, S.; van der Velde, M.; Perger, C.; Foody, G. Using control data to determine the reliability of volunteered geographic information about land cover. *Int. J. Appl. Earth Obs. Geoinf.* **2013**, *23*, 37–48. [CrossRef]

51. Hashemi, P.; Abbaspour, R.A. *Assessment of Logical Consistency in OpenStreetMap Based on the Spatial Similarity Concept, in Openstreetmap in Giscience*; Springer: Berlin/Heidelberg, Germany, 2015; pp. 19–36.

52. Blom, R.; Elachi, C. Multifrequency and multipolarization radar scatterometry of sand dunes and comparison with spaceborne and airborne radar images. *J. Geophys. Res. B* **1987**, *92*, 7877–7889. [CrossRef]

53. Dong, Z.; Wang, X.; Chen, G. Monitoring sand dune advance in the Taklimakan Desert. *Geomorphology* **2000**, *35*, 219–231. [CrossRef]

54. Dong, Z.; Chen, G.; He, X.; Han, Z.; Wang, X. Controlling blown sand along the highway crossing the Taklimakan Desert. *J. Arid Environ.* **2004**, *57*, 329–344. [CrossRef]

 *remote sensing*

Article

Learn to Extract Building Outline from Misaligned Annotation through Nearest Feature Selector

Yuxuan Wang [1,†], Guangming Wu [1,*,†], Yimin Guo [1], Yifei Huang [2] and Ryosuke Shibasaki [1]

1   Center for Spatial Information Science, The University of Tokyo, Kashiwa 277-8568, Japan;
    yuxuan@csis.u-tokyo.ac.jp (Y.W.); guo.ym@csis.u-tokyo.ac.jp (Y.G.); shiba@csis.u-tokyo.ac.jp (R.S.)
2   Institute of Industrial Science, The University of Tokyo, Tokyo 153-8505, Japan; hyf@iis.u-tokyo.ac.jp
*   Correspondence: huster-wgm@csis.u-tokyo.ac.jp; Tel.: +81-04-7136-4390
†   These authors contributed equally to this work.

Received: 4 July 2020; Accepted: 20 August 2020; Published: 23 August 2020

Abstract: For efficient building outline extraction, many algorithms, including unsupervised or supervised, have been proposed over the past decades. In recent years, due to the rapid development of the convolutional neural networks, especially fully convolutional networks, building extraction is treated as a semantic segmentation task that deals with the extremely biased positive pixels. The state-of-the-art methods, either through direct or indirect approaches, are mainly focused on better network design. The shifts and rotations, which are coarsely presented in manually created annotations, have long been ignored. Due to the limited number of positive samples, the misalignment will significantly reduce the correctness of pixel-to-pixel loss that might lead to a gradient explosion. To overcome this, we propose a nearest feature selector (NFS) to dynamically re-align the prediction and slightly misaligned annotations. The NFS can be seamlessly appended to existing loss functions and prevent misleading by the errors or misalignment of annotations. Experiments on a large scale aerial image dataset with centered buildings and corresponding building outlines indicate that the additional NFS brings higher performance when compared to existing naive loss functions. In the classic L1 loss, the addition of NFS gains increments of 8.8% of f1-score, 8.9% of kappa coefficient, and 9.8% of Jaccard index, respectively.

Keywords: deep convolutional networks; outline extraction; misalignments; nearest feature selector

---

1. Introduction

The rooftops of buildings are dominant features in urban satellite or aerial imagery. For many remote sensing applications, such as slum mapping [1], urban planning [2], and solar panel capacity analysis [3], the spatial distributions and temporal renews of buildings are critical. These information are collected from labor-intensive and time-consuming field surveys [4]. For analyses in the city or country scale, especially in developing countries, a robust and cost-efficient method for automatic building extraction is preferred.

Over the past decades, many algorithms have been proposed [5]. These methods are verified by datasets of various types (e.g., imagery or point cloud), scales (e.g., city or country), resolutions (e.g., centimeter or meter), or spectrums (e.g., visible light, or multispectral) [6–10]. Based on whether sampled ground truths are required, existing building outline extraction methods can be classified into two categories: (i) unsupervised and (ii) supervised methods.

1.1. Unsupervised Methods

For most unsupervised methods, building outlines are extracted using thresholding pixel values or histograms [11], edge detectors [12], and region techniques [13,14]. Because of their simplicity, these methods do not require additional training data and are fast. However, when applied to residential areas with complex backgrounds, some artifacts and noises are inevitable in the extracted building outlines.

1.2. Supervised Methods

Unlike unsupervised methods, supervised methods extract building outlines from the images through patterns learned from ground truths. By learning from correct examples, supervised methods typically performed better in terms of both generalization and precision [15–17].

In the early stages, a two-stage approach that combines handcrafted descriptors for feature extraction [18–21] and classifiers for categorizing [22–24] are adopted in supervised methods. Because of the separation, an optimal combination of both the feature descriptor and classifier is difficult to achieve. Rather than the two-stage approach, convolutional neural network (CNN) methods enable a unified feature extraction and classification through sequential convolutional and fully connected layers [25,26]. Initially, CNN-based methods are constructed in a patch-by-patch manner that predicts the class of a pixel through the surrounding patch [27]. Subsequently, fully convolutional networks (FCNs) are introduced to reduce memory costs and improve computational efficiency through sequential convolutional, subsampling, and upsampling operations [28,29]. Because of information loss caused by subsampling and upsampling operations, the prediction results of classic FCN models often present blurred edges. Hence, advanced FCN-based methods using various strategies have been proposed, such as unpooling [30], deconvolution [31], skip connections [32,33], multi-constraints [34], and stacking [35]. Among FCN-based methods, two different approaches exist: (a) indirect and (b) direct approaches.

1.2.1. Indirect Approach

In the indirect approach, instead of extracting the building outline directly from the input aerial or satellite image, semantic maps are first generated. The outlines on top of those maps are computed consequently. Because the outlines are derived from segmentation output, the final accuracy relies significantly on the robustness of semantic segmentation.

In principle, all FCN-based methods mentioned above can be used for indirect building outline extraction. However, owing to the sensitivity of the outline/boundary, training with only semantic information typically results in an inconsistent outline or boundary. To prevent this, BR-Net [36] utilizes a modified U-Net, and a multitask framework to generate predictions for semantic maps and building outlines based on a consistent feature representation from a shared backend.

1.2.2. Direct Approach

Unlike the indirect approach, the direct approach extracts the building outlines directly from the input aerial or satellite images. Compared with the indirect approach, the direct approach learns the extraction pattern directly from the ground truth outline that preserves a higher fidelity. In the direct approach, building outline extraction is considered a segmentation or pixel-level classification problem that involves extremely biased data [37]. In recent years, some advanced FCN-based models, such as RSRCNN [38], ResUNet [39], and D-LinkNet [40] have been proposed for better outline extractions.

However, these models focus on deeper network architectures to better utilize the feature representation capability of hidden layers. Furthermore, regardless of how these models generate predictions, their loss functions are computed directly from the pixel-to-pixel similarity of the ground truth. Owing to the extremely biased distribution of positive and negative pixels, the gradient explosion during training becomes a severe problem. Additionally, because of occasional human errors, several or tens of pixel misalignments will inevitably occur between the annotation and the corresponding aerial image. Owing to the much fewer positive pixels of the building outline, the pixel-to-pixel losses are extremely sensitive to these misalignments.

Hence, we propose a nearest feature selector (NFS) module, enabling a dynamic re-alignment between the ground truth and prediction. A dynamic matching between the ground truth and prediction is performed at every iteration to determine the matched position. Subsequently, the overlapped areas of both the ground truth and prediction are used for further loss computation. Because the NFS is used for the upper stream, it can be seamlessly integrated into all existing loss functions. The effectiveness of the proposed NFS module is demonstrated using a VHR image dataset [36] located in New Zealand (see Section 2.1). In comparative experiments, under different loss functions, the addition of the NFS indicates significantly higher values of the f1-score, Jaccard index [41], and kappa coefficient [42].

The main contributions of this study can are as follows:

- We design a fully convolutional network framework for direct building outline extraction from aerial imagery.
- We propose the nearest feature selector(NFS) module to dynamically re-align the prediction and annotation to avoid misleading by slightly misaligned annotations.
- We analyze the effectiveness of the NFS with different loss functions to understand its effects on the performances of deep CNN models.

The rest of the paper is organized as follows: At first, we introduce the materials and methods used for this research in the Section 2. Then, we present the learning curves and quantitative and qualitative results in the Section 3. Subsequently, we illustrate our discussion and conclusion in the Sections 4 and 5, respectively.

2. Material and Method

2.1. Data

To evaluate the performance of different methods, a research area located in Christchurch, New Zealand, is selected. The original aerial imagery, as well as annotated building polygons, are hosted by the Land Information of New Zealand (LINZ) (https://data.linz.govt.nz/layer/53413-nz-building-outlines-pilot/). The aerial images are in a spatial resolution of 0.075. Prior to performing our experiments, we evenly partition the study area into two areas for training (i.e., Figure 1a, left) and testing (i.e., Figure 1a, right), respectively. The original annotations provided by the LINZ are registered to the corresponding building grounds instead of rooftops (confirmed by visual interpretation uisng QGIS GUI (https://qgis.org/)). For accurate outline extraction, we manually adjust vectorized building outlines to ensure that all building polygons and aerial rooftops are roughly registered (i.e., Figure 1b). Because of the huge amount of buildings and occasional human errors, sub-pixel or several pixel misalignments will be inevitable. Thus, we have to train the models with imperfect "ground truth".

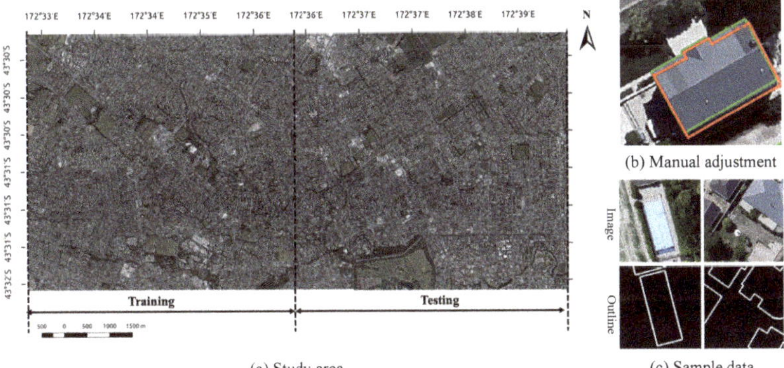

(a) Study area

(b) Manual adjustment

(c) Sample data

Figure 1. (**a**) Aerial imagery of the study area ranging from 172°33′E to 172°40′E and 43°30′S to 43°32′S, encompassing approximately 32 km². (**b**) Manual adjustment of provided annotation (e.g., from Red to Green polygon). (**c**) Sample pairs of the extracted patches.

As shown in Figure 1a, the study area is covered mainly by residential buildings with sparsely distributed factories, trees, and lakes. From training and testing areas, 16,635 and 14,834 patches are extracted. The size of the patch is 224 × 224 pixels. As shown in Figure 1c, within each pair of the patches, there are buildings in the center area.

2.2. Methodology

In this study, we are expected to correctly train and evaluate a model using imperfect annotation. Due to the inevitable misalignments, values of the loss functions or metrics, which are directly computed by the pixel-to-pixel comparison of the prediction and annotation, are inaccurate. To avoid this, we introduce the nearest feature selector (NFS) module to perform similarity selection during training and testing stages.

As shown in Figure 2, at the training phase, the NFS is applied to prediction and imperfect annotation to generate aligned prediction and annotation for accurate loss estimation and proper back-propagation. As for the testing phase, the NFS is applied to prediction and imperfect annotation to generate aligned prediction and annotation that can be used for reliable accuracy analysis. Since the NFS is applied to select the most paired overlap, it can avoid misalignments in the ground truth and produce a more reliable accuracy or prediction error.

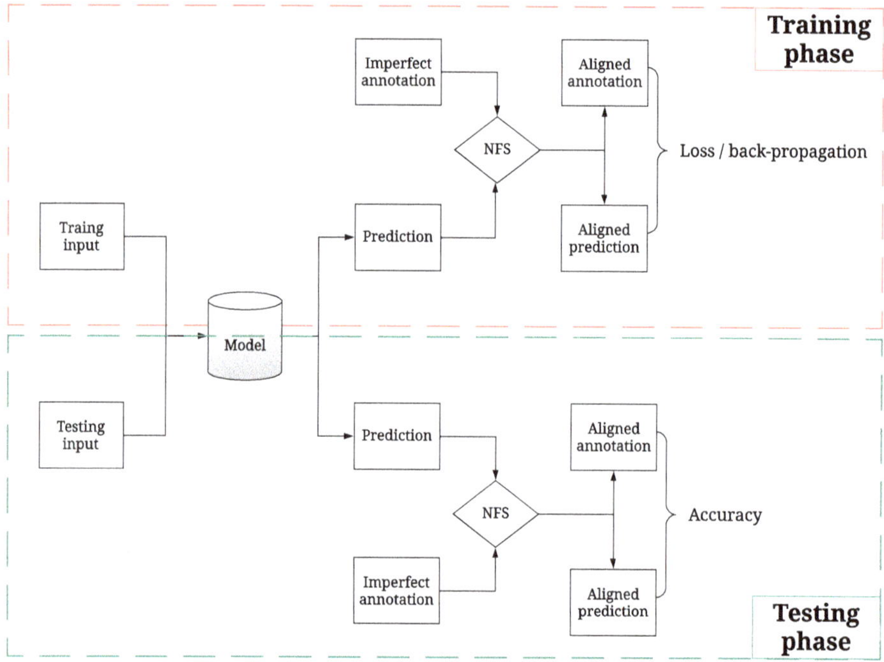

Figure 2. Experimental design for model training and evaluation under imperfect annotation. The proposed nearest feature selector(NFS) is applied to perform similarity selection during training and testing stages.

Figure 3 presents the workflow for building outline extraction. The aerial images and their corresponding building outlines are partitioned into two sets for training and testing. Through several cycles of training and validation, the hyperparameters, including batch size, the number of iterations, random seed, and initial learning rate were determined and optimized using the basic model (i.e., SegNet + L1 loss). Subsequently, the predictions generated by the optimized models are evaluated using the patches within the test set. For performance evaluations, we select three typically used balanced metrics, i.e., the f1-score, Jaccard index, and kappa coefficient. These metrics are computed before the post-processing operations [43,44].

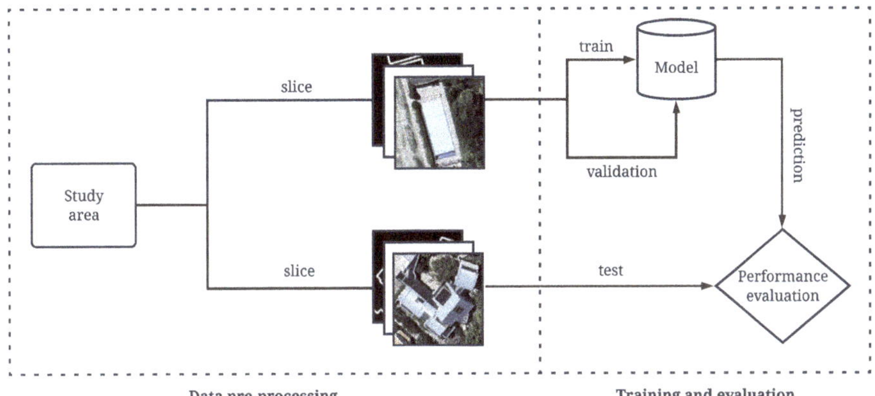

Figure 3. Experimental workflow for buidling outline extraction. Existing loss functions and proposed nearest feature selector are trained and evaluated using 224 × 224 image patches extracted from original dataset.

2.2.1. Data Preprocessing

According to the location and extent of every building polygon, a square window is applied to the centroid of the polygon to extract the corresponding image patch. Later, all patches are resized as 224 × 224 pixels. After data preprocessing, there are 16,635 and 14,834 image patches extracted from training and testing area, respectively. Since we have carefully checked the annotations, there are no negative patches to be discarded. Then, the image patches within the training area are shuffled and partitioned into two groups: training (70%), and validation (30%). Subsequently, the number of patches used for training, validation, and testing are 11,644, 4990, and 14,834, respectively.

2.2.2. Proposed Model

For an efficient building outline extraction, we utilize a modified SegNet [30] for feature extraction and the NFS to achieve a dynamic alignment between the ground truth and prediction (see Figure 4).

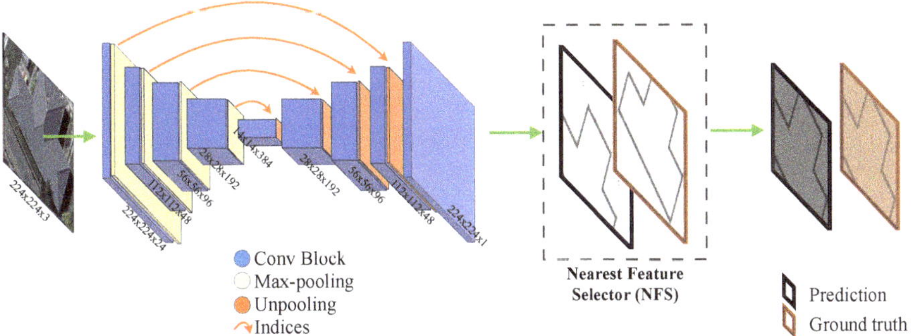

Figure 4. Overview of the proposed model. The model consists of a modified SegNet for feature extraction and the nearest feature selector (NFS) module for dynamic alignment.

- **Feature extraction**

In this study, we utilize a modified SegNet for effective feature extraction from very-high-resolution aerial images. As shown in Figure 4, the modified SegNet comprises sequential operation layers, including convolution, nonlinear activation, batch normalization, subsampling, and unpooling operations.

The convolution operation is an element-wise multiplication within a two-dimensional kernel (e.g., 3×3, or 5×5). The size of the kernel determines the receptive field and computational efficiency of the convolution operation. Owing to the complexity of the task, we set the number of kernels of the corresponding convolutional layers to [24, 48, 96, 192, 384, 192, 96, 48, 24] [34]. Subsequently, the convolution output is managed using a rectified linear unit [45], which treats all values less than zero as zeros. To accelerate network training, a batch normalization [46] layer was appended to every activation function except for the final layer. Max-pooling [47] and the corresponding unpooling [30] were used to reduce and upsample the width and height of intermediate features, respectively.

- **Nearest Feature Selector(NFS)**

Figure 5 shows the mechanisms of the NFS. The center area of the ground truth slides over the corresponding prediction along both the X- and Y-axes to generate overlaps of $X_{i,j}$ and Y_c, respectively, where i and j are the distances from the initial position. To obtain a balance between the computational efficiency and sliding field, we set the maximum values of both i and j to five. Subsequently, they were used for similarity estimation through different criteria according to the number of channels of the output.

For the prediction and ground truth containing a single channel, the classic L1 distance is used. Thus, the distance of the (i,j) overlap can be formulated as:

$$D_{i,j} = \frac{1}{W \times H} \sum_{i=1}^{W} \sum_{j=1}^{H} ||X_{i,j} - Y_c|| \tag{1}$$

where X is the prediction, and Y is the corresponding ground truth. Both X and Y are $\in R^{W \times H}$. W and H are the width and height of the corresponding output, respectively.

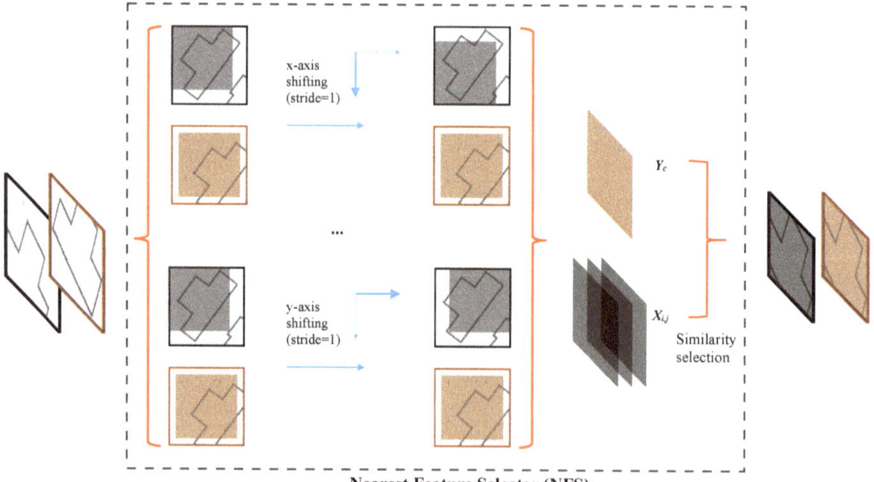

Figure 5. Overview of the nearest feature selector (NFS) module. The center area of ground truth slides over prediction along X- and Y-axes to generate overlaps that are used for similarity selection.

For the prediction and ground truth containing multiple channels, the average cosine similarity along the channels will be calculated. In such cases, the distance of overlaps can be formulated as:

$$D_{i,j} = 1 - \frac{1}{W \times H} \sum_{i=1}^{W} \sum_{j=1}^{H} \frac{X_{i,j} \cdot Y_c}{||X_{i,j}|| \times ||Y_c||} \tag{2}$$

From all overlaps, location indices of the one with the closest distance to the ground truth is determined as:

$$(i_{min}, j_{min}) = \underset{i,j}{\arg\min} D \tag{3}$$

The nearest overlap $(X_{i_{min},j_{min}})$ and corresponding ground truth (Y_c) are selected for further final loss estimation. Four well-known loss functions, namely, L1, mean square error (MSE), binary cross-entropy (BCE) [48], and focal loss [49], are chosen in this study.

$$\mathcal{L}_{L1} = \frac{1}{W \times H} \sum_{m=1}^{W} \sum_{n=1}^{H} ||y_{m,n} - g_{m,n}|| \tag{4}$$

$$\mathcal{L}_{MSE} = \frac{1}{W \times H} \sum_{m=1}^{W} \sum_{n=1}^{H} (y_{m,n} - g_{m,n})^2 \tag{5}$$

where W and H represent the width and hight of the nearest overlap $(X_{i_{min},j_{min}})$ and corresponding ground truth (Y_c). The values of $y_{m,n}$ and $g_{m,n}$ are the predicted probability and ground truth, respectively.

For notational convenience, we define $p_{m,n}$:

$$p_{m,n} = \begin{cases} y_{m,n}, & \text{if } g_{m,n} = 1 \\ 1 - y_{m,n}, & \text{if } g_{m,n} = 0 \end{cases} \tag{6}$$

As compared with traditional cross-entropy, focal loss introduces a scaling factor (γ) to focus on difficult samples. Mathematically, the BCE and focal loss can be formulated as:

$$\mathcal{L}_{BCE} = -\frac{1}{W \times H} \sum_{m=1}^{W} \sum_{n=1}^{H} log(p_{m,n}) \tag{7}$$

$$\mathcal{L}_{focal} = -\frac{1}{W \times H} \sum_{m=1}^{W} \sum_{n=1}^{H} (1 - p_{m,n})^\gamma log(p_{m,n}) \tag{8}$$

Because the NFS is computed dynamically, it can be seamlessly integrated into the existing loss without further modification.

Three typically used balanced metrics, i.e., the f1-score, Jaccard index, and kappa coefficient, are used for the quantitative evaluation. Compared with unbalanced metrics such as precision and recall, the selected metrics provide a more generalized accuracy level by considering both precision and recall.

$$F1 - score = \frac{2 \times TP}{2 \times TP + (FP + FN)} \tag{9}$$

$$Jaccard = \frac{TP}{TP + FP + FN} \tag{10}$$

$$Pe = \frac{(TP + FN) \times (TP + FP) + (FP + TN) \times (FN + TN)}{(TP + FP + FN + FN) \times (TP + FP + FN + FN)} \tag{11}$$

$$Po = \frac{TP + TN}{TP + FP + FN + FN} \tag{12}$$

$$Kappa = \frac{Po - Pe}{1 - Pe} \tag{13}$$

where TP, FP, FN, and TN represent the number of true positives, false positives, false negatives, and true negatives, respectively.

3. Results

Four well-known loss functions, i.e., L1, mean square error (MSE), binary cross-entropy (BCE) [48], and focal loss [49] are used in this study. The L1 and MSE can be regarded as the most classic and typically used criteria for pixel-to-pixel comparisons. The BCE is a typical loss function that increases or decreases exponentially for binary classification. The focal loss introduces a scale factor to the BCE to reduce the importance of the easy example. These loss functions were trained either with or without the NFS, separately. All experiments were performed on the same dataset and processing platform.

Three typically used balanced metrics, i.e., the f1-score, Jaccard index, and kappa coefficient, are used for the quantitative evaluation. Compared with unbalanced metrics such as precision and recall, the selected metrics provide a more generalized accuracy level by considering both precision and recall.

3.1. Learning Curves

Figure 6 shows the relative values of loss from different loss functions under the validation dataset. Among all the loss functions (i.e., L1, MSE, BCE, and focal), the loss with the NFS (i.e., +NFS) indicated a faster converging speed than those without (i.e., −NFS).

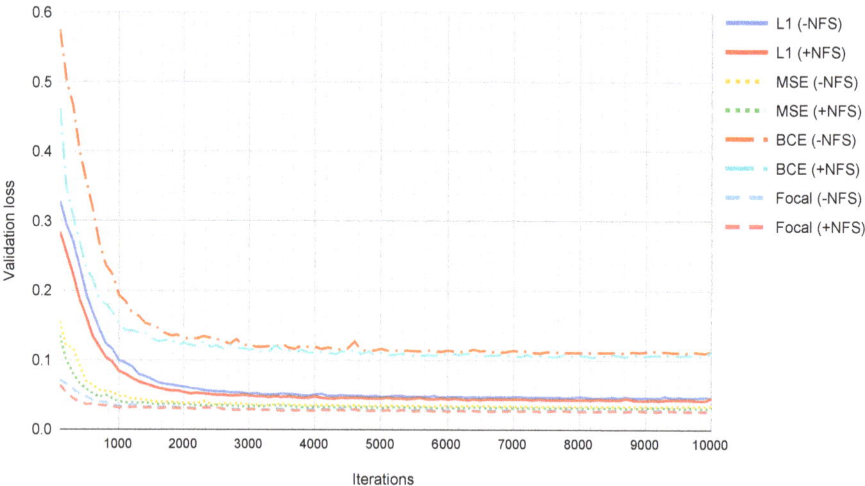

Figure 6. Trends in validation loss values over different iterations.

Figure 7 shows the trend of kappa coefficient values over various iterations from four different loss functions under the validation dataset. Among all the conditions, the focal loss trained with the proposed NFS (i.e., focal + NFS) indicates the highest kappa coefficient values in most of the iterations. By contrast, the L1 loss trained without the NFS (i.e., L1 − NFS) indicated the lowest kappa coefficient values for almost every iteration.

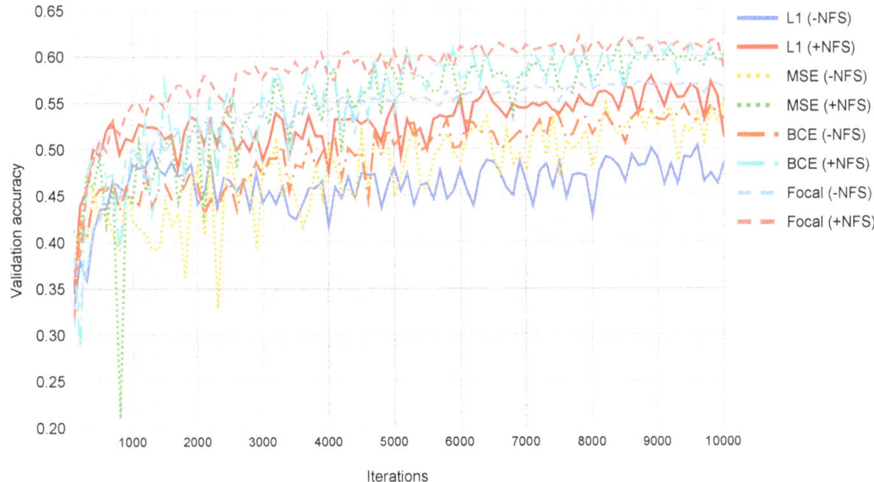

Figure 7. Trends in validation accuracy values over different iterations.

3.2. Quantitative Results

Figure 8a shows the relative performances of different loss functions under the test dataset. Among all loss functions (i.e., L1, MSE, BCE, and focal), the loss with the NFS indicates the higher values for all evaluation metrics.

Figure 8b shows the corresponding values of the evaluation metrics over various loss functions. Among four loss functions, regardless of with or without the NFS, the focal loss is generally better than BCE, MSE, and L1 loss. L1 loss without NFS (L1 − NFS) indicates the lowest values for all metrics in all conditions. The best performance is achieved by focal loss with NFS, i.e., 0.651 for f1-score, 0.490 for the Jaccard index, and 0.626 for the kappa coefficient. Under all loss functions, the addition of the NFS results in significantly higher values for all evaluation metrics. The result indicates that the proposed NFS can effectively manage the slight misalignments from the annotation and achieve better performance. Interestingly, on the weakest L1 loss, the addition of the NFS results in the most significant increments among the three evaluation metrics. The increments of the f1-score, kappa coefficient, and Jaccard index reached 8.8%, 8.9%, and 9.8%, respectively.

3.3. Qualitative Results

Figure 9 presents six representative results of outlines extracted from the model trained by L1 loss with/without the NFS under test dataset. The backgrounds, red lines, and green circles represent the aerial input, predicted outline, and focused area. In general, the addition of the NFS yields a better building outline extraction, particularly on shadowed areas (e.g., green circles in a, b, and e) and turning corners (e.g., green circles in d and f). Additionally, the model trained with the NFS yields a more intact outline (e.g., green circles in c).

Figure 10 shows six representative groups of building outlines extracted from the model trained by the MSE loss with/without the NFS. Generally, the addition of the NFS yields a slightly better building outline extraction. Using the NFS, the extracted outlines contain fewer false positives within buildings (e.g., green circles in a and b) and fewer breakpoints (e.g., green circles c, d, e, and f).

Figure 11 shows six representative groups of outlines extracted from the model trained by BCE loss with or without the NFS. The backgrounds, red lines, and green circles represent the aerial input, predicted outline, and focused area, respectively. As shown in the figure, the addition of the NFS yields a slightly better line extraction at areas shadowed by surrounding trees (e.g., green circles of column a, e, and f). Moreover, the additional NFS results in better line continuity around corners of the buildings (e.g., green circles of column b, c, and d). In general, using the proposed NFS, the building outline extracted from the aerial image is more intact, particularly on building corners and shadowed areas.

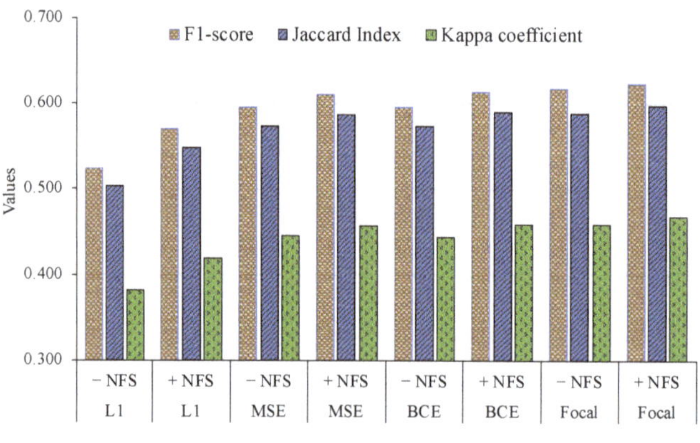

(a) Bar chart

Loss	Condition	F1-score	Jaccard Index	Kappa coefficient
L1	− NFS	0.524	0.503	0.382
L1	+ NFS	**0.571**	**0.548**	**0.419**
MSE	− NFS	0.596	0.573	0.445
MSE	+ NFS	**0.611**	**0.587**	**0.458**
BCE	− NFS	0.596	0.573	0.444
BCE	+ NFS	**0.613**	**0.589**	**0.459**
Focal	− NFS	0.618	0.588	0.459
Focal	+ NFS	**0.624**	**0.597**	**0.468**

(b) Table

Figure 8. Performances of different losses, either with or without nearest feature selector (NFS). (**a**) Bar chart for comparison of relative performances (**b**) Table of performances under different loss functions. For each loss function, the highest values are highlighted in bold.

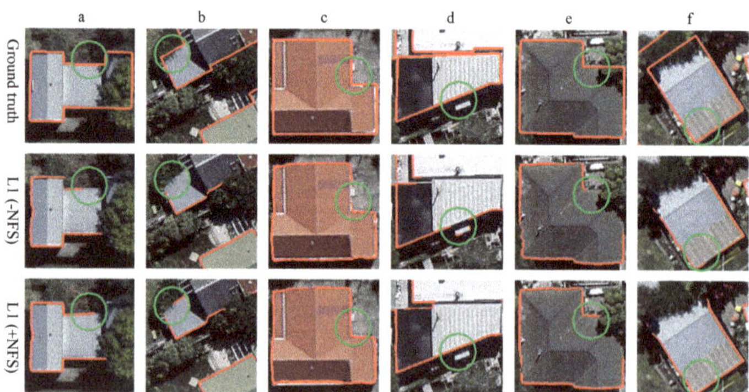

Figure 9. Representative results of extracted outlines from model trained by L1 loss with/without nearest feature selector (NFS). Backgrounds, red lines, and green circles represent aerial input, predicted outline, and focused area, respectively. Selected results are denoted as (**a–f**).

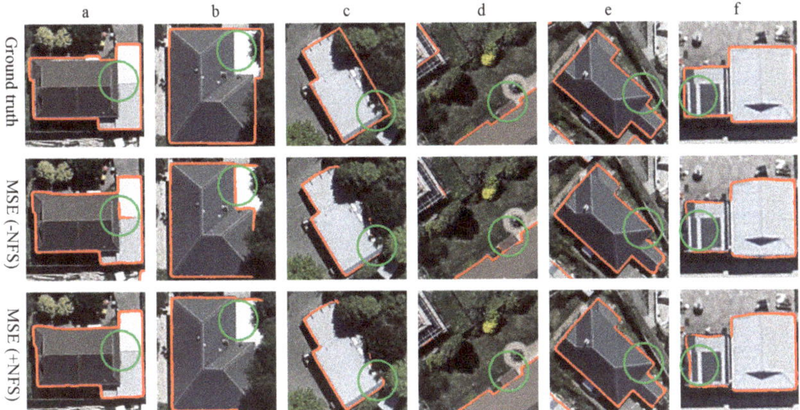

Figure 10. Representative results of outlines extracted from model trained by mean square error (MSE) loss with/without nearest feature selector (NFS). Backgrounds, red lines, and green circles represent aerial input, predicted outline, and focused area, respectively.Selected results are denoted as (**a–f**).

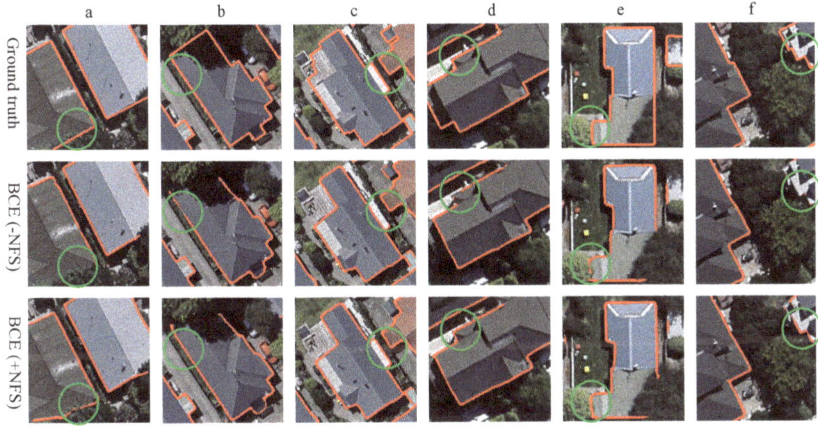

Figure 11. Representative results of outlines extracted from model trained by binary cross-entropy (BCE) loss with/without nearest feature selector (NFS). Backgrounds, red lines, and green circles represent aerial input, predicted outline, and focused area, respectively.Selected results are denoted as (**a–f**).

Figure 12 presents six representative pairs of building outlines extracted from the model trained with the focal loss with or without the NFS. Owing to the robustness of the focal loss, even without the NFS, the model successfully recognizes and extracts the major parts of the building outline from the aerial input (e.g., b, c, and f). However, with the additional NFS, the generated outlines contain fewer false positives around corners with complicated backgrounds (e.g., a, d and e). Compared with L1 loss, the addition of NFS imposes a less significant effect on the model trained with focal loss. This observation is consistent with the quantitative result shown in Figure 8b.

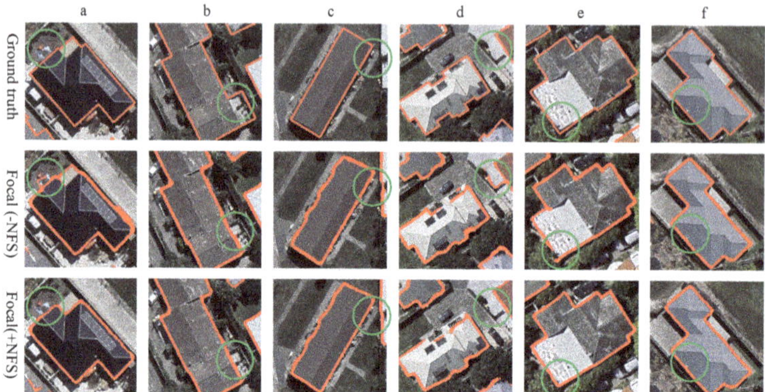

Figure 12. Representative results of outlines extracted from model trained by focal loss with/without nearest feature selector (NFS). Backgrounds, red lines, and green circles represent aerial input, predicted outline, and focused area, respectively.Selected results are denoted as (**a–f**).

Figure 13 presents four representative pairs of failure cases from the model trained with the loss function that combines with or without the nearest feature selector (NFS). As compared with the model trained without NFS, the addition of NFS might lead to un-expected misclassification around corners.

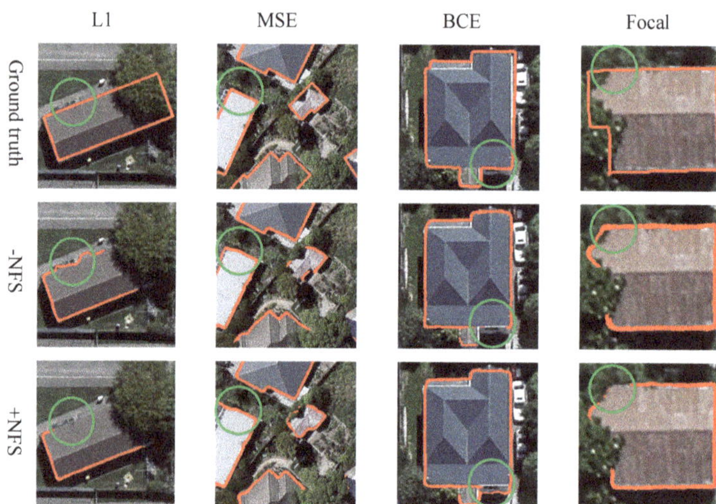

Figure 13. Representative failure cases of outlines extracted from model trained by four losses with/without nearest feature selector (NFS). Backgrounds, red lines, and green circles represent aerial input, predicted outline, and focused area, respectively.

3.4. Computational Efficiency

All experiments are trained and tested on a Sakura "koukakuryoku" Server (https://www.sakura.ad.jp/koukaryoku/) equipped with a 4*times* NVIDIA Tesla V100 GPU (https://www.nvidia.com/en-us/data-center/tesla-v100/) and installed with 64-bit Ubuntu 16.04 LTS. The original SegNet is implemented on Caffe [50] and trained on multi-class scene segmentation tasks, CamVid road scene segmentation [51] and SUN RGB-D indoor scene segmentation [52]. The stochastic gradient descent (SGD) with a fixed learning rate of 0.1 and a momentum of 0.9 is applied to train the model. The implementation of the modified SegNet is based on geoseg (https://github.com/huster-wgm/geoseg) [53], which is built on top of Pytorch(version $\geq$ 0.4.1). To avoid interference by other hyperparameters, all models are trained with a fixed batch size (i.e., 24) and a constant iteration (i.e., 10,000). The Adam stochastic optimizer, which operates at default settings (lr = 2^{-4}, betas = [0.9, 0.999]), is used for training different models.

Table 1 shows the computing speeds of the methods in frames per second (FPS). Among all the loss functions, the additional NFS results in slightly longer processing time during both training and testing. However, the decline in PFS is not significant.

Table 1. Comparison of the computational efficiencies of different loss functions under conditions that with or without NFS.

Loss	Condition	Training FPS	Testing FPS
L1	−NFS	102.3	264.4
L1	+NFS	98.5	236.1
MSE	−NFS	101.9	265.9
MSE	+NFS	98.4	236.2
BCE	−NFS	102.1	266.8
BCE	+NFS	98.7	236.6
Focal	−NFS	101.6	268.5
Focal	+NFS	97.9	236.3

4. Discussion

4.1. Regarding the NFS

In recent years, fully convolutional networks have demonstrated their ability in automatically extracting line features, including roads and building outlines [36,39,54]. However, those studies mainly focused on designing deeper or more complex network architectures to enhance the representation capability for better predictions. The loss functions of fully convolutional networks cannot handle misalignments or rotations between inputs and manually created annotations. Because the building outline occupies a small portion of pixels, misalignments and rotations will severely interfere with the building outline extraction accuracy.

Herein, we propose the NFS module to dynamically re-align the prediction and corresponding annotation. The proposed framework can be easily appended into existing loss functions, such as L1, MSE, and focal loss. Through a dynamic re-alignment, the addition of NFS enables the correct position of the annotation to be located for an appropriate loss calculation. Qualitative and quantitative results based on the testing data demonstrated the effectiveness of our proposed NFS.

4.2. Accuracies, Uncertainties, and Limitations

Among all methods, the focal loss with NFS indicates the highest values for all evaluation metrics. Its values of the f1-score, Jaccard index, and kappa coefficient are 0.624, 0.597, and 0.468. Compared with the naive L1 loss, the addition of the NFS results in significant increments in all evaluation metrics. The increments of the f1-score, kappa coefficient, and Jaccard index reach 8.8%, 8.9%, and 9.8%, respectively. As it is arguable that the kappa coefficient is unsuitable in the assessment and comparison of the accuracy [55], the actual performance gained from the NFS might be less significant (i.e., less than 9.8%). For robust loss functions (e.g., focal, and BCE loss), the improvement afforded by the NFS is less significant (see details in Figure 8b). Owing to the sliding-and-matching mechanism, the proposed NFS cannot be applied to annotations that require rotation correction. Since the methods are designed and trained on image patches with dense buildings, the trained model is not appropriate for evaluating the entire study area where buildings are sparsely presented.

We observe a slight decrease in processing speed when the NFS is applied through the analysis of computational efficiency. Considering the performance gain by the NFS, computational efficiency degradation is negligible. Because the NFS is independent of the aerial characteristic, in principle, it should apply for not only aerial images, but also other data sources (e.g.satellite, SAR, and UAV). The effectiveness of the NFS will be further estimated using publicly available datasets from various sources [56].

Because of the extremely biased negative/positive ratio, complete building outline extraction is still challenging. With the current classification-based scheme, the model is trained to generate pixel-to-pixel prediction using features extracted from sequential convolutional layers. The predicted pixels of the building outline lack of internal connectivity that some pixels might be misclassified as non-outline (e.g., 2nd and 3rd rows in Figure 9).

5. Conclusions

For an accurate building outline extraction, we design a nearest feature selector (NFS) module to dynamically re-align predictions and slightly misaligned annotations. The proposed module can be easily combined with existing loss functions to manage subpixel or pixel-to-level misalignments of the manually created annotations more effectively. For all loss functions, the addition of the proposed NFS yielded significantly better performances in all the evaluation metrics. For the classic L1 loss, the increments gained by using the additional NFS are 8.8%, 8.9%, and 9.8% for the f1-score, kappa coefficient, and Jaccard index, respectively. We plan to improve the similarity selection mechanism and apply it to other data sources to achieve better generalization capacity for large-scale applications.

Author Contributions: Conceptualization, G.W.; Investigation, Y.W.; Project administration, G.W.; Resources, R.S.; Validation, Y.G. and Y.H.; Writing—original draft, Y.W.; Writing—review & editing, G.W. All authors have read and agreed to the published version of the manuscript.

Funding: Part of this work was supported by the JST (Japan Science and Technology Agency) aXis Grant Number JPMJAS2019.

Acknowledgments: We gratefully acknowledge SAKURA Internet Inc. for the provision of the *koukaryoku* GPU server for our experiments.

Conflicts of Interest: The authors declare no conflict of interest.

Abbreviations

The following abbreviations are used in this manuscript:

CNN   Convolutional Neural Network
FCN   Fully Convolutional Networks
NFS   Nearest Feature Selector

References

1. Kuffer, M.; Pfeffer, K.; Sliuzas, R. Slums from space—15 years of slum mapping using remote sensing. *Remote Sens.* **2016**, *8*, 455. [CrossRef]
2. Pham, H.M.; Yamaguchi, Y.; Bui, T.Q. A case study on the relation between city planning and urban growth using remote sensing and spatial metrics. *Landsc. Urban Plan.* **2011**, *100*, 223–230. [CrossRef]
3. Ordóñez, J.; Jadraque, E.; Alegre, J.; Martínez, G. Analysis of the photovoltaic solar energy capacity of residential rooftops in Andalusia (Spain). *Renew. Sustain. Energy Rev.* **2010**, *14*, 2122–2130. [CrossRef]
4. Hamre, L.N.; Domaas, S.T.; Austad, I.; Rydgren, K. Land-cover and structural changes in a western Norwegian cultural landscape since 1865, based on an old cadastral map and a field survey. *Landsc. Ecol.* **2007**, *22*, 1563–1574. [CrossRef]
5. Li, M.; Zang, S.; Zhang, B.; Li, S.; Wu, C. A review of remote sensing image classification techniques: The role of spatio-contextual information. *Eur. J. Remote Sens.* **2014**, *47*, 389–411. [CrossRef]
6. Chen, R.; Li, X.; Li, J. Object-based features for house detection from rgb high-resolution images. *Remote Sens.* **2018**, *10*, 451. [CrossRef]

7. Xu, B.; Jiang, W.; Shan, J.; Zhang, J.; Li, L. Investigation on the weighted ransac approaches for building roof plane segmentation from lidar point clouds. *Remote Sens.* **2015**, *8*, 5. [CrossRef]

8. Huang, Y.; Zhuo, L.; Tao, H.; Shi, Q.; Liu, K. A novel building type classification scheme based on integrated LiDAR and high-resolution images. *Remote Sens.* **2017**, *9*, 679. [CrossRef]

9. Gilani, S.A.N.; Awrangjeb, M.; Lu, G. An automatic building extraction and regularisation technique using lidar point cloud data and orthoimage. *Remote Sens.* **2016**, *8*, 258. [CrossRef]

10. Guo, Z.; Wu, G.; Song, X.; Yuan, W.; Chen, Q.; Zhang, H.; Shi, X.; Xu, M.; Xu, Y.; Shibasaki, R.; et al. Super-Resolution Integrated Building Semantic Segmentation for Multi-Source Remote Sensing Imagery. *IEEE Access* **2019**, *7*, 99381–99397. [CrossRef]

11. Sahoo, P.K.; Soltani, S.; Wong, A.K. A survey of thresholding techniques. *Comput. Vis. Graph. Image Process.* **1988**, *41*, 233–260. [CrossRef]

12. Kanopoulos, N.; Vasanthavada, N.; Baker, R.L. Design of an image edge detection filter using the Sobel operator. *IEEE J. Solid-State Circuits* **1988**, *23*, 358–367. [CrossRef]

13. Wu, Z.; Leahy, R. An optimal graph theoretic approach to data clustering: Theory and its application to image segmentation. *IEEE Trans. Pattern Anal. Mach. Intell.* **1993**, *15*, 1101–1113. [CrossRef]

14. Tremeau, A.; Borel, N. A region growing and merging algorithm to color segmentation. *Pattern Recognit.* **1997**, *30*, 1191–1203. [CrossRef]

15. Gómez-Moreno, H.; Maldonado-Bascón, S.; López-Ferreras, F. Edge detection in noisy images using the support vector machines. In *International Work-Conference on Artificial Neural Networks*; Springer: Berlin/Heidelberg, Germany, 2001; pp. 685–692.

16. Zhou, J.; Chan, K.; Chong, V.; Krishnan, S.M. Extraction of brain tumor from MR images using one-class support vector machine. In Proceedings of the 2005 IEEE Engineering in Medicine and Biology 27th Annual Conference, Shanghai, China, 17–18 January 2006; pp. 6411–6414.

17. Xie, S.; Tu, Z. Holistically-nested edge detection. In Proceedings of the IEEE International Conference on Computer Vision, Santiago, Chile, 7–13 December 2015; pp. 1395–1403.

18. Viola, P.; Jones, M. Rapid object detection using a boosted cascade of simple features. In Proceedings of the 2001 IEEE Computer Society Conference on Computer Vision and Pattern Recognition, CVPR 2001, Kauai, HI, USA, 8–14 December 2001; Volume 1, p. I.

19. Lowe, D.G. Object recognition from local scale-invariant features. In Proceedings of the Seventh IEEE International Conference on Computer Vision, Kerkyra, Greece, 20–27 September 1999; Volume 2, pp. 1150–1157.

20. Ojala, T.; Pietikainen, M.; Maenpaa, T. Multiresolution gray-scale and rotation invariant texture classification with local binary patterns. *IEEE Trans. Pattern Anal. Mach. Intell.* **2002**, *24*, 971–987. [CrossRef]

21. Dalal, N.; Triggs, B. Histograms of oriented gradients for human detection. In Proceedings of the 2005 IEEE Computer Society Conference on Computer Vision and Pattern Recognition (CVPR'05), San Diego, CA, USA, 20–25 June 2005; Volume 1, pp. 886–893.

22. Inglada, J. Automatic recognition of man-made objects in high resolution optical remote sensing images by SVM classification of geometric image features. *ISPRS J. Photogramm. Remote Sens.* **2007**, *62*, 236–248. [CrossRef]

23. Aytekin, Ö.; Zöngür, U.; Halici, U. Texture-based airport runway detection. *IEEE Geosci. Remote Sens. Lett.* **2013**, *10*, 471–475. [CrossRef]

24. Dong, Y.; Du, B.; Zhang, L. Target detection based on random forest metric learning. *IEEE J. Sel. Top. Appl. Earth Obs. Remote Sens.* **2015**, *8*, 1830–1838. [CrossRef]

25. LeCun, Y.; Bengio, Y. Convolutional networks for images, speech, and time series. *Handb. Brain Theory Neural Netw.* **1995**, *3361*, 1995.

26. Ciresan, D.; Giusti, A.; Gambardella, L.M.; Schmidhuber, J. Deep neural networks segment neuronal membranes in electron microscopy images. In *Advances in Neural Information Processing Systems*; Curran Associates: Red Hook, NY, USA, 2012; pp. 2843–2851.

27. Guo, Z.; Shao, X.; Xu, Y.; Miyazaki, H.; Ohira, W.; Shibasaki, R. Identification of village building via Google Earth images and supervised machine learning methods. *Remote Sens.* **2016**, *8*, 271. [CrossRef]

28. Long, J.; Shelhamer, E.; Darrell, T. Fully convolutional networks for semantic segmentation. In Proceedings of the IEEE Conference on Computer Vision and Pattern Recognition, Boston, MA, USA, 7–12 June 2015; pp. 3431–3440.

29. Kampffmeyer, M.; Salberg, A.B.; Jenssen, R. Semantic segmentation of small objects and modeling of uncertainty in urban remote sensing images using deep convolutional neural networks. In Proceedings of the IEEE Conference on Computer Vision and Pattern Recognition Workshops, Las Vegas, NV, USA, 26 June–1 July 2016; pp. 1–9.

30. Badrinarayanan, V.; Kendall, A.; Cipolla, R. Segnet: A deep convolutional encoder-decoder architecture for image segmentation. *IEEE Trans. Pattern Anal. Mach. Intell.* **2017**, *39*, 2481–2495. [CrossRef] [PubMed]

31. Noh, H.; Hong, S.; Han, B. Learning deconvolution network for semantic segmentation. In Proceedings of the IEEE International Conference on Computer Vision, Santiago, Chile, 7–13 December 2015; pp. 1520–1528.

32. Ronneberger, O.; Fischer, P.; Brox, T. U-net: Convolutional networks for biomedical image segmentation. In *International Conference on Medical Image Computing and Computer-Assisted Intervention*; Springer: Berlin/Heidelberg, Germany, 2015; pp. 234–241.

33. Lin, T.Y.; Dollár, P.; Girshick, R.; He, K.; Hariharan, B.; Belongie, S. Feature pyramid networks for object detection. In Proceedings of the IEEE Conference on Computer Vision and Pattern Recognition, Honolulu, HI, USA, 21–26 July 2017; Volume 1, p. 4.

34. Wu, G.; Shao, X.; Guo, Z.; Chen, Q.; Yuan, W.; Shi, X.; Xu, Y.; Shibasaki, R. Automatic Building Segmentation of Aerial Imagery Using Multi-Constraint Fully Convolutional Networks. *Remote Sens.* **2018**, *10*, 407. [CrossRef]

35. Wu, G.; Guo, Y.; Song, X.; Guo, Z.; Zhang, H.; Shi, X.; Shibasaki, R.; Shao, X. A stacked fully convolutional networks with feature alignment framework for multi-label land-cover segmentation. *Remote Sens.* **2019**, *11*, 1051. [CrossRef]

36. Wu, G.; Guo, Z.; Shi, X.; Chen, Q.; Xu, Y.; Shibasaki, R.; Shao, X. A Boundary Regulated Network for Accurate Roof Segmentation and Outline Extraction. *Remote Sens.* **2018**, *10*, 1195. [CrossRef]

37. Mnih, V.; Hinton, G.E. Learning to detect roads in high-resolution aerial images. In *European Conference on Computer Vision*; Springer: Berlin/Heidelberg, Germany, 2010; pp. 210–223.

38. Wei, Y.; Wang, Z.; Xu, M. Road structure refined CNN for road extraction in aerial image. *IEEE Geosci. Remote Sens. Lett.* **2017**, *14*, 709–713. [CrossRef]

39. Zhang, Z.; Liu, Q.; Wang, Y. Road extraction by deep residual u-net. *IEEE Geosci. Remote Sens. Lett.* **2018**, *15*, 749–753. [CrossRef]

40. Zhou, L.; Zhang, C.; Wu, M. D-LinkNet: LinkNet With Pretrained Encoder and Dilated Convolution for High Resolution Satellite Imagery Road Extraction. In Proceedings of the CVPR Workshops, Salt Lake City, UT, USA, 18–22 June 2018; pp. 182–186.

41. Polak, M.; Zhang, H.; Pi, M. An evaluation metric for image segmentation of multiple objects. *Image Vis. Comput.* **2009**, *27*, 1223–1227. [CrossRef]

42. Carletta, J. Assessing agreement on classification tasks: The kappa statistic. *Comput. Linguist.* **1996**, *22*, 249–254.

43. Li, E.; Femiani, J.; Xu, S.; Zhang, X.; Wonka, P. Robust rooftop extraction from visible band images using higher order CRF. *IEEE Trans. Geosci. Remote Sens.* **2015**, *53*, 4483–4495. [CrossRef]

44. Comer, M.L.; Delp, E.J. Morphological operations for color image processing. *J. Electron. Imaging* **1999**, *8*, 279–290. [CrossRef]

45. Nair, V.; Hinton, G.E. Rectified linear units improve restricted boltzmann machines. In Proceedings of the 27th International Conference on Machine Learning (ICML-10), Haifa, Israel, 21–24 June 2010; pp. 807–814.

46. Ioffe, S.; Szegedy, C. Batch normalization: Accelerating deep network training by reducing internal covariate shift. In Proceedings of the International Conference on Machine Learning, Lille, France, 6–11 July 2015; pp. 448–456.

47. Nagi, J.; Ducatelle, F.; Di Caro, G.A.; Cireşan, D.; Meier, U.; Giusti, A.; Nagi, F.; Schmidhuber, J.; Gambardella, L.M. Max-pooling convolutional neural networks for vision-based hand gesture recognition. In Proceedings of the 2011 IEEE International Conference on Signal and Image Processing Applications (ICSIPA), Kuala Lumpur, Malaysia, 16–18 November 2011; pp. 342–347.

48. Shore, J.; Johnson, R. Properties of cross-entropy minimization. *IEEE Trans. Inf. Theory* **1981**, *27*, 472–482. [CrossRef]

49. Lin, T.Y.; Goyal, P.; Girshick, R.; He, K.; Dollár, P. Focal loss for dense object detection. In Proceedings of the IEEE International Conference on Computer Vision, Venice, Italy, 22–29 October 2017; pp. 2980–2988.

50. Jia, Y.; Shelhamer, E.; Donahue, J.; Karayev, S.; Long, J.; Girshick, R.; Guadarrama, S.; Darrell, T. Caffe: Convolutional architecture for fast feature embedding. In Proceedings of the 22nd ACM International Conference on Multimedia, Orlando, FL, USA, 3–7 November 2014; ACM: New York, NY, USA, 2014; pp. 675–678.
51. Brostow, G.J.; Fauqueur, J.; Cipolla, R. Semantic object classes in video: A high-definition ground truth database. *Pattern Recognit. Lett.* **2009**, *30*, 88–97. [CrossRef]
52. Song, S.; Lichtenberg, S.P.; Xiao, J. Sun rgb-d: A rgb-d scene understanding benchmark suite. In Proceedings of the IEEE Conference on Computer Vision and Pattern Recognition, Boston, MA, USA, 7–12 June 2015; pp. 567–576.
53. Wu, G.; Guo, Z.; Shao, X.; Shibasaki, R. Geoseg: A Computer Vision Package for Automatic Building Segmentation and Outline Extraction. In Proceedings of the IGARSS 2019-2019 IEEE International Geoscience and Remote Sensing Symposium, Yokohama, Japan, 28 July–2 August 2019; pp. 158–161.
54. Wu, S.; Du, C.; Chen, H.; Xu, Y.; Guo, N.; Jing, N. Road Extraction from Very High Resolution Images Using Weakly labeled OpenStreetMap Centerline. *ISPRS Int. J. Geo-Inf.* **2019**, *8*, 478. [CrossRef]
55. Foody, G.M. Explaining the unsuitability of the kappa coefficient in the assessment and comparison of the accuracy of thematic maps obtained by image classification. *Remote Sens. Environ.* **2020**, *239*, 111630. [CrossRef]
56. Chen, Q.; Wang, L.; Wu, Y.; Wu, G.; Guo, Z.; Waslander, S.L. Aerial Imagery for Roof Segmentation: A Large-Scale Dataset towards Automatic Mapping of Buildings. *arXiv* **2018**, arXiv:1807.09532.

 *remote sensing*

Article

A High-Performance Spectral-Spatial Residual Network for Hyperspectral Image Classification with Small Training Data

Wijayanti Nurul Khotimah [1,*], **Mohammed Bennamoun** [1], **Farid Boussaid** [2], **Ferdous Sohel** [3] and **David Edwards** [4]

1   Department of Computer Science and Software Engineering, The University of Western Australia, 35 Stirling Highway, Crawley, Perth, WA 6009, Australia; mohammed.bennamoun@uwa.edu.au
2   Department of Electrical, Electronic and Computer Engineering, The University of Western Australia, 35 Stirling Highway, Crawley, Perth, WA 6009, Australia; farid.boussaid@uwa.edu.au
3   Information Technology, Mathematics & Statistics, Murdoch University, 90 South Street, Murdoch, Perth, WA 6150, Australia; F.Sohel@murdoch.edu.au
4   School of Plant Biology and The UWA Institute of Agriculture, The University of Western Australia, 35 Stirling Highway, Crawley, Perth, WA 6009, Australia; dave.edwards@uwa.edu.au
*   Correspondence: wijayantinurul.khotimah@research.uwa.edu.au

Received: 13 August 2020; Accepted: 21 September 2020; Published: 24 September 2020

Abstract: In this paper, we propose a high performance Two-Stream spectral-spatial Residual Network (TSRN) for hyperspectral image classification. The first spectral residual network (sRN) stream is used to extract spectral characteristics, and the second spatial residual network (saRN) stream is concurrently used to extract spatial features. The sRN uses 1D convolutional layers to fit the spectral data structure, while the saRN uses 2D convolutional layers to match the hyperspectral spatial data structure. Furthermore, each convolutional layer is preceded by a Batch Normalization (BN) layer that works as a regularizer to speed up the training process and to improve the accuracy. We conducted experiments on three well-known hyperspectral datasets, and we compare our results with five contemporary methods across various sizes of training samples. The experimental results show that the proposed architecture can be trained with small size datasets and outperforms the state-of-the-art methods in terms of the Overall Accuracy, Average Accuracy, Kappa Value, and training time.

Keywords: hyperspectral image classification; two stream residual network; deep learning; Batch Normalization

---

1. Introduction

Hyperspectral imaging has received much attention in recent years due to its ability to capture spectral information that is not detected by the naked human eye [1]. Hyperspectral imaging provides rich cues for numerous computer vision tasks [2] and a wide range of application areas, including medical [1], military [3], forestry [4], food processing [5], and agriculture [6].

One of the main challenges when analyzing Hyperspectral Images (HSIs) lies in extracting features, which is challenging due to the complex characteristics, i.e., the large size and the large spatial variability of HSIs [7]. Furthermore, HSI is composed of hundreds of spectral bands, in which wavelengths are very close, resulting in high redundancies [7,8]. Traditional machine learning methods are less suitable for HSI analysis because they heavily depend on hand-crafted features, which are commonly designed for a specific task, and are thus not generalizable [9]. In contrast, deep learning techniques can capture characteristic features automatically [9,10], thus constituting a promising avenue for HSI analysis.

Several deep learning architectures have been proposed to classify HSIs. Many architectures, such as one-dimensional convolutional neural network (1D-CNN) [11,12], one-dimensional generative adversarial network (1D-GAN) [13,14], and recurrent neural network (RNN) [15,16], have been proposed to learn spectral features. Other works, e.g., Reference [17–19], have shown that adding spatial features can improve the classification performance. Numerous spectral-spatial network architectures have been proposed for HSIs [19–28].

A number of methods argue that extracting the spectral and spatial features in two separate streams can produce more discriminative features [25,29,30]. Examples of such methods include stacked denoising autoencoder (SdAE) and 2D-CNN [30], plain 1D-CNN and 2D-CNN [25], spectral-spatial long short-term memory (SSLSTMs) [27], and a spectral-spatial unified network (SSUN) [23]. In terms of the spectral stream, the work of Reference [27] used a LSTM, which considers the spectral values of the different channels as a sequence. However, using LSTM on hundreds of a sequence of channels is complex; thus, [23] tried to simplify the sequence by grouping them. One of the grouping strategies is dividing the adjacent band into the same sequence following the spectral orders. The work in Reference [30] considered spectral values as a vector with noise and used a denoising technique, SdAE, to encode the spectral features. These networks, based on LSTM and SdAE, are all shallow. To increase the accuracy, Reference [25] tried to make a deeper network by employing a simpler layer, based on 1D convolution. The work in Reference [31] considered that the HSI bands have a different variance and correlation. Hence, they cluster the bands into some groups based on their similarity, then extracted the spectral features of each cluster using 1D convolution. Different from Reference [31], the study in Reference [32] considered that different objects have different spectral reflection profiles; hence, they used 2D convolution with a kernel size of 1x1 to extract the spectral features. For the spatial stream, Reference [27] also used LSTM, and due to its complexity, thus used a shallow network. Other approaches [23,25,30] used 2D convolution with a plain network, which could be made deeper, while Reference [31,32] used 2D convolution with a multi-scale input to extract multi-scale spatial features.

Other works claim that extracting spectral and spatial features directly using a single stream network can be more beneficial as it leverages the joint spectral-spatial features [28,33,34]. Most that adopt this approach utilize 3D convolutional layers [12,19,34,35] because they are naturally suited to the 3D cube data structure of HSIs. Reported experiments show that deep 3D-CNN produces better performance compared with 2D-CNN [18]. However, 3D-CNN requires large memory size and expensive computation cost [36]. Moreover, 3D-CNN faces over-smoothing phenomena because it fails to take the full advantage of spectral information, which results in misclassification for small objects and boundaries [23]. In addition, the labeling process of HSIs is labor-intensive, time-consuming, difficult, and thus expensive [37]. Using a complex deep learning architecture, in which parameters are in the millions, to learn from a small labeled dataset may also lead to over-fitting [38]. Moreover, adjusting millions of parameters during the deep-learning training process consumes a lot of time. Devising a deep learning architecture, which can work well on complex data of HSI, in which labeled datasets are small, is desirable.

Another issue with HSI classification based on deep learning is the depth of the network. The deeper the layer is, the richer the features will be, where the first layer of the deep network extracts general characteristics, and the deeper layers extract more specific features [39,40]. However, such deep networks are prone to the vanishing/exploding gradient problem, which occurs when the layers are deeper [41,42]. To solve this problem, Reference [40] reformulated the layers as learning residual functions with reference to the layer inputs. This approach is called a residual network (ResNet), which has become popular because of its remarkable performance on image classification [43]. For HSI classification, a single stream ResNet has been used by Reference [19,44–46].

Another problem related to HSI feature extraction is that the spectral values are prone to noise [47]. However, most of the previous research, which focus on the extraction of spectral features with deep-networks, have not taken noise into account. They usually use a pixel vector along the spectral dimension directly as their spectral network input [23,25,27,29,30], without considering that noise can worsen the classification performance.

Considering the aforementioned challenges and the limitations of existing network architectures, associated with HSI feature extraction and classification, we propose an efficient yet high performance two-stream spectral-spatial residual network. The spectral residual network (sRN) stream uses 1D convolutional layers to fit the spectral data structure, and the spatial residual network (saRN) uses 2D convolutional layers to fit the spatial data structure. The residual connection in the sRN and saRN can solve the vanishing/exploding gradient problem. Since proceeding the convolutional layer with Batch Normalization (BN) layer and full pre-activation rectified linear unit (ReLU) generalizes better than the original ResNet [48], in each of our residual unit, we use BN layer and ReLU layer before the convolutional layer. We then combine our sRN and saRN in a parallel pipeline. As shown in Figure 1, given a spectral input cube X_{ij}^{s} of a pixel x_{ij}, the sRN extracts its spectral features. Concurrently, given a spatial input cube X_{ij}^{sa} of a pixel x_{ij}, the saRN will extract its spatial characteristics. Since the sRN and the saRN use different input sizes and different types of convolution layers, they produce different sizes of feature maps. The gap between the number of spectral feature maps and the number of spatial feature maps can worsen the classification accuracy. To make the number of feature maps in each network proportional, we add an identical fully connected layer at the end of each network. Subsequently, we employ a dense layer to fuse the spectral features and the spatial features. Finally, we classify the joint spectral-spatial features using a softmax layer (Figure 1).

In summary, the main contributions of this research are:

- We propose TSRN, a Two-Stream Spectral-Spatial network with residual connections, to extract spectral and spatial features for HSI classification. The identity shortcut in the residual-unit is parameter-free, thus adding shortcut connections into a residual-unit does not increase the number of parameters. Furthermore, the use of 1D convolutional layers in the sRN and 2D convolutional layers in the saRN results in few trainable parameters. We can, therefore, construct a deeper and wider network with fewer parameters, making it particularly suitable for HSI classification when the amount of available labeled data is small.

- We achieve the state-of-the-art performance on HSI classification with various sizes of training samples (4%, 6%, 8%, 10%, and 30%). Moreover, compared to networks based on 3D convolutional layers, our proposed architecture is faster.

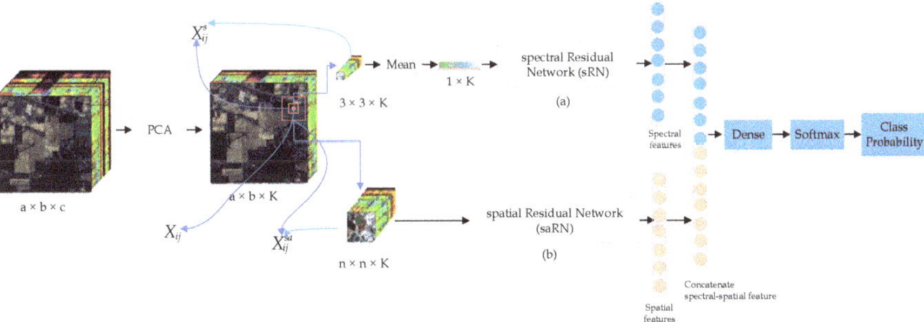

Figure 1. Proposed Two-Stream Spectral-Spatial Residual Network (TSRN) architecture. The details of spectral residual network (sRN) and spatial residual network (saRN) sub-networks are shown in Figure 2.

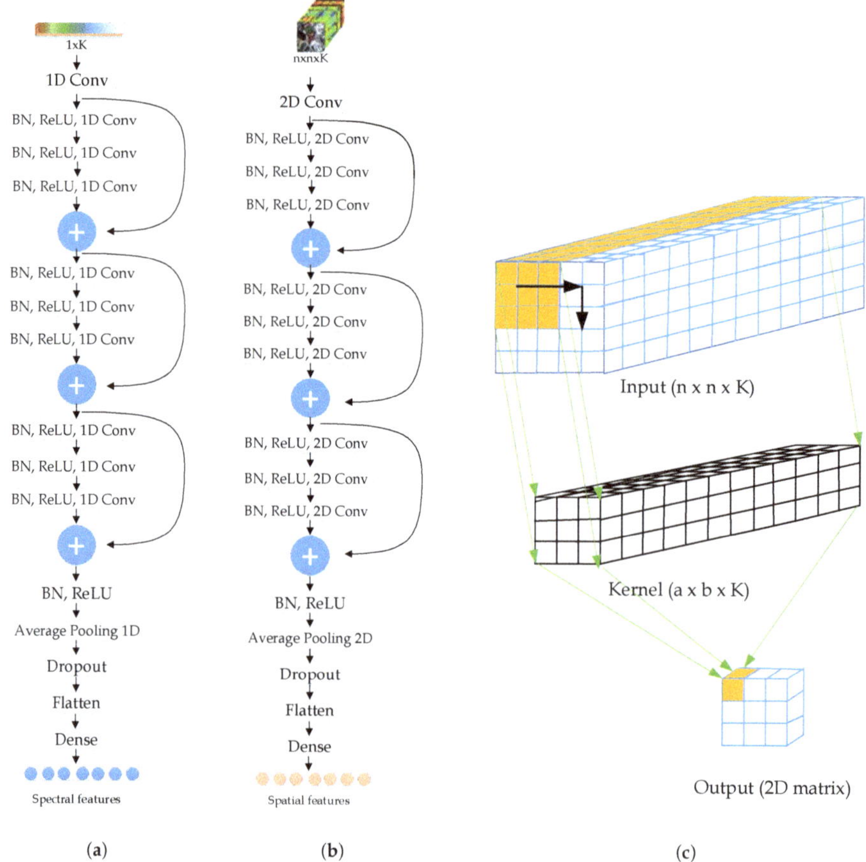

Figure 2. The detailed network of the (**a**) sRN, (**b**) saRN, and (**c**) the detail process of 2D convolution on 3D input.

2. Technical Preliminaries

2.1. CNN

Convolutional Neural Networks (CNNs) have been increasingly used for HSI analysis. A number of works aimed at improving the performance of Deep CNNs (DCNNs) have focused on different aspects, e.g., the network architecture, the type of nonlinear activation function, supervision methods, regularization mechanisms, and optimization techniques [4,49,50]. Based on the network architecture, specifically on the convolutional layers, there are different types of CNNs, namely 1D-CNN, 2D-CNN, and 3D-CNN. The 1D-CNN has one-dimensional filters in its convolutional layers which are naturally fit to the spectral data structure. Consider the case when the size of the input is $K \times 1$, and the kernel size is $B \times 1$, with K representing the number of HSI bands, B is the kernel size, and $B << K$. Wei Hu et al. [11] used 1D convolutional layers to extract the spectral features of HSI. Their network input is an HSI pixel vector, with size $(K, 1)$. This research initiated the use of multiple convolutional layers for HSI classification. Compared to 2D convolution and 3D convolution, the process of 1D convolution, which is shown in Equation (1), is the simplest one.

A 2D-CNN has two-dimensional filters in its convolutional layers. It has been widely used to solve several computer vision problems, such as object detection [51], scene recognition [52], and image classification [51], because of its ability to extract features from a raw image directly. For the HSI classification problem, 2D convolutions have been used to extract spatial features [25,30]. In contrast to RGB images, HSI has a much larger number of channels. Applying 2D convolutions along the hundreds of channels results in more learned parameters [18]; hence, several studies on HSIs, which employ 2D convolutions, do not use all of the channels. Most of them use a dimensionality reduction technique as a preprocessing step with their network [53–55] or use the average of all the images over the spectral bands of the HSI [25].

A 3D-CNN employs 3D convolutions in its convolutional layers. 3D-CNNs are popular for video classification [36], 3D object reconstruction [56], and action recognition [57]. For the case of HSIs, the form of the 3D filter suits the data structure of the HSI cube. Some research papers on HSI classification use 3D convolutional layers to extract the spectral-spatial features directly [18,33]. Their research shows that 3D-CNN outperforms both 1D-CNN and 2D-CNN. However, as shown in Equation (3), the process of 3D convolution requires more parameters, more memory, and requires a higher computational time and complexity compared to 1D convolution in Equation (1) and 2D convolution in Equation (2).

$$v_{ij}^{z} = f\left(\sum_{m}\sum_{b=0}^{B_i-1} k_{ijm}^{b} v_{(i-1)m}^{z+b} + r_{ij}\right), \tag{1}$$

$$v_{ij}^{xy} = f\left(\sum_{m}\sum_{h=0}^{H_i-1}\sum_{w=0}^{W_i-1} k_{ijm}^{wh} v_{(i-1)m}^{(x+h)(y+w)} + r_{ij}\right), \tag{2}$$

$$v_{ij}^{xyz} = f\left(\sum_{m}\sum_{b=0}^{B_i-1}\sum_{h=0}^{H_i-1}\sum_{w=0}^{W_i-1} k_{ijm}^{whb} v_{(i-1)m}^{(x+h)(y+w)(z+b)} + r_{ij}\right), \tag{3}$$

where: i is the layer under consideration, m is the index of feature map, z is the index that corresponds to the spectral dimension, v_{ij}^{z} is the output of the ith layer and the jth feature map at position z, k_{ijm}^{b} is the kernel value at index b on the layer i and feature map j, r_{ij} is the bias at layer i and feature map j. For the 1D convolution in Equation (1), B_i is the size of the 1D filter in layer i, while, for the 3D convolution in Equation (3), B_i is the depth of 3D kernel. W_i and H_i are the width and height of the kernel, respectively, for both 2D and 3D convolutions.

The expensive computational cost and memory demand of 3D convolution has led studies to investigate alternative network architectures based on (2D + 1D) convolutions. For instance, in the case of action recognition, a study in Reference [58] proposed to replace 3D convolution with m parallel streams of n 2D and one 1D convolution. This study empirically showed that their network, which is based on (2D + 1D) convolution, achieves around 40% reduction in model size and yields a drastic reduction in the number of learning parameters compared to another network with 3D convolution. In Reference [36], a simplified 3D convolution was implemented using 2D and 1D convolutions in three different blocks: 2D followed by 1D, 2D and 1D in parallel, and 2D followed by 1D with skip connections. These blocks were subsequently interleaved using a sequence network. The proposed architecture has a depth of 199 and a model size of 261 MB, which is much lighter compared to the 3D-CNN, in which model size is 321 MB when the depth is 11. The architecture was also shown to be faster than its 3D-CNN counterpart [36]. Using (2D + 1D) convolutions instead of 3D convolutions allows the network to be deeper without significantly increasing the number of parameters. Such deep networks can extract richer features and have been shown to outperform 3D-CNN architectures [36,58,59]. Because the model size and the number of parameters grow dramatically as the network becomes deeper, the training of deep 3D-CNNs is extremely difficult with the risk of overfitting.

2.2. Residual Network

Deeper CNN can extract richer features [39]. In some cases, when the networks are deeper, their accuracy degrades because of the vanishing/exploding gradients problem [60]. Hence, He et al. [40] proposed to use a shortcut connection to perform identity mapping without adding extra parameters or extra computational time. The shortcut connection outputs are added to the output of the stacked layers, and a ReLU is applied as the activation function. This network is named ResNet. It has achieved outstanding classification accuracy on some image benchmark datasets, such as ImageNet, ILSVRC 2015, and CIFAR-10.

He et al. [48] followed up their work on ResNet by analyzing the propagation formulation behind the residual unit. Their analysis has shown that a "clean" information path in the skip connection results in the lowest training loss compared to those with scaling, gating, and convolution. Regarding the position of the ReLU activation function in the residual building blocks, they proved that putting ReLU and BN before the add function (full pre-activation) generalizes better than the original ResNet [40], which used ReLU after the add function (post-activation). The difference between pre-activation and post-activation in the residual building blocks is shown in Figure 3.

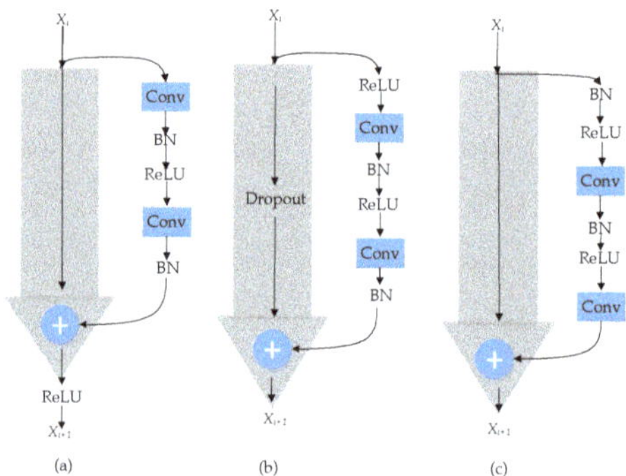

Figure 3. (**a**) Original residual unit with clear short-cut connection. (**b**) rectified linear unit (ReLU)-only pre-activation with dropout short-cut connection. (**c**) Full pre-activation with clear short-cut connection.

For the use of ResNet for HSI classification, Zhong et al. [44] proved that with the same size of convolutional layers, ResNet achieves better recognition accuracy than a plain CNN. Then, they explored ResNet by applying more various kernel sizes to sequentially extract the spectral features and the spatial features [19]. Roy et al. [61] used 3D convolutional layers followed by 2D convolutional layers in their residual network. Their network achieved high performance with 30% training samples. Meanwhile, Reference [45] explored ResNet by implementing a variable number of kernels in each convolutional layer. The kernel number was set to increase gradually in all convolutional layers like a pyramid to increase the diversity of the spectral-spatial features. In contrast to Reference [19,45,61], which focus on exploring the convolutional layer, Reference [46] improved the ResNet architecture by combining it with a dense convolutional network, which helps the ResNet to explore new features. These various works all improve ResNet performance by using a single network to extract both spectral and spatial features. Our proposed architecture extracts the spectral and spatial features from two separate stream networks to produce distinctive spectral and spatial features.

3. Proposed TSRN Network

The flowchart of the proposed TSRN is displayed in Figure 1. From the diagram, we can see that TSRN has two important residual network streams: a sRN and a saRN. Since the number of bands in the spectral dimension is very large (hundreds of channels), and thus comprises much redundancy, we first apply PCA to extract the first K principal components. Then, for each pixel, we take a 3×3 cube alongside the spectral direction, which is centered at that pixel, as the input of the sRN stream to learn the spectral features. Similarly, we take an $n \times n \times K$ cube and feed it to the saRN streams to extract the spatial features. In this method, we propose to use the same number of spectral and spatial features. Hence, we need to ensure that their feature map sizes at the output of the sRN and the saRN networks are the same. To this end, we have applied a dense layer with the same number of units in each stream. Then, we apply a fully connected layer to fuse the spectral features and the spatial features. Finally, we use a softmax layer to classify the features. In the next subsection, we will explain in more detail both spectral and spatial networks.

3.1. Spectral Residual Network

Although we have minimized the data redundancy of the HSI by using PCA, the spectral values can still be noisy. In the sRN, we propose to compute the mean of the reflectance in each spectral band before inputting the spectral cube into the sRN to minimize the effects of the noise. Given a pixel x_{ij}, we choose a spectral cube $X_{ij}^s \in R^{3\times3\times K}$, which is centered at x_{ij} and K is the number of PCA components. Then, we compute the mean reflectance of each band by using Equation (4), where $k \in K$, $\overline{X}_{ij}^s = \{\overline{x}_{ij1}^s, \overline{x}_{ij2}^s, ..., \overline{x}_{ijK}^s\}$, and $\overline{X}_{ij}^s \in R^{1 \times K}$.

$$\overline{x}_{ijk}^s = \frac{\sum_{h=j-1}^{h=j+1} \sum_{g=i-1}^{g=i+1} x_{g,h,k}^s}{9}. \tag{4}$$

Then, we input the mean of the spectral value $(\overline{X}_{ij}^s)$ into the sRN. In this proposed architecture, we use three full pre-activation with clear skip connection residual units inspired by Reference [48]. It has been shown that these residual units are better than the traditional residual units. These residual units consist of three different layers.

- One-dimensional convolutional layers, which perform a dot product between every small window of input data ($1 \times B_i$) and the kernel's weights and biases (see Equation (1)).
- BN layer that normalizes the layer inputs of each training mini-batch to overcome the internal covariate shift problem [62]. The internal covariate shift problem is a condition which occurs when the distribution of each layer's inputs in deep-network changes due to the change of the previous layer's parameters. This situation slows down the training. Normalizing the layer's inputs stabilizes its distribution and thus speeds up the training process.
- ReLU is an activation function, which learns the non-linear representations of each feature map's components [63].

In the full pre-activation residual unit, the BN layer and the ReLU activation layer are established before the convolution layer, as shown in Figure 2a. From that figure, we can see that an Average Pooling layer, a Dropout layer, and a Dense layer have been applied at the end of the sRN. The Average Pooling layer and the Dropout layer are used as a regularizer to minimize the over-fitting problem due to the small number of training samples, while the Dense layer is used to perform the high-level reasoning to produce 128 spectral features.

3.2. Spatial Residual Network

The saRN is devised to extract the spatial features of a pixel x_{ij}. The input is an $n \times n \times K$ cube, centered at pixel x_{ij}. Then, the input is processed by the full pre-activation residual unit. As shown in Figure 2b, the rest of the layers architecture of this saRN are similar to those of sRN. The main difference between them is that the saRN uses 2D convolutional layers, while the sRN uses 1D convolutional layers. In the end of this network, 128 spatial features are extracted.

Figure 2c illustrates the 2D convolution process with a spatial cube in which size is $n \times n \times K$, where K is the number of channels or the input depth. Since the input is 3D then the kernel is also 3D, i.e., the kernel depth must be the same as the input depth. Hence, in this case, the kernel depth must be K. So, we can only select the kernel width and height. The convolution process is performed along the x and y-direction and produces a 2D matrix as output [64].

Given a pixel x_{ij}, the spectral features F_{ij}^s produced by sRN and the spatial features F_{ij}^{sa} produced by saRN are given by Equations (5) and (6), respectively.

$$F_{ij}^s = sRN(\overline{X}_{ij}^s),\qquad(5)$$

$$F_{ij}^{sa} = saRN(X_{ij}^{sa}).\qquad(6)$$

Using F_{ij}^s and F_{ij}^{sa}, we implement a fully connected layer to obtain the joint spectral-spatial features F_{ij}^{ssa} using the formula in Equation (7), where W^{fcl} and b^{fcl} are the weight vector and the bias of the fully connected layer, respectively, and $\oplus$ is the concatenation operation.

$$F_{ij}^{ssa} = f(W^{fcl} \cdot \{F_{ij}^s \oplus F_{ij}^{sa}\} + b^{fcl}).\qquad(7)$$

After obtaining the joint spectral-spatial feature F_{ij}^{ssa}, we use a softmax regression layer (SRL) to predict the class probability distribution of the pixel x_{ij} by using Equation (8). Here, N is the number of classes, and $P(x_{ij})$ is a vector consisting of the probability distribution of each class on pixel x_{ij}. In the end, the label of the pixel x_{ij} is decided using Equation (9).

$$P(x_{ij}) = \frac{1}{\sum_{n=1}^{N} e^{w_n^{srl}.F_{ij}^{ssa}}} \begin{bmatrix} e^{w_1^{srl}.F_{ij}^{ssa}} \\ e^{w_2^{srl}.F_{ij}^{ssa}} \\ \vdots \\ e^{w_N^{srl}.F_{ij}^{ssa}} \end{bmatrix}\qquad(8)$$

$$label(x_{ij}) = argmax P(x_{ij}).\qquad(9)$$

4. Experiments

4.1. Experimental Datasets

We evaluated the proposed architecture on three publicly available HSI datasets, which are frequently used for pixel-wise HSI classification. These datasets are Indian Pine (IP), Pavia University (PU), and Kennedy Space Center (KSC). These datasets were captured by different sensors, thus having different spatial resolutions and a different number of bands. Each dataset has a different category (class), and each class has a different number of instances. The details of the datasets are provided below:

1. IP Dataset: IP dataset was acquired by Airborne Visible/Infrared Imaging Spectrometer (AVIRIS) hyperspectral sensor data on 12 June 1992, over Purdue University Agronomy farm and nearby area, in Northwestern Indiana, USA. The major portions of the area are Indian Creek and Pine Creek watershed; thus, the dataset is known as the Indian Pine dataset. The captured scene contains 145×145 pixels with a spatial resolution of 20 meters per pixel. In other words, the dataset has 21,025 pixels. However, not all of the pixels have ground-truth information. Only 10,249 pixels are categorized between 1 to 16 (as shown in Table 1a), and the remaining pixels remain unknown (labeled with zero in the ground truth). Regarding the spectral information, the entire spectral band of this dataset is 224, with wavelengths ranging from 400 to 2500 nm. Since some of the bands cover the region of water absorption, (104–108), (150–163), and 220, they are removed, so only 200 bands remain Reference [65].

2. KSC Dataset: Same as the IP dataset, the KSC dataset was also collected by AVIRIS sensor in 1996 over the Kennedy Space Center, Florida, USA. Its size is 512×614; thus, this dataset consists of

314,368 pixels, but only 5122 pixels have ground-truth information (as shown in Table 1b). The dataset's spatial resolution is 18 meters per pixel, and its band number is 174.

3. PU Dataset: The PU dataset was gathered during a flight campaign over the campus in Pavia, Northern Italy, using a Reflective Optics System Imaging Spectrometer (ROSIS) hyperspectral sensor. The dataset consists of 610 × 610 pixels, with a spatial resolution 1.3 meters per pixel. Hence 207,400 pixels are available in this dataset. However, only 20% of these pixels have ground-truth information, which are labeled into nine different classes, as shown in Table 1c. The number of its spectral bands is 103, ranging from 430 to 860 nm.

Table 1. Detailed categories and number of instances of Indian Pines dataset (The colours represent the colour labels that are used in the figures of Section 4.3).

Label	Category Name	# Pixel
(a) Indian Pines Dataset		
C1	Alfafa	46
C2	Corn-notil	1428
C3	Corn-mintill	830
C4	Corn	237
C5	Grass-pasture	483
C6	Grass-trees	730
C7	Grass-pasture-mowed	28
C8	Hay-windrowed	478
C9	Oats	20
C10	Soybean-notil	972
C11	Soybean-mintill	2455
C12	Soybean-clean	593
C13	Wheat	205
C14	Woods	1265
C15	Building-Grass-Tress	386
C16	Stone-Steel-Towers	93
(b) KSC Dataset		
C1	Scrub	761
C2	Willow swamp	253
C3	Cabbage palm hammock	256
C4	Cabbage palm	252
C5	Slash pine	161
C6	Oak	229
C7	Hardwood swamp	105
C8	Graminoid marsh	431
C9	Spartina marsh	520
C10	Cattail marsh	404
C11	Salt marsh	419
C12	Mud flats	503
C13	Water	927
(c) Pavia University Dataset		
C1	Asphalt	6631
C2	Meadows	18,649
C3	Gravel	2099
C4	Trees	3064
C5	Painted metal sheets	1345
C6	Bare soil	5029
C7	Bitumen	1330
C8	Self-Blocking Bricks	3682
C9	Shadows	947

4.2. Experimental Configuration

In our proposed model, we do standardization (a technique to rescale the data to have a mean of 0 and a standard deviation of 1) in advance, before dividing the data into training and testing. Hyperparameters are initialized based on previous research or optimized during experiments. We initialized the convolution kernel by using the "He normal optimizer" [66] and applied l2 (0.0001) for the kernel regularizer. We use 1D convolutional kernels of size 5 in the sRN sub-network and 2D convolutional kernels of size 3 × 3 in the saRN sub-network. For the number of filters, we use the same size of filters in each convolution layer, 24. We apply 1D average pooling layer with pool size 2 and 2D average pooling layer with pool size 5 × 5 in the sRN and saRN, respectively. Furthermore, a 50% dropout is applied in both sub-networks. Then, we trained our model using Adam optimizer with a learning rate of 0.0003 [67].

Regarding batch-size, a constant batch-size sometimes results in a tiny mini-batch (see Figure 4a). Meanwhile, in a network with BN layers, there is dependency between the mini-batch elements because BN uses mini-batch statistics to normalize the activations during the learning process [68]. This dependency may decrease the performance if the mini-batch is too small [68,69]. Some approaches can be applied to overcome this problem. The first is to ignore the samples in the last mini-batch. This approach is not viable for the IP dataset because the number of training samples in a category can be very small; for example, with 10% training samples, we only have two training samples in Oats category. Performance will be badly affected if the removed samples are from this category (see Figure 4a). The second approach is by copying other samples from the previous mini-batch. This technique will make some samples appear twice in the training process, and these samples will have more weight. Another approach is by dividing the training size over the intended batch number. For example, if we intend to have three batches so the batch size = training size/3. However, when the training sample is too large, the batch size will be large and thus prone to an out of memory problem. If the training size is too small, the batch size will also be small, having a tiny batch size can decrease the performance. Therefore, in our experiment, we used Equation (10) to compute the batch-size prior to the training process to prevent the occurrence of a tiny mini-batch, where s_b is the standard batch, tr is the training size, and th is the threshold (the allowed smallest mini-batch, see Figure 4b). We used $s_b = 256$ and $th = 64$ in this paper.

$$batch_{size} = \begin{cases} s_b, & \text{if } tr \bmod s_b > th. \\ s_b + \dfrac{tr \bmod s_b}{int(\frac{tr}{s_b})}, & \text{otherwise.} \end{cases} \tag{10}$$

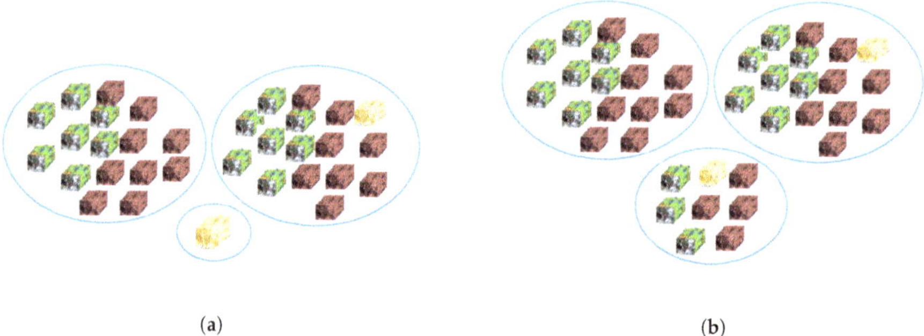

(a)   (b)

Figure 4. Example condition when the batch size cannot divide the training size evenly: (**a**) the latest mini-batch size is one, and (**b**) the latest mini-batch size is more than threshold (if the threshold is seven).

In the sRN, we used a 3 × 3 spectral cube and computed its mean instead of using a pixel vector directly to minimize the effect of spectral noise. In contrast to sRN, saRN focus is to get the spatial features; hence, the region size of the input cube gives an impact on the spatial feature representation.

In this research, in order to find the optimum saRN input region size, n, we experiment on a variable set of n {21, 23, 25, and 27} with the number of PCA components set to 30 by using 10% random training samples and repeat the experiment 10 times. Table 2 shows the results of the Overall Accuracy mean and standard deviation, and from this table, we can conclude that each dataset has a different optimum number n. For the PU, IP, and KSC dataset, the optimum n is 21, 25, and 27, respectively.

We then use the optimum value of n to find the optimum number of PCA components, K. We experiment with different size of K {25, 30, 35, and 40}. The Overall Accuracy (OA)-mean with different values of K and 10% training samples are shown in Table 3. The table shows that the optimum K of KSC dataset is 25, while, for the IP dataset and PU dataset, the optimum K is 35.

Table 2. Overall Accuracy of each dataset based on various patch sizes (SoP). The number in bold is the best Overall Accuracy.

SoP	21	23	25	27
IP	98.73 ± 0.22	98.66 ± 0.29	$\mathbf{98.77 \pm 0.32}$	98.75 ± 0.16
KSC	97.73 ± 0.47	97.95 ± 1.12	96.74 ± 1.43	$\mathbf{98.51 \pm 0.31}$
PU	$\mathbf{99.87 \pm 0.06}$	99.46 ± 0.02	99.65 ± 1.28	99.82 ± 0.39

Table 3. Overall Accuracy based on PCA number. The number in bold is the best Overall Accuracy.

nPCA	25	30	35	40
IP	98.74 ± 0.24	98.77 ± 0.32	$\mathbf{98.82 \pm 0.38}$	98.80 ± 0.15
KSC	$\mathbf{99.15 \pm 0.18}$	98.51 ± 0.31	98.29 ± 0.82	98.10 ± 0.88
PU	99.72 ± 0.50	99.87 ± 0.06	$\mathbf{99.91 \pm 0.02}$	99.72 ± 0.47

Given the optimal parameters for our proposed method, we perform two experiments to understand the impact of each module of our proposed architecture. The first is an experiment to discover the effect of using the mean in the sRN sub-network. Second, we perform an experiment to evaluate the performance of sRN, saRN, and our proposed architecture.

To demonstrate the effectiveness of our proposed method, we compare our method with the state-of-the-art architectures, which focus on exploring the spectral-spatial features of HSI, namely 3D-CNN [34], SSLSTMs [27], SSUN [23], spectral-spatial residual network (SSRN) [19], and hybrid spectral convolutional neural network (HybridSN) [61]. The SSLSTMs and the SSUN explore the spectral and the spatial features using two different streams, while the 3D-CNN, the SSRN, and the HybridSN extract features using a single stream network based on 3D convolutional layer. The implementation codes of the 3D-CNN (https://github.com/nshaud/DeepHyperX), the SSUN (https://github.com/YonghaoXu/SSUN), the SSRN (https://github.com/zilongzhong/SSRN), and the HybridSN (https://github.com/gokriznastic/HybridSN) are publicly available, letting us execute the codes to produce the classification results with all datasets. For the SSLSTMs, even though the implementation code is not accessible, we wrote the code based on their paper architecture and parameters. To confirm that our implemented code is correct, we tested it on 10% of the training dataset and verified our results with the work of Reference [27]. All experiments except the 3D-CNN were conducted on X299 UD4 Pro desktop computer with the GeForce RTX 2080 Ti Graphical Processing Unit (GPU). The experiment of the 3D-CNN was conducted on Google Colab server because 3D-CNN used the Pytorch framework.

To validate the performance of the proposed model with respect to the training size of each compared model, we performed three different experiments. In all of these experiments, we used 10-fold cross-validation. To guarantee that all of the techniques use the same training indices and testing indices, we created a module to generate the training indices and testing indices by using StratifiedShuffleSplit function available in Keras. The input of this function is the training size percentage and the number of the fold/group (k). The output is k fold training indices and testing indices, where each fold is made by preserving the percentage of samples for each class. We then saved the training indices and testing indices of each fold in a file. Those files were read by each method during the experiment. Following the protocol in Reference [19], we use the same number of training epoch, 200, for all of the experiments. Regarding the hyperparameters, we used the optimum parameter of each model that has been provided in their respective paper. For the hyperparameters of this proposed approach, we used the optimum settings that have been optimized on 10% training samples, which were provided by Tables 2 and 3.

In conclusion, we divided the experiments into two groups. The first group (experiments 1 and 2) is an ablation analysis to understand the impact of using the mean, and concatenating sRN and saRN in the proposed method with respect to the overall performance accuracy. The second group (experiments 3, 4, and 5) are experiments to understand the effectiveness of the proposed method compared to other previous studies. The details of these experiments are as follows:

1. To evaluate the effect of the mean operation in the sRN sub-network input, we performed experiments on our proposed architecture with two case scenarios. First, the sRN input is the mean of a 3×3 spectral cube. Second, the sRN input is a spectral vector of a pixel x_{ij} without the mean operation. We performed experiments with 4%, 6%, 8%, 10%, and 30% training samples for IP dataset, PU dataset, and KSC dataset. We use the rest of the data that is not used in training for testing. In each run, we use the same data points for the training of both "with mean" and "without mean" setups.

2. To discover the concatenation effect of the sRN sub-network and the saRN sub-network on the performance accuracy, we performed experiments on three different architectures, the sRN network only, the saRN network only, and our proposed method with 30%, 10%, and 4% training samples. Here, we divided the data into a training set (30%) and a testing set (70%). Then, from the training set, we used all, one-third, and one-seventh point five for training. For testing, in all experiments in this part, we used all data in the testing set. The examples of train and test split on IP dataset with various percentage training samples are shown in Figure 5.

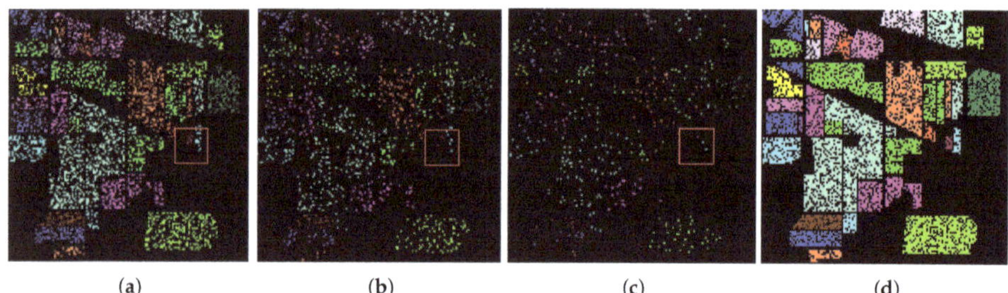

(a)   (b)   (c)   (d)

Figure 5. (**a–c**) The train split with 30%, 10%, and 4% training size on Indian Pine (IP) dataset (**d**) the test split.

3. In our third experiment, this proposed approach is compared to 3D-CNN, SSLSTMs, SSUN, SSRN, and HybridSN by using 10% training samples. We chose 10% training samples because the SSLSTMs and other experiments on SSUN, SSRN, and HybridSN have also been conducted using 10% training samples.

4. In our fourth experiment, we compared all of those methods on the smaller training samples, 4%, 6%, and 8%. Besides, because 3D-CNN has been tested using 4% training samples, the use of small labeled samples during training can be used to investigate over-fitting issues.

5. In the last experiment, we compared all of those methods on large training samples, 30%. Not only because HybridSN [61] had been tested on 30% training samples but also to investigate under-fitting issues with large training samples.

4.3. Experimental Results

Experiment 1: Table 4 shows the OA-mean and standard deviation of this proposed architecture in two different cases. In the first case, the sRN input of our network is a 3×3 cube followed by mean operations (with mean), and the second case, the sRN input of our network is a spectral vector, which was not followed by mean operations (without mean). From the table, we can see that, in 11 cases out of 15, the "with mean" slightly outperform the "without mean". We also found that, in 10 cases out of 15, the "with mean" is more stable than "without mean".

Table 4. Comparison between with mean and without mean in our proposed network (Bold represents the best results in the experiment setup).

Training Percentage	Indian Pines		Pavia University		KSC	
	with Mean	w\o Mean	with Mean	w\o Mean	with Mean	w\o Mean
4%	**95.40 ± 0.79**	95.07 ± 0.81	**99.85 ± 0.06**	99.62 ± 0.55	**96.97 ± 0.86**	95.32 ± 1.16
6%	**97.38 ± 0.58**	97.37 ± 0.63	99.44 ± 1.54	**99.93 ± 0.03**	**98.04 ± 0.65**	96.62 ± 1.46
8%	**98.24 ± 0.50**	98.14 ± 0.43	99.78 ± 0.55	**99.94 ± 0.05**	**99.36 ± 0.31**	99.02 ± 0.28
10%	98.70 ± 0.26	**98.81 ± 0.24**	**99.86 ± 0.27**	99.67 ± 0.74	**99.48 ± 0.34**	99.20 ± 0.42
30%	99.70 ± 0.10	**99.75 ± 0.15**	**99.89 ± 0.20**	99.03 ± 3.04	**99.96 ± 0.02**	99.93 ± 0.06

Experiment 2: Table 5 displays the OA-mean and standard deviation of sRN, saRN, and TSRN with various training samples, where the best performance is shown in bold. The table shows that with 30% training data, saRN's performance is slightly better than others. With 10% training samples, TSRN's performance starts to exceed saRN's performance. TSRN's superiority is clearly shown in 4% training samples. When the training size is large (30%), and the train and test sets are sampled randomly over the whole image, the possibility of the training samples become the testing samples' neighbor is high. Other spatial features, such as line and shape, are clear, too. See Figure 5a, suppose the center of the red window is the testing sample, we can easily predict its label by seeing its spatial features. However, with 10% training samples, predicting the pixel's label only by using its spatial features is slightly difficult (see Figure 5b). The prediction problems are more complicated when the training size is 4%. Figure 5c shows that the spatial features (e.g., neighborhood, shape, line) alone cannot perform well. Therefore, with 4% training samples, the TSRN, which also use spectral features, produces much better performance then saRN. Meanwhile, the low performance of sRN on IP dataset and KSC dataset probably because IP and KSC dataset have significantly low spatial resolution 20 m per pixel and 18 m per pixel, respectively. For example, in IP dataset, where most classes are vegetation, one pixel corresponds to the average reflectance of vegetation in 400 m^2, which results in a mixture of ground materials. As a consequence, classifying the objects based on spectral information only is difficult.

Table 5. Comparison between sRN, saRN, and Proposed (TSRN) with 30%, 10%, and 4% training samples (Bold represents the best results in the experiment setup).

Training	30%			10%			4%		
Dataset	sRN	saRN	TSRN	sRN	saRN	TSRN	sRN	saRN	TSRN
IP	92.16 ± 0.66	**99.75 ± 0.18**	99.69 ± 0.15	86.61 ± 0.86	98.44 ± 0.26	**99.03 ± 0.24**	79.94 ± 1.54	94.20 ± 0.43	**95.50 ± 0.87**
PU	98.96 ± 0.08	**99.97 ± 0.02**	99.95 ± 0.13	98.05 ± 0.19	99.84 ± 0.05	**99.93 ± 0.10**	96.68 ± 0.14	99.56 ± 0.11	**99.82 ± 0.10**
KSC	95.55 ± 0.77	**100 ± 0.02**	99.95 ± 0.05	92.52 ± 0.86	**99.55 ± 0.2**	99.52 ± 0.29	82.42 ± 3.48	94.60 ± 1.25	**95.62 ± 1.27**

Experiment 3: Tables 6–8 show the quantitative evaluations of those compared models with 10% training samples. The tables present three generally used quantitative metrics, i.e., Overall Accuracy (OA), Average Accuracy (AA), Kappa coefficient (K), and the classification accuracy of each class. The first three rows show the OA, AA, and K of each method. The following rows show the classification accuracy of each class. The numbers indicate the mean, followed by the standard deviation of each evaluation with a 10-fold cross-validation. The bold, the underlined, and the

italic numbers represent the first-best performance, the second-best, and the third-best performance, respectively. Subsequently, Figures 6–8 display the false-color image, the ground-truth image, and the classification map of each method on Indian Pine, Pavia University and KSC datasets.

Table 6. Overall Accuracy, Average Accuracy, Kappa Value, and Class Wise Accuracy of our proposed method versus other methods on IP dataset when using 10% training samples. The best performance is in bold, the second-best performance is underlined, and the third-best is in italic.

Label	3D-CNN [34]	SSLSTMs [27]	SSUN [23]	SSRN [19]	HybridSN [61]	Proposed
OA	85.29 ± 7.24	94.65 ± 0.72	96.79 ± 0.36	98.24 ± 0.29	*97.36 ± 0.82*	**98.70 ± 0.25**
AA	81.11 ± 0.12	94.47 ± 1.77	95.78 ± 2.97	91.69 ± 3.46	*95.70 ± 1.08*	**98.71 ± 0.61**
K × 100%	83.20 ± 0.08	93.89 ± 0.82	96.33 ± 0.41	97.99 ± 0.32	*96.99 ± 0.93*	**98.52 ± 0.28**
C1	68.70 ± 0.28	*99.32 ± 2.05*	99.51 ± 0.98	**100 ± 0**	97.80 ± 3.53	98.72 ± 3.83
C2	84.30 ± 0.17	93.09 ± 1.71	94.65 ± 1.33	**98.63 ± 0.82**	96.76 ± 1.44	*98.02 ± 1.08*
C3	75.50 ± 0.11	86.37 ± 1.81	96.09 ± 1.87	96.82 ± 0.70	96.27 ± 2.58	**97.08 ± 1.72**
C4	74.50 ± 0.09	89.35 ± 4.41	94.65 ± 4.90	**99.20 ± 1.37**	96.67 ± 2.44	*99.16 ± 1.22*
C5	88.80 ± 0.07	93.69 ± 3.10	94.89 ± 2.09	96.95 ± 2.27	94.57 ± 3.93	**99.41 ± 0.36**
C6	96.00 ± 0.02	95.46 ± 0.92	99.06 ± 0.78	98.19 ± 1.17	*98.48 ± 0.70*	**99.76 ± 0.26**
C7	61.90 ± 0.27	**99.22 ± 1.57**	*93.20 ± 4.75*	70 ± 45.83	88.80 ± 13.24	99.15 ± 1.71
C8	96.80 ± 0.02	98.16 ± 1.73	99.81 ± 0.49	*99.15 ± 2.13*	99.81 ± 0.39	**99.98 ± 0.07**
C9	60.00 ± 0.24	90.70 ± 19.01	*86.67 ± 13.19*	20.00 ± 40.00	86.11 ± 14.33	**98.50 ± 4.50**
C10	82.70 ± 0.08	96.50 ± 1.42	95.18 ± 1.92	97.35 ± 1.55	97.34 ± 1.29	**98.19 ± 1.10**
C11	87.30 ± 0.07	96.73 ± 0.85	98.13 ± 0.38	98.65 ± 0.40	98.22 ± 0.89	**99.32 ± 0.46**
C12	77.40 ± 0.11	88.92 ± 1.58	92.85 ± 4.39	95.51 ± 1.28	93.82 ± 2.73	**97.87 ± 1.82**
C13	97.70 ± 0.02	93.83 ± 4.27	**99.57 ± 0.22**	98.21 ± 2.38	99.02 ± 0.94	*98.76 ± 2.15*
C14	95.30 ± 0.03	98.07 ± 1.03	98.66 ± 0.54	**99.57 ± 0.42**	99.39 ± 0.40	*98.98 ± 0.91*
C15	69.30 ± 0.05	96.63 ± 3.98	*97.32 ± 2.91*	**99.34 ± 0.82**	95.48 ± 2.81	98.44 ± 1.74
C16	81.50 ± 0.28	*95.48 ± 4.10*	92.26 ± 6.74	**99.48 ± 0.85**	92.74 ± 4.95	98.11 ± 1.20

Table 7. Overall Accuracy, Average Accuracy, Kappa Value, and Class Wise Accuracy of our proposed method versus other methods on Pavia University (PU) dataset with 10% training samples. The best performance is in bold, the second-best performance is underlined, and the third-best is in italic.

Label	3D-CNN [34]	SSLSTMs [27]	SSUN [23]	SSRN [19]	HybridSN [61]	Proposed
OA	94.07 ± 0.86	98.58±0.23	99.53 ± 0.09	*99.59 ± 0.72*	99.73 ± 0.11	**99.86 ± 0.26**
AA	96.54 ± 0.01	98.65 ± 0.16	99.18 ± 29.9	*99.31 ± 1.48*	99.43 ± 0.23	**99.77 ± 0.54**
K × 100%	92.30 ± 0.01	98.11 ± 0.31	99.38 ± 0.12	*99.46 ± 0.95*	99.65 ± 0.15	**99.82 ± 0.34**
C1	96.50 ± 0.01	97.47 ± 0.46	99.29 ± 0.22	*99.85 ± 0.17*	99.91 ± 0.18	**99.94 ± 0.07**
C2	95.20 ± 0.01	98.95 ± 0.34	99.91 ± 0.04	*99.93 ± 0.07*	**100 ± 0.01**	99.94 ± 0.06
C3	92.20 ± 0.03	*98.80 ± 0.59*	97.67 ± 0.74	**99.56 ± 0.56**	99.22 ± 0.81	98.56 ± 4.09
C4	97.20 ± 0.01	98.43 ± 0.46	*99.36 ± 0.29*	99.66 ± 0.33	98.4 ± 0.96	**99.99 ± 0.03**
C5	99.90 ± 0	99.87 ± 0.15	99.93 ± 0.07	*99.91 ± 0.08*	**99.97 ± 0.06**	99.84 ± 0.10
C6	97.60 ± 0.02	98.92 ± 0.51	*99.89 ± 0.12*	99.99 ± 0.05	99.99 ± 0.01	**100 ± 0**
C7	95.50 ± 0.02	*98.36 ± 1.03*	97.87 ± 1.64	95.67 ± 12.94	**100 ± 0**	99.73 ± 0.75
C8	95.40 ± 0.02	97.67 ± 0.60	99.19 ± 0.36	*99.28 ± 1.11*	99.41 ± 0.45	**99.92 ± 0.08**
C9	99.40 ± 0.01	99.44 ± 0.50	*99.54 ± 0.35*	99.92 ± 0.12	97.98 ± 0.95	**100 ± 0**

Table 8. Overall Accuracy, Average Accuracy, Kappa Value, and Class Wise Accuracy of our proposed method versus other methods on Kennedy Space Center (KSC) dataset with 10% training samples. The best performance is in bold, the second-best performance is underlined, and the third-best is in italic.

Label	3D-CNN [34]	SSLSTMs [27]	SSUN [23]	SSRN [19]	HybridSN [61]	Proposed
OA	82.21 ± 2.96	*97.51 ± 0.63*	96.22 ± 0.86	98.77 ± 0.75	91.72 ± 1.52	**99.48 ± 0.32**
AA	71.68 ± 0.12	*97.36 ± 0.69*	94.65 ± 2.87	98.10 ± 1.13	88.94± 1.53	**99.04 ± 0.46**
K × 100%	80.20 ± 0.03	*97.22 ± 0.71*	95.79 ± 0.96	98.63 ± 0.83	90.77 ± 1.69	**99.42 ± 0.36**
C1	92.40 ± 0.03	*96.65 ± 1.73*	97.12 ± 0.98	**100 ± 0**	95.68 ± 4.17	**100 ± 0**
C2	84.20 ± 0.08	97.57 ± 2.39	94.25 ± 4.04	*96.90 ± 7.10*	76.99 ± 5.48	**99.77 ± 0.43**
C3	43.00 ± 0.27	*96.75 ± 2.76*	95.05 ± 3.41	**100 ± 0**	90.35 ± 3.60	97.17 ± 4.41
C4	33.50 ± 0.16	**98.41 ± 1.38**	*89.08 ± 5.15*	88.86 ± 11.12	70.40 ± 5.30	98.39 ± 2.69
C5	34.70 +- 0.19	97.55 ± 2.52	89.93 ± 8.78	96.66 ± 5.93	97.24 ± 3.26	**97.91 ± 4.81**
C6	40.90 ± 0.22	*97.82 ± 2.84*	79.56 ± 3.851	**99.90 ± 0.30**	81.89 ± 6.60	99.19 ± 1.42
C7	59.30 ± 0.31	96.34 ± 4.93	**98.19 ± 1.43**	93.56 ± 10.79	82.23 ± 6.73	*96.02 ± 4.26*
C8	75.60 ± 0.08	*95.27 ± 2.83*	93.61 ± 2.73	99.54 ± 0.76	93.40 ± 3.81	**99.62 ± 0.35**
C9	84.10 ± 0.09	96.93 ± 2.34	*98.25 ± 2.60*	**100 ± 0**	88.29 ± 4.40	99.62 ± 0.42
C10	94.40 ± 0.04	97.28 ± 3.23	96.95 ± 2.45	**100 ± 0**	90.08 ± 4.19	99.89 ± 0.33
C11	97.70 ± 0.02	98.21 ± 1.35	*98.86 ± 1.11*	**99.92 ± 0.24**	97.19 ± 2.55	99.90 ± 0.32
C12	92.10 ± 0.02	97.06 ± 1.52	*99.56 ± 0.83*	99.93 ± 0.20	92.50 ± 3.78	**100 ± 0**
C13	99.90 ± 0	*99.88 ± 0.14*	**100 ± 0**	**100 ± 0**	**100 ± 0**	**100 ± 0**

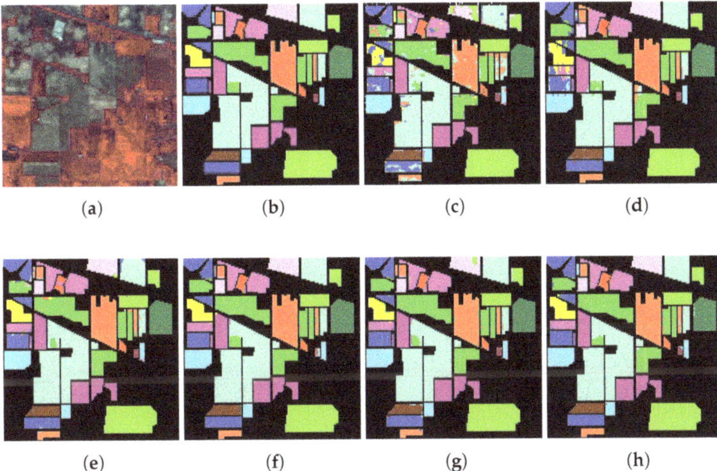

(a)   (b)   (c)   (d)

(e)   (f)   (g)   (h)

Figure 6. The classification map of IP dataset. (**a**) False color image, (**b**) Ground truth, and (**c–h**) Prediction classification maps of 3D-Convolutional Neural Network (CNN) (85.29%), spectral-spatial long short-term memory (SSLSTMs) (95%), spectral-spatial unified network (SSUN) (97.24%), SSRN (98.29%), HybridSN (97.38%), and our proposed architecture (98.69%).

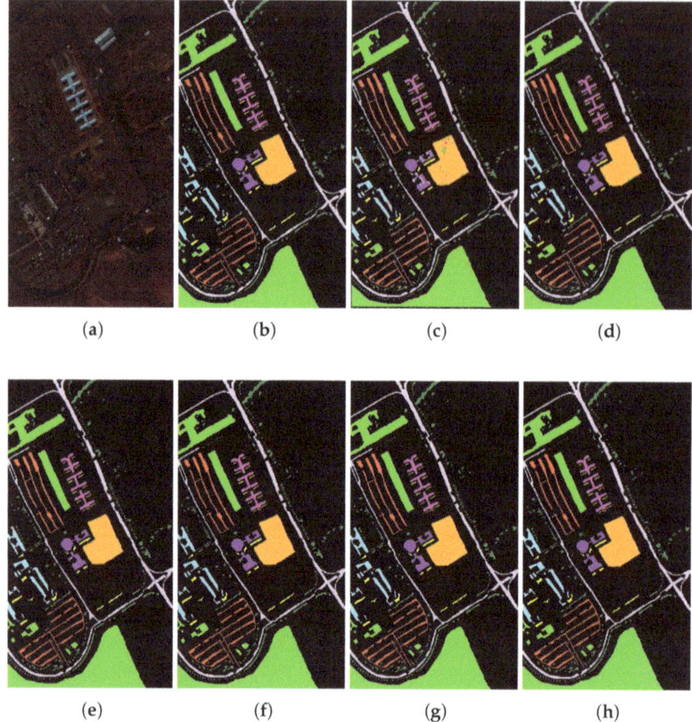

Figure 7. The classification map of Pavia University dataset. (**a**) False color image, (**b**) Ground truth, (**c–h**) Prediction classification maps of 3D-CNN (94.07%), SSLSTMs (98.50%), SSUN (99.52%), SSRN (99.88%), HybridSN (99.85%), and our proposed architecture (99.94%).

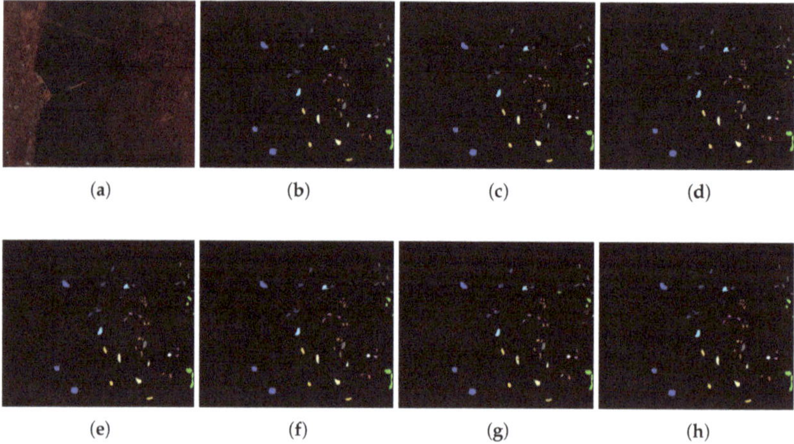

Figure 8. The classification map of KSC dataset. (**a**) False color image, (**b**) Ground truth, (**c–h**) Prediction classification maps of 3D-CNN (82.21%), SSLSTMs (97%), SSUN (97.10%), SSRN (99.27%), HybridSN (87.46%), and our proposed architecture (99.61%).

Experiment 4: Figure 9 presents the graphic of AO-mean obtained from our fifth experiment, where all of those methods are trained on smaller training samples 4%, 6%, and 8%. In the figure, we include the results of our first experiment, where those methods are trained on the 10% samples. The performances of all of the compared methods are displayed using a dotted line, while our proposed method is displayed with a solid line.

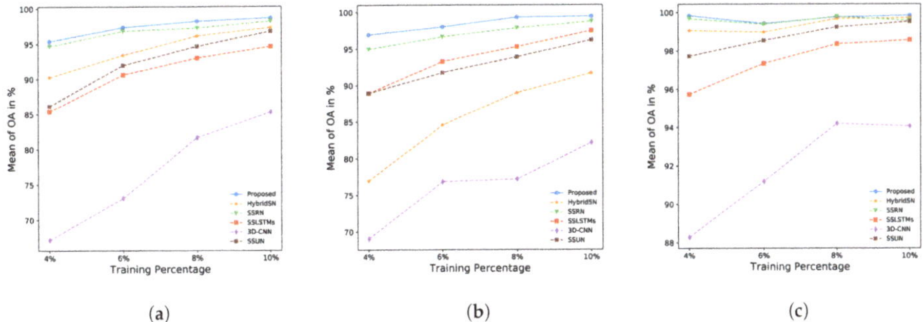

Figure 9. Overall accuracy of each method for different training data sizes of: (**a**) Indian Pine dataset, (**b**) KSC dataset, and (**c**) Pavia University dataset.

Experiment 5: Tables 9–11 show the OA, AA, and K of each method with 30% training samples. On large training samples, almost all of the compared methods produce a high accuracy. The difference is small. Hence, in the table, we report the comparison on each fold for a more detailed comparison. The bold numbers are the best accuracies produced by these methods.

Table 9. Fold Overall Accuracy, Average Accuracy, and Kappa Value on IP dataset with 30% training data. The best performance is in bold, the second-best performance is underlined, and the third-best is in italic.

Fold	1	2	3	4	5	6	7	8	9	10	OA-Mean and Std
						OA					
Proposed	**99.80**	99.71	**99.61**	**99.72**	99.57	*99.54*	99.69	**99.85**	99.68	**99.79**	**99.70 ± 0.10**
HybridSN	*99.53*	99.71	99.57	99.61	**99.68**	**99.75**	**99.83**	*99.53*	99.67	99.64	99.65 ± 0.09
SSRN	98.89	**99.79**	99.53	99.26	99.50	99.68	99.04	99.33	99.46	99.48	99.40 ± 0.26
SSUN	99.60	99.48	99.57	99.46	99.57	99.46	99.67	99.61	**99.74**	99.55	*99.57 ± 0.09*
SSLSTMs	99.16	99.19	99.07	99.15	98.94	99.23	97.09	98.91	99.4	99.32	98.95 ± 0.64
3D-CNN	95.15	95.89	92.42	95.47	94.51	95.55	95.86	95.26	95.41	95.33	95.09 ± 0.96
						AA					
Proposed	**99.89**	**99.66**	**99.68**	**99.73**	99.61	99.58	**99.77**	**99.82**	**99.65**	**99.58**	**99.70 ± 0.10**
HybridSN	*99.04*	*99.50*	98.82	99.44	**99.62**	**99.69**	99.71	*99.03*	99.61	99.07	*91.98 ± 5.19*
SSRN	80.29	99.64	92.98	91.55	93.18	93.20	86.22	91.96	92.79	97.99	91.98 ± 5.19
SSUN	99.60	99.14	99.47	99.26	*99.34*	99.50	99.24	99.61	99.50	99.53	99.42 ± 0.15
SSLSTMs	*99.04*	99.12	98.26	*99.34*	98.77	97.10	97.66	98.65	*99.55*	*99.31*	98.68 ± 0.75
3D-CNN	96.57	97.33	94.04	95.42	95.20	96.75	97.02	96.81	96.57	96.01	96.17 ± 0.96
						K × 100%					
Proposed	**99.78**	99.67	**99.56**	**99.68**	99.51	*99.48*	99.65	**99.83**	99.63	**99.76**	**99.66 ± 0.11**
HybridSN	*99.46*	99.67	99.51	99.56	**99.63**	**99.71**	**99.81**	*99.46*	99.62	99.59	99.60 ± 0.11
SSRN	98.73	**99.76**	99.46	99.16	99.43	99.63	98.90	99.24	99.38	99.41	99.31 ± 0.30
SSUN	99.54	99.41	99.51	*99.38*	99.51	99.38	99.62	99.56	**99.70**	99.49	*99.51 ± 0.10*
SSLSTMs	99.05	99.08	98.94	99.03	98.79	99.13	96.68	98.76	99.32	99.22	98.80 ± 0.73
3D-CNN	94.49	95.33	91.40	94.85	93.75	94.95	95.29	94.61	94.78	94.69	94.41 ± 1.09

Table 10. Fold Overall Accuracy, Average Accuracy, and Kappa Value on PU dataset with 30% training data. The best performance is in bold, the second-best performance is underlined, and the third-best is in italic.

Fold	1	2	3	4	5	6	7	8	9	10	OA-Mean and Std
						OA					
Proposed	99.96	99.96	99.98	99.98	99.99	99.84	99.34	99.99	99.92	99.95	99.89 ± 0.19
HybridSN	99.97	99.96	99.97	99.96	99.99	100	99.96	99.97	99.92	99.98	99.97 ± 0.02
SSRN	99.93	99.83	99.83	99.92	99.81	99.89	99.82	99.86	97.44	99.58	99.59 ± 0.72
SSUN	99.94	99.95	99.92	99.91	99.91	99.94	99.9	99.91	99.94	99.96	99.93 ± 0.02
SSLSTMs	99.85	99.80	99.83	99.76	99.72	99.79	99.68	99.86	99.70	99.82	99.78 ± 0.06
3D-CNN	95.76	95.80	95.80	95.30	95.75	95.76	94.75	95.94	95.92	94.06	95.48 ± 0.58
						AA					
Proposed	99.96	99.94	99.94	99.97	100	99.77	98.70	99.99	99.95	99.97	99.82 ± 0.38
HybridSN	99.97	99.93	99.95	99.94	99.97	100	99.93	99.95	99.81	99.96	99.94 ± 0.05
SSRN	99.93	99.74	99.85	99.86	99.81	99.92	99.84	99.74	94.88	99.48	99.31 ± 1.48
SSUN	99.90	99.91	99.81	99.84	99.82	99.92	99.82	99.79	99.87	99.94	99.86 ± 0.05
SSLSTMs	99.88	99.74	99.81	99.79	99.68	99.73	99.66	99.86	99.67	99.85	99.77 ± 0.08
3D-CNN	97.46	97.69	97.54	97.00	97.68	97.54	96.91	97.77	97.79	94.80	97.22 ± 0.86
						K × 100%					
Proposed	99.95	99.94	99.97	99.98	99.99	99.78	99.12	99.98	99.90	99.93	99.85 ± 0.25
HybridSN	99.96	99.94	99.96	99.95	99.98	100	99.94	99.96	99.90	99.97	99.96 ± 0.03
SSRN	99.91	99.78	99.77	99.90	99.75	99.85	99.77	99.81	96.62	99.45	99.46 ± 0.95
SSUN	99.92	99.93	99.89	99.88	99.88	99.92	99.87	99.88	99.92	99.95	99.90 ± 0.03
SSLSTMs	99.80	99.74	99.77	99.68	99.63	99.72	99.58	99.82	99.60	99.77	99.71 ± 0.08
3D-CNN	94.46	94.50	94.51	93.87	94.45	94.47	93.19	94.69	94.66	92.25	94.11 ± 0.75

Table 11. Fold Overall Accuracy, Average Accuracy, and Kappa Value on KSC dataset with 30% training data. The best performance is in bold, the second-best performance is underlined, and the third-best is in italic.

Fold	1	2	3	4	5	6	7	8	9	10	OA-Mean and Std
						OA					
Proposed	99.95	99.97	100	99.95	99.97	99.97	99.92	99.95	99.92	99.95	99.96 ± 0.02
HybridSN	99.01	99.37	99.23	99.48	98.96	98.93	98.85	98.96	99.59	99.26	99.16 ± 0.24
SSRN	99.67	100	100	100	99.84	99.37	98.27	100	99.97	99.64	99.68 ± 0.51
SSUN	99.10	99.31	99.67	98.68	99.56	99.31	99.29	99.29	99.01	99.40	99.26 ± 0.27
SSLSTMs	99.78	99.95	99.56	99.73	99.95	99.10	99.78	99.42	99.26	99.78	99.63 ± 0.27
3D-CNN	94.60	95.39	94.76	95.34	91.89	94.24	94.57	95.45	94.46	93.15	94.39 ± 1.05
						AA					
Proposed	99.95	99.97	100	99.96	99.97	99.97	99.90	99.95	99.92	99.96	99.96 ± 0.03
HybridSN	98.36	99.18	98.63	99.17	98.33	98.31	98.38	98.42	99.35	98.92	98.71 ± 0.39
SSRN	99.51	100	100	100	99.75	99.09	97.97	100	99.96	99.44	99.57 ± 0.61
SSUN	98.25	98.73	99.27	97.59	99.28	98.61	98.43	98.98	98.26	99.03	98.64 ± 0.50
SSLSTMs	99.74	99.85	99.44	99.62	99.95	98.95	99.66	99.49	99.34	99.78	99.58 ± 0.28
3D-CNN	92.57	93.37	92.01	92.70	87.23	91.71	91.82	93.09	91.58	89.46	91.55 ± 1.77
						K × 100%					
Proposed	99.94	99.97	100	99.94	99.97	99.97	99.91	99.94	99.91	99.94	99.95 ± 0.03
HybridSN	98.90	99.30	99.15	99.42	98.84	98.81	98.72	98.84	99.54	99.18	99.07 ± 0.27
SSRN	99.63	100	100	100	99.82	99.30	98.08	100	99.97	99.60	99.64 ± 0.57
SSUN	98.99	99.24	99.63	98.53	99.51	99.24	99.21	99.21	98.90	99.33	99.18 ± 0.30
SSLSTMs	99.76	99.94	99.51	99.69	99.94	98.99	99.76	99.36	99.18	99.76	99.59 ± 0.30
3D-CNN	93.99	94.88	94.18	94.81	90.98	93.60	93.97	94.94	93.85	92.39	93.76 ± 1.17

4.4. Discussion

According to the highlighted results in Section 4.3, one of the first apparent points is that the proposed method is able to produce a high performance for all ranges of training sizes (4%, 6%, 8%, 10%, and 30% training samples). Its OA, AA, and K values are higher compared to 3D-CNN, SSLSTMs, SSUN, SSRN, and HybridSN. The differences get higher as the training sample size is reduced. With large training samples, e.g., 30% training samples, the performances of these methods are similar.

The quantitative evaluation of those models with 10% training samples are reported in Tables 6–8. These tables show three standard quantitative metrics, i.e., OA, AA, K; and the classification accuracy of each class. More specifically, on Indian Pines dataset (see Table 6), in which class sizes are imbalanced, our proposed method produces the highest OA, AA, and K value, and the proposed approach yields OA 0.46% higher than the second-best method, SSRN. Considering the AA, the difference between the proposed architecture and SSRN is much higher, more than 7%. From Table 6, we can see that TSRN tries to optimize the recognition of each class even though the number of instances in the class is tiny. Hence, it achieves a high accuracy compared to the other methods when classifying C9 (Oats), in which number of instances is 20, which means C9 training samples is 2. For more detailed classification accuracy of each class, TSRN yields the best recognition on eight classes out of 16 classes. Its recognition for the other five classes and two classes are the second- and third-best, respectively.

The results are consistent on the Pavia University dataset, in which characteristics are different from the Indian Pine dataset. In the PU dataset, the number of data for each class is large, with the minimum number of instances on Shadows category equal to 947. As shown in Table 7, our proposed method attains the best OA, AA, and K compared to the other architectures, albeit insignificant disparity. The small gap between TSRN and the second-best method, HybridSN, shows that those methods are very competitive for large training samples. For class recognition, the proposed method achieves the highest accuracy on five out of nine classes in the PU dataset, with an improvement of less than 1%.

In contrast to the IP and PU datasets, the total number of instances of KSC dataset is relatively small. From Table 8, we can see that our proposed approach achieves the best performance. Its OA, AA, and K is ± 0.71, ± 0.94, and ± 0.79 higher compared to the second-best method, SSRN. In contrast, HybridSN yields performance that is not as good as its performance on IP and PU dataset.

The comparison between the proposed architecture and other methods on smaller training samples for IP, KSC, and PU dataset is demonstrated in Figure 9a–c, respectively. These figures reveal that the proposed method achieves the best accuracy even with smaller training samples. The accuracy gap between our method and the second-best method is high on KSC dataset. With 4% training data, our method achieves OA ± 2% higher than the second-best method, SSRN. The difference is smaller on IP dataset and is extremely small on PU dataset. The reason is that the size of the KSC dataset is the smallest compared to other datasets. Four percent training samples in the experiment correspond to 208 training instances on KSC dataset, 410 instances on IP dataset, and 1711 samples on PU dataset.

The performance of TSRN and the other methods with larger training samples, 30%, is shown in Tables 9–11 for IP dataset, PU dataset, and KSC dataset, respectively. In IP dataset, out of ten folds, the proposed method achieves the best OA and K on five-folds, and the best AA on eight folds. Our method also outperforms HybridSN, which presents the best OA on three folds. In PU dataset (see Table 10), the HybridSN shows a slightly better OA than the proposed architecture. HybridSN produces the best OA on six-folds while TSRN produces the best performance within five-folds. In the 2nd fold and 5th fold, their OA is precisely the same. Regarding AA, the proposed method achieves the best AA on six-folds when HybridSN achieves the best AA on four-folds. In terms of K, those two methods yield the best value on five-folds. The same with the result from IP dataset, with KSC dataset, our proposed approach also produces the best performance or the second-best performance in each fold (see Table 11). From these results, we can conclude that on large training samples, those approaches, i.e., TSRN, SSUN, HybridSN, SSRN, SSLSTMs, are very competitive.

Table 12 presents the number of parameters, the model size, the depth, the training time, and the testing time of the other methods with 10% training samples from IP dataset. We do not report the time complexity of 3D-CNN because 3D-CNN was tested on a different machine. However, Reference [61] has shown that 3D-CNN is less efficient compared to HybridSN. Moreover, from Table 12, we perceive that the proposed method is more efficient than HybridSN. In other words, we can conclude that TSRN is much more efficient than 3D-CNN. Compared to 3D convolution-based SSRN [19] and (3D + 2D) convolution-based HybridSN [61], our proposed network has a faster training time and fewer learning parameters (Table 12). Our network model size, in which depth is 24, is 3.3 MB. The size is smaller compared to 61.5 MB of a 7-layer HybridSN and more effective compared to 2.9 MB of a 9-layer SSRN. Note that an increase in the network depth results in a model size increase. From Table 12, we can see that our network, which uses (2D+1D) convolution, can be deeper without increasing the number of parameters by a large number. Such a deeper network can extract richer features. On the other hand, for 3D convolution, the model size and the number of parameters will grow dramatically as the network becomes deeper. As a result, training on very deep 3D-CNN becomes challenging with the risk of overfitting. Our network yields a smaller number of learnable parameters, making it less prone to the overfitting problem especially when small samples are used for training.

Table 12. Number of parameters, model size, depth, training time, and testing time on IP dataset on different methods with 10% training samples.

Method	SSLSTMs	SSUN	SSRN	HybridSN	Proposed
# parameters	343,072	949,648	346,784	5,122,176	239,672
Model Size	6.7 MB	9.6 MB	2.9 MB	61.5 MB	3.3 MB
Depth	4	10	9	7	24
Training Time	300 s	66 s	150 s	120 s	60 s
Testing Time	5.3 s	3.03 s	4.1 s	2.57 s	3.1 s

5. Conclusions

The paper presents a novel two-streams residual network architecture for the classification of hyperspectral data. Our network improves the spectral and the spatial feature extraction by applying a full pre-activation sRN and saRN separately. These two networks are similar in their structure but use a different type of convolutional layer. The convolutional layer of sRN is based on 1D convolution, which best fits the spectral data structure, while the saRN is based on 2D convolution, which best fits the spatial data structure of HSI.

Our experiments were conducted on three well-known hyperspectral datasets, versus five different methods, as well as various sizes of training samples. One of the main conclusions that arises from our experiments is that our proposed method can provide a higher performance versus state-of-the-art classification methods, even with various training samples proportion from 4% training samples up to 30% training samples. The high accuracy of our proposed method on small training samples, 4%, shows that this method does not overfit. Otherwise, the competitive accuracy of our proposed method with large training samples, 30%, explains that this architecture is not under-fitting either.

Author Contributions: W.N.K. proposed the methodology, implemented it, did experiments and analysis, and wrote the original draft manuscript. M.B., F.B., and F.S. supervised the study, directly contributed to the problem formulation, experimental design and technical discussions, reviewed the writing, and gave insightful suggestions for the manuscript. D.E. supervised the study, reviewed the writing, gave insightful suggestions for the manuscript and provided resources. All authors have read and agreed to the published version of the manuscript.

Funding: W.N. Khotimah is supported by a scholarship from Indonesian Endowment Fund for Education, Ministry of Finance, Indonesia. This research is also supported by Australia Research Council Grants DP150100294, DP150104251, and Grains Research and Development Corporation Grant UWA2002-003RTX.

Conflicts of Interest: The authors declare no conflict of interest. The funders had no role in the design of the study; in the collection, analyses, or interpretation of data; in the writing of the manuscript, or in the decision to publish the results.

References

1. Lu, G.; Fei, B. Medical hyperspectral imaging: a review. *J. Biomed. Opt.* **2014**, *19*, 010901. [CrossRef] [PubMed]
2. Wang, L.; Sun, C.; Fu, Y.; Kim, M.H.; Huang, H. Hyperspectral Image Reconstruction Using a Deep Spatial-Spectral Prior. In Proceedings of the IEEE Conference on Computer Vision and Pattern Recognition (CVPR), Long Beach, CA, USA, 16–20 June 2019.
3. Shimoni, M.; Haelterman, R.; Perneel, C. Hypersectral Imaging for Military and Security Applications: Combining Myriad Processing and Sensing Techniques. *IEEE Geosci. Remote Sens. Mag.* **2019**, *7*, 101–117. [CrossRef]
4. Stuart, M.B.; McGonigle, A.J.S.; Willmott, J.R. Hyperspectral Imaging in Environmental Monitoring: A Review of Recent Developments and Technological Advances in Compact Field Deployable Systems. *Sensors* **2019**, *19*, 3071. [CrossRef] [PubMed]
5. Ma, J.; Sun, D.W.; Pu, H.; Cheng, J.H.; Wei, Q. Advanced Techniques for Hyperspectral Imaging in the Food Industry: Principles and Recent Applications. *Annu. Rev. Food Sci. Technol.* **2019**, *10*, 197–220. [CrossRef] [PubMed]
6. Mahlein, A.K.; Kuska, M.; Behmann, J.; Polder, G.; Walter, A. Hyperspectral Sensors and Imaging Technologies in Phytopathology: State of the Art. *Annu. Rev. Phytopathol.* **2018**, *56*, 535–558. [CrossRef] [PubMed]
7. Li, S.; Wu, H.; Wan, D.; Zhu, J. An effective feature selection method for hyperspectral image classification based on genetic algorithm and support vector machine. *Knowl. Based Syst.* **2011**, *24*, 40–48. [CrossRef]
8. Cao, F.; Yang, Z.; Ren, J.; Ling, W.K.; Zhao, H.; Sun, M.; Benediktsson, J.A. Sparse representation-based augmented multinomial logistic extreme learning machine with weighted composite features for spectral–spatial classification of hyperspectral images. *IEEE Trans. Geosci. Remote Sens.* **2018**, *56*, 6263–6279. [CrossRef]
9. Li, S.; Song, W.; Fang, L.; Chen, Y.; Ghamisi, P.; Benediktsson, J.A. Deep Learning for Hyperspectral Image Classification: An Overview. *IEEE Trans. Geosci. Remote Sens.* **2019**, *57*, 6690–6709. [CrossRef]
10. Shah, S.A.A.; Bennamoun, M.; Boussaïd, F. Iterative deep learning for image set based face and object recognition. *Neurocomputing* **2016**, *174*, 866–874. [CrossRef]
11. Hu, W.; Huang, Y.; Wei, L.; Zhang, F.; Li, H. Deep Convolutional Neural Networks for Hyperspectral Image Classification. *J. Sens.* **2015**, *2015*, 1–12. [CrossRef]
12. Haut, J.M.; Paoletti, M.E.; Plaza, J.; Li, J.; Plaza, A. Active learning with convolutional neural networks for hyperspectral image classification using a new bayesian approach. *IEEE Trans. Geosci. Remote Sens.* **2018**, *56*, 6440–6461. [CrossRef]
13. Zhu, L.; Chen, Y.; Ghamisi, P.; Benediktsson, J.A. Generative Adversarial Networks for Hyperspectral Image Classification. *IEEE Trans. Geosci. Remote Sens.* **2018**, *56*, 5046–5063. [CrossRef]
14. Zhan, Y.; Hu, D.; Wang, Y.; Yu, X. Semisupervised Hyperspectral Image Classification Based on Generative Adversarial Networks. *IEEE Geosci. Remote Sens. Lett.* **2018**, *15*, 212–216. [CrossRef]
15. Angelopoulou, T.; Tziolas, N.; Balafoutis, A.; Zalidis, G.; Bochtis, D. Remote Sensing Techniques for Soil Organic Carbon Estimation: A Review. *Remote Sens.* **2019**, *11*, 676. [CrossRef]
16. Yang, X.; Ye, Y.; Li, X.; Lau, R.Y.; Zhang, X.; Huang, X. Hyperspectral image classification with deep learning models. *IEEE Trans. Geosci. Remote Sens.* **2018**, *56*, 5408–5423. [CrossRef]
17. Liu, B.; Gao, G.; Gu, Y. Unsupervised Multitemporal Domain Adaptation With Source Labels Learning. *IEEE Geosci. Remote Sens. Lett.* **2019**, *16*, 1477–1481. [CrossRef]
18. Li, Y.; Zhang, H.; Shen, Q. Spectral–Spatial Classification of Hyperspectral Imagery with 3D Convolutional Neural Network. *Remote Sens.* **2017**, *9*, 67. [CrossRef]
19. Zhong, Z.; Li, J.; Luo, Z.; Chapman, M. Spectral–Spatial Residual Network for Hyperspectral Image Classification: A 3-D Deep Learning Framework. *IEEE Trans. Geosci. Remote Sens.* **2018**, *56*, 847–858. [CrossRef]
20. Wang, W.; Dou, S.; Jiang, Z.; Sun, L. A Fast Dense Spectral–Spatial Convolution Network Framework for Hyperspectral Images Classification. *Remote Sens.* **2018**, *10*, 1068. [CrossRef]
21. PV, A.; Buddhiraju, K.M.; Porwal, A. Capsulenet-Based Spatial–Spectral Classifier for Hyperspectral Images. *IEEE J. Sel. Top. Appl. Earth Obs. Remote Sens.* **2019**, *12*, 1849–1865. [CrossRef]

22. Liu, Q.; Zhou, F.; Hang, R.; Yuan, X. Bidirectional-Convolutional LSTM Based Spectral-Spatial Feature Learning for Hyperspectral Image Classification. *Remote Sens.* **2017**, *9*, 1330. [CrossRef]
23. Xu, Y.; Zhang, L.; Du, B.; Zhang, F. Spectral-Spatial Unified Networks for Hyperspectral Image Classification. *IEEE Trans. Geosci. Remote Sens.* **2018**, *56*, 5893–5909. [CrossRef]
24. Koda, S.; Melgani, F.; Nishii, R. Unsupervised Spectral-Spatial Feature Extraction With Generalized Autoencoder for Hyperspectral Imagery. *IEEE Geosci. Remote Sens. Lett.* **2019**, *17*, 469–473. [CrossRef]
25. Yang, J.; Zhao, Y.Q.; Chan, J.C.W. Learning and Transferring Deep Joint Spectral-Spatial Features for Hyperspectral Classification. *IEEE Trans. Geosci. Remote Sens.* **2017**, *55*, 4729–4742. [CrossRef]
26. Hu, W.S.; Li, H.C.; Pan, L.; Li, W.; Tao, R.; Du, Q. Feature Extraction and Classification Based on Spatial-Spectral ConvLSTM Neural Network for Hyperspectral Images. *arXiv* **2019**, arXiv:1905.03577.
27. Zhou, F.; Hang, R.; Liu, Q.; Yuan, X. Hyperspectral image classification using spectral-spatial LSTMs. *Neurocomputing* **2019**, *328*, 39–47. [CrossRef]
28. Liu, B.; Yu, X.; Zhang, P.; Tan, X.; Wang, R.; Zhi, L. Spectral–spatial classification of hyperspectral image using three-dimensional convolution network. *J. Appl. Remote Sens.* **2018**, *12*, 016005. [CrossRef]
29. Pan, B.; Shi, Z.; Zhang, N.; Xie, S. Hyperspectral Image Classification Based on Nonlinear Spectral-Spatial Network. *IEEE Geosci. Remote Sens. Lett.* **2016**, *13*, 1782–1786. [CrossRef]
30. Hao, S.; Wang, W.; Ye, Y.; Nie, T.; Bruzzone, L. Two-stream deep architecture for hyperspectral image classification. *IEEE Trans. Geosci. Remote Sens.* **2017**, *56*, 2349–2361. [CrossRef]
31. Sun, G.; Zhang, X.; Jia, X.; Ren, J.; Zhang, A.; Yao, Y.; Zhao, H. Deep Fusion of Localized Spectral Features and Multi-scale Spatial Features for Effective Classification of Hyperspectral Images. *Int. J. Appl. Earth Obs. Geoinf.* **2020**, *91*, 102157. [CrossRef]
32. Gao, L.; Gu, D.; Zhuang, L.; Ren, J.; Yang, D.; Zhang, B. Combining t-Distributed Stochastic Neighbor Embedding With Convolutional Neural Networks for Hyperspectral Image Classification. *IEEE Geosci. Remote Sens. Lett.* **2019**, *17*, 1368–1372. [CrossRef]
33. He, M.; Li, B.; Chen, H. Multi-scale 3D deep convolutional neural network for hyperspectral image classification. In Proceedings of the 2017 IEEE International Conference on Image Processing (ICIP), Beijing, China, 17–20 September 2017; pp. 3904–3908. [CrossRef]
34. Ben Hamida, A.; Benoit, A.; Lambert, P.; Ben Amar, C. 3-D deep learning approach for remote sensing image classification. *IEEE Trans. Geosci. Remote Sens.* **2018**, *56*, 4420–4434. [CrossRef]
35. Paoletti, M.E.; Haut, J.M.; Fernandez-Beltran, R.; Plaza, J.; Plaza, A.; Li, J.; Pla, F. Capsule Networks for Hyperspectral Image Classification. *IEEE Trans. Geosci. Remote Sens.* **2019**, *57*, 2145–2160. [CrossRef]
36. Qiu, Z.; Yao, T.; Mei, T. Learning spatio-temporal representation with pseudo-3d residual networks. In Proceedings of the IEEE International Conference on Computer Vision, Honolulu, HI, USA, 21–26 July 2017; pp. 5533–5541.
37. Lin, J.; Zhao, L.; Li, S.; Ward, R.; Wang, Z.J. Active-Learning-Incorporated Deep Transfer Learning for Hyperspectral Image Classification. *IEEE J. Sel. Top. Appl. Earth Obs. Remote Sens.* **2018**, *11*, 4048–4062. [CrossRef]
38. Alam, F.I.; Zhou, J.; Liew, A.W.C.; Jia, X.; Chanussot, J.; Gao, Y. Conditional Random Field and Deep Feature Learning for Hyperspectral Image Classification. *IEEE Trans. Geosci. Remote Sens.* **2019**, *57*, 1612–1628. [CrossRef]
39. Zeiler, M.D.; Fergus, R. Visualizing and understanding convolutional networks. In Proceedings of the European Conference on Computer Vision, Zurich, Switzerland, 6–12 September 2014; pp. 818–833.
40. He, K.; Zhang, X.; Ren, S.; Sun, J. Deep residual learning for image recognition. In Proceedings of the IEEE Conference on Computer Vision and Pattern Recognition, Las Vegas, NV, USA, 26 June 26–1 July 2016; pp. 770–778.
41. Bengio, Y.; Simard, P.; Frasconi, P. Learning long-term dependencies with gradient descent is difficult. *IEEE Trans. Neural Netw.* **1994**, *5*, 157–166. [CrossRef] [PubMed]
42. Glorot, X.; Bengio, Y. Understanding the difficulty of training deep feedforward neural networks. In Proceedings of the Thirteenth International Conference on Artificial Intelligence and Statistics, Sardinia, Italy, 13–15 May 2010; pp. 249–256.
43. Hanif, M.S.; Bilal, M. Competitive residual neural network for image classification. *ICT Express* **2020**, *6*, 28–37. [CrossRef]

44. Zhong, Z.; Li, J.; Ma, L.; Jiang, H.; Zhao, H. Deep residual networks for hyperspectral image classification. In Proceedings of the 2017 IEEE International Geoscience and Remote Sensing Symposium (IGARSS), Fort Worth, TX, USA, 23–28 July 2017; pp. 1824–1827.

45. Paoletti, M.E.; Haut, J.M.; Fernandez-Beltran, R.; Plaza, J.; Plaza, A.J.; Pla, F. Deep pyramidal residual networks for spectral–spatial hyperspectral image classification. *IEEE Trans. Geosci. Remote Sens.* **2018**, *57*, 740–754. [CrossRef]

46. Kang, X.; Zhuo, B.; Duan, P. Dual-path network-based hyperspectral image classification. *IEEE Geosci. Remote Sens. Lett.* **2018**, *16*, 447–451. [CrossRef]

47. Aggarwal, H.K.; Majumdar, A. Hyperspectral Unmixing in the Presence of Mixed Noise Using Joint-Sparsity and Total Variation. *IEEE J. Sel. Top. Appl. Earth Obs. Remote Sens.* **2016**, *9*, 4257–4266. [CrossRef]

48. He, K.; Zhang, X.; Ren, S.; Sun, J. Identity mappings in deep residual networks. In Proceedings of the European Conference on Computer Vision, Amsterdam, The Netherlands, 8–16 October 2016; pp. 630–645.

49. Rawat, W.; Wang, Z. Deep convolutional neural networks for image classification: A comprehensive review. *Neural Comput.* **2017**, *29*, 2352–2449. [CrossRef]

50. Khan, S.H.; Rahmani, H.; Shah, S.A.A.; Bennamoun, M. *A Guide to Convolutional Neural Networks for Computer Vision*; Morgan & Claypool: San Rafael, CA, USA, 2018.

51. Kim, J.; Song, J.; Lee, J.K. Recognizing and Classifying Unknown Object in BIM Using 2D CNN. In Proceedings of the International Conference on Computer-Aided Architectural Design Futures, Daejeon, Korea, 26–28 June 2019; pp. 47–57.

52. Parmar, P.; Morris, B. HalluciNet-ing Spatiotemporal Representations Using 2D-CNN. *arXiv* **2019**, arXiv:1912.04430.

53. Chen, Y.; Zhu, L.; Ghamisi, P.; Jia, X.; Li, G.; Tang, L. Hyperspectral Images Classification with Gabor Filtering and Convolutional Neural Network. *IEEE Geosci. Remote Sens. Lett.* **2017**, *14*, 2355–2359. [CrossRef]

54. He, N.; Paoletti, M.E.; Haut, J.M.; Fang, L.; Li, S.; Plaza, A.; Plaza, J. Feature extraction with multiscale covariance maps for hyperspectral image classification. *IEEE Trans. Geosci. Remote Sens.* **2019**, *57*, 755–769. [CrossRef]

55. Fang, L.; Liu, G.; Li, S.; Ghamisi, P.; Benediktsson, J.A. Hyperspectral Image Classification With Squeeze Multibias Network. *IEEE Trans. Geosci. Remote Sens.* **2019**, *57*, 1291–1301. [CrossRef]

56. Han, X.F.; Laga, H.; Bennamoun, M. Image-based 3D Object Reconstruction: State-of-the-Art and Trends in the Deep Learning Era. *IEEE Trans. Pattern Anal. Mach. Intell.* **2019**. [CrossRef]

57. Hara, K.; Kataoka, H.; Satoh, Y. Can spatiotemporal 3d cnns retrace the history of 2d cnns and imagenet? In Proceedings of the IEEE Conference on Computer Vision and Pattern Recognition, Salt Lake City, UT, USA, 18–22 June 2018; pp. 6546–6555.

58. Gonda, F.; Wei, D.; Parag, T.; Pfister, H. Parallel separable 3D convolution for video and volumetric data understanding. *arXiv* **2018**, arXiv:1809.04096.

59. Tran, D.; Wang, H.; Torresani, L.; Ray, J.; LeCun, Y.; Paluri, M. A closer look at spatiotemporal convolutions for action recognition. In Proceedings of the IEEE Conference on Computer Vision and Pattern Recognition, Salt Lake City, UT, USA, 18–22 June 2018; pp. 6450–6459.

60. He, K.; Sun, J. Convolutional neural networks at constrained time cost. In Proceedings of the IEEE Conference on Computer Vision and Pattern Recognition, Boston, MA, USA, 7–12 June 2015; pp. 5353–5360.

61. Roy, S.K.; Krishna, G.; Dubey, S.R.; Chaudhuri, B.B. Hybridsn: Exploring 3-d-2-d cnn feature hierarchy for hyperspectral image classification. *IEEE Geosci. Remote Sens. Lett.* **2019**, *17*, 277–281. [CrossRef]

62. Ioffe, S.; Szegedy, C. Batch normalization: Accelerating deep network training by reducing internal covariate shift. In Proceedings of the 32nd International Conference on International Conference on Machine Learning, Lille, France, 6–11 July 2015; Volume 37, pp. 448–456.

63. Nair, V.; Hinton, G.E. Rectified linear units improve restricted boltzmann machines. In Proceedings of the 27th international conference on machine learning (ICML-10), Haifa, Israel, 21–24 June 2010; pp. 807–814.

64. Tran, D.; Bourdev, L.; Fergus, R.; Torresani, L.; Paluri, M. Learning spatiotemporal features with 3d convolutional networks. In Proceedings of the IEEE International Conference on Computer Vision, Santiago, Chile, 7–13 December 2015; pp. 4489–4497.

65. Baumgardner, M.F.; Biehl, L.L.; Landgrebe, D.A. 220 Band AVIRIS Hyperspectral Image Data Set: June 12, 1992 Indian Pine Test Site 3. 2015. Available online: https://doi.org/doi:10.4231/R7RX991C (accessed on 20 January 2020).

66. He, K.; Zhang, X.; Ren, S.; Sun, J. Delving deep into rectifiers: Surpassing human-level performance on imagenet classification. In Proceedings of the IEEE International Conference on Computer Vision, Santiago, Chile, 7–13 December 2015; pp. 1026–1034.

67. Kingma, D.P.; Ba, J. Adam: A method for stochastic optimization. *arXiv* **2014**, arXiv:1412.6980.

68. Singh, S.; Krishnan, S. Filter Response Normalization Layer: Eliminating Batch Dependence in the Training of Deep Neural Networks. In Proceedings of the IEEE/CVF Conference on Computer Vision and Pattern Recognition (CVPR), Seattle, WA, USA, 14–19 June 2020.

69. Lian, X.; Liu, J. Revisit Batch Normalization: New Understanding and Refinement via Composition Optimization. In Proceedings of the 22nd International Conference on Artificial Intelligence and Statistics, Okinawa, Japan, 16–18 April 2019.

Article

VddNet: Vine Disease Detection Network Based on Multispectral Images and Depth Map

Mohamed Kerkech [1,*], Adel Hafiane [1] and Raphael Canals [2]

[1] INSA-CVL, University of Orléans, PRISME, EA 4229, F18022 Bourges, France; adel.hafiane@insa-cvl.fr
[2] INSA-CVL, University of Orléans, PRISME, EA 4229, F45072 Orléans, France; raphael.canals@univ-orleans.fr
* Correspondence: mohamed.kerkech@insa-cvl.fr

Received: 15 September 2020; Accepted: 6 October 2020; Published: 11 October 2020

Abstract: Vine pathologies generate several economic and environmental problems, causing serious difficulties for the viticultural activity. The early detection of vine disease can significantly improve the control of vine diseases and avoid spread of virus or fungi. Currently, remote sensing and artificial intelligence technologies are emerging in the field of precision agriculture. They offer interesting potential for crop disease management. However, despite the advances in these technologies, particularly deep learning technologies, many problems still present considerable challenges, such as semantic segmentation of images for disease mapping. In this paper, we present a new deep learning architecture called Vine Disease Detection Network (VddNet). It is based on three parallel auto-encoders integrating different information (i.e., visible, infrared and depth). Then, the decoder reconstructs and retrieves the features, and assigns a class to each output pixel. An orthophotos registration method is also proposed to align the three types of images and enable the processing by VddNet. The proposed architecture is assessed by comparing it with the most known architectures: SegNet, U-Net, DeepLabv3+ and PSPNet. The deep learning architectures were trained on multispectral data from an unmanned aerial vehicle (UAV) and depth map information extracted from 3D processing. The results of the proposed architecture show that the VddNet architecture achieves higher scores than the baseline methods. Moreover, this study demonstrates that the proposed method has many advantages compared to methods that directly use the UAV images.

Keywords: plant disease detection; precision agriculture; UAV multispectral images; machine learning; orthophotos registration; 3D information; orthophotos segmentation

---

1. Introduction

In agricultural fields, the main causes of losing quality and yield of harvest are virus, bacteria, fungi and pest [1]. To prevent these harmful pathogens, farmers generally treat the global crop to prevent different diseases. However, using a large amount of chemicals has a negative impact on human health and ecosystems. This constitutes a significant problem to be solved; precision agriculture presents an interesting alternative.

In recent decades, the precision agriculture [2,3] has introduced many new farming methods to improve and optimize crop yields, constituting a research field in continuous evolution. New sensing technologies and algorithms have enabled the development of several applications such as water stress detection [4], vigour evaluation [5], estimation of evaporate-transpiration and harvest coefficient [6], weed localization [7,8], and disease detection [9,10].

Disease detection in vine is an important topic in precision agriculture [11–22]. The aim is to detect and treat the infected area at the right place and the right time, and with the right dose of phytosanitary products. At the early stage, it is easier to control diseases with small amounts of chemical products. Indeed, intervention before infection spreads offers many advantages, such as preservation of vine, grap production and environment, and reducing the economics losses. To achieve this goal, frequent monitoring of the parcel is necessary. Remote sensing (RS) methods are among the most widely used for that purpose and essential in the precision agriculture. RS images can be obtained at leaf- or parcel-scale. At the leaf level, images are acquired using a photo sensor, either held by a person [23] or mounted on a mobile robot [24]. At the parcel level, satellite was the standard RS imaging system [25,26]. Recently, drones or UAVs have gained popularity due to their low cost, high-resolution images, flexibility, customization and easy data access [27]. In addition, unlike satellite imaging, UAV does not have the cloud problem, which has helped to solve many remote sensing problems.

Parcels monitoring generally requires orthophotos building from geo-referenced visible and infrared UAV images. However, two separated sensors generate a spatial shift between images of the two sensors. This problem also occurred after building the orthophotos. It has been established that it is more interesting to combine the information from the two sensors to increase the efficiency of disease detection. Therefore, image registration is required.

The existent algorithms of registration rely on an approach based on either the area or feature methods. The most commonly used in the precision agriculture are feature-based methods, which are based on matching features between images [28]. In this study, we adopted the feature-based approach to align orthophotos of the visible and infrared ranges. Then, the two are combined for the disease detection procedure, where the problem consists of assigning a class-label to each pixel. For that purpose, the deep learning approach is currently the most preferred approach for solving this type of problem.

Deep learning methods [29] have achieved a high level of performance in many applications, in which different network architectures have been proposed. For instance, R-CNN [30], Siamese [31], ResNet [32], SegNet [33] are architectures used for object detection, tracking, classification and segmentation, respectively, which operate in most cases in visible ranges. However, in certain situations, the input data are not only visible images but can be combined with multispectral or hyperspectral images [34], and even depth information [35]. In these contexts, the architectures can undergo modification for improving the methods [36]. Thus, in some studies [37–40], depth information is used as input data. These data generally provide precious information about scene or environment.

Depth or height information is extracted from the 3D reconstruction or photogrammetry processing. In UAV remote sensing imagery, the photogrammetry processing can to build a digital surface model (DSM) before creating the orthophoto. The DSM model can provide much information about the parcel, such as the land variation and objects on its surface. Certain research works have shown the ability to extract vinerows by generating a depth map from the DSM model [41–43]. These solutions have been proposed to solve the vinerows misextraction resulting from the NDVI vegetation index. Indeed, in some situations, the NDVI method cannot be used to extract vinerows when the parcel has a green grassy soil. The advantage of the depth map is its ability to separate areas above-ground from the ground, even if the color is the same for all zones. To date, there has been no work on the vine disease detection that combines depth and multispectral information with a deep learning approach.

This paper presents a new system for vine disease detection using multispectral UAV images. It combines a highly accurate orthophotos registration method, a depth map extraction method and a deep learning network adapted to the vine disease detection data.

The article is organized as follows. Section 2 presents a review of related works. Section 3 describes the materials and methods used in this study. Section 4 details the experiments. Section 5 discusses the performances and limitations of the proposed method. Finally, Section 6 concludes the paper and introduces ideas to improve the method.

2. Related Work

Plant disease detection is an important issue in precision agriculture. Much research has been carried out and a large survey has been realised by Mahlein (2016) [44], Kaur et al. (2018) [45], Saleem et al. (2019) [46], Sandhu et al. (2019) [47] and Loey et al. (2020) [48]. Schor et al. (2016) [49] presented a robotic system for detecting powdery mildew and wilt virus in tomato crops. The system is based on an RGB sensor mounted on a robotic arm. Image processing and analysis were developed using the principal component analysis and the coefficient of variation algorithms. Sharif et al. (2018) [50] developed a hybrid method for disease detection and identification in citrus plants. It consists of lesion detection on the citrus fruits and leaves, followed by a classification of the citrus diseases. Ferentinos (2018) [51] and Argüeso et al. (2020) [52] built a CNN model to perform plant diagnosis and disease detection using images of plant leaves. Jothiaruna et al. (2019) [53] proposed a segmentation method for disease detection at the leaf scale using a color features and region-growing method. Pantazi et al. (2019) [54] presented an automated approach for crop disease identification on images of various leaves. The approach consists of using a local binary patterns algorithm for extracting features and performing classification into disease classes. Abdulridha et al. (2019) [55] proposed a remote sensing technique for the early detection of avocado diseases. Hu et al. (2020) [56] combined an internet of things (IoT) system with deep learning to create a solution for automatically detecting various crop diseases and communicating the diagnostic results to farmers.

Disease detection in vineyards has been increasingly studied in recent years [11–22]. Some works are realised at the leaf scale, and others at the crop scale. MacDonald et al. (2016) [11] used a Geographic Information System (GIS) software and multispectral images for detecting the leafroll-associated virus in vine. Junges et al. (2018) [12] investigated vine leaves affected by the esca in hyperspectral ranges and di Gennaro et al. (2016) [13] worked at the crop level (UAV images). Both studies concluded that the reflectance of healthy and diseased leaves is different. Albetis et al. (2017) [14] studied the Flavescence dorée detection in UAV images. The results obtained showed that the vine disease detection using aerial images is feasible. The second study of Albetis et al. (2019) [15] examined of the UAV multispectral imagery potential in the detection of symptomatic and asymptomatic vines. Al-Saddik has conducted three studies on vine disease detection using hyperspectral images at the leaf scale. The aim of the first one (Al-Saddik et al. 2017) [16] was to develop spectral disease indices able to detect and identify the Flavescence dorée on grape leaves. The second one (Al-Saddik et al. 2018) [17] was performed to differentiate yellowing leaves from leaves diseased by esca through classification. The third one (Al-saddik et al., 2019) [18] consisted of determining the best wavelengths for the detection of the Flavescence dorée disease. Rançon et al. (2019) [19] conducted a similar study for detecting esca disease. Image sensors were embedded on a mobile robot. The robot moved along the vinerows to acquire images. To detect esca disease, two methods were used: the scale Invariant Feature Transform (SIFT) algorithm and the MobileNet architecture. The authors concluded that the MobileNet architecture provided a better score than the SIFT algorithm. In the framework of previous works, we have realized three studies on vine disease detection using UAV images. The first one (Kerkech et al. 2018) [20] was devoted to esca disease detection in the visible range using the LeNet5 architecture combined with some color spaces and vegetation indices. In the second study (Kerkech et al. 2019) [21], we used near-infrared images and visible images. Disease detection was considered as a semantic segmentation problem performed by the SegNet architecture. Two parallel SegNets were applied for each imaging modality and the results obtained were merged to generate a disease map. In (Kerkech et al. 2020) [22], a correction process using a depth map was added to the output of the previous method. Post-processing with these depth information demonstrated the advantage of this approach in reducing detection errors.

3. Materials and Methods

This section presents the materials and each component of the vine disease detection system. Figure 1 provides an overview of the methods. It includes the following steps: data acquisition, orthophotos registration, depth map building and orthophotos segmentation (disease map generation). The next sections detail these different steps.

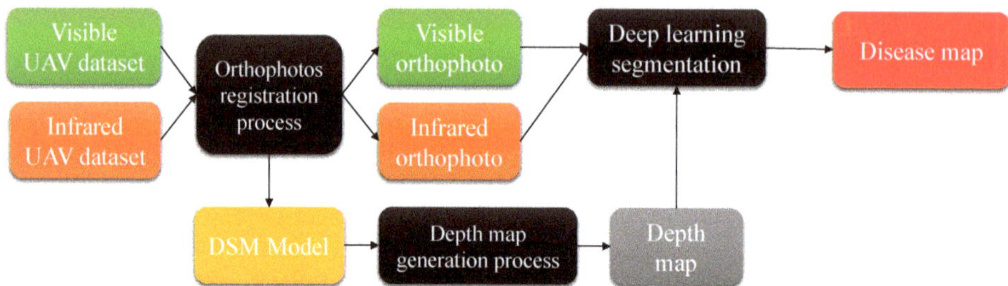

Figure 1. The proposed vine disease detection system.

3.1. Data Acquisition

Multispectral images are acquired using a quadricopter UAV that embeds a MAPIR Survey2 camera and a Global Navigation Satellite System (GNSS) module. This camera integrates two sensors in the visible and infrared ranges with a resolution of 16 megapixels (4608 × 3456 pixels). The visible sensor captures the red, green, and blue (RGB) channels and the infrared sensor captures the red, green, and near-infrared (R-G-NIR) channels. The wavelength of the near-infrared channel is 850 nm. The accuracy of the GNSS module is approximately 1 m.

The acquisition protocol consists of a drone flying over vines at an altitude of 25 m and at an average speed of 10 km/h. During flights, the sensors acquire an image every 2 s. Each image has a 70% overlap with the previous and the next ones. Each point of the vineyard has six different viewpoints (can be observed on six different images). The flight system is managed by a GNSS module. The flight plans include topographic monitoring aimed at guaranteeing a constant distance from the sol. Images are recorded with their GNSS position. Flights are performed at the zenith to avoid shadows, and with moderate weather conditions (light wind and no rain) to avoid UAV flight problems.

3.2. Orthophotos Registration

The multispectral acquisition protocol using two sensors causes a shift between visible and infrared images. Hence, a shift in multispectral images automatically implies a shift in orthophotos. Usually, the orthophotos registration is performed manually using the QGIS software. The manual method is time-consuming, requires a high focus to select many key points between visible and infrared orthophotos, and the result is not very accurate. To overcome this problem, a new method for automatic and accurate orthophotos registration is proposed.

The proposed orthophotos registration method is illustrated in Figure 2 and is divided into two steps. The first one concerns the UAV multispectral images registration and the second permits the building of registered multispectral orthophotos. In this study, the first step uses the optimized multispectral images registration method proposed in [21]. Based on the Accelerated-KAZE (AKAZE) algorithm, the registration method uses feature-matching between visible and infrared images to match key points extracted from the two images and compute the homographic matrix for geometric correction. In order to increase accuracy, the method uses an iterative process to reduce the Root Mean Squared Error (RMSE) of the registration. The second step consists of using the Agisoft Metashape software to build and obtain the registered visible and infrared orthophotos. The Metashape software is based on the Structure from motion (SfM) algorithm for the photogrammetry processing. Building orthophotos requires the UAV images and the digital surface model (DSM). To obtain this DSM model, the software must go through a photogrammetry processing and perform the following steps: alignment of the images to build a sparse point cloud, then a dense point cloud, and finally the DSM. The orthophotos building is carried out by the option "build orthomosaic" process in the software.

To build the visible orthophotos, it is necessary to use the visible UAV images and the DSM model, while, to build a registered infrared orthophoto, it is necessary to use the registered infrared UAV images and the same DSM model of the visible orthophoto. The parameters used in the Metashape software are detailed in Table 1.

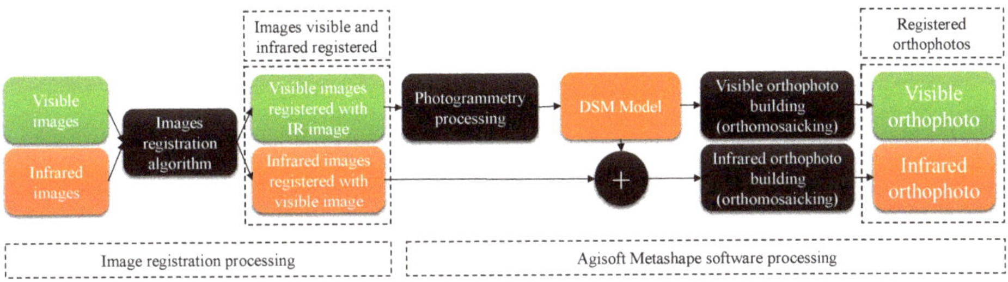

Figure 2. The proposed orthophotos registration method.

3.3. Depth Map

The DSM model previously built in the orthophotos registration process is used here to obtain the depth map. In fact, the DSM model represents the terrain surface variation and includes all objects found here (in this case, objects are vine trees). Therefore, some processings are required to determine only the vine height. To extract the depth map from the DSM, the method proposed in [41] is used. It consists of applying the following processings: the DSM is first filtered using a low-pass filter of size 20×20; this filter is chosen to smooth the image and keep only the terrain surface variations, also called the digital terrain model (DTM). The DTM is thereafter subtracted from the DSM to eliminate the terrain variations and retain only the vine height. Due to the weak contrast of the result, an enhancement processing was necessary. The contrast is enhanced here by using a histogram-based (histogram normalization) method. The obtained result is an image with a good difference in grey levels between vines and non-vines. Once the contrast is corrected, an automatic thresholding using the Otsu's algorithm is applied to obtain a binary image representing the depth map.

3.4. Segmentation and Classification

The last stage of the vine disease detection system concerns the data classification. This step is performed using a deep learning architecture for segmentation. Deep learning has proven its performance in numerous research studies and in various domains. Many architectures have been developed, such as SegNet [33], U-Net [57], DeepLabv3+ [58], and PSPNet [59]. Each architecture can provide good results in a specific domain and be less efficient in others. These architectures are generally used for the segmentation of complex indoor/outdoor scenes, medical ultrasound images, or even in agriculture. One channel is generally used for greyscale medical imaging or three channels for visible RGB color images. Hence, they are not always adapted to a specific problem. Indeed, for this study, multispectral and depth map data offer additional information. This can improve the segmentation representation and the final disease map result. For this purpose, we have designed our deep learning architecture adapted to the vine disease detection problem, and we have compared it to the most well known deep learning architectures. In the following sections, we describe the proposed deep learning architecture and the training process.

3.4.1. VddNet Architecture

Vine Disease Detection Network (VddNet), Figure 3 is inspired by VGG-Net [60], SegNet [33], U-Net [57] and the parallel architectures proposed in [37,61–63]. VddNet is a parallel architecture based on the VGG encoder; it has three types of data as inputs: visible a RGB image, a near-infrared image and a depth map. VddNet is dedicated to segmentation, so the output has the same input, with a number of channels equal to the number of classes (4). It is designed with three parallel encoders and one decoder. Each encoder can typically be considered as a convolutional neural network without the fully connected layers. The convolutional operation is repeated twice using a 3×3 mask, a rectified linear unit (ReLU), a batch normalization and a subsampling using a max pooling function of 2×2 size and a stride of 2. The number of feature map channels is doubled at each subsampling step. The idea of VddNet is to encode each type of data separately and, at the same time, concatenate the near-infrared and the features map of the depth map with the visible features map before each subsampling. Hence, the central encoder preserves the features of the near-infrared and the depth map data merged with the visible features map, and concatenated at the same time. The decoder phase consists of upsampling and convolution with a 2×2 mask. It is then followed by two convolution layers with a 3×3 mask, a rectified linear unit, and a batch normalization. In contrast to the encoder phase, after each upsampling operation, the number of features map channels is halved. Using the features map concatenation technique of near-infrared and depth map, the decoder retrieves features lost during the merging and the subsample process. The decoder follows the same steps until it reaches the final layer, which is a convolution with a 1×1 mask and a softmax providing classes probabilities, at pixel-wise.

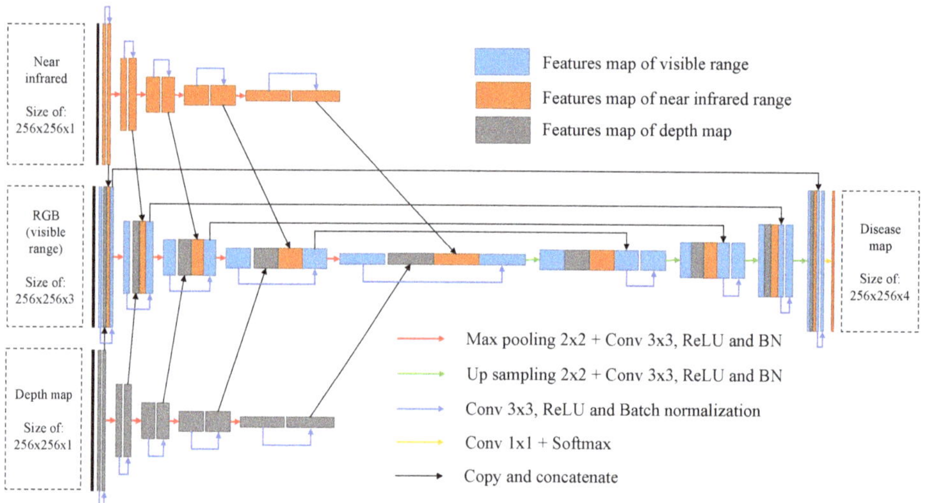

Figure 3. VddNet architecture.

3.4.2. Training Dataset

In this study, one crop is used for model training and validation, and two crops for testing. To build the training dataset, four steps are required: data source selection, classes definition, data labelling, and data augmentation.

The first step is probably the most important one. Indeed, to allow a good learning, the data source for feeding models must represent the global data in terms of richness, diversity and classes. In this study, a particular area was chosen that contains a slight shadow area, brown ground (soil) and a vine partially affected by mildew.

Once the data source has been selected, it is necessary to define the different classes present in these data. For that purpose, each type of data (visible, near-infrared and depth map) is important in this step. In visible and near-infrared images, four classes can be distinguished. On the other hand, the depth map contains only two distinct classes, which are the vine canopy and the non-vine. Therefore, the choice of classes must match all data types. Shadow is the first class; it is any dark zone. It can be either on the vine or on the ground. This class was created to avoid confusion and misclassification on a non-visible pattern. Ground is the second class; from one parcel to another, ground is generally different. Indeed, the ground can have many colors: brown, green, grey, etc. To solve this color confusion, the ground is chosen as any pixels in the non-vine zone from the depth map data. Healthy vine is the third class; it is the green leaves of the vine. Usually, it is easy to classify these data, but when ground is also green, this leads to confusion between vine and ground in 2D images. To avoid that, the healthy class is defined as the green color in the visible spectrum and belonging to the vine canopy according to the depth map. The fourth and last class corresponds to diseased vine. Disease symptoms can present several colors in the visible range: yellow, brown, red, golden, etc. In the near-infrared, it is only possible to differentiate between healthy and diseased reflectances. In general, diseased leaves have a different reflectance than healthy leaves [17], but some confusion between disease and ground classes may occur when the two colors are similar. Ground must also be eliminated from the disease class using the depth map.

Data labelling was performed with the semi-automatic labelling method proposed in [21]. The method consists of using automatic labelling in a first step, followed by manual labelling in a second step. The first step is based on the deep learning LeNet-5 [64] architecture, where the classification is carried out using a 32×32 sliding window and a 2×2 stride. The result is equivalent to a coarse image segmentation which contains some misclassifications. To refine the segmentation, output results were manually corrected using the Paint.Net software. This task was conducted based on the ground truth (realized in the crop by a professional reporting occurred diseases), and observations in the orthophotos.

The last stage is the generation of a training dataset from the labelled data. In order to enrich the training dataset and avoid an overfitting of networks, data augmentation methods [65] are used in this study. A 256×256 pixels patches dataset is generated from the data source matrix and its corresponding labelled matrix. The data source consists of multimodal and depth map data and has a size of $4626 \times 3904 \times 5$. Four data augmentation methods are used: translation, rotation, under and oversampling, and brightness variation. Translation was performed with an overlap of 50% using a sliding window in the horizontal and vertical displacements. The rotation angle was set at $30°$, $60°$ and $90°$. Under- and oversampling were parametrized to obtain 80% and 120% of the original data size. Brightness variation is only applied to multispectral data. Pixel values are multiplied by the coefficients of 0.95 and 1.05, which introduce a brightness variation of $\pm5\%$. Each method brings an effect on the data (translation, rotation, etc.), allowing the networks to learn, respectively, transition, vinerows orientations, acquisition scale variation and weather conditions. At the end, the data augmentation generated 35.820 patches.

4. Experimentations and Results

This section presents the different experimental devices, as well as qualitative and quantitative results. The experiments are performed on Python 2.7 software, using the Keras 2.2.0 library for the development of deep learning architectures, and GDAL 3.0.3 for the orthophotos management. The Agisoft Metashape software version 1.6.2 is also used for photogrammetry processing. The codes were developed under the Linux Ubuntu 16.04 LTS 64-bits operating system and run on a hardware with an Intel Xeon 3.60 GHz $\times$ 8 processor, 32 GB RAM, and a NVidia GTX 1080 Ti graphics card with 11 GB of internal RAM. The cuDNN 7.0 library and the CUDA 9.0 Toolkit are used for deep learning processing on GPU.

4.1. Orthophotos Registration and Depth Map Building

To realize this study, multispectral and depth map orthophotos were required. Two parcels were selected and data were aquired at two different times to construct the orthophotos dataset. Each parcel had one or more of the following characteristics: with or without shadow, green or brown ground, healthy or partially diseased. Registered visible and infrared orthophotos were built from multispectral images using the optimized image registration algorithm [21] and the Agisoft Metashape software version 1.6.2. Orthophotos were saved in the geo-referenced file format "TIFF". The parameters used in the Metashape software are listed in Table 1.

To evaluate the registration and depth map quality, we chose a chessboard test pattern. Figure 4 presents an example of visible and infrared orthophotos registration. As can be seen, the alignment between the two orthophotos is accurate. The registration of the depth map with the visible range also provides good results (Figure 6).

Table 1. The parameters used for the orthophotos building process in the Agisoft Metashape software.

Sparse point cloud	
Accuracy :	High
Image pair selection :	Ground control
Constrain features by mask :	No
Maximum number of feature points :	40,000
Dense point cloud	
Quality :	High
Depth filtering :	Disabled
Digital Surface Model	
Type :	Geographic
Coordinate system :	WGS 84 (EPSG::4326)
Source data :	Dense cloud
Orthomosaic	
Surface :	DSM
Blending mode :	Mosaic

4.2. Training and Testing Architectures

In order to determine the best parameters for each deep learning architecture, four cross-optimizers with two loss functions were compared. Architectures were compiled using either the loss function "cross entropy" or "mean squared error", and with one among the four optimizers: SGD [66], Adadelta [67], Adam [68], or Adamax [69]. Once the best parameters were defined for each architecture, a final fine-tuning was performed on the "learning rate" parameter to obtain the best results (to achieve a good model without overfitting). The best parameters found for each architecture are presented in Table 2.

Table 2. The parameters used for the different deep learning architectures. LR means learning rate.

Network	Base Model	Optimizer	Loss Function	LR	Learning Ate Decrease Parameters
SegNet	VGG-16	Adadelta	Categorical cross entropy	1.0	rho = 0.95, epsilon = 1×10^{-7}
U-Net	VGG-11	SGD	Categorical cross entropy	0.1	decay = 1×10^{-6}, momentum = 0.9
PSP-Net	ResNet-50	Adam	Categorical cross entropy	0.001	beta1 = 0.9, beta2 = 0.999, epsilon = 1×10^{-7}
DeepLabv3+	Xception	Adam	Categorical cross entropy	0.001	beta1 = 0.9, beta2 = 0.999, epsilon = 1×10^{-7}
VddNet	Parallel VGG-13	SGD	Categorical cross entropy	0.1	decay = 1×10^{-6}, momentum = 0.9

For training the VddNet model, data from visible, near-infrared and depth maps were incorporated separately in the network inputs. For the other architectures, a multi-data matrix consists of five channels with a size of 256×256. The first three channels correspond to the visible spectrum, the 4th channel to the near-infrared data and the 5th channel to the depth map. Each multi-data matrix has a corresponding labelled matrix. Models training is an iterative process that is fixed at 30.000 epochs for each model. For each iteration, a batch of five multi-data matrices with their corresponding labelled matrices are randomly selected from the dataset and sent to feed the model. In order to check the convergence of the model, a test using validation data is performed each 10 iterations.

A qualitative study was conducted for determining the importance of depth map information. For this purpose, an experience was conducted by training the deep learning models with only multispectral data and with a combination of both (multispectral and depth maps). The comparison results are shown in Figures 7 and 8.

To test the deep learning models, test areas are segmented using a 256×256 sliding window (without overlap). For each position of the sliding window, the visible, near-infrared and depth maps are sent to the network inputs (respecting the data order for each architecture) in order to perform segmentation. The output of the networks is a matrix of size of $256 \times 256 \times 4$. The results are saved after an application of the Argmax function. They are then stitched together to obtain the original size of the orthophoto tested data.

4.3. Segmentation Performance Measurements

Segmentation performance measurements are expressed in terms of recall, precision, F1-Score/Dice and accuracy (using Equations (1)–(5)) for each class (shadow, ground, healthy and diseased) at grapevine-scale. Grapevine-scale assessment was chosen because pixel-wise evaluation is not suitable for providing disease information. Moreover, imprecision of the ground truth, small surface of the disease and difference of deep learning segmentation results do not allow for a good evaluation of the different architectures, at pixel-wise. These measurements use a sliding window equivalent to the average size of a grapevine (in this study, approximatively 64×64 pixels). For each step of the sliding window, the class evaluated is the dominant class in the ground truth. The window is considered "true positive" if the dominant class is the same as the ground truth, otherwise it is a "false positive". The confusion matrix is updated for each step. Finally, the score is given by

$$Recall = \frac{TP}{TP+FN} \tag{1}$$

$$Precision = \frac{TP}{TP+FP} \tag{2}$$

$$F1\text{-}Score = 2\frac{Recall \times Precision}{Recall + Precision} = \frac{2TP}{FP+2TP+FN} \tag{3}$$

$$Dice = \frac{2|X \cap Y|}{|X|+|Y|} = \frac{2(TP)}{(FP+TP)+(TP+FN)} = \frac{2TP}{FP+2TP+FN} \tag{4}$$

$$Accuracy = \frac{TP+TN}{TP+TN+FP+FN} \tag{5}$$

where TP, TN, FP and FN are the number of samples for "true positive", "true negative", "false positive" and "false negative", respectively. Dice equation is defined by X (set of ground truth pixels) and Y (set of the classified pixels).

5. Discussion

To validate the proposed vine disease detection system, it is necessary to evaluate and compare qualitative and quantitative results for each block of the whole system. For this purpose, several experiments were conducted at each step of the disease detection procedure. The first experience was carried out on the multimodal orthophotos registration. Figure 4 shows the obtained results. As can be seen, the continuity of the vinerows is highly accurate and the continuity is respected between the visible and infrared ranges. However, if image acquisition is incorrectly conducted, this results in many registration errors. To avoid these problems, two rules must be followed. The first one regards the overlapping between visible and infrared images acquired in the same position, which must be greater than 85%. The second rule is that the overlapping between each acquired image must be greater than 70%; this rule must be respected in both ranges. Non-compliance with the first rule affects the building of the registered infrared orthophoto. Indeed, this latter may present some black holes (this means that there are no data available to complete these holes). Non-compliance with the second rule affects the photogrammetry processing and the DSM model. This can lead to deformations in the orthophoto patterns (as can be seen on the left side of the visible and infrared orthophotos in Figure 5). In case the DSM model is impacted, the depth map automatically undergoes the same deformation (as can be seen in the depth map in Figure 5). The second quality evaluation is the building of the depth map (Figure 6). Despite the slight deformation in the left side of the parcel, the result of the depth map is consistent and well aligned with the visible orthophotos, and can be used in the segmentation process.

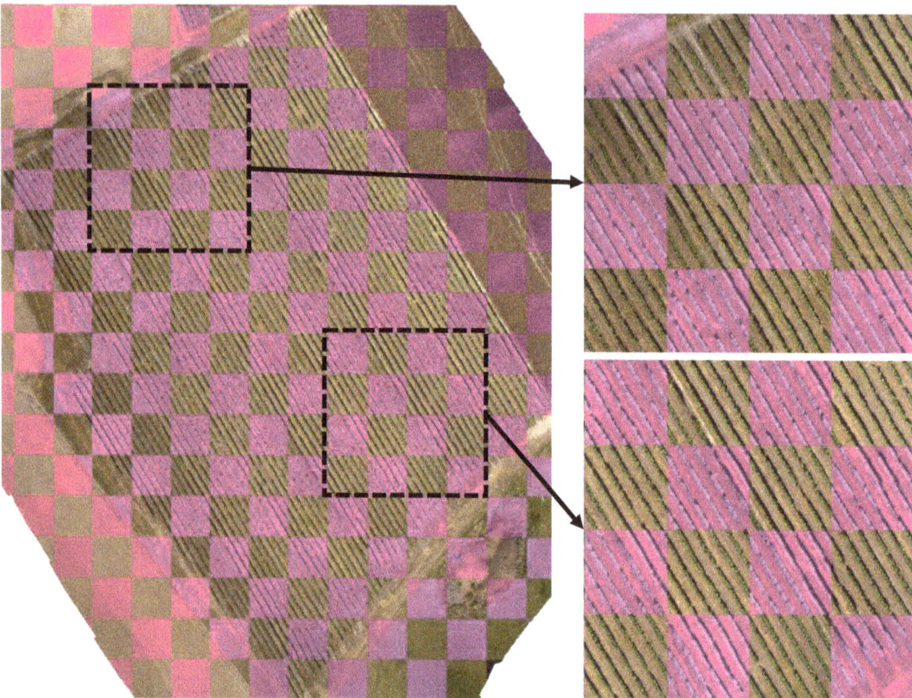

Figure 4. Qualitative results of orthophotos registration using a chessboard pattern.

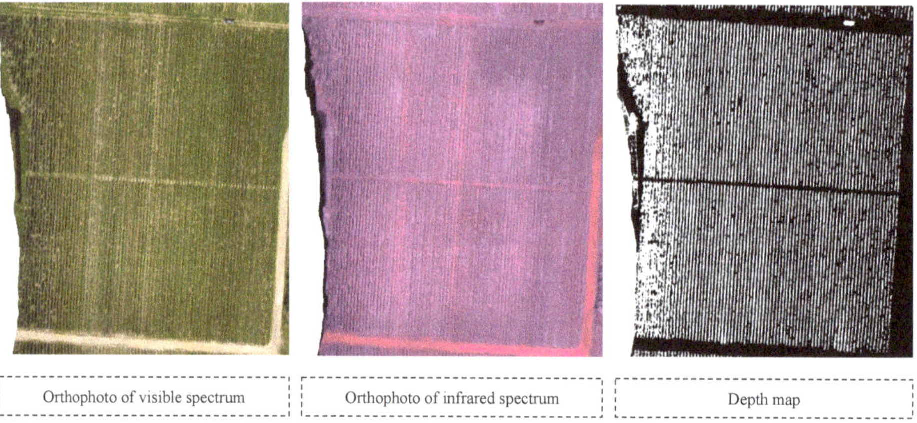

Figure 5. Qualitative results of orthophotos and depth map.

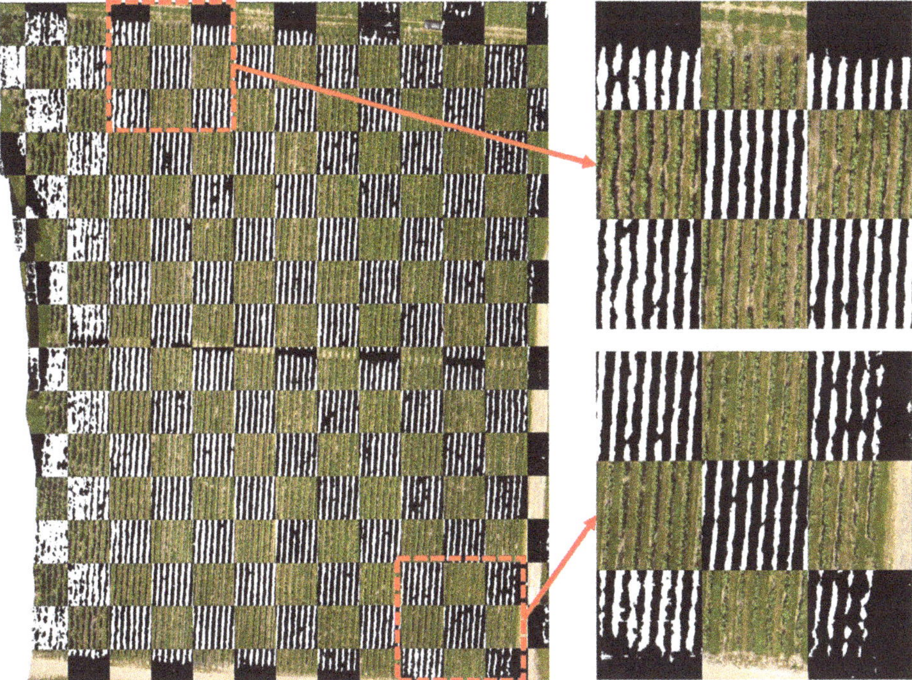

Figure 6. Evaluation of the depth map alignment using a chessboard pattern.

In order to assess the added value of depth map information, two training sessions were performed on the SegNet [33], U-Net [57], DeepLabv3+ [58] and PSPNet [59] networks. The first training session was conducted only on multispectral data, and the second one on multispectral data combined with depth map information. Figures 7 and 8 illustrate the qualitative test results of the comparison between the two trainings. The left side of Figure 7 shows an example of a parcel with a green ground. The center of the figure presents the segmentation result of the SegNet model trained only on multispectral data. As can be seen, in some areas of the parcel, it is difficult to dissociate vinerows. The right side of the figure depicts the segmentation result of the SegNet model trained on multispectral data combined with depth map information. This result is better than the previous one and it can easily separate vinerows. This is due to additional depth map information that allows a better learning of the scene environment and distinction between classes. Figure 7 illustrates other examples realised under the same conditions as above. On the first row, we observe an area composed of green ground. The segmentation results using the first and second models are displayed in the centre and on the right side, respectively. We can notice in this example a huge confusion between ground and healthy vine classes. This is mainly due to the fact that the ground color is similar to the healthy vine one. This problem has been solved by adding depth map information in the second model, the result of which is shown on the right side. The second row of Figure 8 presents an example of a partially diseased area. The first segmentation result reveals the detection of the disease class on the ground. The brown color (original ground color) merged with a slight green color (grass color) on the ground confused the first model and led it to misclassifying the ground. This confusion does not exist in the second segmentation result (right side). From these results, it can be concluded that the second model learned that the diseased vine class could not be detected on "no-vine" when this one was trained on multispectral and depth map information. Based on these results, the following experiments were conducted using multispectral data and the depth map information.

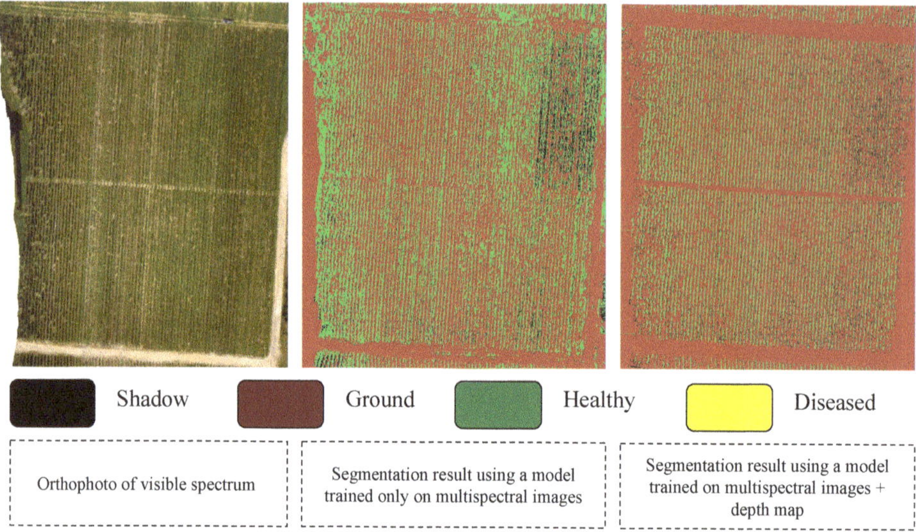

Figure 7. Difference between a SegNet model trained only on multispectral data and the same trained on multispectral data combined with depth map information. The presented example is on an orthophoto of a healthy parcel with a green ground.

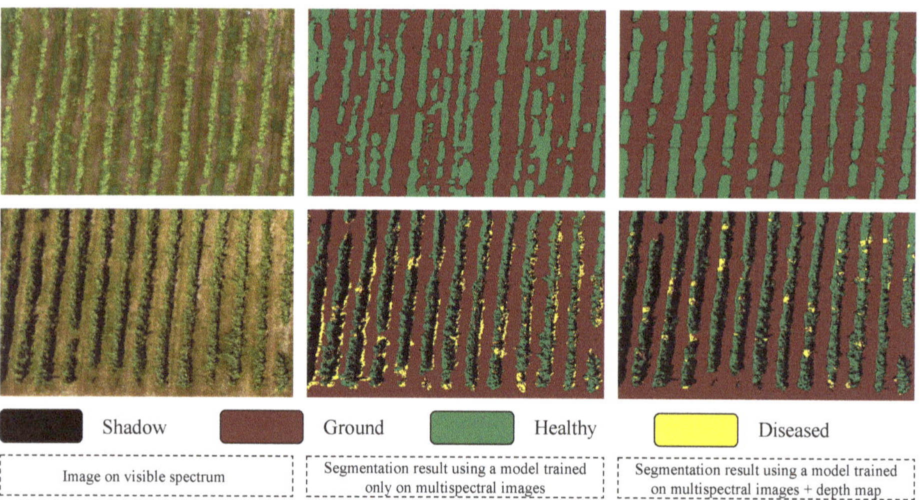

Figure 8. Difference between a SegNet model trained only on multispectral data and the same trained on multispectral data combined with depth map information. Two examples are presented here, the first row is an example on a healthy parcel with a green ground. The second one is an example on a partially diseased parcel with a brown ground.

In order to validate the proposed architecture, a comparative study was conducted on the most well-known deep learning architectures, SegNet [33], U-Net [57], DeepLabv3+ [58] and PSPNet [59]. All architectures were trained and tested on the following classes: shadow, ground, healthy and diseased, with the same data (same training and test). Table 3 lists the segmentation results of the different architectures. The quantitative evaluations are based on the F1-score and the global accuracy.

As can be seen, the shadow and ground classes have obtained an average scores of 94% and 95%, respectively, with all architectures. The high scores are due to the easy detection of these classes. The healthy class scored between 91% and 92% for VddNet, SegNet, U-Net and DeepLabv3+. However, PSPNet obtained the worst result of 73.96%, due to a strong confusion between the ground and healthy classes. PSPNet was unable to generate a good segmentation model although the training dataset was rich. The diseased vine class is the most important class in this study. VddNet obtained the best result for this class with a score of 92.59%, followed by SegNet with a score of 88.85%. The scores of the other architectures are 85.78%, 81.63% and 74.87% for U-Net, PSPNet and DeepLabv3+, respectively. VddNet achieved the best result because the feature extraction was performed separately. Indeed, in [21] it was proven that merging visible and infrared segmentations (with two separate trained models) provides a better detection than visible or infrared separately. The worst result of the diseased class was obtained with DeepLabv3+; this is due to a insensibility in the color variation. In fact, the diseased class can correspond to the yellow, brown or golden color and these colors are usually between the green color of healthy neighbour leaves. This situation led classifiers to be insensitive to this variation. The best global segmentation accuracy was achieved by VddNet, with an accuracy of 93.72%. This score can be observed on the qualitative results of Figures 9 and 10. Figure 9 presents an orthophoto of a parcel (on the left side) partially contaminated with mildew. The right side shows the segmentation result by VddNet. It can be seen that it correctly detects the diseased areas. Figure 10 is an example of parcel without disease; here, VddNet also performs well in detecting true negatives.

Figure 9. Qualitative result of VddNet on a parcel partially contaminated with mildew and with green ground. The visible orthophoto of the healthy parcel is in the left side, and its disease map in the right side.

Table 3. Quantitative results with measurement of recall (Rec.), precision (Pre.), F1-Score/Dice (F1/D.) and accuracy (Acc.) for the performances of VddNet, SegNet, U-Net, DeepLabv3+ and PSPNet networks, using multispectral and depth map data. Values are presented as a percentage.

Class name	Shadow			Ground			Healthy			Diseased			Total
Measure	Rec.	Pre.	F1/D.	Rec.	Pre.	F1/D.	Rec.	Pre.	F1/D.	Rec.	Pre.	F1/D.	Acc.
VddNet	94.88	94.89	94.88	94.84	95.11	94.97	87.96	94.84	91.27	90.13	95.19	**92.59**	**93.72**
SegNet	94.97	94.60	94.79	95.16	94.99	**95.07**	90.14	94.81	**92.42**	83.45	95.00	88.85	92.75
U-Net	95.09	94.70	**94.90**	94.99	95.07	95.03	89.09	94.74	91.83	78.27	94.90	85.78	90.69
DeepLabV3+	94.90	94.68	94.79	95.21	94.90	95.06	88.78	95.16	91.86	61.78	94.98	74.87	88.58
PSPNet	95.07	94.25	94.66	94.94	87.29	90.95	60.54	95.04	73.96	71.70	94.75	81.63	84.63

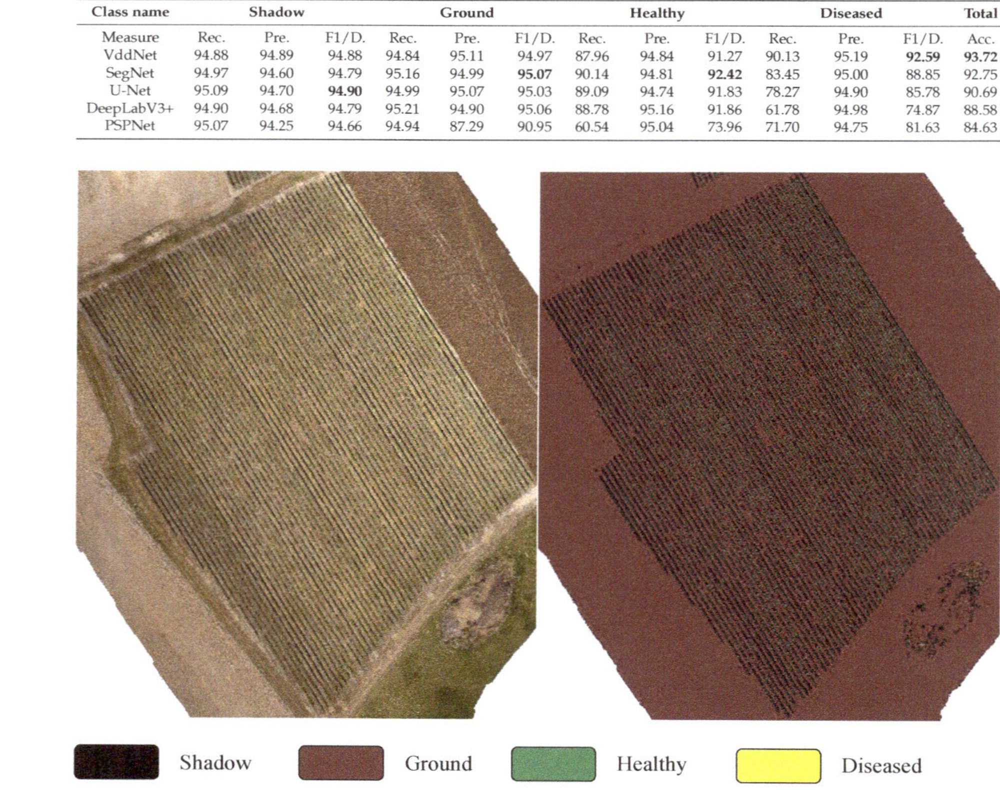

Figure 10. Qualitative result of VddNet on a healthy parcel with brown ground. The visible orthophoto of the healthy parcel is in the left side, and its disease map in the right side.

6. Conclusions

The main goal of this study is to propose a new method that improves vine disease detection in UAV images. A new deep learning architecture for vine disease detection (VddNet), and automatic multispectral orthophotos registration have been proposed. UAV images in the visible and near-infrared spectra are the input data of the detection system for generating a disease map. UAV input images were aligned using an optimized multispectral registration algorithm. Aligned images were then used in the process of building registered orthophotos. During this process, a digital surface model (DSM) was generated to built a depth map. At the end, VddNet generated the disease map from visible, near-infrared and depth map data. The proposed method brought many benefits to the whole process. The automatic multispectral orthophotos registration provides high precision and fast processing compared to conventional procedures. A 3D processing enables the building of the depth map, which is relevant for the VddNet training and segmentation process. Depth map data reduce misclassification and confusion between close color classes. VddNet improves disease detection and global segmentation compared to the state-of-the-art architectures. Moreover, orthophotos are georeferenced with GNSS coordinates, making it easier to locate diseased vines for traitment. In future work, it would be interesting to acquire new multispectral channels to enhance disease detection and improve the VddNet architecture.

Author Contributions: M.K. and A.H. conceived and designed the method; M.K. implemented the method and performed the experiments; M.K., A.H. and R.C. discussed the results and revised the manuscript. All authors have read and agreed to the published version of the manuscript.

Funding: This research received no external funding.

Acknowledgments: This work is part of the VINODRONE project supported by the Region Centre-Val de Loire (France). We gratefully acknowledge Region Centre-Val de Loire for its support.

Conflicts of Interest: The authors declare no conflict of interest.

References

1. Oerke, E.C. Crop losses to pests. *J. Agric. Sci.* **2006**, *144*, 31–43. [CrossRef]
2. Patrício, D.I.; Rieder, R. Computer vision and artificial intelligence in precision agriculture for grain crops: A ystematic review. *Comput. Electron. Agric.* **2018**, *153*, 69–81. [CrossRef]
3. Mogili, U.R.; Deepak, B.B. Review on Application of Drone Systems in Precision Agriculture. *Procedia Comput. Sci.* **2018**, *133*, 502–509. [CrossRef]
4. Bellvert, J.; Zarco-Tejada, P.J.; Girona, J.; Fereres, E. Mapping crop water stress index in a 'Pinot-noir' vineyard: Comparing ground measurements with thermal remote sensing imagery from an unmanned aerial vehicle. *Precis. Agric.* **2014**, *15*, 361–376. [CrossRef]
5. Mathews, A.J. Object-based spatiotemporal analysis of vine canopy vigor using an inexpensive unmanned aerial vehicle remote sensing system. *J. Appl. Remote Sens.* **2014**, *8*, 085199. [CrossRef]
6. Vanino, S.; Pulighe, G.; Nino, P.; de Michele, C.; Bolognesi, S.F.; D'Urso, G. Estimation of evapotranspiration and crop coefficients of tendone vineyards using multi-sensor remote sensing data in a mediterranean environment. *Remote Sens.* **2015**, *7*, 14708–14730. [CrossRef]
7. Bah, M.D.; Hafiane, A.; Canals, R. CRowNet: Deep Network for Crop Row Detection in UAV Images. *IEEE Access* **2020**, *8*, 5189–5200. [CrossRef]
8. Dian Bah, M.; Hafiane, A.; Canals, R. Deep learning with unsupervised data labeling for weed detection in line crops in UAV images. *Remote Sens.* **2018**, *10*, 1690. [CrossRef]
9. Tichkule, S.K.; Gawali, D.H. Plant diseases detection using image processing techniques. In Proceedings of the 2016 Online International Conference on Green Engineering and Technologies (IC-GET 2016), Coimbatore, India, 19 November 2016; pp. 1–6. [CrossRef]
10. Pinto, L.S.; Ray, A.; Reddy, M.U.; Perumal, P.; Aishwarya, P. Crop disease classification using texture analysis. In Proceedings of the 2016 IEEE International Conference on Recent Trends in Electronics, Information and Communication Technology, RTEICT 2016—Proceedings, Bangalore, India, 20–21 May 2016; pp. 825–828. [CrossRef]

11. MacDonald, S.L.; Staid, M.; Staid, M.; Cooper, M.L. Remote hyperspectral imaging of grapevine leafroll-associated virus 3 in cabernet sauvignon vineyards. *Comput. Electron. Agric.* **2016**, *130*, 109–117. [CrossRef]

12. Junges, A.H.; Ducati, J.R.; Scalvi Lampugnani, C.; Almança, M.A.K. Detection of grapevine leaf stripe disease symptoms by hyperspectral sensor. *Phytopathol. Mediterr.* **2018**, *57*, 399–406. [CrossRef]

13. di Gennaro, S.F.; Battiston, E.; di Marco, S.; Facini, O.; Matese, A.; Nocentini, M.; Palliotti, A.; Mugnai, L. Unmanned Aerial Vehicle (UAV)-based remote sensing to monitor grapevine leaf stripe disease within a vineyard affected by esca complex. *Phytopathol. Mediterr.* **2016**, *55*, 262–275. [CrossRef]

14. Albetis, J.; Duthoit, S.; Guttler, F.; Jacquin, A.; Goulard, M.; Poilvé, H.; Féret, J.B.; Dedieu, G. Detection of Flavescence dorée grapevine disease using Unmanned Aerial Vehicle (UAV) multispectral imagery. *Remote Sens.* **2017**, *9*, 308. [CrossRef]

15. Albetis, J.; Jacquin, A.; Goulard, M.; Poilvé, H.; Rousseau, J.; Clenet, H.; Dedieu, G.; Duthoit, S. On the potentiality of UAV multispectral imagery to detect Flavescence dorée and Grapevine Trunk Diseases. *Remote Sens.* **2019**, *11*, 23. [CrossRef]

16. Al-Saddik, H.; Simon, J.; Brousse, O.; Cointault, F. Multispectral band selection for imaging sensor design for vineyard disease detection: Case of Flavescence Dorée. *Adv. Anim. Biosci.* **2017**, *8*, 150–155. [CrossRef]

17. Al-Saddik, H.; Laybros, A.; Billiot, B.; Cointault, F. Using image texture and spectral reflectance analysis to detect Yellowness and Esca in grapevines at leaf-level. *Remote Sens.* **2018**, *10*, 618. [CrossRef]

18. Al-saddik, H. Assessment of the optimal spectral bands for designing a sensor for vineyard disease detection: The case of ' Flavescence dorée '. *Precis. Agric.* **2018**. [CrossRef]

19. Rançon, F.; Bombrun, L.; Keresztes, B.; Germain, C. Comparison of SIFT encoded and deep learning features for the classification and detection of esca disease in Bordeaux vineyards. *Remote Sens.* **2019**, *11*, 1. [CrossRef]

20. Kerkech, M.; Hafiane, A.; Canals, R. Deep learning approach with colorimetric spaces and vegetation indices for vine diseases detection in UAV images. *Comput. Electron. Agric.* **2018**, *155*, 237–243. [CrossRef]

21. Kerkech, M.; Hafiane, A.; Canals, R. Vine disease detection in UAV multispectral images using optimized image registration and deep learning segmentation approach. *Comput. Electron. Agric.* **2020**, *174*, 105446. [CrossRef]

22. Kerkech, M.; Hafiane, A.; Canals, R.; Ros, F. *Vine Disease Detection by Deep Learning Method Combined with 3D Depth Information*; Lecture Notes in Computer Science; Springer International Publishing: Cham, Switzerland, 2020; Volume 12119, pp. 1–9. [CrossRef]

23. Singh, V.; Misra, A.K. Detection of plant leaf diseases using image segmentation and soft computing techniques. *Inf. Process. Agric.* **2017**, *4*, 41–49. [CrossRef]

24. Pilli, S.K.; Nallathambi, B.; George, S.J.; Diwanji, V. EAGROBOT—A robot for early crop disease detection using image processing. In Proceedings of the 2nd International Conference on Electronics and Communication Systems (ICECS 2015), Coimbatore, India, 26–27 February 2015; pp. 1684–1689. [CrossRef]

25. Abbas, M.; Saleem, S.; Subhan, F.; Bais, A. Feature points-based image registration between satellite imagery and aerial images of agricultural land. *Turk. J. Electr. Eng. Comput. Sci.* **2020**, *28*, 1458–1473. [CrossRef]

26. Ulabhaje, K. Survey on Image Fusion Techniques used in Remote Sensing. In Proceedings of the 2018 Second International Conference on Electronics, Communication and Aerospace Technology (ICECA), Coimbatore, India, 29–31 March 2018; pp. 1860–1863.

27. Mukherjee, A.; Misra, S.; Raghuwanshi, N.S. A survey of unmanned aerial sensing solutions in precision agriculture. *J. Netw. Comput. Appl.* **2019**, *148*, 102461. [CrossRef]

28. Xiong, Z.; Zhang, Y. A critical review of image registration methods. *Int. J. Image Data Fusion* **2010**, *1*, 137–158. [CrossRef]

29. Unal, Z. Smart Farming Becomes even Smarter with Deep Learning—A Bibliographical Analysis. *IEEE Access* **2020**, *8*, 105587–105609. [CrossRef]

30. Girshick, R.; Donahue, J.; Darrell, T.; Malik, J. Rich Feature Hierarchies for Accurate Object Detection and Semantic Segmentation. In Proceedings of the 2014 IEEE Conference on Computer Vision and Pattern Recognition, Columbus, OH, USA, 23–28 June 2014; Volume 1, pp. 580–587. [CrossRef]

31. Bertinetto, L.; Valmadre, J.; Henriques, J.F.; Vedaldi, A.; Torr, P.H.S. Fully-Convolutional Siamese Networks for Object Tracking. In *Lecture Notes in Computer Science (Including Subseries Lecture Notes in Artificial Intelligence and Lecture Notes in Bioinformatics)*; Elsevier B.V.: Amsterdam, The Netherlands, 2016; Volume 9914 LNCS, pp. 850–865._56. [CrossRef]

32. He, K.; Zhang, X.; Ren, S.; Sun, J. Deep residual learning for image recognition. In Proceedings of the IEEE Computer Society Conference on Computer Vision and Pattern Recognition, Las Vegas, NV, USA, 27–30 June 2016; pp. 770–778. [CrossRef]

33. Badrinarayanan, V.; Kendall, A.; Cipolla, R. SegNet: A Deep Convolutional Encoder-Decoder Architecture for Image Segmentation. *IEEE Trans. Pattern Anal. Mach. Intell.* **2017**, *39*, 2481–2495. [CrossRef]

34. Polder, G.; Blok, P.M.; de Villiers, H.A.; van der Wolf, J.M.; Kamp, J. Potato virus Y detection in seed potatoes using deep learning on hyperspectral images. *Front. Plant Sci.* **2019**, *10*, 209. [CrossRef] [PubMed]

35. Naseer, M.; Khan, S.; Porikli, F. Indoor Scene Understanding in 2.5/3D for Autonomous Agents: A Survey. *IEEE Access* **2019**, *7*, 1859–1887. [CrossRef]

36. Sa, I.; Chen, Z.; Popovic, M.; Khanna, R.; Liebisch, F.; Nieto, J.; Siegwart, R. WeedNet: Dense Semantic Weed Classification Using Multispectral Images and MAV for Smart Farming. *IEEE Robot. Autom. Lett.* **2018**, *3*, 588–595. [CrossRef]

37. Ren, X.; Du, S.; Zheng, Y. Parallel RCNN: A deep learning method for people detection using RGB-D images. In Proceedings of the 2017 10th International Congress on Image and Signal Processing, BioMedical Engineering and Informatics (CISP-BMEI 2017), Shanghai, China, 14–16 October 2017; pp. 1–6. [CrossRef]

38. Gené-Mola, J.; Vilaplana, V.; Rosell-Polo, J.R.; Morros, J.R.; Ruiz-Hidalgo, J.; Gregorio, E. Multi-modal deep learning for Fuji apple detection using RGB-D cameras and their radiometric capabilities. *Comput. Electron. Agric.* **2019**, *162*, 689–698. [CrossRef]

39. Bezen, R.; Edan, Y.; Halachmi, I. Computer vision system for measuring individual cow feed intake using RGB-D camera and deep learning algorithms. *Comput. Electron. Agric.* **2020**, *172*, 105345. [CrossRef]

40. Aghi, D.; Mazzia, V.; Chiaberge, M. Local Motion Planner for Autonomous Navigation in Vineyards with a RGB-D Camera-Based Algorithm and Deep Learning Synergy. *Machines* **2020**, *8*, 27. [CrossRef]

41. Burgos, S.; Mota, M.; Noll, D.; Cannelle, B. Use of very high-resolution airborne images to analyse 3D canopy architecture of a vineyard. *Int. Arch. Photogramm. Remote Sens. Spat. Inf. Sci. ISPRS Arch.* **2015**, *40*, 399–403. [CrossRef]

42. Matese, A.; Di Gennaro, S.F.; Berton, A. Assessment of a canopy height model (CHM) in a vineyard using UAV-based multispectral imaging. *Int. J. Remote Sens.* **2017**, *38*, 2150–2160. [CrossRef]

43. Weiss, M.; Baret, F. Using 3D Point Clouds Derived from UAV RGB Imagery to Describe Vineyard 3D Macro-Structure. *Remote Sens.* **2017**, *9*, 111. [CrossRef]

44. Mahlein, A.K. Plant disease detection by imaging sensors—Parallels and specific demands for precision agriculture and plant phenotyping. *Plant Dis.* **2016**, *100*, 241–254. [CrossRef]

45. Kaur, S.; Pandey, S.; Goel, S. Plants Disease Identification and Classification Through Leaf Images: A Survey. *Arch. Comput. Methods Eng.* **2019**, *26*, 507–530. [CrossRef]

46. Saleem, M.H.; Potgieter, J.; Arif, K.M. Plant disease detection and classification by deep learning. *Plants* **2019**, *8*, 468. [CrossRef]

47. Sandhu, G.K.; Kaur, R. Plant Disease Detection Techniques: A Review. In Proceedings of the 2019 International Conference on Automation, Computational and Technology Management (ICACTM 2019), London, UK, 24–26 April 2019; pp. 34–38. [CrossRef]

48. Loey, M.; ElSawy, A.; Afify, M. Deep learning in plant diseases detection for agricultural crops: A survey. *Int. J. Serv. Sci. Manag. Eng. Technol.* **2020**, *11*, 41–58. [CrossRef]

49. Schor, N.; Bechar, A.; Ignat, T.; Dombrovsky, A.; Elad, Y.; Berman, S. Robotic Disease Detection in Greenhouses: Combined Detection of Powdery Mildew and Tomato Spotted Wilt Virus. *IEEE Robot. Autom. Lett.* **2016**, *1*, 354–360. [CrossRef]

50. Sharif, M.; Khan, M.A.; Iqbal, Z.; Azam, M.F.; Lali, M.I.U.; Javed, M.Y. Detection and classification of citrus diseases in agriculture based on optimized weighted segmentation and feature selection. *Comput. Electron. Agric.* **2018**, *150*, 220–234. [CrossRef]

51. Ferentinos, K.P. Deep learning models for plant disease detection and diagnosis. *Comput. Electron. Agric.* **2018**, *145*, 311–318. [CrossRef]

52. Argüeso, D.; Picon, A.; Irusta, U.; Medela, A.; San-Emeterio, M.G.; Bereciartua, A.; Alvarez-Gila, A. Few-Shot Learning approach for plant disease classification using images taken in the field. *Comput. Electron. Agric.* **2020**, *175*. [CrossRef]

53. Jothiaruna, N.; Joseph Abraham Sundar, K.; Karthikeyan, B. A segmentation method for disease spot images incorporating chrominance in Comprehensive Color Feature and Region Growing. *Comput. Electron. Agric.* **2019**, *165*, 104934. [CrossRef]

54. Pantazi, X.E.; Moshou, D.; Tamouridou, A.A. Automated leaf disease detection in different crop species through image features analysis and One Class Classifiers. *Comput. Electron. Agric.* **2019**, *156*, 96–104. [CrossRef]

55. Abdulridha, J.; Ehsani, R.; Abd-Elrahman, A.; Ampatzidis, Y. A remote sensing technique for detecting laurel wilt disease in avocado in presence of other biotic and abiotic stresses. *Comput. Electron. Agric.* **2019**, *156*, 549–557. [CrossRef]

56. Hu, W.j.; Fan, J.I.E.; Du, Y.X.; Li, B.S.; Xiong, N.N.; Bekkering, E. MDFC—ResNet: An Agricultural IoT System to Accurately Recognize Crop Diseases. *IEEE Access* **2020**, *8*, 115287–115298. [CrossRef]

57. Ronneberger, O.; Fischer, P.; Brox, T. U-net: Convolutional networks for biomedical image segmentation. In *Lecture Notes in Computer Science (Including Subseries Lecture Notes in Artificial Intelligence and Lecture Notes in Bioinformatics)*; Elsevier B.V.: Amsterdam, The Netherlands, 2015; Volume 9351, pp. 234–241. [CrossRef]

58. Chen, L.C.; Zhu, Y.; Papandreou, G.; Schroff, F.; Adam, H. Encoder-Decoder with Atrous Separable Convolution for Semantic Image Segmentation. *Pertanika J. Trop. Agric. Sci.* **2018**, *34*, 137–143.

59. Zhao, H.; Shi, J.; Qi, X.; Wang, X.; Jia, J. Pyramid scene parsing network. In Proceedings of the 30th IEEE Conference on Computer Vision and Pattern Recognition (CVPR 2017), Honolulu, HI, USA, 21–26 July 2017; Volume 2017, pp. 6230–6239. [CrossRef]

60. Sermanet, P.; Eigen, D.; Zhang, X.; Mathieu, M.; Fergus, R.; LeCun, Y. Overfeat: Integrated recognition, localization and detection using convolutional networks. In Proceedings of the 2nd International Conference on Learning Representations (ICLR 2014), Conference Track Proceedings, Banff, AB, Canada, 14–16 April 2014.

61. Liu, Y.; Chen, X.; Peng, H.; Wang, Z. Multi-focus image fusion with a deep convolutional neural network. *Inf. Fusion* **2017**, *36*, 191–207. [CrossRef]

62. Adhikari, S.P.; Yang, H.; Kim, H. Learning Semantic Graphics Using Convolutional Encoder–Decoder Network for Autonomous Weeding in Paddy. *Front. Plant Sci.* **2019**, *10*, 1404. [CrossRef]

63. Dunnhofer, M.; Antico, M.; Sasazawa, F.; Takeda, Y.; Camps, S.; Martinel, N.; Micheloni, C.; Carneiro, G.; Fontanarosa, D. Siam-U-Net: Encoder-decoder siamese network for knee cartilage tracking in ultrasound images. *Med Image Anal.* **2020**, *60*, 101631. [CrossRef]

64. LeCun, Y.; Bottou, L.; Bengio, Y.; Haffner, P. Gradient-based learning applied to document recognition. *Proc. IEEE* **1998**, *86*, 2278–2323. [CrossRef]

65. Dellana, R.; Roy, K. Data augmentation in CNN-based periocular authentication. In Proceedings of the 6th International Conference on Information Communication and Management (ICICM 2016), Hatfield, UK, 29–31 October 2016; pp. 141–145. [CrossRef]

66. Hoffman, M.D.; Blei, D.M.; Wang, C.; Paisley, J. Stochastic variational inference. *J. Mach. Learn. Res.* **2013**, *14*, 1303–1347.

67. Zeiler, M.D. ADADELTA: An Adaptive Learning Rate Method. *arXiv* **2012**, arXiv:1212.5701.

68. Kingma, D.P.; Ba, J.L. Adam: A method for stochastic optimization. In Proceedings of the 3rd International Conference on Learning Representations (ICLR 2015), Conference Track Proceedings, San Diego, CA, USA, 7–9 May 2015; pp. 1–15.

69. Zeng, X.; Zhang, Z.; Wang, D. AdaMax Online Training for Speech Recognition. 2016, pp. 1–8. Available online: http://cslt.riit.tsinghua.edu.cn/mediawiki/images/d/df/Adamax_Online_Training_for_Speech_Recognition.pdf (accessed on 10 September 2020).

 *remote sensing*

Article

Wildfire-Detection Method Using DenseNet and CycleGAN Data Augmentation-Based Remote Camera Imagery

Minsoo Park [1], Dai Quoc Tran [1], Daekyo Jung [2] and Seunghee Park [1,3,*]

[1]   School of Civil, Architectural Engineering & Landscape Architecture, Sungkyunkwan University, Suwon 16419, Korea; pms5343@skku.edu (M.P.); daitran@skku.edu (D.Q.T.)
[2]   Department of Convergence Engineering for Future City, Sungkyunkwan University, Suwon 16419, Korea; jdaekyo@skku.edu
[3]   Technical Research Center, Smart Inside Co., Ltd., Suwon 16419, Korea
*   Correspondence: shparkpc@skku.edu

Received: 21 October 2020; Accepted: 11 November 2020; Published: 12 November 2020

Abstract: To minimize the damage caused by wildfires, a deep learning-based wildfire-detection technology that extracts features and patterns from surveillance camera images was developed. However, many studies related to wildfire-image classification based on deep learning have highlighted the problem of data imbalance between wildfire-image data and forest-image data. This data imbalance causes model performance degradation. In this study, wildfire images were generated using a cycle-consistent generative adversarial network (CycleGAN) to eliminate data imbalances. In addition, a densely-connected-convolutional-networks-based (DenseNet-based) framework was proposed and its performance was compared with pre-trained models. While training with a train set containing an image generated by a GAN in the proposed DenseNet-based model, the best performance result value was realized among the models with an accuracy of 98.27% and an F1 score of 98.16, obtained using the test dataset. Finally, this trained model was applied to high-quality drone images of wildfires. The experimental results showed that the proposed framework demonstrated high wildfire-detection accuracy.

Keywords: wildfire detection; convolutional neural networks; densenet; generative adversarial networks; CycleGAN; data augmentation

---

1. Introduction

Wildfires cause significant harm to humans and damage to private and public property; they pose a constant threat to public safety. More than 200,000 wildfires occur globally every year, with a combustion area of 3.5–4.5 million km^2 [1]. In addition, climate change is gradually accelerating the effects of these wildfires; there is thus considerable interest in wildfire management [2–4]. As wildfires are difficult to control once they spread over a certain area, early detection is the most important factor in minimizing wildfire damage. Traditionally, wildfires were primarily detected by human observers, but a deep learning-based automatic wildfire detection system with real-time surveillance cameras has the advantage of the possibility of constant and accurate monitoring, compared to human observers. The available methods for the early detection of wildfires can be categorized as a sensor-based technology and image-processing-based technology, using a camera. Sensors that detect changes in smoke, pressure, humidity, and temperature are widely used for fire detection. However, this method has several disadvantages, such as high initial cost and high false-alarm rates, as the performance of sensors is significantly affected by the surrounding environment [5–7].

With the rapid development of digital-cameras and image-processing technologies, traditional methods are replaced by video- and image-data-based methods [8]. Using these methods, a large area of a forest can be monitored, where fires and smoke can be detected immediately after the outbreak of a wildfire. In addition, owing to intelligent image-analysis technology, image-based methods can be used to address the problem of the inflexibility of sensing technology to new environments [9]. Such early approaches include the use of support vector machines (SVM) [10,11] for classifying wildfire images, and fuzzy c-means clustering [12] for identifying potential fire regions. Recently, convolutional neural networks (CNNs), which provide excellent image classification and object detection by extracting features and patterns from images, made many contributions to the wildfire-detection field [13–16]. CNN is one of the most popular neural networks and was successfully used in many research and industry applications, such as computer vision and image processing [17,18]. These networks were developed and successfully applied to many challenging image-classification problems, such as for improving a model's performance [19,20]. Muhammad et al. [21] developed a modified model from GoogleNet Architecture for fire detection, to increase the model's accuracy, and proposed a framework for fire detection in closed-circuit television surveillance systems. Jung et al. [22] developed a decision support system concept architecture for wildfire management and evaluated CNN-based fire-detection technology from the Fire dataset. As noted by Jain et al. in their review of machine-learning applications in wildfire detection [23], Zhang et al. found that CNN outperforms the SVM-based method [24], and Cao et al. reported a 97.8% accuracy rate for smoke detection, using convolutional layers [25]. Recently, advances in mobile communication technology made it possible to use unmanned aerial vehicles (UAVs), which are more flexible than fixed fire-monitoring towers; images obtained from UAVs are used to learn fire-detection models [26,27].

Despite the contributions of these successful studies, some issues still need to be resolved in order to apply this technology in the field. Mountain-image data are easy to obtain, owing to the availability of various built-up datasets. However, not only is there a dearth of fire or smoke images of wildfires in datasets, but such data are also relatively difficult to obtain because they require the use of installed surveillance cameras or operational drones at the site of the wildfire [28,29]. Therefore, research on damage detection is frequently faced with a data imbalance problem, which causes overfitting; overfitting results in the deterioration of the model performance [30]. In order to solve this data imbalance problem, in a recent study, synthetic images were generated and used to expand the fire/smoke dataset [24,31]. In early studies, the data were increased using indoor artificially generated smoke and flames or artificial images that comprised cut-and-pasted images of flames in their background. However, this requires considerable manpower, and it is difficult to emulate the characteristics of wildfire images using indoor images. Generative adversarial networks (GANs) [32] are models that create new images using two networks—a generator and a discriminator. The generator creates similar data using the training set, and the discriminator distinguishes between the real data and the fake data created by the generator. The image rotation and image cropping data augmentation method can also be used to expand the training dataset; however, GANs can be used to increase dataset diversity as well as to increase the amount of data. They recently exhibited impressive photorealistic-image-creation results [33–36]. GANs were proven to improve performance when learning the classifier, mainly in areas where it is difficult to obtain damage data [37–39]. However, there are relatively few related studies in the field of wildfire detection. Namozov et al. used GANs to create fire photographs with winter and evening backgrounds in the original photographs, and added a variety of seasons and times [28]. However, it is difficult to provide various types of fire scenarios in various places as the resultant image retains not only the background of the original photo, but also the shape of the flame and smoke. To apply the early wildfire detection model to the field, it is necessary to learn various types of wildfire images using new backgrounds, such that wildfire detection can be actively performed even in a new environment.

With the development of the CNN model and the deepening of neural networks, problems such as vanishing gradients arise, which causes overfitting and deterioration of the model performance. An algorithm constructed using the latest neural network architecture of DenseNet [40] could be used to address this issue. DenseNet improves the performance of a model by connecting the feature maps of the previous layers to the inputs of the next layer using concatenation to maximize the information flow between layers.

Inspired by recent works, we generated synthetic wildfire images using GANs to change the image of a fire-free mountain to that of a mountain with a wildfire. The k-folds (k = 5) cross validation scheme was used on the models, and the train set was separated, train sets A and B, consisting of only the original images and of the original and generated images, respectively. Each dataset was divided to obtain the training data and test data, and was used to train a model that was developed based on DenseNet; this facilitated the comparison of the performance with two pre-trained models, VGG-16 [19] and ResNet-50 [20]. This paper is organized as follows. Section 2 describes the architecture of cycle-consistent adversarial networks (CycleGANs) [41], which is one of the main GANs algorithms used for data augmentation, and DenseNet [40], which is used for wildfire-image classification (wildfire detection). The experiment results obtained using both the models and the classification performance comparison with those of the pre-trained models are presented in Section 3. Section 4 presents the conclusion of this study.

2. Materials and Methods

2.1. Data Collection

The wildfire and non-fire images that were used for training the GAN model and CNN classification models were collected. The mountain datasets were obtained from eight scene-categories databases [42] and a Korean tourist spot database [43]. However, there is no open data benchmark available for fire or smoke images of wildfires [28]. The collection was, thus, solely obtained using web crawling; this limitation resulted in a data imbalance. Considering that the early fire-detection model is intended for application in drones and surveillance cameras for the purpose of monitoring, both categories of datasets were crawled from images or videos obtained using a drone. The sample of the dataset is presented in Figure 1. A total of 4959 non-wildfire images and 1395 wildfire images were set up in our original dataset and resized to 224 × 224 for the network input.

(a)   (b)   (c)   (d)   (e)

Figure 1. Sample mountain and wildfire images from conducted data collection. (**a**) Mountain images from eight scene categories database. (**b**) Mountain images from Korean tourist spot database. (**c**) Drone-captured mountain images obtained via web image crawling. (**d**) Drone-captured wildfire images obtained via web image crawling. (**e**) Drone-captured wildfire images obtained via web video crawling.

2.2. CycleGAN Image-to-Image Translation

To generate wildfire images, CycleGAN [41] was used, which is a method used for image-to-image translation from the reference image domain (X) to the target image domain (Y), without relying on paired images. As illustrated in Figure 2, in the CycleGAN, two loss functions called the adversary loss [33] and cycle-consistency loss [41] were used.

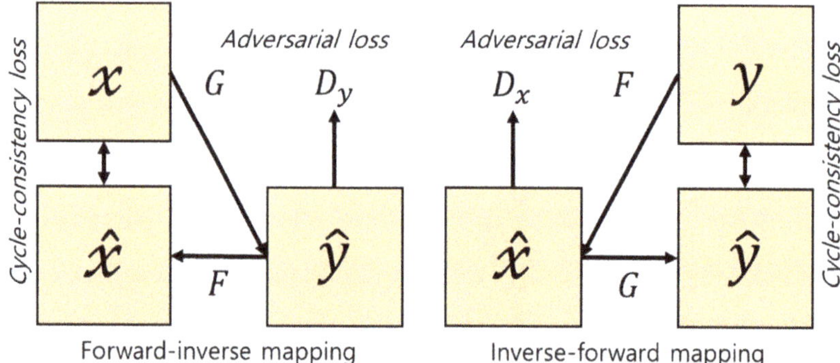

Figure 2. Architecture of CycleGAN; mapping between two image domains x and y. The model training is performed as the forward and inverse mappings are learned simultaneously using the adversarial loss and cycle-consistency loss.

Our objective was to train $G_{x \to y}$ such that the discriminator D_y cannot distinguish the image data distribution from $G_{x \to y}$ and the image data distribution from domain Y. This objective can be written as follows:

$$\mathscr{L}_{GAN}\big(G_{x \to y}, D_Y, X, Y\big) = \mathbb{E}_{y \sim p_{data}(y)}\big[logD_Y(y)\big] + \mathbb{E}_{x \sim p_{data}(x)}\big[\log(1 - D_Y(G_{x \to y}(x)))\big]. \tag{1}$$

$$\mathcal{L}_{GAN}\big(G_{y \to x}, D_x, X, Y\big) = \mathbb{E}_{x \sim p_{data}(x)}\big[logD_x(x)\big] + \mathbb{E}_{y \sim p_{data}(y)}\big[\log(1 - D_x(G_{y \to x}(y)))\big]. \tag{2}$$

However, in a general GAN, the model is not trained over the entire distribution of actual data; it is only trained for reducing the loss. Therefore, a mode collapsing problem occurs in which the optimization fails, as the generator cannot find the entire data distribution, and all input images are mapped to the same output image. To solve this problem, in the CycleGAN, inverse mapping and cycle-consistency loss ($\mathcal{L}_{cyc}$) were applied to Equations (1) and (2), respectively, and various outputs were thus produced. The equations of the cycle-consistency loss were as follows:

$$\mathcal{L}_{cyc}\big(G_{x \to y}, G_{y \to x}\big) = \mathbb{E}_{x \sim p_{data}(x)}\big[\|G_{y \to x}\big(G_{x \to y}(x)\big) - x\|_1\big] + \mathbb{E}_{y \sim p_{data}(y)}\big[\|G_{x \to y}\big(G_{y \to x}(y)\big) - y\|_1\big]. \tag{3}$$

In addition, by converting the X domain into $G_{y \to x}$ while adding an identity loss ($\mathcal{L}_{im}$) that regularized the generator, such that the calculated output was the same as the input, the converted image could be generated, while minimizing the damage to the original image.

$$\mathcal{L}_{im}\big(G_{x \to y}, G_{y \to x}\big) = \mathbb{E}_{y \sim p_{data}(y)}\big[\|G_{x \to y}(y) - y\|_1\big] + \mathbb{E}_{x \sim p_{data}(x)}\big[\|G_{y \to x}(x) - x\|_1\big]. \tag{4}$$

The final loss combined with all losses was as follows. Using CycleGAN with this method, it was possible to create various wildfire images, while maintaining the shape and background color of the forest site.

$$\mathcal{L}\big(G_{x \to y}, G_{y \to x}, D_x, D_y\big) = \mathcal{L}_{GAN}\big(G_{x \to y}, D, X, Y\big) + \mathcal{L}_{GAN}\big(G_{y \to x}, D, X, Y\big) + \\ \lambda\mathcal{L}_{cyc}\big(G_{x \to y}, G_{y \to x}\big) + \mathcal{L}_{im}\big(G_{x \to y}, G_{y \to x}\big). \tag{5}$$

2.3. DenseNet

The early wildfire-detection algorithm was constructed using the state-of-the-art net architecture, DenseNet, which is known to perform well in wildfire detection, while alleviating the vanishing gradient problem and reducing the training time [40]. It is a densely connected CNN structure that has a connection strategy. Figure 3 illustrates the original dense block architecture. The network comprises layers, each of which contain a non-linear transformation, and includes functions such as batch normalization, rectified linear unit (ReLU), and convolution. X_0 is a single image, and the network output of the $(l-1)^{th}$ layer after passing through a convolution is X_{l-1}. The lth layer receives the feature maps of all preceding layers as its input (Equation (6)).

$$X_l = H_l([X_0, X_1, X_2, \ldots, X_{l-1}]) \tag{6}$$

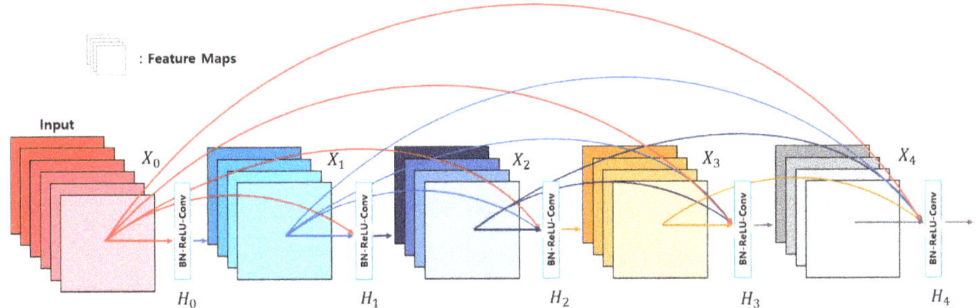

Figure 3. Architecture of five-layer densely connected convolution networks.

2.4. Performance Evaluation Metrics

To compare the performance of the models, five commonly used metrics were calculated—accuracy, precision, sensitivity, specificity, and F1-Score [44–46]. Accuracy is the ratio of accurately predicted observations to the total number of observations and is the most intuitive performance measurement. Precision is the ratio of correctly predicted positive observations to the total predicted positive observations. Sensitivity is the ratio of correctly predicted positive observations to the actual true observations. Specificity is the ratio of correctly predicted negative observations to the total number of predicted negative observations. The F1 score is the harmonic average of precision and sensitivity, which is generally useful for determining the performance of a model in terms of accuracy. The expressions for the evaluation metrics are presented as follows.

$$\text{Accuracy} = \frac{\text{TP} + \text{TN}}{\text{TP} + \text{TN} + \text{FP} + \text{FN}}. \tag{7}$$

$$\text{Precision} = \frac{\text{TP}}{\text{TP} + \text{FP}}. \tag{8}$$

$$\text{Sensitivity} = \frac{\text{TP}}{\text{TP} + \text{FN}}. \tag{9}$$

$$\text{Specificity} = \frac{\text{TN}}{\text{TN} + \text{FP}}. \tag{10}$$

$$\text{F1} - \text{score} = \frac{2 \times \text{precision} \times \text{Sensitivity}}{\text{precision} + \text{Sensitivity}}. \tag{11}$$

In the aforementioned equations, the number of true positives that the model predicts, i.e., the number of wildfire images predicted as wildfires and the number of true negatives that model the predicts, i.e., the number of non-fire images identified as non-fire, are denoted by true positive (TP) and true negative (TN), respectively. In addition, the number of false positives that the model predicts, i.e., the non-fire images predicted as wildfires, and the number of false negatives that model predicts, i.e., the wildfire images predicted as non-fire, are denoted as false positive (FP) and false negative (FN), respectively. These four types of data are defined using a confusion matrix in the binary classification. The overall performance-evaluation metrics were evaluated using the wildfire and non-wildfire testing sets.

3. Experimental Results

The following sections present the obtained results of the dataset balancing and wildfire detection models. The experiment environment was CentOS (Community enterprise operating system) Linux release 8.2.2004, which was constructed as an artificial intelligence server. The hardware configuration of the server consists of an Intel(R) Xeon(R) Gold 6240 central processing unit, 2.60 GHz, with an Nvidia Tesla V100 GPU, 32 GB memory. The experiences were conducted using the PyTorch deep learning framework [47] with Python language. The result and the example experiment code is available online at Github repository (https://github.com/pms5343/pms5343-WildfireDetection_by_DenseNet).

3.1. Dataset Augmentation Using GAN

To alleviate the data imbalance of the collected images, new wildfire images were generated using the CycleGAN as a data augmentation strategy. The objective of using the image-generation model is to convert non-wildfire images from a part of the collected data into wildfire images. A total of 1294 wildfire images (Domain A) and 2311 non-wildfire images (Domain B) from our original dataset were used.

As can be observed from Figure 4, the training was performed by increasing the number of epochs until there was a slight change in each loss, in order to improve the model. The generator loss was learned in the direction of increasing loss as the number of epochs increased because the objective of the generators was to create a fake image such that the discriminator could not determine whether the generated image was real or fake. Conversely, the discriminator losses were trained to reduce the loss, in order to distinguish between the generated and original images. Figure 4b shows that the cycle consistency loss added for the purpose of increasing the diversity of the generated image and the identity mapping loss added for the purpose of minimizing changes in the background of the generated image were also trained in the direction of decreasing exposure. After 650 epochs, there was no significant change in loss, and the training was thus terminated.

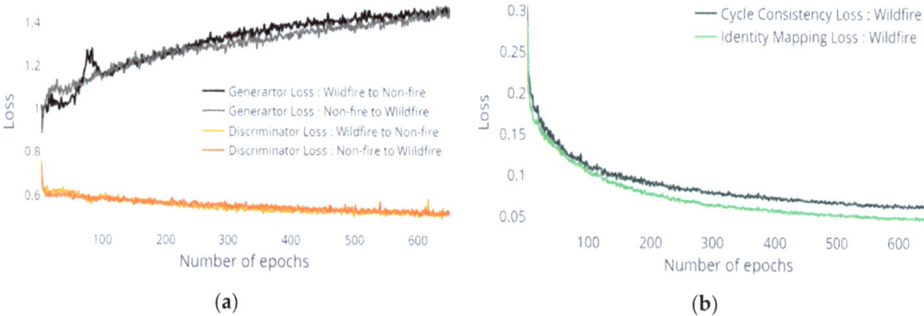

Figure 4. Training loss curve of CycleGAN-based non-fire–wildfire image converter. (**a**) Adversarial loss curve for generator and discriminator by the number of epochs. (**b**) Cycle-consistency and identity mapping loss curve by the number of epochs.

Figure 5 illustrates the overall process of the model and an example of when the images of domains A and B undergo the model-training process. The mountain image without a fire in domain B was converted into a wildfire image through the generator G_{BA} and then compared with the image of domain A (original wildfire image), by discriminator A (D_A) (①→②→③ process in Figure 5). The converted image was the image reconstructed by generator G_{AB}, and the result was not significantly different from that of domain B (①→②→④ process in Figure 5). In addition, it was confirmed that there was no difference in the image converted by generator G_{AB} from domain B (①→⑤ process in Figure 5). Conversely, the process was conducted in the same manner, and 1195 new 224 × 224-pixel fire images were created from domain B (Figure 6) and included in the wildfire dataset.

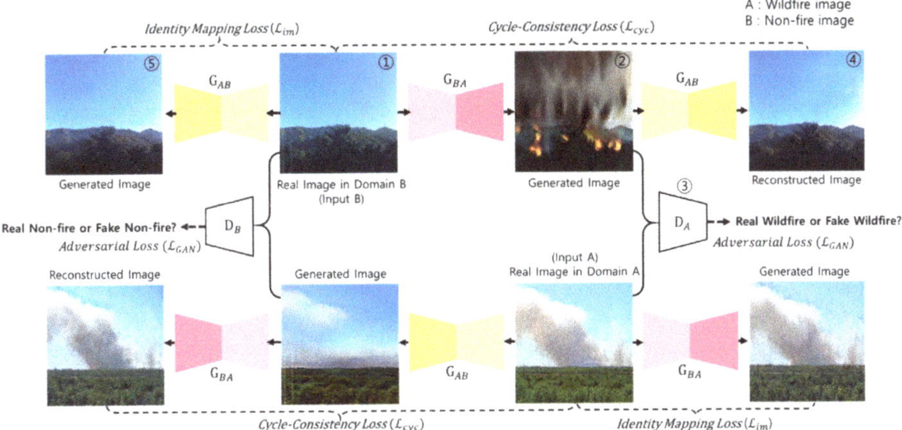

Figure 5. CycleGAN-based wildfire-image-generation architecture.

Figure 6. Sample of the wildfire images converted from non-fire mountain images.

3.2. Wildfire Detection

The wildfire detection was realized through the use of a DenseNet-based classification network model consisting of three dense blocks and two transition layers to identify the fire with 224 × 224-pixel-size image inputs. The architecture of the simple network is illustrated in Figure 7.

The dense block included a two-kernel filter. One filter was a 1 × 1 size convolution, which was used to decrease the number of input feature map channels, and the other was a 3 × 3 size convolution. After the dense block, the feature maps passed through a phase layer consisting of batch normalization, ReLU, 1 × 1 convergence, and 2 × 2 average pooling, which reduced the width and length of the feature map and the number of feature maps. Finally, after three dense block sessions, the result was drawn after the linear layer at the end, after passing through the global average pooling and softmax classifier sequentially, as in the case of a traditional CNN.

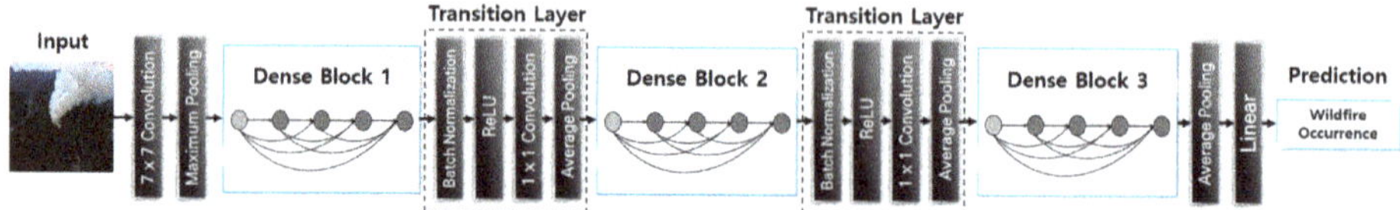

Figure 7. DenseNet-based wildfire-detection architecture.

The following section presents the results of the wildfire-detection performance obtained using the deep learning classification model based on DenseNet, as compared to the pre-trained model. Two results were derived for each model—one for train set A and the other for train set B.

3.2.1. Dataset Partition

The train and test set partition are specified in the following section. From the collected original dataset, several images were used to generate new images. The forest image used as the GAN domain was deleted from the dataset for the classification model; however, the wildfire domain was not eliminated because it was used as a reference; it was thus not deleted from the dataset. A total of horizontal flip and random crop (by 200 pixel) were used to expand the number of samples of the training sets. The train sets were divided into trainset A, consisting only of photographs taken, and trainset B, consisting of wildfire images generated by the GAN. Many precedent research showed that accuracy becomes lower when the number of data points is imbalanced [48]. In order to avoid the disadvantages of already well-known data imbalances, Train set A kept the data ratio between the two classes similar, even if the total number of data is set less than B. The test set only contains the original photograph and not the generated image. Twenty percent of the total collected original image dataset was selected as the test dataset. Partition of the datasets are shown in Table 1.

Table 1. Image datasets for wildfire-detection model.

	Original Non-Fire Images	Original Wildfire Images	Generated Wildfire Images
Train set A [Real database]	3165	2427	
Train set B [Real + synthetic database]	6309	2427	3585
Test set	545	486	

3.2.2. Model Training and Comparison of the Models

To demonstrate the performance of the proposed method, two train sets were used in the proposed model and well-known pre-trained models, ResNet-16 and VGG-50, for the performance evaluation. To improve the models' performance of each model, the learning rate and optimizer were tested. Ten values of the initial learning rate between 0.1 and 0.00001 were tested, while changing three representative optimizers—stochastic gradient descent (SGD), Adam [49], and PMSprop [50]. The number of epochs was fixed at 250, and batch size was fixed at 64. The best hyperparameter combination was found based on the average accuracy from the k-folds (k = 5) cross-validation process; presented in Table 2.

Table 2. Selected hyperparameters for CNN architectures.

	VGG-16	**ResNet-50**	**DenseNet**
Batch Size	60	60	60
Initial Learning Rate	0.0002	0.0002	0.01
Number of Training Epochs	250	250	250
Optimizer	Adam	Adam	SGD

The training process of each model using the selected hyperparameter combination is illustrated in Figure 8. The training accuracy curve obtained as the number of epochs increased is presented in Figure 8a. The accuracy of the six models increased most significantly between epochs 1 and 10 and then increased steadily until epoch 250.

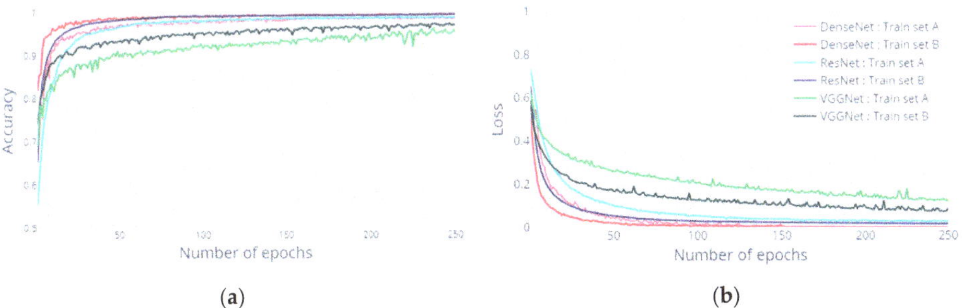

Figure 8. Learning curve of training process over epochs. (**a**) Accuracy curve. Final accuracy: VGG-16, trainset A (0.954); VGG-16, trainset B (0.969); ResNet-50, trainset A (0.989); ResNet-50, trainset B (0.995); DenseNet trainset A (0.985); and DenseNet trainset B (0.995). (**b**) Loss curve. Final loss: VGG-16, trainset A (0.123); VGG-16 trainset B (0.085); ResNet-50, trainset A (0.028); ResNet-50, trainset B (0.016); DenseNet, trainset A (0.0003; SGD); and DenseNet, trainset B (0.00006; SGD).

The DenseNet-based proposed model demonstrated the highest training accuracy, with an approximate accuracy of 99% in the final learning approach, followed by ResNet-50 and then VGG-16. In addition, it was demonstrated that the accuracy performance of trainset B, which included generated images, was greater than that of trainset A for all three models. The training loss curve obtained as the number of epochs increased is presented in Figure 8b. The DenseNet and ResNet-16 losses rapidly decreased until epoch 20, whereas the loss of VGG-16 continued to decrease steadily. The training loss also exhibited a better performance for trainset B than that for trainset A in the case of both the initial and final losses.

The classifier models were evaluated based on the performance results, using the five metrics presented in Table 3. DenseNet yielded the best results in terms of all five metrics. Although the VGG-50 model exhibited a slightly lower accuracy, sensitivity, and F1-score, the results obtained on using trainset B were at a similar level as (or better than) those obtained with trainset A. For example, in the case of DenseNet, the accuracy increased from 96.734% to 98.271%, the precision increased from 96.573% to 99.380%, sensitivity increased from 96.573% to 96.976, specificity increased from 96.881% to 99.450%, and the F1-score increased from 96.573 to 98.163. The experimental results showed that a new image created by changing a normal image of a mountain into an image of a mountain on which a fire had occurred could maintain the performance of the CNN and also improve the model performance via the input of various data as training.

Table 3. Comparisons of performance evaluation.

	VGG-16		ResNet-50		Proposed Method	
	Train Set A	Train Set B	Train Set A	Train Set B	Train Set A	Train Set B
Accuracy (%)	93.756	93.276	96.734	96.926	96.734	**98.271**
Precision (%)	93.890	97.973	97.727	97.934	96.573	**99.380**
Sensitivity (%)	92.944	87.702	95.363	95.565	96.573	**96.976**
Specificity (%)	94.495	98.349	97.982	98.165	96.881	**99.450**
F1-Score	93.414	92.553	96.531	96.735	96.573	**98.163**

The bold is the best result among other methods.

3.2.3. Influence of Data Augmentation Methods

In this section, proposed model performance is compared with and without using CycleGAN-based data augmentation, to verify the influence of the proposed method. Horizontal flip, random zoom (200 pixel), rotation (original images were rotated by 10° and 350°), and random brightness (two values were selected arbitrarily from $l_{min} = 0.8$ to $l_{max} = 1.2$) methods were used in this section, as traditional data augmentation without GAN. The F1-score was obtained from a combination of training sets consisting of various augmentation methods.

Based on the experimental results, it could be seen that data augmentation from CycleGAN improved the accuracy of wildfire detection models. As can be seen from Table 4, the F1 score trained from data combination including the GAN method was higher by 1.154, 0.902, and 0.821, respectively, than the model trained from traditional method without GAN.

Table 4. F1-scores for model trained by various combination sets.

Data Augmentation Method	Training Images	F1-Score
Original + GAN + Horizontal flip + Zoom (200)	6312	98.163
Original + GAN + Rotation (10° and 350°)	6312	97.911
Original + GAN + Random brightness (from l_{min} to l_{max})	6312	97.830
Original + Traditional augmentation (Without GAN)	6363	97.009

3.2.4. Visualization of the Contributed Features

In order to visualize the output result of the model that exhibits the best performance, a class activation map (CAM) [51] was used to determine the features of the image that were extracted to detect the wildfire. As can be observed from the example of the CAM results in Figure 9, the detection was made primarily based on the presence of smoke or flames in the image, and the elements used for the classification as wildfires were found even in the early stages of the fire, with no flame and little smoke.

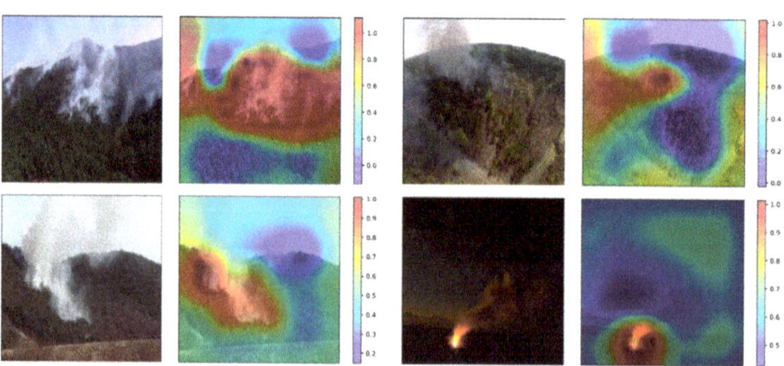

Figure 9. Sample of CAM results of the wildfire images.

The smoke in the part of the image that comprises the forest could be detected well, but the smoke in the part that comprises sky was not judged as a factor. It is hypothesized that this occurred because the model confused smoke with clouds or fog, and the smoke near the sky background could thus not be treated as a powerful factor for classifying the features.

3.3. Model Application

To apply the learned model to on-site drones or surveillance cameras used to monitor forests, a method of application for higher-resolution images than the model input-image size (224 × 224) is required. There is also a method used for resizing a remote camera image to a lower resolution; however, the method proposed in this study comprises cropping high-resolution images at regular intervals—considering that surveillance cameras are generally used to observe large areas—to derive the result values for each image.

Figures 10–12 present an example of a model application based on a drone-tested forest video [52]. This is a 1280 × 720-size drone video of a wildfire that occurred in Daejeon, Korea, in 2015. The white and jade green boxes denote the cropped areas of size 224 × 224 and are indicated in alternate colors for visualization convenience. The cropped images were cut to overlap each other at a certain interval, and 28 images per video frame were cut and input to the classification model. The text in the square box indicates the value derived from the softmax layer of the model, which was the final layer of the model (as it was trained using two classes; if the softmax value of the model was greater than 0.5, it was determined that the range comprised a fire, otherwise, it was determined that the range did not comprise a fire.)

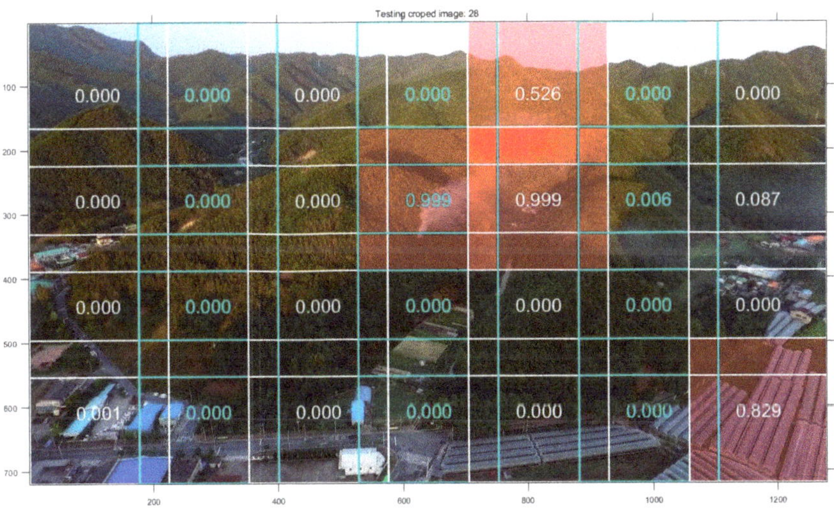

Figure 10. Example of model application with softmax result for early wildfire (with error).

Figure 10 presents the result of the application of the model to the image captured approximately 1 min after the wildfire occurred. The photos include not only the forest, but also parts of the nearby villages. The model detected the smoke generated in the forest and determined the location at which the fire had occurred. However, a greenhouse at the bottom right of the photo was falsely detected as a wildfire (0.829). It was suggested that this was a problem caused by the error of not properly taking into consideration specific images like cities, roads, and farmland, when training the initial model. This phenomenon was also found when applied to other sites.

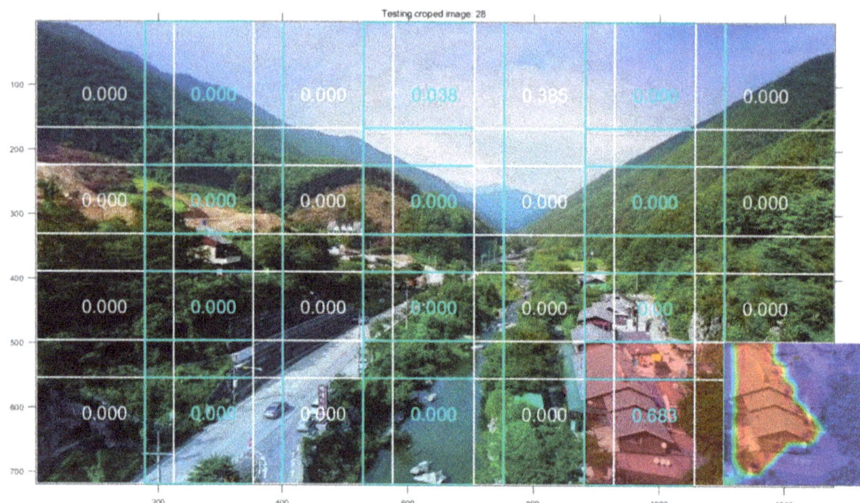

Figure 11. Example of model application with softmax result for non-wildfire (with error).

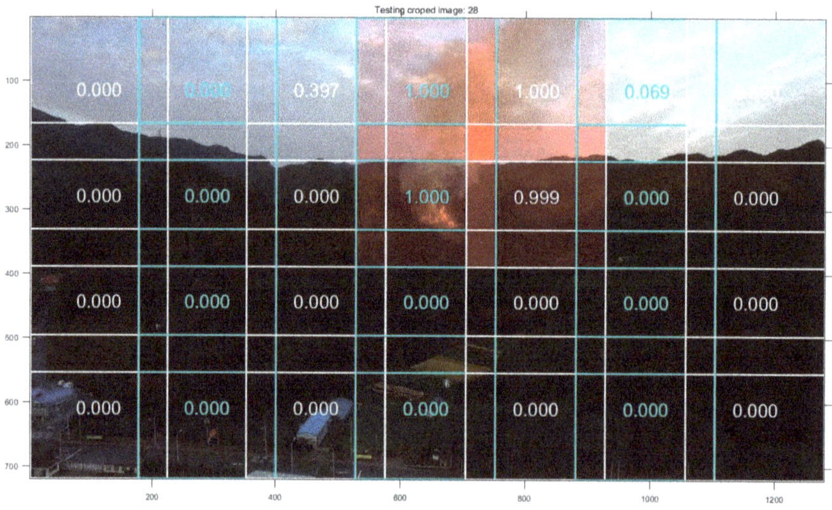

Figure 12. Example of model application with softmax result after 10-min wildfire progress.

As can be seen from the class activation map in Figure 11, the model mistook the building feature. Although it could not be judged that this was falsely detected by all artifacts, it was confirmed that false positives might occur when more than half of the cropped images were not natural objects. Conversely, there were no false positives caused by natural objects, such as confusion of distinguishing between clouds and smoke.

Figure 12 presents the result of the application of the model, approximately 10 min after the wildfire progression. As the fire was accompanied by flames after the fire had grown to some extent, the softmax layer provided a prediction with 100% probability, and the fire could be detected more easily than at the beginning of the fire. After applying the method of cropping without resizing the image, damaging the original image becomes unnecessary. As each cropped image is discriminated individually, the location of the fire can be tracked, while continuously obtaining real-time video footage, using a surveillance camera.

4. Conclusions

With the development of remote camera sensing technology, many researchers attempted to improve existing wildfire-detection systems using CNN-based deep learning. In the damage-detection field, it is difficult to obtain a sufficient amount of the necessary data for training models; data imbalance or overfitting problems have thus caused the deterioration of the models' performance. To solve these problems, traditional image transformation methods such as image rotation were primarily used. A method of increasing the learning data was also adopted, wherein the flame image was artificially cut and pasted over a forest background. However, these two methods have their respective weaknesses—failure to increase the diversity of images and the necessity of manual labor, while providing unnatural images. The results of this study addressed this issue.

Our study had several advantages. First, a data augmentation method based on the same rules as those of artificial intelligence was used. It could also generate data while requiring minimal manpower. Using adversary, cycle-consistency, and identity losses, the optimized model could be used to produce various flame scenarios. The model could also be pre-trained for various wildfire scenarios in new environments, prior to the management of the forest; higher detection accuracy could, thus, be expected. Second, we improved the detection accuracy by applying a dense block based on DenseNet in the model. The training history and test results showed that the proposed methods facilitated good model performance. Third, it was proposed that the model could be applied to high-resolution images to overcome the limitations that depend primarily on the use of small-sized images, as inputs to the model. This allows us to identify the approximate location of the wildfire from a wide range of photographs.

There were also several limitations to our study. The model training was conducted using a limited forest class. Although during the experiment conducted with drone images the model identified the cloud and wildfire areas well (the upper part of the cropped photos in Figure 11), the smoke in the part of the image comprising the sky was not captured as a feature when the test data was obtained using CAM. This could be adjusted by increasing the class range or by learning additional models using images that are likely to confuse the model. Another potential problem was that the model performance for detection of wildfires in the nighttime was not considered. This temporal variable was excluded from the study because the purpose of this study was to check the efficiency of the data augmentation from artificial intelligence method and the efficiency of dense block in wildfire detection models. However, these details should be considered in further studies because of the different characteristics in the nighttime detection and in the day-time detection.

By improving upon the achievements and limitations of this study, in a future study, we intend to implement a forest-fire detection model in the field, by installing real-time surveillance cameras in Gangwon-do, Korea, which is exposed to the risk of wildfires every year.

In addition, by developing a technology that calculates the location of fires using image processing to measure fire area distance from camera and displays it on a map user interface, we intend to provide disaster-response support information for decision makers to realize a quick response in the event of the occurrence of a wildfire.

Author Contributions: Idea development and original draft writing, M.P.; draft review and editing, D.Q.T.; project administration, D.J.; supervision and funding acquisition, S.P. All authors have read and agreed to the published version of the manuscript.

Funding: This research was supported by a grant [2019-MOIS31-011] from the Fundamental Technology Development Program for Extreme Disaster Response funded by the Ministry of Interior and Safety, Korea.

Conflicts of Interest: The authors declare no conflict of interest.

References

1. Meng, Y.; Deng, Y.; Shi, P. Mapping forest wildfire risk of the world. In *World Atlas of Natural Disaster Risk*; Springer: Berlin/Heidelberg, Germany, 2015; pp. 261–275.
2. Jolly, W.M.; Cochrane, M.A.; Freeborn, P.H.; Holden, Z.A.; Brown, T.J.; Williamson, G.J.; Bowman, D.M. Climate-induced variations in global wildfire danger from 1979 to 2013. *Nat. Commun.* **2015**, *6*, 7537. [CrossRef]
3. Williams, A.P.; Allen, C.D.; Macalady, A.K.; Griffin, D.; Woodhouse, C.A.; Meko, D.M.; Swetnam, T.W.; Rauscher, S.A.; Seager, R.; Grissino-Mayer, H.D. Temperature as a potent driver of regional forest drought stress and tree mortality. *Nat. Clim. Chang.* **2012**, *2*, 1–6. [CrossRef]
4. Solomon, S.; Matthews, D.; Raphael, M.; Steffen, K. *Climate Stabilization Targets: Emissions, Concentrations, and Impacts over Decades to Millennia*; National Academies Press: Washington, DC, USA, 2011.
5. Mahmoud, M.A.I.; Ren, H. Forest Fire Detection Using a Rule-Based Image Processing Algorithm and Temporal Variation. *Math. Probl. Eng.* **2018**, *2018*, 1–8. [CrossRef]
6. Kim, Y.J.; Kim, E.G. Image based fire detection using convolutional neural network. *J. Korea Inst. Inf. Commun. Eng.* **2016**, *20*, 1649–1656. [CrossRef]
7. Celik, T.; Demirel, H. Fire detection in video sequences using a generic color model. *Fire Saf. J.* **2009**, *44*, 147–158. [CrossRef]
8. Wang, Y.; Dang, L.; Ren, J. Forest fire image recognition based on convolutional neural network. *J. Algorithms Comput. Technol.* **2019**, *13*. [CrossRef]
9. Souza, M.; Moutinho, A.; Almeida, M. Wildfire detection using transfer learning on augmented datasets. *Expert Syst. Appl.* **2020**, *142*, 112975. [CrossRef]
10. Ko, B.C.; Cheong, K.H.; Nam, J.Y. Fire detection based on vision sensor and support vector machines. *Fire Saf. J.* **2009**, *44*, 322–329. [CrossRef]
11. Zhao, J.; Zhang, Z.; Han, S.; Qu, C.; Yuan, Z.; Zhang, D. SVM based forest fire detection using static and dynamic features. *Comput. Sci. Inf. Syst.* **2011**, *8*, 821–841. [CrossRef]
12. Tung, T.X.; Kim, J. An effective four-stage smoke-detection algorithm using video image for early fire-alarm systems. *Fire Saf. J.* **2011**, *5*, 276–282. [CrossRef]
13. Gomes, P.; Santana, P.; Barata, J. A Vision-Based Approach to Fire Detection. *Int. J. Adv. Robot. Syst.* **2014**, *11*, 149. [CrossRef]
14. Xu, G.; Zhang, Y.; Zhang, Q.; Lin, G.; Wang, Z.; Jia, Y.; Wang, J. Video Smoke Detection Based on Deep Saliency Network. *Fire Saf. J.* **2019**, *105*, 277–285. [CrossRef]
15. Zhang, Q.; Xu, J.; Xu, L.; Guo, H. Deep convolutional neural networks for forest fire detection. In Proceedings of the 2016 International Forum on Management, Education and Information Technology Application, Guangzhou, China, 30–31 January 2016.
16. Pan, H.; Diaa, B.; Ahmet, E.C. Computationally efficient wildfire detection method using a deep convolutional network pruned via Fourier analysis. *Sensors* **2020**, *20*, 2891. [CrossRef]
17. Krizhevsky, A.; Sutskever, I.; Hinton, G.E. ImageNet classification with deep convolutional neural networks. *Adv. Neural Inf. Process. Syst.* **2012**, *1*, 1097–1105. [CrossRef]
18. Md, Z.A.; Aspiras, T.; Taha, T.M.; Asari, V.K.; Bowen, T.J.; Billiter, D.; Arkell, S. Advanced deep convolutional neural network approaches for digital pathology image analysis: A comprehensive evaluation with different use cases. *arXiv* **2019**, arXiv:1904.09075.
19. Simonyan, K.; Zisserman, A. Very deep convolutional networks for large-scale image recognition. *arXiv* **2014**, arXiv:1409.1556.
20. He, K.; Zhang, X.; Ren, S.; Sun, J. Deep residual learning for image recognition. In Proceedings of the IEEE Conference on Computer Vision and Pattern Recognition, Las Vegas, NV, USA, 27–30 June 2016; pp. 770–778.
21. Muhammad, K.; Ahmad, J.; Mehmood, I.; Rho, S. Convolutional Neural Networks based Fire Detection in Surveillance Videos. *IEEE Access* **2018**, *6*, 18174–18183. [CrossRef]
22. Jung, D.; Tuan, V.T.; Tran, D.Q.; Park, M.; Park, S. Conceptual Framework of an Intelligent Decision Support System for Smart City Disaster Management. *Appl. Sci.* **2020**, *10*, 666. [CrossRef]
23. Jain, p.; Coogan, S.C.; Subramanian, S.G.; Crowley, M.; Taylor, S.; Flannigan, M.D. A review of machine learning applications in wildfire science and management. *arXiv* **2020**, arXiv:2003.00646v2.

24. Zhang, Q.X.; Lin, G.H.; Zhang, Y.M.; Xu, G.; Wang, J.J. Wildland forest fire smoke detection based on faster r-cnn using synthetic smoke images. *Procedia Eng.* **2018**, *211*, 411–466. [CrossRef]

25. Cao, Y.; Yang, F.; Tang, Q.; Lu, X. An attention enhanced bidirectional LSTM for early forest fire smoke recognition. *IEEE Access* **2019**, *7*, 154732–154742. [CrossRef]

26. Zhao, Y.; Ma, J.; Li, X.; Zhang, J. Saliency Detection and Deep Learning-Based Wildfire Identification in UAV Imagery. *Sensors* **2018**, *18*, 712. [CrossRef]

27. Alexandrov, D.; Pertseva, E.; Berman, I.; Pantiukhin, I.; Kapitonov, A. Analysis of Machine Learning Methods for Wildfire Security Monitoring with an Unmanned Aerial Vehicles. In Proceedings of the 2019 24th Conference of Open Innovations Association (FRUCT), Moscow, Russia, 8–12 April 2019.

28. Namozov, A.; Cho, Y.I. An efficient deep learning algorithm for fire and smoke detection with limited data. *Adv. Electr. Comput. Eng.* **2018**, *18*, 121–129. [CrossRef]

29. Zhikai, Y.; Leping, B.; Teng, W.; Tianrui, Z.; Fen, W. Fire Image Generation Based on ACGAN. In Proceedings of the 2019 Chinese Control and Decision Conference (CCDC), Nanchang, China, 3–5 June 2019; pp. 5743–5746.

30. Li, T.; Zhao, E.; Zhang, J.; Hu, C. Detection of Wildfire Smoke Images Based on a Densely Dilated Convolutional Network. *Electronics* **2019**, *8*, 1131. [CrossRef]

31. Xu, G.; Zhang, Y.; Zhang, Q.; Lin, G.; Wang, J. Domain adaptation from synthesis to reality in single-model detector for video smoke detection. *arXiv* **2017**, arXiv:1709.08142.

32. Goodfellow, I.; Pouget-Abadie, J.; Mirza, M.; Xu, B.; Warde-Farley, D.; Ozair, S.; Courville, A.; Bengio, Y. Generative adversarial nets. In *Advances in Neural Information Processing Systems*; ACM: New York, NY, USA, 2014; pp. 2672–2680.

33. Zheng, K.; Wei, M.; Sun, G.; Anas, B.; Li, Y. Using Vehicle Synthesis Generative Adversarial Networks to Improve Vehicle Detection in Remote Sensing Images. *ISPRS Int. J. Geo-Inf.* **2019**, *8*, 390. [CrossRef]

34. Ledig, C.; Theis, L.; Huszár, F.; Caballero, J.; Cunningham, A.; Acosta, A.; Aitken, A.; Tejani, A.; Totz, J.; Wang, Z.; et al. Photo-realistic single image super-resolution using a generative adversarial network. In Proceedings of the IEEE Conference on Computer Vision and Pattern Recognition, Honolulu, HI, USA, 21–26 July 2017; pp. 105–114.

35. Isola, P.; Zhu, J.Y.; Zhou, T.; Efros, A.A. Image-To-Image Translation with Conditional Adversarial Networks. In Proceedings of the IEEE Conference on Computer Vision and Pattern Recognition (CVPR), Honolulu, HI, USA, 21–26 July 2017.

36. Zhang, H.; Xu, T.; Li, H.; Zhang, S.; Wang, X.; Huang, X.; Metaxas, D.N. StackGAN++: Realistic Image Synthesis with Stacked Generative Adversarial Networks. *IEEE Trans. Pattern Anal. Mach. Intell.* **2019**, *41*, 1947–1962. [CrossRef]

37. Sandfort, V.; Yan, K.; Pickhardt, P.J.; Summers, R.M. Data augmentation using generative adversarial networks (CycleGAN) to improve generalizability in CT segmentation tasks. *Sci. Rep.* **2019**, *9*, 16884. [CrossRef]

38. Uzunova, H.; Ehrhardt, J.; Jacob, F.; Frydrychowicz, A.; Handels, H. Multi-scale GANs for memory-efficient generation of high resolution medical images. In Proceedings of the International Conference on Medical Image Computing and Computer-Assisted Intervention (MICCAI), Shenzhen, China, 13–17 October 2019; pp. 112–120.

39. Hu, Q.; Wu, C.; Wu, Y.; Xiong, N. UAV Image High Fidelity Compression Algorithm Based on Generative Adversarial Networks Under Complex Disaster Conditions. *IEEE Access* **2019**, *7*, 91980–91991. [CrossRef]

40. Huang, G.; Liu, Z.; Van der Maaten, L.; Weinberger, K.Q. Densely connected convolutional networks. In Proceedings of the IEEE Conference on Pattern Recognition and Computer Vision 2017, Honolulu, HI, USA, 21–26 July 2017; pp. 4700–4708.

41. Zhu, J.Y.; Park, T.; Isola, P.; Efros, A.A. Unpaired Image-to-Image Translation using Cycle-Consistent Adversarial Networks. In Proceedings of the 2017 IEEE International Conference on Computer Vision (ICCV), Venice, Italy, 22–29 October 2017; pp. 2242–2251.

42. Oliva, A.; Torralba, A. Modeling the shape of the scene: A holistic representation of the spatial envelope. *Int. J. Comput. Vis.* **2001**, *42*, 145–175. [CrossRef]

43. Jeong, C.; Jang, S.-E.; Na, S.; Kim, J. Korean Tourist Spot Multi-Modal Dataset for Deep Learning Applications. *Data* **2019**, *4*, 139. [CrossRef]

44. Sokolova, M.; Japkowicz, N.; Szpakowicz, S. Beyond accuracy, F-Score and ROC: A family of discriminant measures for performance evaluation. In *AI 2006: Advances in Artificial Intelligence*; Springer: Berlin/Heidelberg, Germany, 2006; pp. 1015–1021.

45. Powers, D.M.W. Evaluation: From precision, recall and F-measure to ROC, informedness, markedness and correlation. *J. Mach. Learn. Technol.* **2011**, *2*, 37–63.

46. Tharwat, A. Classification assessment methods. *Appl. Comput. Inform.* **2018**, *13*. [CrossRef]

47. Paszke, A.; Gross, S.; Massa, F.; Lerer, A.; Bradbury, J.; Chanan, G.; Killeen, T.; Lin, Z.; Gimelshein, N.; Antiga, L. PyTorch: An imperative style, high-performance deep learning library. In Proceedings of the Advances in Neural Information Processing Systems, Vancouver, BC, Canada, 8–14 December 2019; pp. 8024–8035.

48. Johnson, J.M.; Khoshgoftaar, T.M. Survey on deep learning with class imbalance. *J. Big Data* **2019**, *6*, 27. [CrossRef]

49. Kingma, D.; Ba, J. Adam: A method for Stochastic Optimization. In Proceedings of the 3rd International Conference for Learning Representations, San Diego, CA, USA, 7–9 May 2015.

50. Hinton, G.; Srivastava, N.; Swersky, K. Neural networks for machine learning lecture 6a overview of mini-batch gradient descent. *Cited* **2012**, *14*, 1–31.

51. Zhou, B.; Khosla, A.; Lapedriza, A.; Oliva, A.; Torralba, A. Learning deep features for discriminative localization. In Proceedings of the IEEE Conference on Computer Vision and Pattern Recognition, Las Vegas, NV, USA, 27–30 June 2016; pp. 2921–2929.

52. Drone Center. Wildfire Video of Nangwol-dong. December 2015. Available online: http://www.dronecenter.kr/bbs/s5_4/3266 (accessed on 13 October 2020).

Publisher's Note: MDPI stays neutral with regard to jurisdictional claims in published maps and institutional affiliations.

Article

Development of an Automated Visibility Analysis Framework for Pavement Markings Based on the Deep Learning Approach

Kyubyung Kang [1], Donghui Chen [2], Cheng Peng [2], Dan Koo [1], Taewook Kang [3,*] and Jonghoon Kim [4]

1   Department of Engineering Technology, Indiana University-Purdue University Indianapolis (IUPUI), Indianapolis, IN 46038, USA; kyukang@iu.edu (K.K.); dankoo@iu.edu (D.K.)
2   Department of Computer and Information Science, Indiana University-Purdue University Indianapolis (IUPUI), Indianapolis, IN 46038, USA; dch1@iu.edu (D.C.); cp16@iu.edu (C.P.)
3   Korea Institute of Civil Engineering and Building Technology, Goyang-si 10223, Korea
4   Department of Construction Management, University of North Florida, Jacksonville, FL 32224, USA; jongkim@unf.edu
*   Correspondence: ktw@kict.re.kr; Tel.: +82-10-3008-5143

Received: 29 September 2020; Accepted: 13 November 2020; Published: 23 November 2020

Abstract: Pavement markings play a critical role in reducing crashes and improving safety on public roads. As road pavements age, maintenance work for safety purposes becomes critical. However, inspecting all pavement markings at the right time is very challenging due to the lack of available human resources. This study was conducted to develop an automated condition analysis framework for pavement markings using machine learning technology. The proposed framework consists of three modules: a data processing module, a pavement marking detection module, and a visibility analysis module. The framework was validated through a case study of pavement markings training data sets in the U.S. It was found that the detection model of the framework was very precise, which means most of the identified pavement markings were correctly classified. In addition, in the proposed framework, visibility was confirmed as an important factor of driver safety and maintenance, and visibility standards for pavement markings were defined.

Keywords: pavement markings; deep learning; visibility; framework

---

1. Introduction

Pavement markings play a critical role in reducing crashes and improving safety on public roads. They do not only convey traffic regulations, road guidance, and warnings for drivers, but also supplement other traffic control devices such as signs and signals. Without good visibility conditions of pavement markings, the safety of drivers is not assured. Therefore, it is important for transportation agencies and other stakeholders to establish a systematic way of frequently inspecting the quality of pavement markings before accidents occur.

State highway agencies in the U.S. invest tremendous resources to inspect, evaluate, and repair pavement markings on nearly nine million lane-miles [1]. One of the challenges in pavement marking inspection and maintenance is the variable durability of pavement markings. Conditions of pavement markings vary even if they were installed at the same time. Such conditions are highly dependent on the material characteristics, pavement characteristics, traffic volumes, weather conditions, etc. Unfortunately, inspecting all pavement markings at the right time is very challenging due to the lack of available human resources. Hence, an automated system for analyzing the condition of pavement markings is critically needed. This paper discusses a study that developed an automated condition analysis framework for pavement markings using machine learning technology. The proposed framework consists of three modules: a data processing module, a pavement marking detection module, and a visibility analysis module. The data processing module includes data acquisition and data annotation, which provides a clear and accurate dataset for the detection module to train. In the pavement marking detection module, a framework named YOLOv3 is used for training to detect and localize pavement markings. For the visibility analysis module, the contour of each pavement marking is clearly marked, and each contrast intensity value is also provided to measure visibility. The framework was validated through a case study of pavement markings training data sets in the U.S.

2. Related Studies

With the remarkable improvements in cameras and computers, pavement conditions can now be analyzed remotely using image processing technologies. Unlike traditional manual inspection, remote analysis does not require on-site operations and closed traffic, yet has high inspection accuracy and efficiency, which greatly reduces the management costs of the government's transportation department. With the continuous development of computer vision technology, more and more researchers are exploring how to use videos or images to complete the analysis of pavement systems. Ceylan et al. summarized the recent computer vision-based pavement engineering applications into seven categories: estimation of pavement conditions and performance, pavement management and maintenance, pavement distress prediction, structural evaluation of pavement systems, pavement image analysis and classification, pavement materials modeling, and other transportation infrastructure applications [2,3]. The increasing number of publications and technologies in these fields in recent years undoubtedly demonstrates that more and more researchers are interested in exploring the use of computer vision technology to study pavement engineering problems [4–10].

A complete pavement system consists mainly of the pavement and the painted markings or lanes. Intuitively, most studies of pavement systems focused on analyzing the pavements and markings. Regarding pavements, researchers pay more attention to exploring how to efficiently and precisely detect cracks on roads. Traditional techniques start mainly from pattern matching or texture analysis to help locate cracks. However, due to the diversity of cracks and their unfixed shapes, such traditional techniques have been found wanting. Studies have been conducted on the automatic identification of pavement cracks using neural network algorithms, due to their powerful learning and representing capabilities. In this new technique, the characteristic information on the road images is first extracted, and then the neural network is trained to recognize it. For sample pavements with known characteristics, the neural network can automatically learn and memorize them, whereas for unknown pavement samples, the neural network can automatically make inferences based on previously learned information.

Meignen et al. directly flattened all the pixels of each image into one-dimensional feature vectors, which were taken as the inputs to a neural network [11]. This method did not work very well, as different roads had different crack characteristics, and the training input data set was too large. Therefore, it is wise to first extract the features that are meaningful for recognizing pavement cracks, and then process the features using a neural network. Xu et al. proposed a modified neural network structure to improve the recognition accuracy [8]. First, the collected pavement images were segmented into several parts, after which the features were extracted from each part. For each segment, the probability that it could have cracks was inferred with the neural network model. The regional division strategy reduced the differences between the samples and effectively improved the performance of the network. Zhang et al. trained a supervised convolutional neural network (CNN) to decide if a patch represents a crack on a pavement [10]. The authors used 500 road system images taken with a low-cost smartphone to inspect the performance of the proposed model. The experiment results showed that the automatically learned features of the deep CNN provided a superior crack recognition capability compared with the features extracted from the hand-craft approaches. Li et al. explored the possibility that the size of the reception field in the CNN structure influences its performance [6]. They trained four CNNs with different reception field sizes and compared them. The results showed that the smaller reception fields had slightly better model accuracy, but also had a more time-consuming training process. Therefore, a good trade-off between effectiveness and efficiency was needed. In the study of Zhang et al., a recurrent neural network (RNN) named CrackNet-R was modeled to perform automatic pixel-wise crack detection for three-dimensional asphalt pavements [12]. In this model, the authors applied a new structure, a gated recurrent multilayer perceptron network, which showed a better memorization ability than other recurrent schemes. Relying on such a memorization ability, the CrackNet-R first searched the image sequence with the highest probability of having a crack pattern. Then an output layer was adopted to transform the timely probabilities of the sequence into pixel-wise probabilities. This novel pixel-wise pavement crack detection model provided a new orientation for the development of the field.

For pavement markings, many publications have also focused on the detection and classification of road signs or lanes, which is an important task for pavement system maintenance or autonomous driving. Most previous studies on this problem were developed with the image processing theory and the hand-craft pattern functions, which made it very difficult to generalize in various situations. Chen et al. proposed a two-stage framework for road markings detection and classification based on machine learning [13]. The first-stage detection model was carried out with the binarized normed gradient (BING) approach, and the second-stage classification model was realized with the principal component analysis network (PCANet). Both BING and PCANet are popular techniques in the field of machine learning. Yamamoto et al. adopted a simple neural network to recognize pavement markings on road surfaces [9]. The authors first extracted the candidate road areas based on the edge information, and then fed them to the neural network to accomplish the recognition. Gurghian et al. proposed a novel method called DeepLanes to directly estimate, from images taken with a side-mounted camera, the position of the lanes using a deep neural network [14]. Besides the ability of the proposed model to determine the existence of lane markings, it could also predict the position of the lane markings with an average speed of 100 frames per second at the centimeter level, without any auxiliary processing. This algorithm can provide significant support for the driver-assistant system that depends on the lanes. The aforementioned models mainly treated pavement markings and lanes as different objects for processing and analysis, until the emergence of the vanishing point guided network (VPGNet), which Lee et al. proposed [15]. VPGNet was an end-to-end deep CNN inspired by the multi-task network structure that can simultaneously detect road markings and lanes. It introduced the vanishing point prediction task into the network to guide lane detection, which improved the performance of the network in some bad situations such as rainy days or at night. The authors also provided a public image dataset for lane and road marking detection tasks with pixel-wise annotations.

3. Methodology

Figure 1 shows an overview of the proposed framework with its three modules: a data processing module, a pavement marking detection module, and a visibility analysis module.

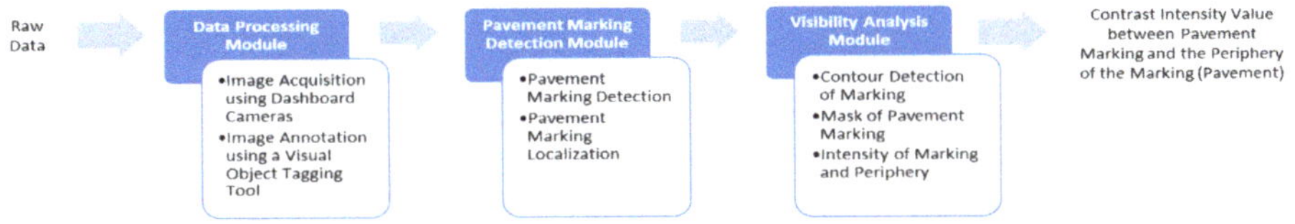

Figure 1. Overview of the Framework for Condition Analysis of Pavement Markings.

3.1. Data Processing Module

3.1.1. Data Acquisition

Since deep learning is a kind of data-driven algorithm, a training dataset must be prepared for the network model. Due to the quick development of autonomous driving, many public datasets for each driving situation have been collected, such as BDD100K, KITTI, Caltech Lanes, and VPGNet [15–18]. However, these datasets mainly focus on lane detection rather than the pavement markers. The VPGNet dataset provides annotations for lanes and pavement markers, but its pixel-wise annotations are inappropriate for the detection module used in this study. Thus, a system for automatically gathering images or videos of pavement systems must be set up. An action camera mounted behind the front windshield of a car driving on the roadways of Indianapolis, U.S.A. was used to record high-definition (HD) videos. Generally, the camera can capture 90% of the view in front of the moving vehicle, including pavements, transportation systems, and nearby street views, but only the data on pavements were used in this study. For this study, several trips were taken to record plenty of video data. The collected dataset covered various weather conditions, such as daytime, nighttime, sunny days, and rainy days, and different regions such as highways and urban areas. After screening all the video data, more than 1000 high-quality pictures were intercepted, of which about 200 were used for testing, and the remaining pictures were used for training, which maintained a good training-testing ratio.

3.1.2. Data Annotation

Since the primary goal of this study was to make the computer recognize the pavement markings in the road view videos or images, a labeled dataset had to be prepared for the model training process. After comparing multiple open-source labeling software, the visual object tagging tool (VoTT) was chosen to perform the data annotation. VoTT is a powerful open-source labeling software released by Microsoft [19]. This software provides a technique for automatic labeling based on the pre-trained network, which can significantly reduce the workload for annotations. It also supports many formats of the exported annotation results, which make the labeled sample set suitable for various deep learning development frameworks. Figure 2 shows an example of the labeling process.

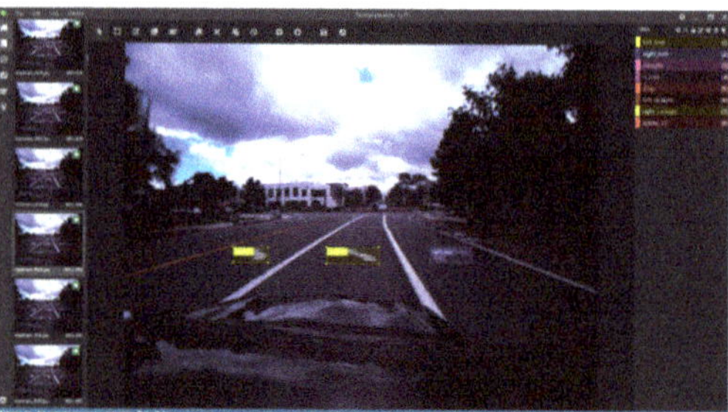

Figure 2. The software interface for data annotation tasks.

The VOC XML file format was chosen to generate annotations for each imported image. The key step in this procedure is to develop categories for the pavement markings. This study mainly focused on arrow-like pavement markings, such as those for left turn, right turn, etc. However, up to 10 categories of pavement markings were additionally captured with rectangular boxes for future research. Those categories are described in Figure 3. The annotated data were divided into a training dataset and a testing dataset at a ratio of 0.9:0.1.

Figure 3. Types of labeled pavement markings.

3.2. Pavement Markings Detection Module

In the field of computer vision, many novel object recognition frameworks have been studied in recent years. Among these frameworks, the most studied frameworks are deep learning-based models. Generally, according to the recognition principle, existing object detection models can be divided into two categories: two-stage frameworks and one-stage frameworks [20].

In two-stage frameworks, the visual target is detected in mainly two steps. First, abundant candidate regions that can possibly cover the targets are proposed, and then the validity of such regions is determined. R-CNN, Fast-RCNN, and Faster-RCNN are the representative two-stage frameworks, all of which have high a detection precision [21–23]. However, since they take time to generate candidate regions, their detection efficiency is relatively unpromising, which makes them unsuitable for real-time applications. To make up for this deficiency, researchers proposed the one-stage framework.

Compared to the two-stage framework, the one-stage framework gets rid of the phase for proposing candidate regions, and simultaneously performs localization and classification by treating the object detection task as a regression problem. Moreover, with the help of CNN, the one-stage framework can be constructed as an end-to-end network so that inferences can be made with simple matrix computations. Although this type of framework is slightly inferior to the two-stage framework in detection accuracy, its detection speed is dozens of times better. One of the representative one-stage frameworks, You Only Look Once (YOLO), achieves a balance between detection accuracy and speed [24]. After continuous updating and improvement, the detection accuracy of YOLOv3 has already caught up with that of most two-stage frameworks. This is why YOLOv3 was chosen as the pavement markings detection model in this study.

3.2.1. Demonstration of the YOLO Framework

Previous studies, such as on Region-CNN (R-CNN) and its derivative methods, used multiple steps to complete the detection, and each independent stage had to be trained separately, which slowed down the execution and made optimizing the training process difficult. YOLO uses an end-to-end design idea to transform the object detection task into a single regression problem, and directly obtains the coordinates and classification probabilities of the targets from raw image data. Although Faster-RCNN also directly takes the entire image as an input, it still uses the idea of the proposal-and-classifier of the R-CNN model. The YOLO algorithm brings a new solution to the object detection problem. It only scans the sample image once and uses the deep CNNs to perform both the classification and the localization. The detection speed of YOLO can reach 45 frames per second, which basically meets the requirement of real-time video detection applications.

YOLO divides the input image into $S * S$ sub-cells, each of which can detect objects individually. If the center point of an object falls in a certain sub-cell, the possibility of including the object in that sub-cell is higher than the possibility of including it in the adjacent sub-cells. In other words, this sub-cell should be responsible for the object. Each sub-cell needs to predict B bounding boxes and the confidence score that corresponds to each bounding box. In detail, the final prediction is a five-dimensional array, namely, $(x, y, w, h, c)^{T}$, where (x, y) is the offset that compares the center point of the bounding box with the upper left corner of the current sub-cell; (w, h) is the aspect ratio of the bounding box relative to the entire image; and c is the confidence value. In the YOLO framework, the confidence score has two parts: the possibility that there is an object in the current cell, and the Intersection over Union (IoU) value between the predicted box and the reference one. Suppose the possibility of the existence of the object is $Pr(Obj)$, and the IoU value between the predicted box and the reference box is $IoU(pred, truth)$, the formula for the confidence score is shown as Equation (1).

$$Confidence = Pr(Obj) * IoU(pred, truth) \tag{1}$$

Suppose that box_p is the predicted bounding box, and box_t is the reference bounding box. Then the IoU value can be calculated using the following formula.

$$IoU_p^t = \frac{box_p \cap box_t}{box_p \cup box_t} \tag{2}$$

In addition, YOLO outputs the individual conditional probability of C object categories for each cell. The final output of the YOLO network is a vector with $S * S * (5 * B + C)$ nodes.

YOLO adopted the classic network structure of CNN, which first extracted spatial features through convolutional layers, and then computed predictions by fully connected layers. This type of architecture limits the number of predictable target categories, which makes the YOLO model insufficient for multi-object detection. Moreover, since YOLO randomly selects the initial prediction boxes for each cell, it cannot accurately locate and capture the objects. To overcome the difficulties of YOLO and enhance its performance, Redmon et al. further modified its structure, applied novel features, and proposed improved models such as YOLOv2 and YOLOv3 [25,26].

The YOLOv2 network discarded the fully connected layers of YOLO, transformed YOLO into a fully convolutional network, and used the anchor boxes to assist in the prediction of the final detection bounding boxes. It predefined a set of anchor boxes with different sizes and aspect ratios in each cell to cover different positions and multiple scales of the entire image. These anchor boxes were used as initial candidate regions, which were distinguished according to the presence or absence of the targets inside them through the network. The position of the predicted bounding boxes was also continuously fine-tuned [27]. To fit the characteristics of the training samples, YOLOv2 used the k-means clustering algorithm to automatically learn the best initial anchor boxes from the training dataset. Moreover, YOLOv2 applied the Batch Normalization (B.N.) operation to the network structure. B.N. decreased the shift in the unit value in the hidden layer, and thus improved the stability of the neural network [28]. The B.N. regularization can prevent overfitting of the model, which makes the YOLOv2 network easier to converge.

Compared to YOLOv2, YOLOv3 mainly integrated some advanced techniques. While maintaining the fast detection, it further improved the detection accuracy and the ability to recognize small targets. YOLOv3 adopted a novel framework called Darknet-53 as its main network. Darknet-53 contained a total of 53 convolutional layers and adopted the skip-connection structure inspired by ResNet [29]. The much deeper CNN helped improve feature extraction. Motivated by the idea of multilayer feature fusion, YOLOv3 used the up-sampling method to re-extract information from the previous feature maps, and performed feature fusion with different-scale feature maps. In this way, more fine-grained information can be obtained, which improved the accuracy of the detection of small objects.

3.2.2. Structure of YOLOv3

Figure 4 shows the YOLOv3 network structure, which has two parts: Darknet-53 and the multi-scale prediction module. Darknet-53 is performed to extract features from the input image, the size of which is set at 416×416. It consists of two $1 * 1$ and $3 * 3$ convolutional layers, without any fully connected layers. Each convolutional layer is followed by a B.N. layer and a LeakyReLU activation function, which is regarded as the DBL block. In addition, Darknet-53 applies residual blocks in some layers. The main distinction of the residual block is that it adds a direct connection from the block entrance to the block exit, which helps the model to converge more easily, even if the network is very deep. When the feature extraction step is completed, feature maps are used for multi-scale object detection. In this part, YOLOv3 extracts three feature maps of different scales in the middle, middle-bottom, and bottom layers. In these layers, the concatenation operations are used to fuse the multi-scale features. In the end, three predictions of different scales will be obtained, each of which will contain the information on the three anchor boxes. Each anchor box is represented as a vector of $(5 + num_{class})$ dimensions, in which the former five values indicate the coordinates and the confidence score, and num_{class} refers to the category number of the objects. In this study, five kinds of arrow-like pavement markings were considered.

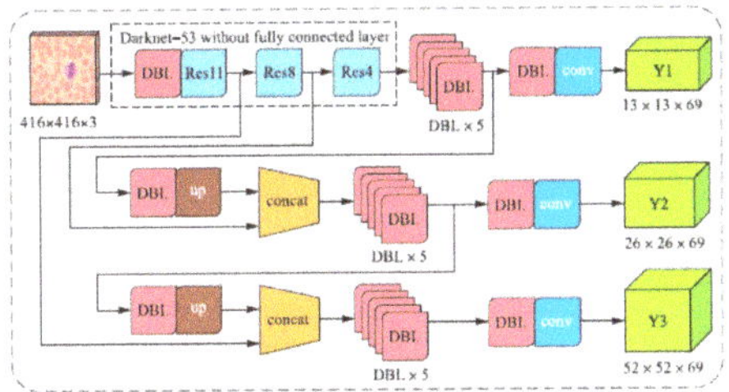

Figure 4. Structure of the YOLOv3 network. (https://plos.figshare.com/articles/YOLOv3_architecture_/8322632/1).

3.3. Visibility Analysis Module

After the data collected by the dashboard camera are labeled, a YOLOv3-based pavement marking detection module can be constructed and trained. The target pavement markings can be extracted and exported as small image patches. Thus, the next step is to design a visibility analysis module to help determine the condition of the pavement markings.

Pavement markings are painted mainly to give notifications to drivers in advance. As such, a significant property of pavement markings is their brightness. However, brightness is an absolute value affected by many factors, such as the weather and the illumination. Since the human visual system is more sensitive to contrast than to absolute luminance, the intensity contrast is chosen as the metric for the visibility of pavement markings [30]. In this study, contrast was defined as the difference between the average intensity of the pavement marking and the average intensity of the surrounding pavement. The main pipeline of this visibility analysis module is shown in Figure 5.

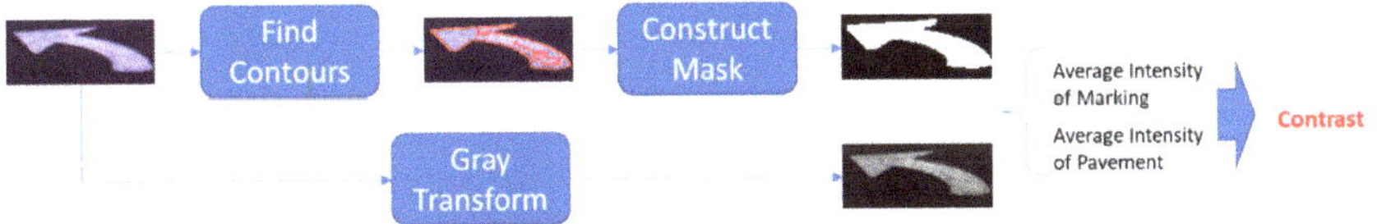

Figure 5. A demonstration of the pipeline of the visibility analysis module.

3.3.1. Finding Contours

As the pavement markings are already exported as image patches, the first step is to separate the pavement markings from the pavement. Since only arrow-like markings were considered in this study, the portion with the marking can be detached easily from the image, for as long as the outer contour of the marking is found. The contour can be described as a curve that joins all the continuous points along the boundary with the same color or intensity.

The contour tracing algorithm used in this part was proposed by Suzuki et al. [30] It was one of the first algorithms to define the hierarchical relationships of the borders and to differentiate the outer bounders from the hole bounders. This method has been integrated into the OpenCV Library [31].

The input image should be a binary image, which means the image has only two values: 0 and 1, with 0 representing the black background, and 1, the bright foreground or object. Thus, the border should mainly serve as the edge.

Assume that p_{ij} denotes the pixel value at position (i, j) in the image. Two variables, Newest Border Number (*NBD*), Last Newest Border Number (*LNBD*), are created to record the relationship between the pixels during the scanning process. The algorithm uses the row-by-row and left-to-right scanning schemes to process each *NBD* and *LNBD*, where $p_{ij} > 0$.

Step 1. If $p_{ij} = 1$ and $p_{i,j-1} = 0$, which indicate that this point is the starting point of an outer border, increment *NBD* by 1 and set $(i_2, j_2) \leftarrow (i, j-1)$. If $p_{ij} \geq 1$ and $p_{i,j+1} = 0$, which means it leads a hole border, increment *NBD* by 1 and set $(i_2, j_2) \leftarrow (i, j+1)$ and $LNBD \leftarrow p_{ij}$ in case $p_{ij} > 1$. Otherwise, jump to Step 3.

Step 2. From this starting point (i, j), perform the following operations to trace the border.

    2.1. Starting from pixel (i_2, j_2), traverse the neighborhoods of pixel (i, j) in a clockwise direction. In this study, the 4-connected case is selected to determine the neighborhoods, which means only the points connected horizontally and vertically are regarded as the neighborhoods. If a non-zero value exists, denote such pixel as (i_1, j_1). Otherwise, let $p_{ij} = -NBD$ and jump to Step 3.

    2.2. Assign $(i_2, j_2) \leftarrow (i_1, j_1)$ and $(i_3, j_3) \leftarrow (i, j)$.

    2.3. Taking pixel (i_3, j_3) as the center, traverse the neighborhoods in a counterclockwise direction from the next element (i_2, j_2) to find the first non-zero pixel, and assign it as (i_4, j_4).

    2.4. Update the value p_{i_3, j_3} according to Step 2.4 in Figure 6.

    2.5. If $p_{i_3, j_3+1} = 0$, update $p_{i_3, j_3} \leftarrow -NBD$.

    2.6. If $p_{i_3, j_3+1} \neq 0$ and $p_{i_3, j_3} = 1$, update $p_{i_3, j_3} \leftarrow NBD$.

    2.7. If the current condition satisfies $(i_4, j_4) = (i, j)$ and $(i_3, j_3) = (i_1, j_1)$, which means it goes back to the starting point, jump to Step 3. Otherwise, assign $(i_2, j_2) \leftarrow (i_3, j_3)$ and $(i_3, j_3) \leftarrow (i_4, j_4)$ and return to Sub-step 2.3.

Step 3. If $p_{ij} \neq 1$, update $LNBD \leftarrow |p_{ij}|$. Let $(i, j) \leftarrow (i, j+1)$ and return to Step 1 to process the next pixel. This algorithm stops after the most bottom-right pixel of the input image is processed.

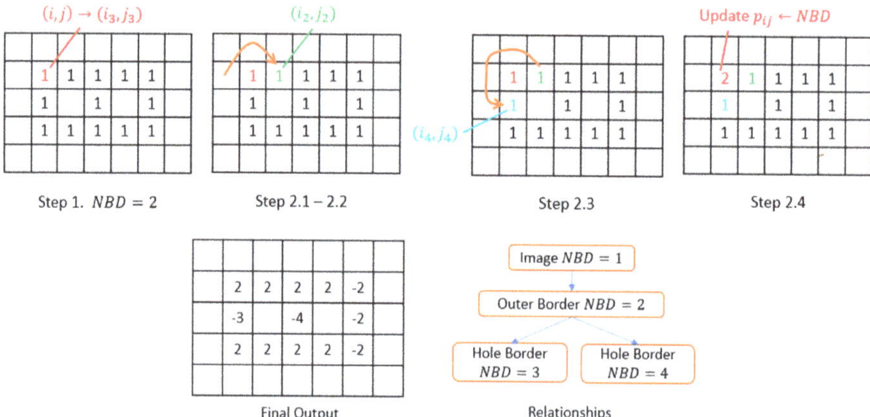

Figure 6. The introduction of Step 1, 2.1–2.4 and the introduction of the final output to the contour tracing algorithm.

Figure 6 show the contour tracing algorithm. By using this approach, the outer border or the contour of the arrow-like pavement marking can be found. However, due to uneven lighting or faded markings, the detected contours are not closed curves, as shown in Figure 7b. The incomplete contours cannot help separate the pavement marking portion. To solve this problem, the dilation operation is performed before the contours are traced.

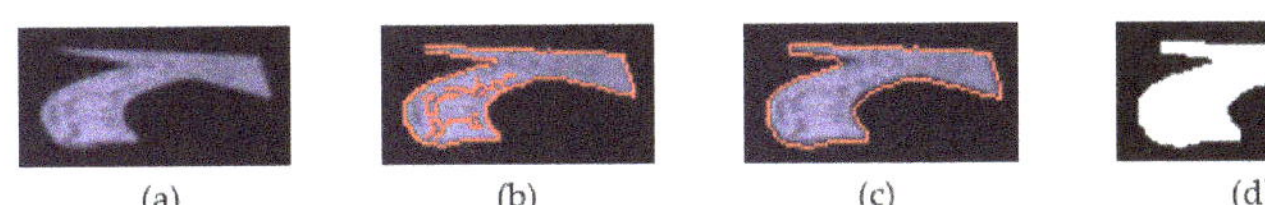

Figure 7. Results of the visibility analysis module. (**a**) Original patch, including the pavement marking; (**b**) Found contours without the dilation operation; (**c**) Found contours with the dilation operation; (**d**) Generated image mask for the marking; and (**e**) Generated image mask for the pavement.

Dilation is one of the morphological image processing methods, opposite to erosion [32]. The basic effect of the dilation operator on a binary image is the gradual enlargement of the boundaries of the foreground pixels so that the holes in the foreground regions would become smaller. The dilation operator takes two pieces of data as inputs. The first input is the image to be dilated, and the second input is a set of coordinate points known as a kernel. The kernel determines the precise effect of the dilation on the input image. It presumes that the kernel is a 3×3 square, with the origin at its center. To compute the dilation output of a binary image, each background pixel (i.e., 0-value) should be processed in turns. For each background pixel, if at least one coordinate point inside the kernel coincides with a foreground pixel (i.e., 1-value), the background pixel must be flipped to the foreground value. Otherwise, the next background pixel must be continually processed. Figure 8 shows the effect of a dilation using a 3×3 kernel. By using the dilation method before detecting the contours for the pavement marking patches, the holes in the markings are significantly eliminated, and the outer border becomes consistent and complete, which can be easily observed in Figure 7b,c.

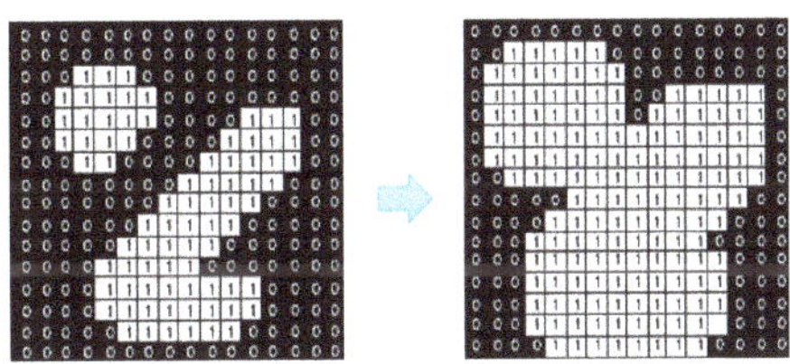

Figure 8. An example of the effect of the dilation operation (https://homepages.inf.ed.ac.uk/rbf/HIPR2/dilate.htm).

3.3.2. Construct Masks

Once the complete outer border of the pavement marking is obtained, the next step is to detach the pavement marking from the surrounding pavement. In practical scenarios, the pavement marking cannot be physically separated from the image patch due to its arbitrary shape. The most common way to achieve the target is to use masks to indicate the region segmentation. Since there are only two categories of objects, i.e., the pavement markings and the pavement, in this study, two masks had to be generated for each image patch.

Image masking is a non-destructive process of image editing that is universally employed in graphics software such as Photoshop to hide or reveal some portions of an image. Masking involves setting some of the pixel values in an image to 0 or another background value. Ordinary masks have only 1 and 0 values, and areas with a 0 value should be hidden (i.e., masked). Examples of masks generated for pavement markings are shown in Figure 7d,e.

3.3.3. Computing the Intensity Contrast

According to the pipeline of the visibility analysis module, the final step is to calculate the contrast between the pavement markings and the surrounding pavement. The straightforward way to determine the contrast value is to simply compute the difference between the average intensities of the markings and the pavement. However, this procedure does not adapt to the changes in the overall luminance. For instance, a luminance difference of 60 grayscales in a dark scenario (e.g., at night) should be more significant than the same luminance difference in a bright scenario (e.g., a sunny day). The human eyes sense brightness approximately logarithmically over a moderate range, which means the human visual system is more sensitive to intensity changes in dark circumstances than in bright environments [33]. Thus, in this study, the intensity contrast was computed using the Weber contrast method, the formula for which is:

$$
Contrast(M, P) = \frac{\overline{I_M} - \overline{I_P}}{\overline{I_P}}, \quad \overline{I_M} = \frac{\sum_{v \in Marking} I_v}{N_{Marking}}, \quad \overline{I_P} = \frac{\sum_{v \in Pavement} I_v}{N_{Pavement}} \tag{3}
$$

where $\overline{I_M}$ and $\overline{I_P}$ are the average intensity values of the pavement marking and the surrounding pavement, respectively, and the N_{region} is the number of pixels in the specific region.

4. Experimental Validation of the Framework

4.1. Experiment Settings

Regarding the pavement marking detection model, it needs to be trained with a labelled dataset to enhance its performance. In this study, a Windows 10 personal computer with an Nvidia GeForce RTX 2060 Super GPU and a total memory of 16 GB was used to perform the training and validation procedures. The deep learning framework that was used to build, train, and evaluate the detection network is the TensorFlow platform, which is one of the most popular software libraries used for machine learning tasks [34].

On actual roads, left-turn markings are much more common than right-turn markings. This leads to an imbalanced ratio of the proportions of these two kinds of pavement markings in the training dataset. If a classification network is trained without fixing this problem, the model could be completely biased [35]. Thus, in this study, data augmentation was performed before the model was trained. Specifically, for each left-turn (right-turn) marking, the image was flipped along the horizontal axis to make the left-turn (right-turn) marking a new right-turn (left-turn) marking. By applying this strategy to the whole training dataset, the numbers of the two markings should be the same. An example of this data augmentation method is shown in Figure 9.

Figure 9. An example of data augmentation.

4.2. Model Training

The neural network is trained by first calculating the loss through a forward inference, and then updating related parameters based on the derivative of loss to make the predictions as accurate as possible. Therefore, the design of loss functions is significant. In the YOLOv3 algorithm, the loss function has mainly three parts: the location offset of the predicted boxes, the deviation of the target confidence score, and the target classification error. The formula for the loss function is:

$$L(l, g, O, o, C, c) = \lambda_1 L_{loc}(l, g) + \lambda_2 L_{conf}(o, c) + \lambda_3 L_{cla}(O, C),\qquad(4)$$

where $\lambda_1 \sim \lambda_3$ refers to the scaling factors.

The location loss function uses the sum of the square errors between the true offset and the predicted offset, which is formulated as:

$$L_{loc}(l, g) = \sum_{m \in \{x, y, w, h\}} (\hat{l}^m - \hat{g}^m)^2 \qquad(5)$$

where $\hat{l}$ and $\hat{g}$ represent the coordinate offsets of the predicted bounding box and the referenced bounding box, respectively. Both $\hat{l}$ and $\hat{g}$ have four parameters: x for the offset along the x-axis, y for the offset along the y-axis, w for the box width, and h for the box height.

The target confidence score indicates the probability that the predicted box contains the target, which is computed as:

$$L_{conf}(o, c) = -\sum (o_i ln(\hat{c}_i) + (1 - o_i)ln(1 - \hat{c}_i)). \qquad(6)$$

The function L_{conf} uses the binary cross-entropy loss, where $o_i \in \{0, 1\}$ indicates whether the target actually exists in the predicted rectangle i. The 1 value means yes, and the 0 value means no. $c_i \in [0, 1]$ denotes the estimated probability that there is a target in the rectangle i.

The formulation of the target classification error in this study slightly differs from that in the YOLOv3 network. In the YOLOv3 network, the authors still used the binary cross-entropy loss function, as the author thought the object was possibly classified into more than one category in complicated reality scenes. However, in this study, the categories of the pavement markings were mutually exclusive. Thus, the multi-class cross-entropy loss function was used to measure the target classification error, the mathematical expression of which is:

$$L_{cla}(O, C) = -\sum_{i \in pos} \sum_{j \in cla} (O_{ij} ln(\hat{C}_{ij}) + (1 - O_{ij})ln(1 - \hat{C}_{ij})), \qquad(7)$$

where $O_{ij} \in \{0, 1\}$ indicates if the predicted box i contains the object j, and $\hat{C}_{ij} \in [0, 1]$ represents the estimated probability occurring in the aforementioned event.

Pan and Yang (2010) found that in the machine learning field, the knowledge gained while solving one problem can be applied to another different but related problem, which is called transfer learning [36]. For instance, the knowledge obtained while learning to recognize cars could be useful for recognizing trucks. In this study, the pavement marking detection network was not trained from scratch. Instead, a pre-trained model learning to recognize objects in the MS COCO dataset was used for the initialization. The MS COCO dataset, published by Lin et al., contains large-scale object detection data and annotations [37]. The model pre-trained from the COCO dataset can provide the machine with some general knowledge on object detection tasks. Starting from the pre-trained network, a fine-tuned procedure is conducted by feeding the collected data to the machine to make it capable of recognizing pavement markings. The total training process runs for a total of 50 epochs.

With the help of the TensorBoard integrated into the TensorFlow platform, users can monitor the training progress in real time. It can export figures to indicate the trends of specific parameters or predefined metrics. Figure 10 shows the trend of three different losses during the training process. The figure shows a decreasing trend for all the losses.

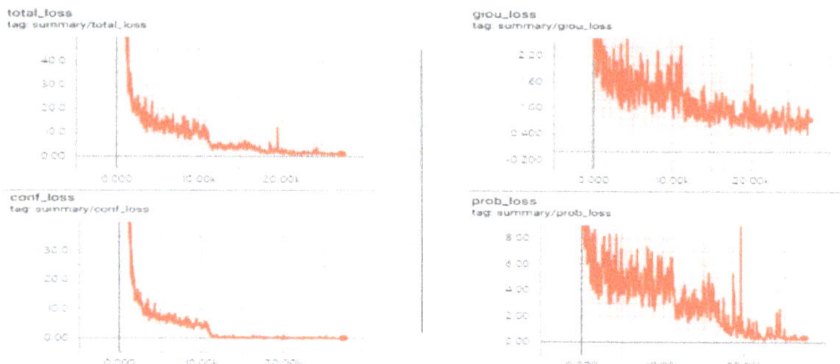

Figure 10. The trends of various loss functions during the training process monitored by TensorBoard.

4.3. Model Inference and Performance

After the training, the produced model is evaluated on the testing dataset. At the end of each training epoch, the network structure and the corresponding parameters are stored as the checkpoint file. For the evaluation, the checkpoint file with the least loss is chosen to be restored. The testing sample images are directly fed to the model as the inputs, and then the machine will automatically detect and locate the pavement markings in the image. Once the arrow-like pavement markings are recognized in the image, the detected areas are extracted to perform the visibility analysis. In this study, the function of the visibility analysis module was integrated into the evaluation of the pavement marking detection module. Thus, for each input image, the model drew the predicted bounding boxes, and added text to indicate the estimated category, the confidence score, and the contrast score on the image. Some examples of the evaluation of testing images are shown in Figure 11.

Figure 11. Visual results were evaluated on the testing samples.

From the figure, it can be seen that most of the pavement markings are correctly located and classified, and the contrast value provides a good measure of the visibility of the markings. The two subfigures on the left belong to the cloudy scenario, and the two on the right represent the sunny case. The pavements in the two subfigures on the left are both dark; but due to the poor marking condition, the contrast values of the top subfigure (i.e., 1.0, 0.4) are much lower than those at the bottom (i.e., 2.1, 1.9, 2.3). It can be observed that the pavement markings in the bottom subfigure are much more recognizable than those in the top subfigure, which validates the effectiveness of the contrast value for analyzing the visibility of pavement markings. Similarly, for the two subfigures on the right, all the detected pavement markings are in good condition; nevertheless, the contrast value of the bottom subfigure (i.e., 0.9, 1.1) is higher than that of the top subfigure (i.e., 0.2, 0.3), because the pavement in the bottom image is darker. This means the markings in the bottom-right subfigure are easier to identify than those in the top-right subfigure. The high brightness of the pavement could reduce the visibility of the markings on it, as the markings are generally painted white.

For the quantitative evaluation of the performance of the pavement marking detection model in this study, the mean average precision (mAP) was used. The results of the object detection system were divided into the following four categories by comparing the estimation with the reference label: true positives (TP), true negatives (TN), false positives (FP), and false negatives (FN). Table 1 defines these four metrics.

Table 1. Four categories of the metrics.

	Positive Predication	**Negative Prediction**
Positive Label	TP	FN
Negative Label	FP	TN

The precision refers to the proportion of the correct results in the identified positive samples, and the recall denotes the ratio of the correctly identified positive samples to all the positive samples. The formulas for these two metrics are as follows.

$$Precision = \frac{TP}{TP + FP} \qquad\qquad Recall = \frac{TP}{TP + FN} \tag{8}$$

To determine if a prediction box correctly located the target, an IoU threshold was predefined before the model was evaluated. For as long as the IoU value between the estimated bounding box and the ground truth was bigger than the threshold, this prediction was considered a correct detection. When the threshold value was adjusted, both the precision and the recall changed. As the threshold decreased, the recall value continued to increase, and the accuracy decreased after reaching a certain level. According to this pattern, the precision-recall curve, i.e., the PR curve, was drawn [38]. The AP value refers to the area under the PR curve, and the mAP value indicates the average AP among the multiple categories.

Figure 12 shows the results of the quantitative validation of the detection model on the testing dataset. As shown in the top-left subfigure, there are 203 sample images and 223 pavement marking objects included in the evaluation dataset. It can be seen that the distribution of different pavement markings is imbalanced. Thus, collecting more images and enlarging the dataset are the future study orientations for this proposal. The bottom-left subfigure demonstrates the number of true/false predictions upon the testing samples for each category, where the red portion represents the false predictions and the green potion refers to the true predictions. Given the number shown in the figure, it can be surmised that the detection model is working properly since most of the identified pavement markings were correctly classified. The right subfigure provides the average precision values for each category. The mAP value can reflect the overall performance of the detection module. However, the low mAP value indicates that there are some spaces to further improve the model.

From the validation results on testing samples, it is observed that some left-turn and right-turn markings are misclassified as the other category. By exploring the whole project, the reason causing this issue is finally found: a code issue. Since YOLOv3 is a representative framework in the object detection field, there are many open-source implementation codes. In this project, the detection model upon pavement markings is also trained with the open-source codes. Within the data preprocessing step of the codes, the author randomly chooses some training samples and flips them horizontally to enhance the diversity of the training data. Actually, this is a common and useful operation to achieve data augumentation. However, it does not fit for this pavement marking detection task. For general objects, the horizontal flipping would not change its category so that this operation is valid. But in terms of pavement markings, the flip process may transform the marking into another type, e.g., left-turn to right-turn. Thus, the flip operation within the codes generates wrong training samples, misleads the machine and hinders the performance of the model.

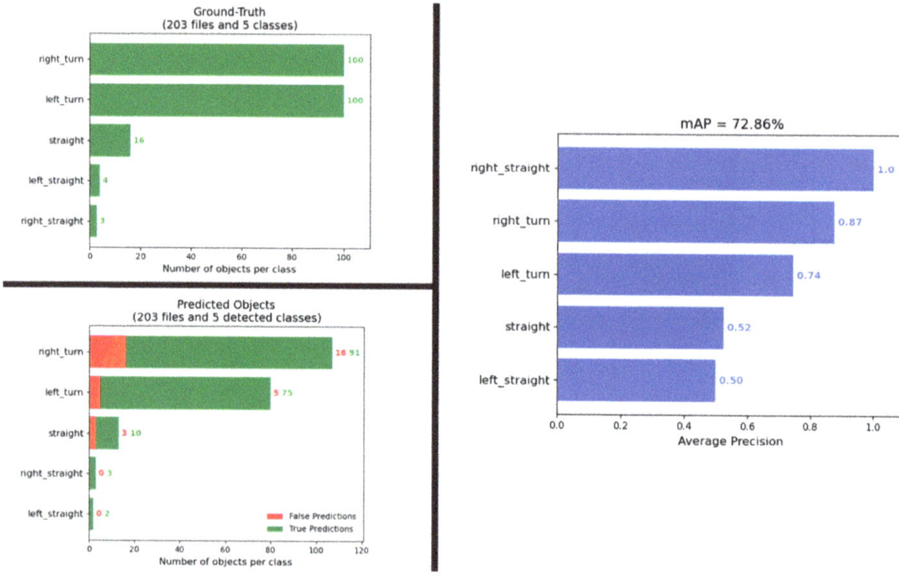

Figure 12. The quantitative evaluation information on the testing dataset of the trained model.

By removing the codes and re-training the model, the new quantitative validation results are shown in Figure 13. Comparing the Figures 12 and 13, the performance of the model is greatly enhanced, i.e., there is a 24% increment on the mAP value. The evaluation results fully prove the effectiveness of the YOLOv3 model in the pavement marking recognition task.

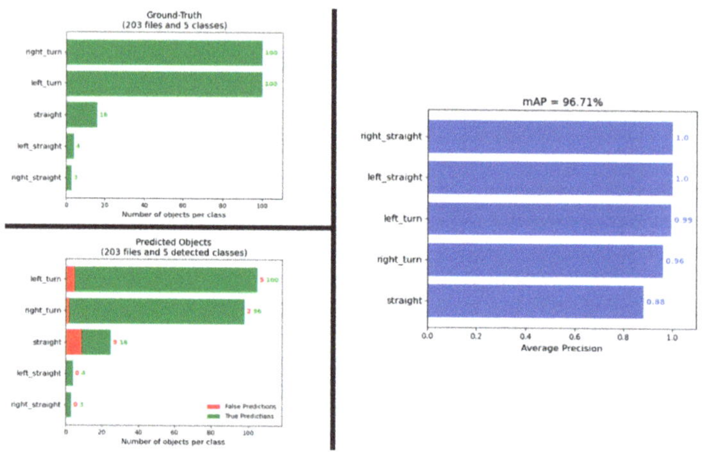

Figure 13. The quantitative evaluation information on the testing dataset of the improved model.

5. Conclusions

To identify issues with the detection and visibility of pavement markings, relevant studies were reviewed. The automated condition analysis framework for pavement markings using machine learning technology was proposed. The framework has three modules: a data processing module, a pavement marking detection module, and a visibility analysis module. The framework was validated through a case study of pavement marking training data sets in the U.S. From the quantitative results in the experimental section, the precision of the pavement marking detection module was pretty high, which fully validates the effectiveness of the YOLOv3 framework. Meanwhile, observing the visual results, all the pavement markings are correctly detected with the rectangle boxes and classified with the attached text in the road-scene images. In addition, the visibility metric of pavement markings was defined and the visibility within the proposed framework was confirmed as an important factor of driver safety and maintenance. The computed visibility values were also attached besides the detected pavement markings in the images. If the proposed study is used properly, pavement markings can be detected accurately, and their visibility can be analyzed to quickly identify places with safety concerns.

From the distribution of the testing samples, it can be inferred that the proportions of the straight markings, the right straight markings, and the left straight markings could be very low. Enlarging and enriching the training dataset could be a goal for future research.

Author Contributions: Conceptualization, K.K., T.K. and J.K.; Data curation, K.K.; Formal analysis, K.K., D.C. and C.P.; Investigation, D.K. and J.K.; Methodology, K.K., D.K., T.K. and J.K.; Project administration, T.K.; Resources, T.K.; Software, K.K., D.C., C.P. and T.K.; Writing—review & editing, K.K. and T.K. All authors have read and agreed to the published version of the manuscript.

Funding: This research was supported by a grant (KICT 2020-0559) from the Remote Scan and Vision Platform Elementary Technology Development for Facility and Infrastructure Management funded by KICT (Korea Institute of Civil Engineering and Building Technology) and a grant (20AUDP-B127891-04) from the Architecture and Urban Development Research Program funded by the Ministry of Land, Infrastructure, and Transport of the Korean government.

Conflicts of Interest: The authors declare no conflict of interest.

References

1. U.S. Department of Transportation Federal Highway Administration. Available online: https://www.fhwa.dot.gov/policyinformation/statistics/2018/hm15.cfm (accessed on 27 August 2020).
2. Ceylan, H.; Bayrak, M.B.; Gopalakrishnan, K. Neural networks applications in pavement engineering: A recent survey. *Int. J. Pavement Res. Technol.* **2014**, *7*, 434–444.
3. El-Basyouny, M.; Jeong, M.G. Prediction of the MEPDG asphalt concrete permanent deformation using closed form solution. *Int. J. Pavement Res. Technol.* **2014**, *7*, 397–404. [CrossRef]
4. Eldin, N.N.; Senouci, A.B. A pavement condition-rating model using backpropagation neural networks. *Comput. Aided Civ. Infrastruct. Eng.* **1995**, *10*, 433–441. [CrossRef]
5. Hamdi; Hadiwardoyo, S.P.; Correia, A.G.; Pereira, P.; Cortez, P. Prediction of Surface Distress Using Neural Networks. In *AIP Conference Proceedings*; AIP Publishing LLC: College Park, MD, USA, 2017; Volume 1855, p. 040006. [CrossRef]
6. Li, B.; Wang, K.C.P.; Zhang, A.; Yang, E.; Wang, G. Automatic classification of pavement crack using deep convolutional neural network. *Int. J. Pavement Eng.* **2020**, *21*, 457–463. [CrossRef]
7. Nhat-Duc, H.; Nguyen, Q.L.; Tran, V.D. Automatic recognition of asphalt pavement cracks using metaheuristic optimized edge detection algorithms and convolution neural network. *Autom. Constr.* **2018**, *94*, 203–213. [CrossRef]
8. Xu, G.; Ma, J.; Liu, F.; Niu, X. Automatic Recognition of Pavement Surface Crack Based on B.P. Neural Network. In Proceedings of the 2008 International Conference on Computer and Electrical Engineering, Phuket, Thailand, 20–22 December 2008; pp. 19–22. [CrossRef]
9. Yamamoto, J.; Karungaru, S.; Terada, K. Road Surface Marking Recognition Using Neural Network. In Proceedings of the 2014 IEEE/SICE International Symposium on System Integration; SII, Tokyo, Janpan, 13–15 December 2014; pp. 484–489. [CrossRef]

10. Zhang, L.; Yang, F.; Yimin, D.Z.; Zhu, Y.J. Road Crack Detection Using Deep Convolutional Neural Network. In Proceedings of the 2016 IEEE international conference on image processing (ICIP), Phoenix, AZ, USA, 25–28 September 2016; Volume 2016, pp. 3708–3712. [CrossRef]

11. Meignen, D.; Bernadet, M.; Briand, H. One Application of Neural Networks for Detection of Defects Using Video Data Bases: Identification of Road Distresses. In Proceedings of the International Conference on Database and Expert Systems Applications, 8th International Conference, Toulouse, France, 1–5 September 1997; pp. 459–464. [CrossRef]

12. Zhang, A.; Wang, K.C.P.; Fei, Y.; Liu, Y.; Chen, C.; Yang, G.; Li, J.Q.; Yang, E.; Qiu, S. Automated pixel-level pavement crack detection on 3D asphalt surfaces with a recurrent neural network. *Comput.-Aided Civ. Infrastruct. Eng.* **2019**, *34*, 213–229. [CrossRef]

13. Chen, T.; Chen, Z.; Shi, Q.; Huang, X. Road marking detection and classification using machine learning algorithms. In Proceedings of the 2015 IEEE Intelligent Vehicles Symposium (IV), Seoul, Korea, 28 June–1 July 2015; Volume 2015, pp. 617–621. [CrossRef]

14. Gurghian, A.; Koduri, T.; Bailur, S.V.; Carey, K.J.; Murali, V.N. DeepLanes: End-To-End Lane Position Estimation Using Deep Neural Networks. In Proceedings of the 2016 IEEE Computer Society Conference on Computer Vision and Pattern Recognition Workshops, Las Vegas, NV, USA, 26 June–1 July 2016. [CrossRef]

15. Lee, S.; Kim, J.; Yoon, J.S.; Shin, S.; Bailo, O.; Kim, N.; Lee, T.H.; Hong, H.S.; Han, S.H.; Kweon, I.S. VPGNet: Vanishing Point Guided Network for Lane and Road Marking Detection and Recognition. In Proceedings of the 2017 IEEE International Conference on Computer Vision, Venice, Italy, 22–29 October 2017. [CrossRef]

16. Geiger, A.; Lenz, P.; Stiller, C.; Urtasun, R. Vision meets robotics: The KITTI dataset. *Int. J. Robot. Res.* **2013**, *32*, 1231–1237. [CrossRef]

17. Griffin, G.; Holub, A.; Perona, P. Caltech-256 object category dataset. *Caltech Mimeo* **2007**, *11*. Available online: http://authors.library.caltech.edu/7694 (accessed on 27 August 2020).

18. Yu, F.; Chen, H.; Wang, X.; Xian, W.; Chen, Y.; Liu, F.; Madhavan, V.; Darrell, T. BDD100K: A Diverse Driving Dataset for Heterogeneous Multitask Learning. In Proceedings of the 2020 IEEE/CVF Conference on Computer Vision and Pattern Recognition (CVPR), Seattle, WA, USA, 13–19 June 2020. [CrossRef]

19. GitHub Repository. Available online: https://github.com/microsoft/VoTT (accessed on 27 August 2020).

20. Liu, L.; Ouyang, W.; Wang, X.; Fieguth, P.; Chen, J.; Liu, X.; Pietikäinen, M. Deep learning for generic object detection: A survey. *Int. J. Comput. Vis.* **2020**, *128*, 261–318. [CrossRef]

21. Girshick, R.; Donahue, J.; Darrell, T.; Malik, J. Rich Feature Hierarchies for Accurate Object Detection and Semantic Segmentation. In Proceedings of the 2014 IEEE Computer Society Conference on Computer Vision and Pattern Recognition, Columbus, OH, USA, 23–28 June 2014; pp. 580–587. [CrossRef]

22. Girshick, R. Fast R-CNN. In Proceedings of the 2015 IEEE International Conference on Computer Vision, Santiago, Chile, 7–13 December 2015; pp. 1440–1448.

23. Ren, S.; He, K.; Girshick, R.; Sun, J. Faster R-CNN: Towards real-time object detection with region proposal networks. *IEEE Trans. Pattern Anal. Mach. Intell.* **2017**, *39*, 1137–1149. [CrossRef] [PubMed]

24. Redmon, J.; Divvala, S.; Girshick, R.; Farhadi, A. You Only Look Once: Unified, Real-Time Object Detection. In Proceedings of the 2016 IEEE Computer Society Conference on Computer Vision and Pattern Recognition, Vegas, NV, USA, 27–30 June 2016; Volume 2016. [CrossRef]

25. Redmon, J.; Farhadi, A. YOLO9000: Better, Faster, Stronger. Available online: http://pjreddie.com/yolo9000/ (accessed on 23 November 2020).

26. Redmon, J.; Farhadi, A. YOLOv3: An Incremental Improvement. Available online: http://arxiv.org/abs/1804.02767 (accessed on 23 November 2020).

27. Ioffe, S.; Szegedy, C. Batch Normalization: Accelerating Deep Network Training by Reducing Internal Covariate Shift. In Proceedings of the 32nd International Conference on Machine Learning, Lille, France, 6–11 July 2015; Volume 1, pp. 448–456. Available online: https://arxiv.org/abs/1502.03167v3 (accessed on 23 November 2020).

28. Zhong, Y.; Wang, J.; Peng, J.; Zhang, L. Anchor Box Optimization for Object Detection. In Proceedings of the 2020 IEEE Winter Conference on Applications of Computer Vision, Snowmass Village, CO, USA, 1–5 March 2020. [CrossRef]

29. He, K.; Zhang, X.; Ren, S.; Sun, J. Deep Residual Learning for Image Recognition. In Proceedings of the 2016 IEEE Computer Society Conference on Computer Vision and Pattern Recognition, Vegas, NV, USA, 27–30 June 2016; Volume 2016, pp. 770–778. [CrossRef]

30. Suzuki, S.; Be, K.A. Topological structural analysis of digitized binary images by border following. *Comput. Vis. Graph. Image Process.* **1985**, *30*, 32–46. [CrossRef]

31. Bradski, G. The OpenCV Library. *Dr. Dobbs J. Softw. Tools* **2000**, *25*, 120–125. [CrossRef]

32. Liang, J.I.; Piper, J.; Tang, J.Y. Erosion and dilation of binary images by arbitrary structuring elements using interval coding. *Pattern Recognit. Lett.* **1989**, *9*, 201–209. [CrossRef]

33. Klein, S.A.; Carney, T.; Barghout-Stein, L.; Tyler, C.W. Seven models of masking. In *Human Vision and Electronic Imaging II.*; International Society for Optics and Photonics: Bellingham, WA, USA, 1997; Volume 3016, pp. 13–24.

34. Abadi, M.; Agarwal, A.; Barham, P.; Brevdo, E.; Chen, Z.; Citro, C.; Corrado, G.S.; Davis, A.; Dean, J.; Devin, M.; et al. TensorFlow: Large-Scale Machine Learning on Heterogeneous Distributed Systems; Distributed, Parallel and Cluster Computing. Available online: http://arxiv.org/abs/1603.04467 (accessed on 23 November 2020).

35. Longadge, R.; Dongre, S. Class imbalance problem in data mining review. *Int. J. Comput. Sci. Netw.* **2013**, *2*. Available online: http://arxiv.org/abs/1305.1707 (accessed on 23 November 2020).

36. Pan, S.J.; Yang, Q. A survey on transfer learning. In *IEEE Transactions on Knowledge and Data Engineering*; IEEE: Piscataway, NJ, USA, 2010; Volume 22, pp. 1345–1359. [CrossRef]

37. Lin, T.Y.; Maire, M.; Belongie, S.; Hays, J.; Perona, P.; Ramanan, D.; Dollár, P.; Zitnick, C.L. Microsoft COCO: Common objects in context. In *European Conference on Computer Vision*; Springer: Cham, Switzerland, 2014; 8693 LNCS (PART 5); pp. 740–755. [CrossRef]

38. Davis, J.; Goadrich, M. The relationship between precision-recall and ROC curves. In *ACM International Conference Proceeding Series*; ACM: New York, NY, USA, 2006; Volume 148, pp. 233–240. [CrossRef]

Publisher's Note: MDPI stays neutral with regard to jurisdictional claims in published maps and institutional affiliations.

Article

Intelligent Mapping of Urban Forests from High-Resolution Remotely Sensed Imagery Using Object-Based U-Net-DenseNet-Coupled Network

Shaobai He [1,2,3], **Huaqiang Du** [1,2,3,*], **Guomo Zhou** [1,2,3], **Xuejian Li** [1,2,3], **Fangjie Mao** [1,2,3], **Di'en Zhu** [4], **Yanxin Xu** [1,2,3], **Meng Zhang** [1,2,3], **Zihao Huang** [1,2,3], **Hua Liu** [1,2,3] and **Xin Luo** [1,2,3]

[1] State Key Laboratory of Subtropical Silviculture, Zhejiang A & F University, Hangzhou 311300, China; 2018103241005@stu.zafu.edu.cn (S.H.); zhougm@zafu.edu.cn (G.Z.); 2017303661004@stu.zafu.edu.cn (X.L.); maofj@zafu.edu.cn (F.M.); xuyanxin@stu.zafu.edu.cn (Y.X.); 2017103242004@stu.zafu.edu.cn (M.Z.); 2018103241008@stu.zafu.edu.cn (Z.H.); 2018103242003@stu.zafu.edu.cn (H.L.); 2019602041026@stu.zafu.edu.cn (X.L.)
[2] Key Laboratory of Carbon Cycling in Forest Ecosystems and Carbon Sequestration of Zhejiang Province, Zhejiang A & F University, Hangzhou 311300, China
[3] School of Environmental and Resources Science, Zhejiang A & F University, Hangzhou 311300, China
[4] The College of Forestry, Beijing Forestry University, Beijing 100083, China; 2015116021018@stu.zafu.edu.cn
[*] Correspondence: duhuaqiang@zafu.edu.cn

Received: 26 October 2020; Accepted: 25 November 2020; Published: 30 November 2020

Abstract: The application of deep learning techniques, especially deep convolutional neural networks (DCNNs), in the intelligent mapping of very high spatial resolution (VHSR) remote sensing images has drawn much attention in the remote sensing community. However, the fragmented distribution of urban land use types and the complex structure of urban forests bring about a variety of challenges for urban land use mapping and the extraction of urban forests. Based on the DCNN algorithm, this study proposes a novel object-based U-net-DenseNet-coupled network (OUDN) method to realize urban land use mapping and the accurate extraction of urban forests. The proposed OUDN has three parts: the first part involves the coupling of the improved U-net and DenseNet architectures; then, the network is trained according to the labeled data sets, and the land use information in the study area is classified; the final part fuses the object boundary information obtained by object-based multiresolution segmentation into the classification layer, and a voting method is applied to optimize the classification results. The results show that (1) the classification results of the OUDN algorithm are better than those of U-net and DenseNet, and the average classification accuracy is 92.9%, an increase in approximately 3%; (2) for the U-net-DenseNet-coupled network (UDN) and OUDN, the urban forest extraction accuracies are higher than those of U-net and DenseNet, and the OUDN effectively alleviates the classification error caused by the fragmentation of urban distribution by combining object-based multiresolution segmentation features, making the overall accuracy (OA) of urban land use classification and the extraction accuracy of urban forests superior to those of the UDN algorithm; (3) based on the Spe-Texture (the spectral features combined with the texture features), the OA of the OUDN in the extraction of urban land use categories can reach 93.8%, thereby the algorithm achieved the accurate discrimination of different land use types, especially urban forests (99.7%). Therefore, this study provides a reference for feature setting for the mapping of urban land use information from VHSR imagery.

Keywords: urban forests; OUDN algorithm; deep learning; object-based; high spatial resolution remote sensing

1. Introduction

Urban land use mapping and the information extraction of urban forest resources are significant, yet challenging tasks in the field of remote sensing and have great value for urban environment monitoring, planning, and designing [1–3]. In addition, smart cities are now an irreversible trend in urban development in the world, and urban forests constitute "vital," "green," and indispensable infrastructure in cities. Therefore, the intelligent mapping of urban forest resources from remote sensing data is an essential component of smart city construction.

Over the past few decades, multispectral (such as the Thematic Mapper (TM)) [4–7], hyperspectral, and LiDAR [8–10] techniques have played important roles in the monitoring of urban forest resources. Currently, with the rapid development of modern remote sensing technologies, a very large amount of VHSR remotely sensed imagery (such as WorldView-3) is commercially available, creating new opportunities for the accurate extraction of urban forests at a very detailed level [11–13]. The application of VHSR images in urban forest resource monitoring has attracted increasing attention because of the rich and fine properties in these images. However, the ground objects of VHSR images are highly complex and confusing. For one thing, numerous land use types (such as Agricultural Land and grassland) have the same spectrum and texture characteristics [14], resulting in strong homogeneity in different categories [15], that is, the phenomenon of "same spectrum with different objects." For another, rich detailed information gives similar objects (such as building composed of different construction materials) strong heterogeneity in the spectral and structural properties [16], resulting in the phenomenon of "same object with different spectra". In addition, traditional statistical classification methods encounter these problems in the extraction of urban forests from VHSR remote sensing images. Additionally, urban forests with fragmented distributions are composed of scattered trees, street trees, and urban park forest vegetation. This also creates very large challenges for urban land use classification and the accurate mapping of urban forests [17].

Object-based classification first aggregates adjacent pixels with similar spectral and texture properties into complementary and overlapping objects through the image segmentation method to achieve image classification, and the processing units are converted from conventional pixels to image objects [18]. This classification method is based on homogeneous objects. In addition to applying the spectral information of images, this method fully exploits spatial features such as geometric shapes and texture details. The essence of object-based classification is to break through the limitations of traditional pixel-based classification and reduce the phenomena of the "same object with different spectra" and the "salt-and-pepper" phenomenon caused by distribution fragmentation. Therefore, object-based classification methods often yield better results than traditional pixel-based classification methods [19]. Recently, the combination of object-based and machine learning (ML) is widely used to detect features in the forest such as damage detection, landslide detection, and insect-infested forests [20–23]. In terms of ML, deep learning (DL) uses a large amount of data to train the model and can simulate and learn high-level features [24], making deep learning a new popular topic in the current research on the intelligent extraction of VHSR remote sensing information [15,25,26].

For DL, DCNNs and semantic segmentation algorithms are widely used in the classification of VHSR images, providing algorithmic support for accurate classification and facilitating great progress [27–38]. Among them, DCNNs are the core algorithms for the development of deep learning [39]. These networks learn abstract features through multiple layers of convolutions, conduct network training and learning, and finally, classify and predict images. DenseNet is a classic convolutional neural network framework [40]. This network can extract abstract features while combining the information features of all previous layers, so it has been widely applied in the classification of remote sensing images [41–44]. However, this network has some problems such as the limited extraction of abstract features. Semantic segmentation places higher requirements on the architectural design of convolutional networks, classifying each pixel in the image into a corresponding category, that is, achieving pixel-level classification. A typical representation of semantic segmentation is U-net [45], which combines upsampling with downsampling. U-net can not only extract deeper features but also achieve accurate classification [46,47]. Therefore, U-net and DenseNet can be integrated to address the problem of the limited extraction of abstract features in DenseNet, and this combination may facilitate more accurate extraction from VHSR images.

In summary, object-based multiresolution segmentation offers obvious advantages in dealing with the problems of "same object with different spectra" and the "salt-and-pepper" phenomenon caused by distribution fragmentation [48–53], and deep learning is an important method for the intelligent mapping of VHSR remote sensing images. Consequently, this research proposes the novel classification method of the object-based U-net-DenseNet-coupled network (OUDN) to realize the intelligent and accurate extraction of urban land use and urban forest resources. This study takes subregion of the Yuhang District of Hangzhou City as the study area, with WorldView-3 images as the data source. First, the DenseNet and U-net network architectures are integrated; then, the network is trained according to the labeled data sets, and land use classification results are obtained based on the trained model. Finally, the object boundaries derived by object-based multiresolution segmentation are combined with the classification results of deep learning to optimize the classification results with the majority voting method.

2. Materials and Methods

2.1. Study Area

In this research, a subregion of the Yuhang District (YH) of Hangzhou, in Zhejiang Province in Southeast China, was chosen as the study area (Figure 1). WorldView-3 images of the study area were captured on 28 October 2018. The images contain four multispectral bands (red, green, blue, and near infrared (NIR)) with a spatial resolution of 2 m and a panchromatic band with a spatial resolution of 0.5 m. According to the USGS land cover classification system [54] and the FROM-GLC10 [55,56], the land use categories were divided into six classes, including Forest, Built-up, Agricultural Land, Grassland, Barren Land, and Water. As shown in Figure 1b, due to shadows in VHSR image, this study added a class of Others, including Shadow of trees and buildings. The detailed descriptions of each land use class and its corresponding subclasses are listed in Table 1.

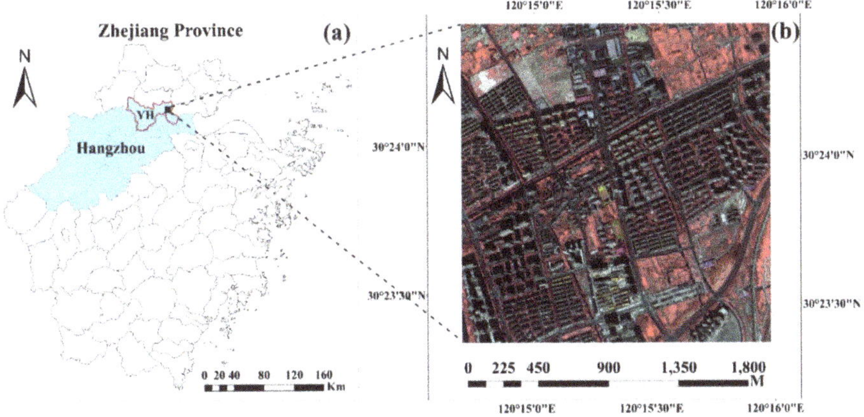

Figure 1. Location of the study area: (**a**) Zhejiang Province, and the blue polygons represent Hangzhou, (**b**) the subregion of the Yuhang District (YH) of Hangzhou.

Table 1. The urban land use classes and the corresponding subclass components.

Land Use Classes	Subclass Components
Forest	Deciduous Forest, Evergreen Forest
Build-up	Residential, Commercial and Services, Industrial, Transportation
Agricultural Land	Cropland, Nurseries, Other Agricultural Land
Grassland	Nature Grassland, Managed Grassland
Barren Land	Dry Salt Flats, Sandy Areas other than Beaches, Bare Exposed Rock
Water	Streams, River, Pond
Others	Shadow of trees and buildings

2.2. Data Processing

The image preprocessing including radiation correction and atmosphere correction was first performed using ENVI 5.3. Then, this label maps of the actual land use categories were made by eCognition software based on the results of the field survey combined with the method of visual interpretation. Due to the limitations on the size of the processed images from the GPU as well as to obtain more training images and to better extract image features, this study used the overlapping cropping method (Figure 2) to segment the images in the sample set into 4761 subimage blocks using 128 × 128 pixel windows for the minibatch training of the DL algorithms.

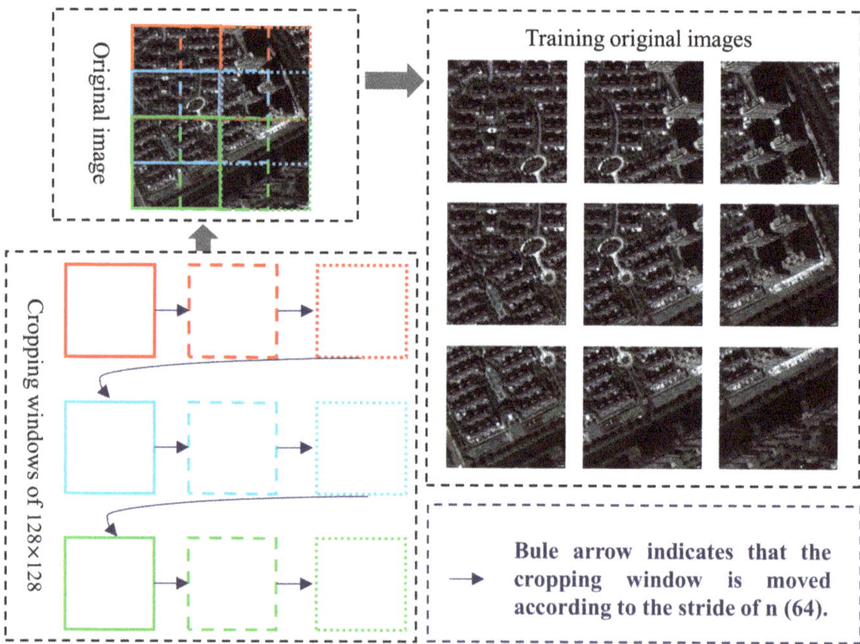

Figure 2. The overlapping cropping method for training the deep learning (DL) network. The size of the cropping windows is set to 128 × 128 pixels, where n is defined as half of 128.

2.3. Feature Setting

In this study, the classification features are divided into three groups: (1) the original R, G, B, and NIR bands, namely, the spectral features (Spe), the spectral features combined with the vegetation index features (Spe-Index), and the spectral features combined with the texture features (Spe-Texture). Based on these three groups of features, the performance of the OUDN algorithm in the mapping of urban land use and urban forest information is evaluated. Descriptions of the spectral features, vegetation index, and texture are given in Table 2. The texture features based on the gray-level co-occurrence matrix (GLCM) [57] include mean, variance, entropy, angular second moment, homogeneity, contrast, dissimilarity, and correlation [58–60] with different calculation windows (3×3, 5×5, 7×7, 9×9, 11×11 and 13×13) [61].

Table 2. All the involved features are listed in detail in this paper, including original bands of WorldView-3 data, vegetation indices, and texture features based on the gray-level co-occurrence matrix (GLCM).

Feature Types	Feature Names	Details	Remarks
Original bands	Blue band (B) Green band (G) Red band (R) Near infrared band (NIR)	450–510 nm 510–580 nm 630–690 nm 770–1040 nm	WorldView-3 data
Vegetation indices	Difference vegetation index (DVI) Ratio vegetation index (RVI) Normalized difference vegetation index (NDVI) Optimized soil adjusted vegetation index (OSAVI) Soil adjusted vegetation index (SAVI) Triangular vegetation index (TVI)	$NIR - R$ NIR/R $(NIR - R)/(NIR + R)$ $(NIR - R)/(NIR + R + X)$ $(NIR - R)(1 + L)/(NIR + R + L)$ $0.5\,[120\,(NIR - G) - 200\,(R - G)]$	X take value for 0.16 L take value for 0.5 [62]
Texture features based on the gray-level co-occurrence matrix (GLCM)	Mean (ME)	$\sum_{i=0}^{N-1}\sum_{j=0}^{N-1} iP(i,j)$	$P(i,j) = V(i,j) / \sum_{i=0}^{N-1}\sum_{j=0}^{N-1} V(i,j)$
	Variance (VA)	$\sum_{i=0}^{N-1}\sum_{j=0}^{N-1}(i - mean)^2 P(i,j)$	$V(i,j)$ is the ith row of the jth column in the Nth moving window
	Entropy (EN)	$-\sum_{i=0}^{N-1}\sum_{j=0}^{N-1} P(i,j)\log(P(i,j))$	
	Angular second moment (SE)	$\sum_{i=0}^{N-1}\sum_{j=0}^{N-1} P(i,j)^2$	$u_x = \sum_{j=0}^{N-1} j \sum_{i=0}^{N-1} P(i,j)$
	Homogeneity (HO)	$\sum_{i=0}^{N-1}\sum_{j=0}^{N-1} \frac{P(i,j)}{1+(i-j)^2}$	$u_y = \sum_{i=0}^{N-1} i \sum_{j=0}^{N-1} P(i,j)$
	Contrast (CON)	$\sum_{\lvert i-j\rvert=0}^{N-1}\lvert i-j\rvert^2\left\{\sum_{i=1}^{N}\sum_{j=1}^{N} P(i,j)\right\}$	$\sigma_x = \sum_{j=0}^{N-1}(j - u_i)^2 \sum_{i=0}^{N-1} P(i,j)$
	Dissimilarity (DI)	$\sum_{\lvert i-j\rvert=0}^{N-1}\lvert i-j\rvert\left\{\sum_{i=1}^{N}\sum_{j=1}^{N} P(i,j)\right\}$	$\sigma_y = \sum_{i=0}^{N-1}(i - u_j)^2 \sum_{j=0}^{N-1} P(i,j)$
	Correlation (COR)	$\frac{\sum_{i=0}^{N-1}\sum_{j=0}^{N-1} P(i,j)^2 - \mu_x\mu_y}{\sigma_x\sigma_y}$	

2.4. Methodology

The DenseNet architecture takes the output of all the previous layers as input and combines the previous information features to extract abstract features that are fairly limited. U-net performs deep feature extraction on the basis of the previous layer. Therefore, this study first improves the U-net and DenseNet networks and deeply couples them into the U-net-DenseNet-coupled network (UDN). Then, this network is combined with object-based multiresolution segmentation methods to construct the OUDN algorithm for intelligent and accurate extraction of urban land use and urban forest resources from VHSR images. The following introduces the DL algorithms in detail based on a brief introduction of DCNNs.

2.4.1. Brief Introduction of CNNs

Convolutional neural networks (CNNs) are the core algorithms of DL in the field of computer vision (CV) applications (such as image recognition) because of their ability to obtain hierarchically abstract representations with local operations [63]. This network structure was first inspired by biological vision mechanisms. There are four key ideas behind CNNs that take advantage of the properties of natural signals: local connections, shared weights, pooling, and the use of many layers [24], which are fully utilized.

As shown in Figure 3, the CNN structure consists of four basic processing layers: the convolution layer (Conv), nonlinear activation layer (such as ReLU), normalization layer (such as batch normalization (BN)), and pooling layer (Pooling) [63,64]. The first few layers are composed of two types of layers:

convolutional layers and pooling layers. The units in a convolutional layer are organized in feature maps, within which each unit is connected to local patches in the feature maps of the previous layer through a set of weights called a filter bank, and all units in a feature map share the same filter bank. Different feature maps in every layer use different filter banks, so different features can be learned. The result of this local weighted sum is then passed through a nonlinear activation function such as a ReLU, and the output results are pooled and nonlinearly processed through normalization (such as BN). In addition, nonlinear activation and nonlinear normalization are nonlinear blocks of processing that leads to a bigger boost in model training, so they play a significant role in CNN architecture. After multiple convolutions (combining a convolutional layer and a pooling layer is called a convolution), the results are flattened as the input of the fully connected layer, namely, the artificial neural network (ANN). Thus, the prediction result is finally obtained. Specifically, the major operations performed in the CNNs can be summarized by Equations (1)–(5):

$$S^{[l]} = pool_p\left(\varphi\left(S^{[l-1]} * W^{[l]} + b^{[l]}\right)\right) \tag{1}$$

$$\varphi(Z) = R = \begin{cases} Z; & if\ Z \geq 0 \\ 0; & Z < 0 \end{cases} \tag{2}$$

$$\mu = \frac{1}{m}\sum_{i=1}^{m} R^{(i)} \tag{3}$$

$$\sigma^2 = \frac{1}{m}\sum_{i=1}^{m}\left(R^{(i)} - \mu\right)^2 \tag{4}$$

$$R_{norm}^{(i)} = \frac{R^{(i)} - \mu}{\sqrt{\sigma^2 + \varepsilon}}, \tag{5}$$

where $S^{[l]}$ indicates the feature map at the lth layer [25], $S^{[l-1]}$ denotes the input feature map to the lth layer, and $W^{[l]}$ and $b^{[l]}$ represent the weights and biases of the layer, respectively, that convolve the input feature map through linear convolution $*$. These steps are often followed by a max-pooling operation with $p \times p$ window size ($pool_p$) to aggregate the statistics of the features within specific regions, which forms the output feature map $S^{[l]}$. The $\varphi(Z)$, R, indicates the nonlinearity function outside the convolution layer and corrects the convolution result of each layer, Z denotes the result of the convolution operation by calculating $S^{[l-1]} * W^{[l]} + b^{[l]}$, m represents the batch size (the number of samples required for a single training iteration), μ represents the mean, σ^2 represents the variance, ε is a constant set to keep the value stable to prevent $\sqrt{\sigma^2 + \varepsilon}$ from being 0, and $R_{norm}^{(i)}$ is the normalized value.

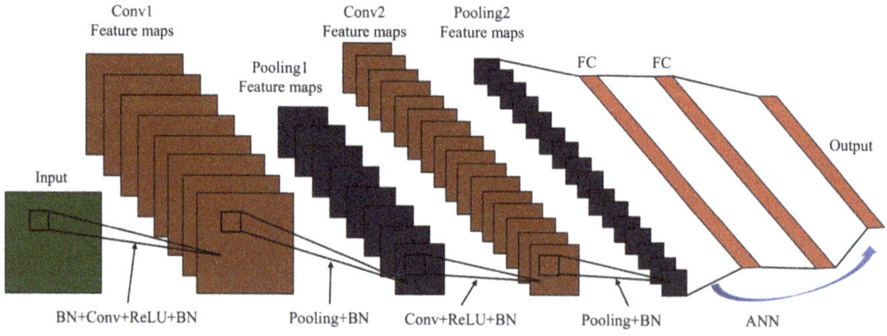

Figure 3. The classical structure of convolutional neural networks (CNNs). Batch normalization (BN) is a technique for accelerating network training by reducing the offset of internal covariates.

2.4.2. DL Algorithms

Improved DenseNet (D): DenseNet is based on ResNet [65], and its most important characteristic is that the feature maps of all previous networks are used as input for each layer of the network. Additionally, the feature maps are used as input by the following network layer, so the problem of gradient disappearance can be alleviated and the number of parameters can be reduced. The improved DenseNet network structure in this study is shown in Figure 4. Figure 4a is the complete structure, which adopts 3 Dense Blocks and 2 Translation layers. Before the first Dense Block, two convolutions are used. In this study, the bottle layer (1×1 convolution) in the Translation layer is converted to a 3×3 convolution operation, followed by an upsampling layer and finally the prediction result. The specific Dense Block structure is shown in Figure 4b and summarized by Equation (6):

$$X_\updownarrow = H_\updownarrow\big([X_0\,,\, X_1\,,\ldots,\, X_{\updownarrow-1}]\big), \tag{6}$$

where $[X_0\,,\, X_1\,,\ldots,\, X_{\updownarrow-1}]$ denotes the feature maps with layers of X_0, X_1 …, $X_{\updownarrow-1}$ and $H_\updownarrow\big([X_0\,,\, X_1\,,\ldots,\, X_{\updownarrow-1}]\big)$ indicates that the $\updownarrow$ layer takes all feature maps of the previous layers (X_0, X_1 …, $X_{\updownarrow-1}$) as input. In this study, all the convolution operations in the Dense Block use 3×3 convolution kernels, and the number of output feature maps (K) in each layer is set to 32.

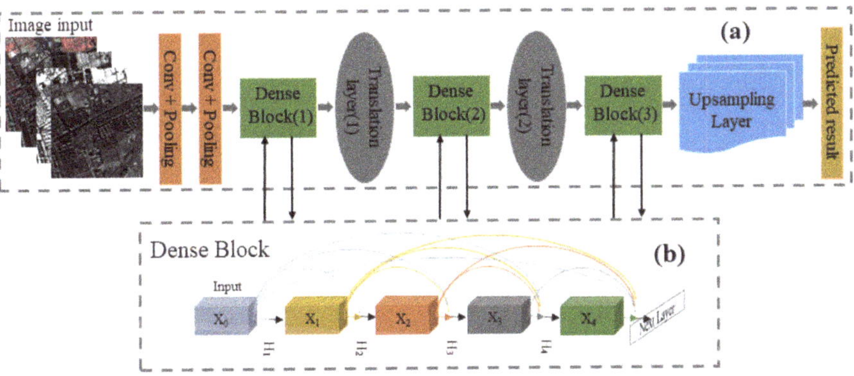

Figure 4. The improved DenseNet structure composed of three dense blocks: (**a**) the complete structure and (**b**) the Dense Block composed of five feature map layers.

Improved U-net (U): U-net is an improved fully convolutional network (FCN) [66]. This network has attracted extensive attention because of its clear structure and excellent performance on small data sets. U-net is divided into a contracting path (to effectively capture contextual information) and an expansive path (to achieve a more precise position for the pixel boundary). Considering the characteristics of urban land use categories and the rich details of WorldView-3 images, the improved structure in this study mainly increases the number of network layers to 11 layers, and each layer increases the convolution operations, thereby obtaining increasingly abstract features. The network is constructed around convolution filters to obtain images with different resolutions, so the structural features of the image can be detected on different scales. More importantly, BN is performed before the convolutional layer and pooling layer, and the details are shown in Figure 5.

(1) The left half of the bottom layer is the contracting path. With the input of a 128×128 image, each layer uses three 3×3 convolution operations. After each convolution, followed by the ReLU activation function, max-pooling with a step of 2 is applied for downsampling. In each downsampling stage, the number of feature channels is doubled. Five downsamplings are applied, followed by two 3×3 convolutions in the bottom layer of the network architecture.

The size of the feature maps is eventually reduced to 4 × 4 pixels, and the number of feature map channels is 1024.

(2) The right half of the network, that is, the expansive path, mainly restores the feature information of the original image. First, a deconvolution kernel with a size of 2 × 2 is used to perform upsampling. In this process, the number of the feature map channels is halved, while the feature maps of the symmetrical position generated by the downsampling and the upsampling are merged; then, three 3 × 3 convolution operations are performed on the merged features, and the above operations are repeated until the image is restored to the size of input image; ultimately, four 3 × 3 and one 1 × 1 convolution operations and a Softmax activation function are used to complete the category prediction of each pixel in the image. The Softmax activation function is defined as Equations (7):

$$p_k(X) = \frac{\exp(a_k(X))}{\left(\sum_{k'=1}^{K} \exp(a_{k'}(X))\right)},\tag{7}$$

where $a_k(X)$ represents the activation value of the kth channel at the position of pixel X. K indicates the number of categories, and $p_k(X)$ denotes the function with the approximate maximum probability. If $a_k(X)$ is the largest activation value in the kth channel, $p_k(X)$ is approximately equal to 1; in contrast, $p_k(X)$ is approximately equal to zero for other k values.

UDN: The detailed coupling process of the improved U-net and DenseNet is shown in Figure 6. (a) The first two layers use the same convolutional layer and pooling layer to obtain abstract feature maps; (b) then, the feature maps obtained by the above operations are input into the Combining Block structure to realize the coupling of the convolution results from the two structures. After two convolution operations are performed on the coupling result, max-pooling is used to perform downsampling, followed by two Combining Block operations; (c) after the downsampling, two convolutions are performed on the coupling result to obtain 1024 feature maps of 4 × 4; (d) the smallest feature maps (4 × 4 × 1024) are restored to the size of the original image after 5 upsamplings; (e) finally, the classification result is output based on the front feature maps through the 1 × 1 convolution operations and the Softmax function.

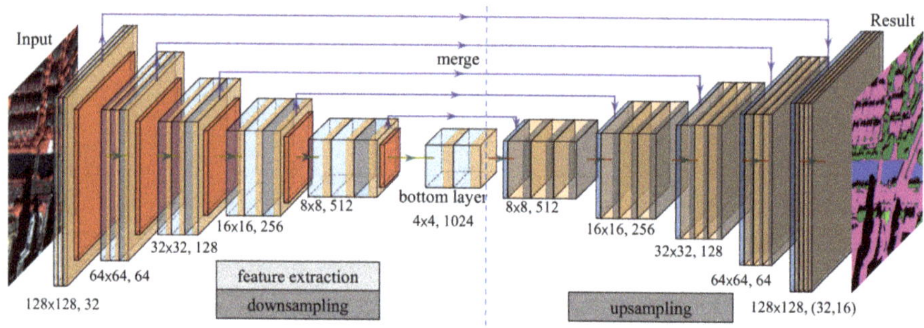

Figure 5. The improved U-net structure is composed of eleven convolution layers.

OUDN: The boundary information of the categories is the basis of the accurate classification of VHSR images. In this study, the OUDN algorithm combines the category objects obtained by object-based multiresolution segmentation [18] with the classification results of the UDN algorithm to constrain and optimize the classification results. Four multispectral bands (red, green, blue, and near infrared) together with vegetation indices and texture features, useful for differentiating urban land use objects with complex information, are incorporated as multiple input data sources for the image segmentation using eCognition software. Then, all the image objects are transformed into GIS vector polygons with distinctive geometric shapes, which are combined with the classification results of the UDN algorithm. Based on the Spatial Analysis Tools of ArcGIS, the category with the largest statistics is taken as the category of the object by counting the number of pixels in each object and using the majority voting method. Thereby, final classification results of the OUDN algorithm are obtained. The segmentation scale directly affects the boundary accuracy of the categories. Therefore, according to the selection method of the optimal segmentation scale [67], this study gains the segmentation results by setting different segmentation scales, and determines the final segmentation scale 50.

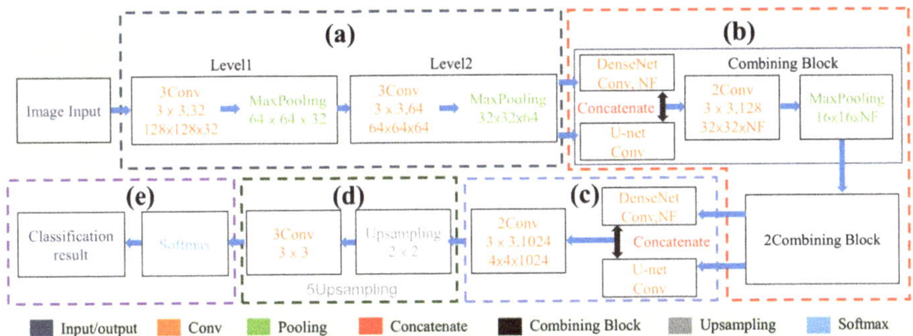

Figure 6. The network structure of the UDN algorithm, where NF represents the number of convolutional filters. (**a**) The first two layers (Level1 and Level2) including convolutional layers and pooling layers; (**b**) the coupling of U-net and DenseNet algorithms; (**c**) the bottom layer of the network; (**d**) Upsampling layers; (**e**) predicted classification result.

Finally, the template for training the minibatch neural network based on the above algorithms in this research is shown in Algorithm 1 [68]. The network uses the loss function of categorical cross entropy and the adaptive optimization algorithm of Adam. Additionally, the number of iterations is set to 50, and the learning rate (lr) is set to 0.0001. In each iteration, b images are sampled to compute the gradients, and then the network parameters are updated. The training of the network stops after K passes through the data set.

Algorithm 1 Train a neural network with the minibatch Adam optimization algorithm.

---

initialize (*net*)
for *epoch* = 1, ... , *K* **do**
    for *batch* = 1, ... , # *images* / *b* **do**
        images ← uniformly sample *batch − size* images
        X, y ← preprocess(images)
        z ← forward (net, X)
        l ← loss (z, y)
        lr, grad ← background (l)
        update (*net, lr, grad*)
    end for
end for

2.5. Experiment Design

The flowchart of steps is shown in Figure 7. The WorldView-3 image with 15.872 × 15.872 pixels is first preprocessed by image fusion, radiometric calibration, and atmospheric correction (Figure 7a). According to this preprocessed image, a 3968 × 3968 pixel subimage with various categories is cropped for model prediction, and other representative subimages are cropped as the sample set including training set and validation set for model training; then, labeled maps are made based on the sample set, followed by image cropping (Figure 7b); the cropped original images and the corresponding labeled maps are used to train the DL models (Figure 7c); the image with 3968 × 3968 pixels is classified by the trained model (Figure 7d); finally, objects of multiresolution segmentation are applied to optimize the classification results of the UDN algorithm to obtain the classification results of the OUDN, followed by detailed comparisons of the results from all algorithms including U, D, UDN, and OUDN (Figure 7e).

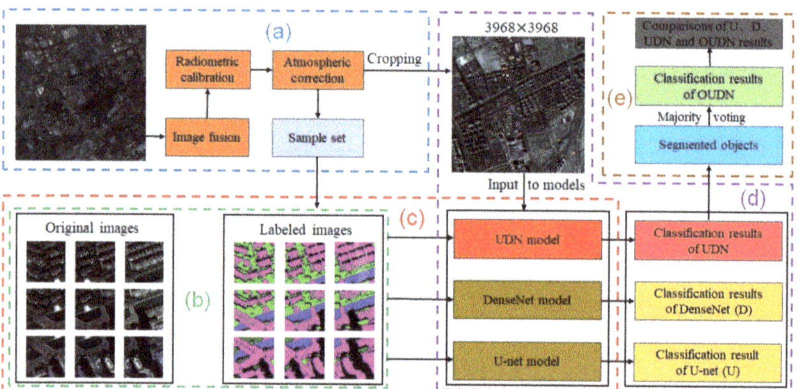

Figure 7. A flowchart of the experimental method in this paper, including five major steps: (**a**) image preprocessing; (**b**) image labeling and image cropping; (**c**) model training; (**d**) model prediction; (**e**) object-based optimization of the UDN results and comparisons of the results from all algorithms.

3. Results and Analysis

The tests of the proposed OUDN algorithm were presented in this section, and the classification results are compared with those of UDN, improved U-net (U), and improved DenseNet (D). To evaluate the proposed algorithm, the classification results in this study were assessed with the overall accuracy (OA), kappa coefficient (Kappa), producer accuracy (PA), and user accuracy (UA) [69]. The detailed results and analysis of the model training and classification results are clarified as follows.

3.1. Training Results of U, D and UDN Algorithms

There were a total of 4761 image blocks with 128 × 128 pixels in the sample set. Additionally, 3984 of these blocks were selected for the training, and the remaining blocks were used for the validation. Then, the cropped original image blocks and the corresponding labeled maps were used to train the minibatch network model according to the template of Algorithm 1. Based on the three feature groups of Spe, Spe-Index, and Spe-Texture, the overall model accuracies including training accuracy (TA) and validation accuracy (VA) of the U, D, and UDN algorithms were demonstrated in Table 3. In all feature combinations, the UDN algorithm obtained the highest training accuracies (98.1%, 98%, and 98.4%). However, for the U and U algorithms, the training accuracies of the Spe-Texture were the lowest (96.3% and 96%) compared with those of the Spe and Spe-Index. The UDN algorithm achieved the highest model accuracies (TA of 98.4% and VA of 93.8%, respectively) based on Spe-Texture.

Table 3. The overall training and validation accuracies of the improved U-net (U), improved DenseNet (D), and U-net-DenseNet-coupled network (UDN) algorithms based on the three feature groups of Spe, Spe-Index, and Spe-Texture. The algorithms in this table did not include object-based U-net-DenseNet-coupled network (OUDN), since the OUDN algorithm was based on UDN algorithm to optimize classification results.

Feature	TA			VA		
	U	**D**	**UDN**	**U**	**D**	**UDN**
Spe	0.975	0.971	0.981	0.914	0.923	0.936
Spe-Index	0.969	0.977	0.980	0.935	0.916	0.920
Spe-Texture	0.963	0.960	0.984	0.927	0.929	0.938

3.2. Classification Results

3.2.1. Classification Results Based on Four Algorithms

The classification accuracies of the U, D, UDN, and OUDN algorithms on the three feature groups of Spe, Spe-Index, and Spe-Texture are demonstrated in Tables 4–6, respectively. In general, among the three feature combinations, the U and D algorithms yielded the lowest OA and Kappa, followed by UDN; in contrast, the OUDN algorithm achieved the highest OA (92.3%, 92.6%, and 93.8%) and Kappa (0.910, 0.914, and 0.928). The average accuracy of the OUDN algorithm was much higher (approximately 3%) than those of the U and D algorithms. As shown in Table 4, the UDN algorithm obtained better accuracies for Agricultural Land and Grassland than the U and D algorithms. For example, the PA values of Agricultural Land were 89%, 88%, and 90.3% for the U, D, and UDN algorithms, respectively, and the PA values of Grassland were 64.3%, 73%, and 74%, respectively. Compared with those of the UDN algorithm, the OUDN algorithm obtained better PA values for Agricultural Land, Grassland, Barren Land, and Water. Table 5 shows that the PA of Agricultural Land of the UDN algorithm was 5% and 3.3% higher than those of the U and D algorithms, respectively. In addition, the OUDN algorithm mainly yielded improvements in the PA values of Forest, Built-up, Agricultural Land, Grassland, and Barren Land. As shown in Table 6, the UDN algorithm yielded higher accuracies for Forest, Built-up, Grassland, Barren Land, and Water than the U and D algorithms, and in particular, the PA value of Grassland was significantly higher, by 15.6% and 17.3%, respectively. Meanwhile, the OUDN algorithm yielded the accuracies superior to those of the UDN algorithm in some categories. In summary, the OUDN algorithm obtained high extraction accuracies for urban land use types, and coupling object-based segmentation effectively addressed the fragmentation problem of classification with high-resolution images, thereby improving the image classification accuracy. Therefore, the OUDN algorithm offered great advantages for urban land-cover classification.

Table 4. The classification accuracies of the U, D, UDN, and OUDN algorithms based on the Spe, including the accuracies (user accuracy (UA) and producer accuracy (PA)) of every class, overall accuracy (OA) and kappa coefficient (Kappa).

Algorithms	OA	Kappa		Forest	Build-UP	Agricultural Land	Grassland	Barren Land	Water	Others
U	0.903	0.887	UA	0.834	0.892	0.756	0.951	0.963	0.997	0.990
			PA	0.990	0.963	0.890	0.643	0.873	0.987	0.973
D	0.905	0.889	UA	0.863	0.855	0.781	0.920	0.975	0.997	0.993
			PA	0.990	0.980	0.880	0.730	0.787	0.990	0.983
UDN	0.920	0.907	UA	0.908	0.910	0.755	0.961	0.973	1.000	0.987
			PA	0.990	0.980	0.903	0.740	0.853	0.987	0.993
OUDN	0.923	0.910	UA	0.911	0.909	0.767	0.958	0.974	0.993	0.993
			PA	0.990	0.970	0.910	0.753	0.857	0.993	0.990

Table 5. The classification accuracies of the U, D, UDN, and OUDN algorithms based on the Spe-Index, including the accuracies (UA and PA) of every class, OA and Kappa.

Algorithms	OA	Kappa		Forest	Build-Up	Agricultural Land	Grassland	Barren Land	Water	Others
U	0.913	0.899	UA	0.878	0.872	0.809	0.930	0.977	1.000	0.958
			PA	0.983	0.977	0.873	0.757	0.867	0.950	0.987
D	0.917	0.903	UA	0.870	0.857	0.842	0.927	0.977	0.997	0.980
			PA	0.983	0.980	0.890	0.760	0.850	0.973	0.983
UDN	0.923	0.910	UA	0.891	0.892	0.817	0.957	0.978	0.997	0.958
			PA	0.983	0.963	0.923	0.750	0.883	0.960	0.997
OUDN	0.926	0.914	UA	0.892	0.901	0.822	0.974	0.982	0.997	0.955
			PA	0.987	0.973	0.937	0.753	0.887	0.957	0.990

Table 6. The classification accuracies of the U, D, UDN, and OUDN algorithms based on the Spe-Texture, including the accuracies (UA and PA) of every class, OA and Kappa.

Algorithms	OA	Kappa		Forest	Build-Up	Agricultural Land	Grassland	Barren Land	Water	Others
U	0.898	0.881	UA	0.864	0.840	0.787	0.914	0.955	1.000	0.961
			PA	0.993	0.963	0.860	0.677	0.857	0.947	0.987
D	0.897	0.879	UA	0.897	0.824	0.750	0.943	0.980	0.993	0.971
			PA	0.987	0.970	0.930	0.660	0.797	0.943	0.990
UDN	0.932	0.921	UA	0.857	0.873	0.913	0.954	0.985	1.000	0.970
			PA	0.997	0.983	0.877	0.833	0.873	0.977	0.983
OUDN	0.938	0.928	UA	0.877	0.866	0.932	0.970	0.985	1.000	0.967
			PA	0.997	0.987	0.913	0.853	0.857	0.973	0.987

The classification maps of the four algorithms based on the Spe, Spe-Index, and Spe-Texture are presented in Figures 8–10, respectively, with the correct or incorrect classification results marked in black or red circles, respectively. In general, the classification results of the UDN and OUDN algorithms were better than those of the other methods, and there was no obvious "salt-and-pepper" effect in the classification results of the four algorithms. However, due to the splicing in the U, D, and UDN algorithms, the ground object boundary exhibited discontinuities, whereas the proposed OUDN algorithm addressed this problem to a certain extent.

Classification maps of different algorithms based on Spe: Based on the Spe, the proposed method in this paper better identified the ground classes that are difficult to distinguish, including Built-up, Barren Land, Agricultural Land, and Grassland. However, the recognition effect of the U and D algorithms was undesirable. As shown in Figure 8, the U and D algorithms confused Built-up and Barren Land (red circle (1)), while the UDN and OUDN algorithms correctly distinguished them (black circle (1)); for the U algorithm, Built-up was misclassified as Barren Land (red circle (2)), while the other algorithms accurately identified these classes (black circle (2)); the D algorithm did not identify Barren Land (red circle (3)), in contrast, the recognition effect of the other methods was favorable (black circle (3)); for the U and D algorithms, Grassland was misclassified as Agricultural Land (red circle (4)), while other algorithms precisely distinguished them (black circle (4)); the four algorithms mistakenly classified some Agricultural Land as Grassland and confused them (red circle (5)).

Classification maps of different algorithms based on Spe-Index: Based on the Spe-Index, the proposed method in this paper better recognized Built-up, Barren Land, Agricultural Land, and Grassland. However, the recognition effect of the U and D algorithms was poor. As demonstrated by Figure 9, U and D algorithms confused Built-up and Barren Land (red circle (1)), whereas the UDN and OUDN algorithms correctly distinguished them (black circle (1)); the U algorithm incorrectly identified Barren Land (red circle (2)), while the classification results of other algorithms were superior (black circle (2)); the U and D algorithms mistakenly classified Barren Land as Agricultural Land (red circle (3)), in contrast, the UDN and OUDN better identified them (black circle (3)); for all four algorithms, some Agricultural Land was misclassified as Grassland (red circle (4)).

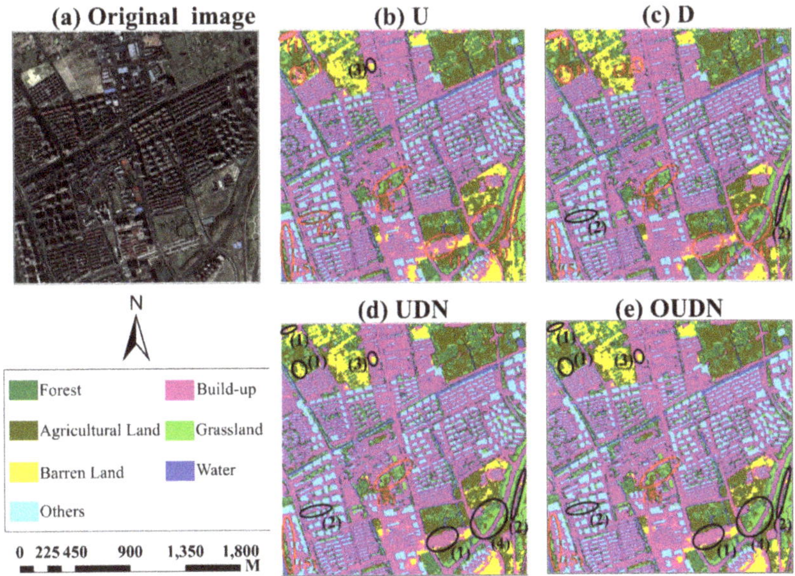

Figure 8. (**a**) Original image; (**b**) the classification map of the U algorithm based on the Spe; (**c**) the classification map of the D algorithm based on the Spe; (**d**) the classification map of the UDN algorithm based on the Spe; (**e**) the classification map of the OUDN algorithm based on the Spe; and the red and black circles denote incorrect and correct classifications, respectively.

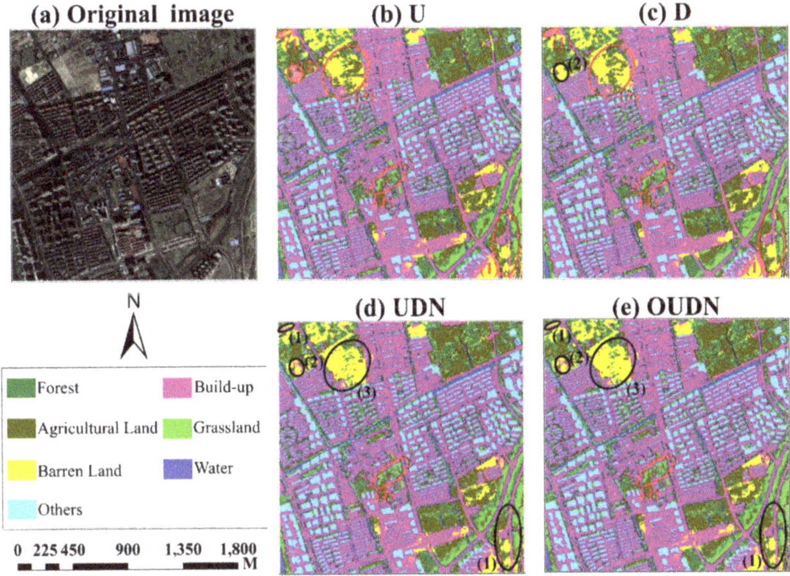

Figure 9. (**a**) Original image; (**b**) the classification map of the U algorithm based on the Spe-Index; (**c**) the classification map of the D algorithm based on the Spe-Index; (**d**) the classification map of the UDN algorithm based on the Spe-Index; (**e**) the classification map of the OUDN algorithm based on the Spe-Index; and the red and black circles denote incorrect and correct classifications, respectively.

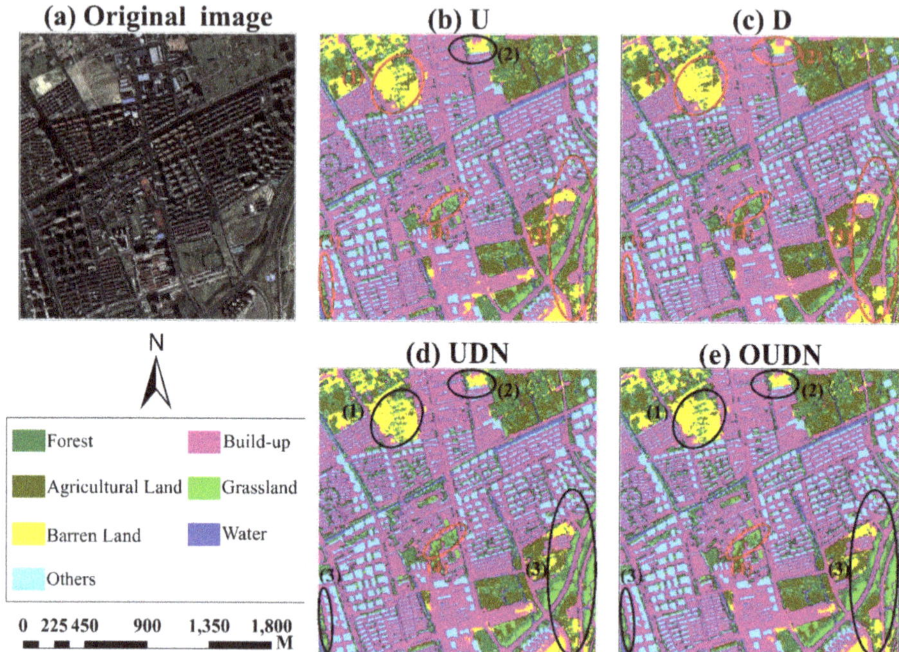

Figure 10. (**a**) Original image; (**b**) the classification map of the U algorithm based on the Spe-Texture; (**c**) the classification map of the D algorithm based on the Spe-Texture; (**d**) the classification map of the UDN algorithm based on the Spe-Texture; (**e**) the classification map of the OUDN algorithm based on the Spe-Texture; and the red and black circles denote incorrect and correct classifications, respectively.

Classification maps of the different algorithms based on Spe-Texture: Based on the Spe-Texture, the proposed method in this paper better identified each category, especially Grassland, yielding the best recognition result; nevertheless, the recognition effect of the U and D algorithms was worse. As shown in Figure 10, the U and D algorithms incorrectly classified much Barren Land as Agricultural Land (red circle (1)), whereas the UDN and OUDN algorithms identified these types better (black circle (1)); the D algorithm confused Built-up and Barren Land (red circle (2)), while the other algorithms better distinguished them (black circle (2)); the extraction effects for Grassland of the UND and OUDN algorithms (black circle (3)) were better than those of the U and D algorithms (red circle (3)); all the algorithms mistakenly classified some Agricultural Land as Grassland (red circle (4)).

3.2.2. Extraction Results of Urban Forests

This section focuses on the analysis of urban forest extraction based on the Spe, Spe-Index, and Spe-Texture with the four algorithms. As shown in Tables 4–6, the PA values of the urban forest information extraction for all algorithms were above 98%, which indicated that the DL algorithms used in this study offered obvious advantages in the extraction of urban forests. Additionally, for the OUDN algorithm, the average PA (99.1%) and UA (89.3%) of urban forest extraction were better than those of the other algorithms based on the three groups of features. This demonstrated that the OUDN algorithm exhibited fewer errors from urban forest leakage and misclassification errors between urban forests and other land use types.

The classification results for urban forests, including scattered trees and street trees, of the different algorithms based on the Spe-Texture are presented in Figure 11. In this study, two representative subregions (subset (1) and subset (2)) were selected for the analysis of the results of the different algorithms, with the correct or incorrect classification results marked in black or blue circles, respectively. In general, the urban forest extraction effect of the OUDN algorithm was the best. According to the classification results of subset (1), the U and D algorithms mistakenly identified some street trees (blue circles), while UDN and OUDN better extracted these trees (black circles). As shown in the results of subset (2), the extraction results for some scattered trees of the U and D algorithms were not acceptable (blue circles); nevertheless, UDN and OUDN accurately distinguished them (black circles). Additionally, the U and D algorithms misclassified some Forest as Grassland and Built-up (blue circles), whereas UDN and OUDN correctly identified the urban forests (black circles).

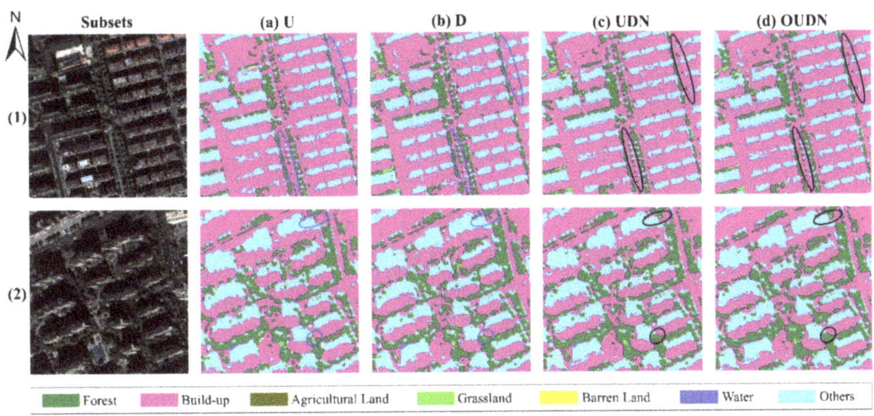

Figure 11. There are two subsets (Subset (**1**) and Subset (**2**)) dominated by urban forests. (**a**) the classification maps of the U algorithm in the subsets; (**b**) the classification maps of the D algorithm in the subsets; (**c**) the classification maps of the UDN algorithm in the subsets; and (**d**) the classification maps of the OUDN algorithm in the subsets.

3.2.3. Result Analysis

According to the classification results of the four algorithms on the Spe, Spe-Index, and Spe-Texture, a confusion matrix is constructed, which is shown in Figure 12. In general, regardless of the feature combinations, the classification accuracies of each algorithm for Forest, Built-up, Water, and Others were relatively high, and the recognition accuracy was above 95%. In particular, the classification accuracy of the Forest was above 98%, whereas the classification accuracies of the other categories varied greatly. As demonstrated by Figure 12, (1) based on the Spe, the extraction accuracies of the OUDN algorithm for Agricultural Land and Grassland were significantly superior to those of the U and D algorithms, whereas the U and D algorithms misclassified Agricultural Land and Grassland as Forest at a higher rate. Compared with that of the U algorithm, the OUDN algorithm yielded better Grassland classification accuracy (75%, an increase in 11%) while optimizing the extraction accuracy of UDN (74%). The D algorithm misclassified 15% of the Barren Land as Built-up, whereas only 8% was incorrectly predicted by the UDN and OUDN algorithms. Therefore, the OUDN algorithm offered obvious advantages in urban land-cover classification. (2) For the Spe-Index, compared with those of the U and D algorithms (87% and 89%, respectively), the OUDN algorithm yielded higher extraction accuracies of Agricultural Land (94%) and optimized the classification accuracy of UDN (92%). The U and D algorithms misclassified 12% of the Barren Land as Built-up, whereas only 11% and 10% were incorrectly predicted by the UDN and OUDN algorithms, so the OUDN algorithm captured the best classification effect. (3) For the Spe-Texture, the extraction accuracies of the UDN and OUDN algorithms for Grassland were very high (83% and 85%, respectively), and the accuracies were the highest among all the Grassland classification results. Compared with the classification accuracies of the U and D algorithms (68% and 66%), the accuracies of UDN and OUDN were 15–19%

higher. Figure 12 showed that the U and D algorithms misclassified 21% and 25% of the Grassland as Agricultural Land, respectively, whereas the misclassification rates of UDN and OUDN were fairly low (7% and 6%, respectively).

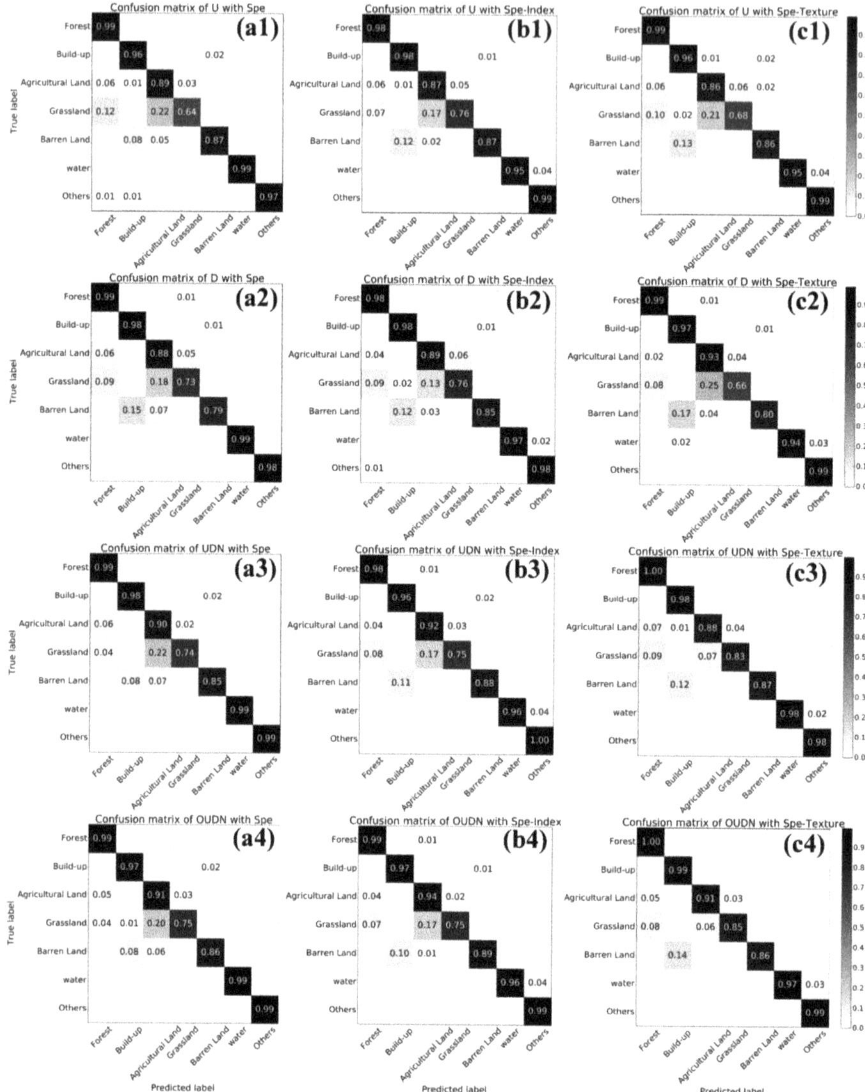

Figure 12. (**a1**) Confusion matrix of U algorithm based on Spe; (**b1**) Confusion matrix of U algorithm based on Spe-Index; (**c1**) Confusion matrix of U algorithm based on Spe-Texture; (**a2**) Confusion matrix of D algorithm based on Spe; (**b2**) Confusion matrix of D algorithm based on Spe-Index; (**c2**) Confusion matrix of D algorithm based on Spe-Texture; (**a3**) Confusion matrix of UDN algorithm based on Spe; (**b3**) Confusion matrix of UDN algorithm based on Spe-Index; (**c3**) Confusion matrix of UDN algorithm based on Spe-Texture; (**a4**) Confusion matrix of OUDN algorithm based on Spe; (**b4**) Confusion matrix of OUDN algorithm based on Spe-Index; (**c4**) Confusion matrix of OUDN algorithm based on Spe-Texture.

For urban forests, as demonstrated by Figure 12, (1) based on the Spe, the extraction accuracy of urban forests was 99% for each algorithm, however, these algorithms misclassified Agricultural Land and Grassland as Forest generally. Compared with those of the U and D algorithms, OUDN's rate of misclassification of Agricultural Land and Grassland as Forest was the lowest (5% and 4%); (2) based on the Spe-Index, the OUDN algorithm obtained the highest urban forest extraction accuracy (99%) and the lowest rate of Agricultural Land and Grassland misclassified as Forest (4% and 7%); (3) based on the Spe-Texture, the urban forest extraction accuracy of the OUDN algorithm was the highest (approximately 100%).

Through the above analysis, it was concluded that (1) the classification results of the OUDN algorithm were significantly better than those of the other algorithms for confusing ground categories (such as Agricultural Land, Grassland, and Barren Land); (2) the accuracy of the UDN algorithm was improved through object constraints; (3) especially for Spe-Texture, the OUDN algorithm achieved the highest OA (93.8%), which was 4% and 4.1% higher than those of the U and D algorithms, respectively; (4) the UDN and OUDN algorithms had obvious advantages regarding the accurate extraction of urban forests, and they not only accurately extracted the street trees but also identified the scattered trees ignored by the U and D algorithms.

4. Discussion

The UDN and OUDN algorithms constructed by this study achieved higher accuracies in the extraction of urban land use information from VHSR imagery than the U and D algorithms. The UDN algorithm applied the coupling of the improved 11-layer U-net network and the improved DenseNet to train the network and realize prediction using the learned deep level features. With the advantages of both networks, the accurate extraction of urban land use and urban forests was ensured. Meanwhile, the UDN algorithm addressed the problems of common misclassifications and omissions in the classification process (Tables 4–6) and dealt with the confusion of Agricultural Land, Grassland, Barren Land and Built-up (Figure 12), thereby improving the classification accuracies of urban land use and urban forest. In all feature combinations, especially for the Spe-Texture, the classification accuracies of the UDN algorithm were 3.4% and 3.5% higher than those of the U and D algorithms, respectively. This study chose 50 as the optimal segmentation scale, and the phenomenon of misclassification with UDN was corrected by the constraints of the segmentation objects (Tables 4–6). The OUDN algorithm not only alleviated the distribution fragmentation of ground objects and the common "salt-and-pepper" phenomenon in the classification process but also dealt with the problem of discontinuous boundaries during the splicing process of the classification results of segmented image blocks (Figures 8–10). Compared with previous studies about classifications using U-net and DenseNet [41,46], this study fully combined the advantages of U-net and DenseNet network and achieved more high classification accuracies. Compared with previous object-based DL classification methods [49], in this study, object-based multiresolution segmentations were used to constrain and optimize the UND classification results and not used to participate in the UND classification. It is necessary to further study in this respect.

The overall classification accuracies (OA) of different features based on different algorithms are shown in Figure 13. (1) In terms of the UDN and OUDN algorithms, accuracies of the Spe-Texture were the highest (93.2% and 93.8%), followed by those of the Spe-Index (92.3% and 92.6%). As demonstrated by Figure 12, for Grassland, Built-up and Water, the classification accuracies of the Spe-Texture were significantly higher than those of the Spe-Index. For example, the Grassland accuracies of the Spe-Texture were 8% and 10% higher than those of the Spe-Index, and the Built-up accuracies of the Spe-Texture were 2% and 2% higher than those of the Spe-Index. It can be concluded from Table 3 that the TA and VA of Spe-Texture are higher. Thus, the classification results of the Spe-Texture were better than those of the Spe-Index and Spe. (2) In terms of the U and D algorithms, classification accuracies of the Spe-Texture were the lowest, 21%, 25% of the Grassland, and 13% and 17% of the Barren Land were misclassified as Forest and Built-up, respectively. In contrast, the accuracies of Spe-Index were the highest.

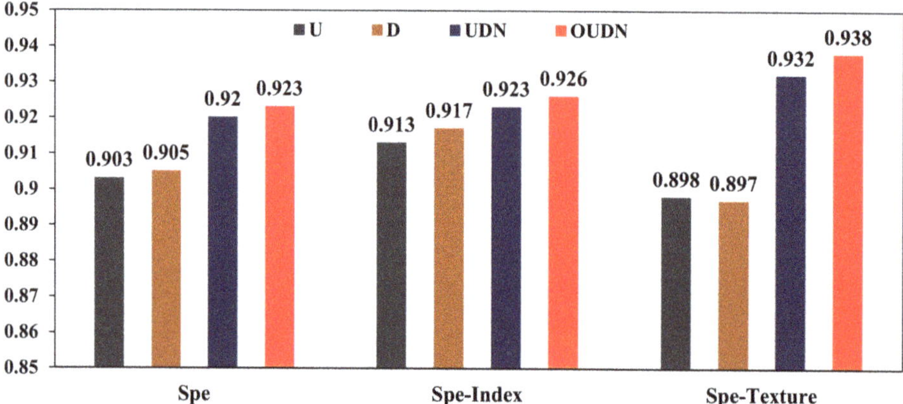

Figure 13. Overall classification accuracies (OA) of different features based on different algorithms.

For urban forests, after texture was added to the Spe, that is, based on the Spe-Texture, the UDN and OUDN algorithms achieved the highest classification accuracies (approximately 100%) for extracting the information of urban forests from VHSR imagery. Similarly, the U and D algorithms also offered relatively obvious advantages for extracting urban forest information based on the feature. As shown in Figure 12, (1) for the U algorithm, the Spe-Index yielded the lowest urban forest extraction accuracy. However, based on Spe-Texture, the accuracy is the highest (99%), and the ratio of Grassland misclassified as Forest was lower (10%) than that of Spe (12%), so the Spe-Texture offered advantages in the extraction of urban forests. (2) For the D algorithm, the urban forest extraction accuracy with the Spe-Texture, compared with those with the Spe and Spe-Index features, was the highest; meanwhile, the ratios of Agricultural Land and Grassland misclassified as Forest were the lowest (2% and 8%). (3) For the UDN and OUDN algorithms, although the urban forest extraction accuracy based on the Spe-Texture was the highest, the ratios of Agricultural Land and Grassland that were misclassified as Forest, compared with those of the other features, were the highest (5% and 8%), thereby resulting in confusion between urban forests and other land use categories.

5. Conclusions

Urban land use classification using VHSR remotely sensed imagery remains a challenging task due to the extreme difficulties in differentiating complex and confusing land use categories. This paper proposed a novel OUDN algorithm for the mapping of urban land use information from VHSR imagery, and the information of urban land use and urban forest resources was extracted accurately. The results showed that the OA of the UDN algorithm for urban land use classification was substantially higher than those of the U and D algorithms in terms of Spe, Spe-Index, and Spe-Texture. Object-based image analysis (OBIA) can address the problem of the "salt-and-pepper" effect encountered in VHSR image classification to a certain extent. Therefore, the OA of urban land use classification and the urban forest extraction accuracy were improved significantly based on the UDN algorithm combined with object-based multiresolution segmentation constraints, which indicated that the OUDN algorithm offered dramatic advantages in the extraction of urban land use information from VHSR imagery. The OA of spectral features combined with texture features (Spe-Texture) in the extraction of urban land use information was as high as 93.8% with the OUDN algorithm, and different land use classes were identified accurately. Especially for urban forests, the OUDN algorithm achieved the highest classification accuracy of 99.7%. Thus, this study provided a reference for the feature setting of urban forest information extraction from VHSR imagery. However, for the OUDN algorithms, the ratios of Agricultural Land and Grassland misclassified as Forest were higher based on Spe-Texture, which led to confusion between urban forests and other categories. This issue will be further studied in future research.

Author Contributions: Conceptualization, H.D.; data curation, D.Z., M.Z., Z.H., H.L., and X.L. (Xin Luo); formal analysis, S.H., X.L. (Xuejian Li), F.M., and H.L.; investigation, S.H., D.Z., Y.X., M.Z., Z.H., and X.L. (Xin Luo); methodology, S.H.; software, S.H. and Z.H.; supervision, H.D. and F.M.; validation, G.Z.; visualization, X.L. (Xuejian Li) and Y.X.; writing—original draft, S.H.; writing—review and editing, H.D. All authors have read and agreed to the published version of the manuscript.

Funding: This research was funded by the National Natural Science Foundation (No. U1809208, 31670644, and 31901310), the State Key Laboratory of Subtropical Silviculture (No. ZY20180201), and the Zhejiang Provincial Collaborative Innovation Center for Bamboo Resources and High-efficiency Utilization (No. S2017011).

Acknowledgments: The authors gratefully acknowledge the supports of various foundations. The authors are grateful to the editor and anonymous reviewers whose comments have contributed to improving the quality.

Conflicts of Interest: The authors declare that they have no competing interests.

References

1. Voltersen, M.; Berger, C.; Hese, S.; Schmullius, C. Object-based land cover mapping and comprehensive feature calculation for an automated derivation of urban structure types at block level. *Remote Sens. Environ.* **2014**, *154*, 192–201. [CrossRef]
2. Wu, C.; Zhang, L.; Du, B. Kernel Slow Feature Analysis for Scene Change Detection. *IEEE Trans. Geosci. Remote Sens.* **2017**, *55*, 2367–2384. [CrossRef]
3. Lin, J.; Kroll, C.N.; Nowak, D.J.; Greenfield, E.J. A review of urban forest modeling: Implications for management and future research. *Urban. For. Urban. Green.* **2019**, *43*, 126366. [CrossRef]
4. Ren, Z.; Zheng, H.; He, X.; Zhang, D.; Yu, X.; Shen, G. Spatial estimation of urban forest structures with Landsat TM data and field measurements. *Urban. For. Urban. Green.* **2015**, *14*, 336–344. [CrossRef]
5. Guang-Rong, S.; Wang, Z.; Liu, C.; Han, Y. Mapping aboveground biomass and carbon in Shanghai's urban forest using Landsat ETM+ and inventory data. *Urban. For. Urban. Green.* **2020**, *51*, 126655. [CrossRef]
6. Zhang, M.; Du, H.; Mao, F.; Zhou, G.; Li, X.; Dong, L.; Zheng, J.; Zhu, D.; Liu, H.; Huang, Z.; et al. Spatiotemporal Evolution of Urban Expansion Using Landsat Time Series Data and Assessment of Its Influences on Forests. *ISPRS Int. J. Geo-Inf.* **2020**, *9*, 64. [CrossRef]
7. Zhang, Y.; Shen, W.; Li, M.; Lv, Y. Assessing spatio-temporal changes in forest cover and fragmentation under urban expansion in Nanjing, eastern China, from long-term Landsat observations (1987–2017). *Appl. Geogr.* **2020**, *117*, 102190. [CrossRef]
8. Alonzo, M.; McFadden, J.P.; Nowak, D.J.; Roberts, D.A. Mapping urban forest structure and function using hyperspectral imagery and lidar data. *Urban. For. Urban. Green.* **2016**, *17*, 135–147. [CrossRef]
9. Alonzo, M.; Bookhagen, B.; Roberts, D.A. Urban tree species mapping using hyperspectral and lidar data fusion. *Remote Sens. Environ.* **2014**, *148*, 70–83. [CrossRef]
10. Liu, L.; Coops, N.C.; Aven, N.W.; Pang, Y. Mapping urban tree species using integrated airborne hyperspectral and LiDAR remote sensing data. *Remote Sens. Environ.* **2017**, *200*, 170–182. [CrossRef]
11. Pu, R.; Landry, S. A comparative analysis of high spatial resolution IKONOS and WorldView-2 imagery for mapping urban tree species. *Remote Sens. Environ.* **2012**, *124*, 516–533. [CrossRef]
12. Pu, R.; Landry, S.M. Mapping urban tree species by integrating multi-seasonal high resolution pléiades satellite imagery with airborne LiDAR data. *Urban. For. Urban. Green.* **2020**, *53*, 126675. [CrossRef]
13. Puissant, A.; Rougier, S.; Stumpf, A. Object-oriented mapping of urban trees using Random Forest classifiers. *Int. J. Appl. Earth Obs. Geoinf.* **2014**, *26*, 235–245. [CrossRef]
14. Pan, G.; Qi, G.; Wu, Z.; Zhang, D.; Li, S. Land-Use Classification Using Taxi GPS Traces. *IEEE Trans. Intell. Transp. Syst.* **2012**, *14*, 113–123. [CrossRef]
15. Zhao, W.; Du, S. Learning multiscale and deep representations for classifying remotely sensed imagery. *ISPRS J. Photogramm. Remote Sens.* **2016**, *113*, 155–165. [CrossRef]
16. Moser, G.; Serpico, S.B.; Benediktsson, J.A. Land-Cover Mapping by Markov Modeling of Spatial–Contextual Information in Very-High-Resolution Remote Sensing Images. *Proc. IEEE* **2012**, *101*, 631–651. [CrossRef]

17. Hu, F.; Xia, G.-S.; Hu, J.; Zhang, L. Transferring Deep Convolutional Neural Networks for the Scene Classification of High-Resolution Remote Sensing Imagery. *Remote Sens.* **2015**, *7*, 14680–14707. [CrossRef]

18. Han, N.; Du, H.; Zhou, G.; Xu, X.; Ge, H.; Liu, L.; Gao, G.; Sun, S. Exploring the synergistic use of multi-scale image object metrics for land-use/land-cover mapping using an object-based approach. *Int. J. Remote Sens.* **2015**, *36*, 3544–3562. [CrossRef]

19. Sun, X.; Du, H.; Han, N.; Zhou, G.; Lu, D.; Ge, H.; Xu, X.; Liu, L. Synergistic use of Landsat TM and SPOT5 imagery for object-based forest classification. *J. Appl. Remote Sens.* **2014**, *8*, 083550. [CrossRef]

20. Hamdi, Z.M.; Brandmeier, M.; Straub, C. Forest Damage Assessment Using Deep Learning on High Resolution Remote Sensing Data. *Remote Sens.* **2019**, *11*, 1976. [CrossRef]

21. Shirvani, Z.; Abdi, O.; Buchroithner, M.F. A Synergetic Analysis of Sentinel-1 and -2 for Mapping Historical Landslides Using Object-Oriented Random Forest in the Hyrcanian Forests. *Remote Sens.* **2019**, *11*, 2300. [CrossRef]

22. Stubbings, P.; Peskett, J.; Rowe, F.; Arribas-Bel, D. A Hierarchical Urban Forest Index Using Street-Level Imagery and Deep Learning. *Remote Sens.* **2019**, *11*, 1395. [CrossRef]

23. Abdi, O. Climate-Triggered Insect Defoliators and Forest Fires Using Multitemporal Landsat and TerraClimate Data in NE Iran: An Application of GEOBIA TreeNet and Panel Data Analysis. *Sensors* **2019**, *19*, 3965. [CrossRef] [PubMed]

24. LeCun, Y.; Bengio, Y.; Hinton, G. Deep learning. *Nature* **2015**, *521*, 436–444. [CrossRef] [PubMed]

25. Romero, A.; Gatta, C.; Camps-Valls, G. Unsupervised Deep Feature Extraction for Remote Sensing Image Classification. *IEEE Trans. Geosci. Remote Sens.* **2015**, *54*, 1349–1362. [CrossRef]

26. Dong, L.; Du, H.; Mao, F.; Han, N.; Li, X.; Zhou, G.; Zhu, D.; Zheng, J.; Zhang, M.; Xing, L.; et al. Very High Resolution Remote Sensing Imagery Classification Using a Fusion of Random Forest and Deep Learning Technique—Subtropical Area for Example. *IEEE J. Sel. Top. Appl. Earth Obs. Remote Sens.* **2019**, *13*, 113–128. [CrossRef]

27. Zhao, W.; Du, S.; Wang, Q.; Emery, W. Contextually guided very-high-resolution imagery classification with semantic segments. *ISPRS J. Photogramm. Remote Sens.* **2017**, *132*, 48–60. [CrossRef]

28. Sherrah, J. Fully convolutional networks for dense semantic labelling of high-resolution aerial imagery. *arXiv* **2016**, arXiv:1606.02585.

29. Liu, Y.; Fan, B.; Wang, L.; Bai, J.; Xiang, S.; Pan, C. Semantic labeling in very high resolution images via a self-cascaded convolutional neural network. *ISPRS J. Photogramm. Remote Sens.* **2018**, *145*, 78–95. [CrossRef]

30. Sun, Y.; Zhang, X.; Xin, Q.; Huang, J. Developing a multi-filter convolutional neural network for semantic segmentation using high-resolution aerial imagery and LiDAR data. *ISPRS J. Photogramm. Remote Sens.* **2018**, *143*, 3–14. [CrossRef]

31. Chen, G.; Zhang, X.; Wang, Q.; Dai, F.; Gong, Y.; Zhu, K. Symmetrical Dense-Shortcut Deep Fully Convolutional Networks for Semantic Segmentation of Very-High-Resolution Remote Sensing Images. *IEEE J. Sel. Top. Appl. Earth Obs. Remote Sens.* **2018**, *11*, 1633–1644. [CrossRef]

32. Chen, K.; Weinmann, M.; Sun, X.; Yan, M.; Hinz, S.; Jutzi, B. Semantic Segmentation of Aerial Imagery via Multi-Scale Shuffling Convolutional Neural Networks with Deep Supervision. *ISPRS Ann. Photogramm. Remote Sens. Spat. Inf. Sci.* **2018**, *4*, 29–36. [CrossRef]

33. Huang, B.; Zhao, B.; Song, Y. Urban land-use mapping using a deep convolutional neural network with high spatial resolution multispectral remote sensing imagery. *Remote Sens. Environ.* **2018**, *214*, 73–86. [CrossRef]

34. Xu, J.; Feng, G.; Zhao, T.; Sun, X.; Zhu, M. Remote sensing image classification based on semi-supervised adaptive interval type-2 fuzzy c-means algorithm. *Comput. Geosci.* **2019**, *131*, 132–143. [CrossRef]

35. Mi, L.; Chen, Z. Superpixel-enhanced deep neural forest for remote sensing image semantic segmentation. *ISPRS J. Photogramm. Remote Sens.* **2020**, *159*, 140–152. [CrossRef]

36. Maggiori, E.; Tarabalka, Y.; Charpiat, G.; Alliez, P. High-resolution semantic labeling with convolutional neural networks. *arXiv* **2016**, arXiv:1611.01962.

37. Liu, Y.; Piramanayagam, S.; Monteiro, S.T.; Saber, E. Dense Semantic Labeling of Very-High-Resolution Aerial Imagery and LiDAR with Fully-Convolutional Neural Networks and Higher-Order CRFs. In Proceedings of the 2017 IEEE Conference on Computer Vision and Pattern Recognition Workshops (CVPRW), Honolulu, HI, USA, 21–26 July 2017; pp. 1561–1570.

38. Zhang, B.; Zhao, L.; Zhang, X. Three-dimensional convolutional neural network model for tree species classification using airborne hyperspectral images. *Remote Sens. Environ.* **2020**, *247*, 111938. [CrossRef]

39. Dumoulin, V.; Visin, F. A guide to convolution arithmetic for deep learning. *arXiv* **2016**, arXiv:1603.07285.

40. Huang, G.; Liu, Z.; van der Maaten, L.; Weinberger, K.Q. Densely Connected Convolutional Networks. In Proceedings of the IEEE Conference on Computer Vision and Pattern Recognition (CVPRW), Honolulu, HI, USA, 21–26 July 2017; pp. 2261–2269.

41. Wang, W.; Dou, S.; Jiang, Z.; Sun, L. A Fast Dense Spectral–Spatial Convolution Network Framework for Hyperspectral Images Classification. *Remote Sens.* **2018**, *10*, 1068. [CrossRef]

42. Li, R.; Duan, C. LiteDenseNet: A Lightweight Network for Hyperspectral Image Classification. *arXiv* **2020**, arXiv:2004.08112.

43. Bai, Y.; Zhang, Q.; Lu, Z.; Zhang, Y. SSDC-DenseNet: A Cost-Effective End-to-End Spectral-Spatial Dual-Channel Dense Network for Hyperspectral Image Classification. *IEEE Access* **2019**, *7*, 84876–84889. [CrossRef]

44. Li, G.; Zhang, C.; Lei, R.; Zhang, X.; Ye, Z.; Li, X. Hyperspectral remote sensing image classification using three-dimensional-squeeze-and-excitation-DenseNet (3D-SE-DenseNet). *Remote Sens. Lett.* **2020**, *11*, 195–203. [CrossRef]

45. Ronneberger, O.; Fischer, P.; Brox, T. U-net: Convolutional networks for biomedical image segmentation. In Proceedings of the International Conference Medical Image Computing and Computer-Assisted Intervention, Munich, Germany, 5–9 October 2015.

46. Flood, N.; Watson, F.; Collett, L. Using a U-net convolutional neural network to map woody vegetation extent from high resolution satellite imagery across Queensland, Australia. *Int. J. Appl. Earth Obs. Geoinfor.* **2019**, *82*, 101897. [CrossRef]

47. Diakogiannis, F.I.; Waldner, F.; Caccetta, P.; Wu, C. Resunet-a: A deep learning framework for semantic segmentation of remotely sensed data. *ISPRS J. Photogramm. Remote Sens.* **2020**, *162*, 94–114. [CrossRef]

48. Zhao, W.; Du, S.; Emery, W.J. Object-Based Convolutional Neural Network for High-Resolution Imagery Classification. *IEEE J. Sel. Top. Appl. Earth Obs. Remote Sens.* **2017**, *10*, 3386–3396. [CrossRef]

49. Zhang, C.; Sargent, I.; Pan, X.; Li, H.; Gardiner, A.; Hare, J.S.; Atkinson, P.M. An object-based convolutional neural network (OCNN) for urban land use classification. *Remote Sens. Environ.* **2018**, *216*, 57–70. [CrossRef]

50. Liu, T.; Abd-Elrahman, A. Deep convolutional neural network training enrichment using multi-view object-based analysis of Unmanned Aerial systems imagery for wetlands classification. *ISPRS J. Photogramm. Remote Sens.* **2018**, *139*, 154–170. [CrossRef]

51. Martins, V.S.; Kaleita, A.L.; Gelder, B.K.; Da Silveira, H.L.; Abe, C.A. Exploring multiscale object-based convolutional neural network (multi-OCNN) for remote sensing image classification at high spatial resolution. *ISPRS J. Photogramm. Remote Sens.* **2020**, *168*, 56–73. [CrossRef]

52. Zhang, C.; Yue, P.; Tapete, D.; Shangguan, B.; Wang, M.; Wu, Z. A multi-level context-guided classification method with object-based convolutional neural network for land cover classification using very high resolution remote sensing images. *Int. J. Appl. Earth Obs. Geoinf.* **2020**, *88*, 102086. [CrossRef]

53. Tong, X.-Y.; Xia, G.-S.; Lu, Q.; Shen, H.; Li, S.; You, S.; Zhang, L. Land-cover classification with high-resolution remote sensing images using transferable deep models. *Remote Sens. Environ.* **2020**, *237*, 111322. [CrossRef]

54. Anderson, J.R. *A Land Use and Land Cover Classification System for Use with Remote Sensor Data*; USGS Professional Paper, No. 964; US Government Printing Office: Washington, DC, USA, 1976. [CrossRef]

55. Gong, P.; Liu, H.; Zhang, M.; Li, C.; Wang, J.; Huang, H.; Clinton, N.; Ji, L.; Li, W.; Bai, Y.; et al. Stable classification with limited sample: Transferring a 30-m resolution sample set collected in 2015 to mapping 10-m resolution global land cover in 2017. *Sci. Bull.* **2019**, *64*, 370–373. [CrossRef]

56. Sun, J.; Wang, H.; Song, Z.; Lu, J.; Meng, P.; Qin, S. Mapping Essential Urban Land Use Categories in Nanjing by Integrating Multi-Source Big Data. *Remote Sens.* **2020**, *12*, 2386. [CrossRef]

57. Bharati, M.H.; Liu, J.; MacGregor, J.F. Image texture analysis: Methods and comparisons. *Chemom. Intell. Lab. Syst.* **2004**, *72*, 57–71. [CrossRef]

58. Li, Y.; Han, N.; Li, X.; Du, H.; Mao, F.; Cui, L.; Liu, T.; Xing, L. Spatiotemporal Estimation of Bamboo Forest Aboveground Carbon Storage Based on Landsat Data in Zhejiang, China. *Remote Sens.* **2018**, *10*, 898. [CrossRef]

59. Fatiha, B.; Abdelkader, A.; Latifa, H.; Mohamed, E. Spatio Temporal Analysis of Vegetation by Vegetation Indices from Multi-dates Satellite Images: Application to a Semi Arid Area in Algeria. *Energy Procedia* **2013**, *36*, 667–675. [CrossRef]

60. Taddeo, S.; Dronova, I.; Depsky, N. Spectral vegetation indices of wetland greenness: Responses to vegetation structure, composition, and spatial distribution. *Remote Sens. Environ.* **2019**, *234*, 111467. [CrossRef]
61. Zhang, M.; Du, H.; Zhou, G.; Li, X.; Mao, F.; Dong, L.; Zheng, J.; Liu, H.; Huang, Z.; He, S. Estimating Forest Aboveground Carbon Storage in Hang-Jia-Hu Using Landsat TM/OLI Data and Random Forest Model. *Forests* **2019**, *10*, 1004. [CrossRef]
62. Ren, H.; Zhou, G.; Zhang, F. Using negative soil adjustment factor in soil-adjusted vegetation index (SAVI) for aboveground living biomass estimation in arid grasslands. *Remote Sens. Environ.* **2018**, *209*, 439–445. [CrossRef]
63. Hadji, I.; Wildes, R.P. What do we understand about convolutional networks? *arXiv* **2018**, arXiv:1803.08834.
64. Ioffe, S.; Szegedy, C. Batch normalization: Accelerating deep network training by reducing internal covariate shift. In Proceedings of the 32nd International Conference on Machine Learning, Lille, France, 6–11 July 2015; pp. 448–456.
65. He, K.; Zhang, X.; Ren, S.; Sun, J. Deep Residual Learning for Image Recognition. *arXiv* **2015**, arXiv:1512.03385.
66. Long, J.; Shelhamer, E.; Darrell, T. Fully convolutional networks for semantic segmentation. In Proceedings of the IEEE Conference on Computer Vision and Pattern Recognition, Boston, MA, USA, 7–12 June 2015; pp. 3431–3440.
67. Yin, R.; Shi, R.; Li, J. Automatic Selection of Optimal Segmentation Scale of High-resolution Remote Sensing Images. *J. Geo-inf. Sci.* **2013**, *15*, 902–910. [CrossRef]
68. He, T.; Zhang, Z.; Zhang, H.; Zhang, Z.; Xie, J.; Li, M. Bag of Tricks for Image Classification with Convolutional Neural Networks. In Proceedings of the 2019 IEEE/CVF Conference on Computer Vision and Pattern Recognition (CVPR), Long Beach, CA, USA, 15–20 June 2019; pp. 558–567.
69. Olofsson, P.; Foody, G.M.; Herold, M.; Stehman, S.V.; Woodcock, C.E.; Wulder, M.A. Good practices for estimating area and assessing accuracy of land change. *Remote Sens. Environ.* **2014**, *148*, 42–57. [CrossRef]

Publisher's Note: MDPI stays neutral with regard to jurisdictional claims in published maps and institutional affiliations.

 *remote sensing*

Article

Post-Disaster Building Damage Detection from Earth Observation Imagery Using Unsupervised and Transferable Anomaly Detecting Generative Adversarial Networks

Sofia Tilon *, Francesco Nex, Norman Kerle and George Vosselman

Faculty of Geo-Information Science and Earth Observation (ITC), University of Twente, 7514 AE Enschede, The Netherlands; f.nex@utwente.nl (F.N.); n.kerle@utwente.nl (N.K.); george.vosselman@utwente.nl (G.V.)
* Correspondence: s.m.tilon@utwente.nl

Received: 13 November 2020; Accepted: 16 December 2020; Published: 21 December 2020

Abstract: We present an unsupervised deep learning approach for post-disaster building damage detection that can transfer to different typologies of damage or geographical locations. Previous advances in this direction were limited by insufficient qualitative training data. We propose to use a state-of-the-art Anomaly Detecting Generative Adversarial Network (ADGAN) because it only requires pre-event imagery of buildings in their undamaged state. This approach aids the post-disaster response phase because the model can be developed in the pre-event phase and rapidly deployed in the post-event phase. We used the xBD dataset, containing pre- and post- event satellite imagery of several disaster-types, and a custom made Unmanned Aerial Vehicle (UAV) dataset, containing post-earthquake imagery. Results showed that models trained on UAV-imagery were capable of detecting earthquake-induced damage. The best performing model for European locations obtained a recall, precision and F1-score of 0.59, 0.97 and 0.74, respectively. Models trained on satellite imagery were capable of detecting damage on the condition that the training dataset was void of vegetation and shadows. In this manner, the best performing model for (wild)fire events yielded a recall, precision and F1-score of 0.78, 0.99 and 0.87, respectively. Compared to other supervised and/or multi-epoch approaches, our results are encouraging. Moreover, in addition to image classifications, we show how contextual information can be used to create detailed damage maps without the need of a dedicated multi-task deep learning framework. Finally, we formulate practical guidelines to apply this single-epoch and unsupervised method to real-world applications.

Keywords: deep learning; Generative Adversarial Networks; post-disaster; building damage assessment; anomaly detection; Unmanned Aerial Vehicles (UAV); satellite; xBD

---

1. Introduction

Damage detection is a critical element in the post-disaster response and recovery phase [1]. Therefore, it has been a topic of interest for decades. Recently, the popularity of deep learning has sparked a renewed interest in this topic [2–4].

Remote sensing imagery is a critical tool to analyze the impacts of a disaster in both the pre- and post-event epoch [4]. Such imagery can be obtained from different platforms: satellites, Unmanned Aerial Vehicles (UAV's) and manned aircrafts [5,6]. Each contains characteristics that need to be considered before deciding on which to use for disaster analysis. Manned airplanes or UAV's can be flexibly deployed and fly at relatively low heights compared to satellites and, therefore, have relatively small ground sampling distances (GSD) [7]. UAV's can fly lower than manned airplanes and in addition, depending on the type of drone, they can hover and maneuver in between obstacles.

Both platforms can be equipped with a camera in oblique mounts, meaning that vital information can be derived from not only the top but also the sides of objects [8]. However, data acquisitions using these platforms have to be carried out and instigated by humans, which makes them time and resource costly. The spatial coverage of these platforms is also typically restricted to small areas of interests (AIO) and biased towards post-event scenarios when new information is required. Therefore, pre-event data from UAV or aerial platforms are less likely to exist. Satellites on the other hand, depending on the type of satellite, have a high coverage and return rate, especially of build-up areas. Therefore, pre-event data from satellites are more likely to exist. Moreover, satellite systems that provide information to Emergency Mapping Services are able to (re)visit the disaster location only hours after an event, enabling fast damage mapping [9]. A disadvantage of satellite imagery is that it has larger GSD footprints. Moreover, excluding the ones that are freely available, obtaining satellite imagery is generally more costly than UAV imagery.

Damage mapping using Earth observation imagery and automatic image analysis is still a challenge for various reasons despite decades of dedicated research. Traditional image analysis remains sensitive to imaging conditions. Shadows, varying lighting conditions, temporal variety of objects, camera angles or distortions of 3D objects that have been reduced to a 2D plane have made it difficult to delineate damage. Moreover, the translation of found damage features to meaningful damage insights have prevented many methods from being implemented in real-world scenarios. Deep learning has made a major contribution towards solving these challenges by allowing the learning of damage features instead of handcrafting them. Several studies have been carried out on post-disaster building damage detection using remote sensing imagery and deep learning [6,10–13]. Adding 3D information, prior cadastral information or multi-scale imagery has contributed towards some of these challenges [11,14–16]. Despite these efforts, persistent problems related to vegetation, shadows or damage interpretation remain. More importantly, a lesser addressed aspect of deep learning-based post-disaster damage detection remains—the transferability of models to other locations or disasters. Models that can generalize and transfer well to other tasks constitute the overarching objective for deep learning applications. Specifically, in the post-disaster management domain, such a model would remove the need to obtain specific training data to address detection tasks for a particular location or disaster. By removing this time-costly part of post-disaster damage detection, resources are saved and fast post-disaster response and recovery is enabled. However, a persisting issue keeping this goal out of reach is the availability of sufficient qualitative training data [13].

Because disasters affect a variety of locations and objects, damage induced by disasters similarly shows a large variety in visual appearances [13]. Obtaining a number of images that sufficiently cover the range of visual appearances is difficult and impractical. In fact, it is impossible to sample the never before seen damage, making supervised deep learning models inherently ad hoc [17]. Moreover, it is challenging to obtain qualitative annotations. Ideally, images are labelled by domain experts. However, the annotation process is time-costly, which critical post-disaster scenarios cannot tolerate. Finally, the process is subjective. Especially in a multi-classification task, two experts are unlikely to annotate all samples with the same label [18]. Questionable quality of the input data makes it difficult to trust the resulting output. The problem of insufficient qualitative training data drives most studies to make use of data from other disaster events with damage similar to the one of interest, to apply transfer learning or to apply unsupervised learning [19].

Most unsupervised methods for damage detection are not adequate for post-disaster applications where time and data are scarce. Principal Component Analysis (PCA) or multi-temporal deep learning frameworks are used for unsupervised change detection [20,21]. Besides the disadvantage of PCA that it is slow and yields high computational overhead, a major disadvantage of change detection approaches in general is that pre-event imagery is required, which is not always available in post-disaster scenarios. Methods such as One-Class Support Vector Machines (OCSVM) make use of a single epoch; however, these methods cannot be considered unsupervised because the normal class, in this case the undamaged class, still needs to be annotated in order to distinguish anomalies such as damage [22]. Moreover, earlier work has shown that OCSVM underperforms in the building damage detection task compared to supervised methods [23].

Anomaly detecting Generative Adversarial Networks (ADGANs), a recently proposed unsupervised deep learning principle used for anomaly detection, have the potential to overcome the aforementioned limitations and, therefore, to improve model transferability. ADGANs have been applied to detect anomalies in images that are less varied in appearance to address problems in constrained settings. For example, reference [17], reference [24] and reference [25] have applied ADGANs to detect prohibited items in x-rays of luggage. Reference [26] and reference [27] have applied ADGANs to detect masses in ultrasounds or disease markers in retina images. Until recently, ADGANs had not been applied to detect anomalies in images that are visually complex, such as remote sensing images, to address a problem that exists in a variety of settings, such as damage detection from remote sensing images.

The fundamental principle of an ADGAN is to view the damaged state as anomalous, and the undamaged state as normal. It only requires images that depict the normal, undamaged state. This principle poses several advantages. First, obtaining images from the undamaged state is less challenging, assuming that this state is the default. Second, data annotations are not required, thus eliminating the need of qualitative annotated training data. Finally, the never before seen damage is inherently considered since it deviates from the norm. This makes ADGAN an all-encompassing approach. The aforementioned advantages have made ADGANs appealing for a variety of applications, and especially appealing for post-disaster damage detection. The main advantage for post-disaster applications is that a model can be trained pre-disaster using only pre-event imagery. It can be instantly applied after the occurrence of a disaster using post-event imagery and thus aid post-disaster response and recovery. ADGANs output binary damage classifications and, therefore, a disadvantage is that they are unable to distinguish between damage severity levels. However, we argue that the practical advantages listed above outweigh this disadvantage, especially considering how the method provides rapid information to first responders in post-disaster scenes.

In earlier work, we showed how an ADGAN could be used under certain pre-processing constraints to detect post-earthquake building damage from imagery obtained from a manned aircraft [23]. Considering these results, and in addition the characteristics of the different remote sensing platforms explained above, we extend the preliminary work by investigating the applicability of ADGAN to detect damage from different remote sensing platforms. By training the ADGAN on a variety of pre-disaster scenes, we expect it to transfer well to different geographical locations or typologies of disasters. Special attention is given to satellite imagery because of its advantages explained above. We aim to provide practical recommendation on how to use this method in operational scenarios.

The contribution of this paper is threefold:

- First, we show how an ADGAN can be applied in a completely unsupervised manner to detect post-disaster building damage from different remote sensing platforms using only pre-event imagery.
- Second, we show how sensitive this method is against different types of pre-processing or data selections to guide practical guidelines for operational conditions.
- Lastly, we show whether this method can generalize over different typologies of damage or locations to explain the usability of the proposed method to real world scenarios.

The goal of this research is the fast detection of damage enabling fast dissemination of information to end-users in a post-disaster scenario. Therefore, it is beyond the scope of this study to examine the link between the proposed method and pre-event building vulnerability estimations or fragility curves. Our main aim is to investigate the applicability of ADGANs for unsupervised damage detection. Based on our results, we present a conclusion regarding the applicability and transferability of this method from an end-user's perspective.

Related work can be found in Section 2; the experiments are detailed in Section 3; results are described in Section 4; the discussion and conclusion can be found in Sections 5 and 6, respectively.

2. Related Work

2.1. Deep Learning for Post-disaster Damage Detection

Deep learning using optical remote sensing imagery has been a widely researched topic to address various aspects in the post-disaster research domain. Reference [2] used a SqueezeNet based Convolutional Neural Net (CNN) to make a distinction between collapsed and non-collapsed buildings after an earthquake event. Reference [28] addressed the combined use of satellite and airborne imagery at different resolutions to improve building damage detection. Reference [12] proposed a method to detect different proxies of damage, such as roof damage, debris, flooded areas, by using transfer learning and airborne imagery. Similarly, Reference [3] aimed to detect blue tarp covered buildings, a proxy for building damage, by utilizing aerial imagery and building footprints. Various researchers focused on utilizing pre- and post-event imagery to its best advantage. Reference [29] showed how fusion of multi-temporal features improved damage localization and classification. Similarly, reference [30] aimed to detect different building damage degrees by evaluating the use of popular CNNs and multi-temporal satellite imagery. Reference [11] proposed an efficient method to update building databases by using pre-disaster satellite imagery and building footprints to train a CNN, which was fine-tuned using post-disaster imagery. Reference [31] proposed a U-Net-based segmentation model to segment roads and buildings from pre- and post-disaster satellite imagery, specifically to update road networks. Progress has also been made towards real-time damage detection. Reference [32] made use of a lightweight CNN that was placed on board an UAV to detect forest fires in semi-real time. Reference [7] developed a similar near-real time low-cost UAV-based system which was able to stream building damage to end-users on the ground. Their approach was one of the first to validate such a system in large-scale projects. Finally, reference [14] showed how adding 3D information to UAV imagery aided the detection of minor damage on building facades from oblique UAV imagery.

Most deep learning methods towards post-disaster damage mapping, including the ones mentioned above, are supervised. However, a persistent issue in supervised learning is the lack of labelled training data [4]. The issue of unbalanced datasets or the lack of qualitative datasets is mentioned by most [2,12,28–30]. As mentioned earlier, researchers bypass this issue by using training datasets from other projects that resembles the data that are needed for the task-at-hand, or by applying transfer learning to boost performance. Despite these solutions, the main weakness of these solutions is that these models generally do not transfer well to other datasets. Reference [13] compared the transferability of different CNNs that were trained on UAV and satellite data from different geographic locations, and concluded that the data used for training a model strongly influences the model its ability to transfer to other datasets. Therefore, especially in data scarce regions, the application of damage detection methodologies in operative scenarios remains limited.

2.2. Generative Adversarial Networks

Generative Adversarial Networks (GANs) were developed by reference [33] and gained popularity due to their applicability in a variety of fields. Applications include augmented reality, data generation and data augmentation [34–36]. A comprehensive review of research towards GANs from recent years can be found in reference [37].

A GAN consists of two Convolutional Neural Nets (CNNs): the Generator and the Discriminator. The Generator receives as input an image dataset with data distribution p_{data}. The Generator aims to produce a new image ($\hat{x}$) that fits within the distribution p_{data}. Therefore, the Generator aims to learn a distribution of p_g that approaches p_{data}. The Discriminator receives as input an image (x) from the original dataset as well as the image ($\hat{x}$) generated by the Generator. The goal of the Discriminator is to distinguish the generated images from the original input data. If the Discriminator wins, the Generator loses and vice versa [33]. The Generator (G) and Discriminator (D) are locked in the two-player zero-sum principle. The discriminator aims to minimize the function $D(G(x))$ and the Generator tries to maximize it according to the function $\log(1 - D(G(x)))$.

2.3. Anomaly detecting Generative Adversarial Networks.

GANs are also applied to detect defects or damage in the medical or manufacturing domain. Similar to post-disaster damage detection, a common limitation for these kind of applications is data imbalance. Therefore, GANs are used to synthesize more data of the underrepresented class. Reference [38] synthesized medical imagery to boost liver lesion detection and reference [39] synthesized road defects samples, which led to a F1-score increase of up to 5 percent. The main limitation of synthesizing data is that examples are required. Moreover, it is unclear to what extent the generated samples are restricted to the data distribution of the input data, inhibiting diversity of the generated images [40,41]. ADGANs provide a better solution, since no examples are needed.

ADGANs are only trained using normal, non-damaged input data. The resulting trained model is proficient in reproducing images that do not show damage, and less proficient in reproducing images that depict damage. Therefore, the distance between the input image and the generated image is large when inference is done using an image that contains damage, which subsequently can be used to produce anomaly scores [24].

The first examples of ADGANs are Efficient GAN-Based Anomaly Detection (EGBAD), which was developed using curated datasets such as MNIST, and AnoGAN, which was geared towards anomaly detection in medical imagery [27,42]. Reference [26] applied an EGBAD-based method to detect malign masses in mammograms. The main limitation in AnoGAN was its low inference speed. This was resolved in f-AnoGAN [43]. The latter was outperformed by GANomaly, which successfully detected prohibited items in x-rays of luggage [17], although it was shown to be less capable of reconstructing visually complex images [23,44]. Using a U-Net as the Generator, the reconstruction of complex imagery was resolved by its successor Skip-GANomaly [24]. Both f-AnoGAN and Skip-GANomaly serve as the basis for ongoing developments [25,44–46].

Considering that Skip-GANomaly outperformed f-AnoGAN, and in addition, how it is proficient in generating visually complex imagery, this architecture was used in this research.

3. Materials and Methods

3.1. ADGAN

The architecture of Skip-GANomaly is shown in Figure 1. The Generator and the Discriminator consist of a U-net and an encoder architecture, respectively [47]. In earlier work, we showed how substituting the Generator for an encoder–decoder architecture without skip-connections—e.g., GANomaly [17]—does not always result in well-reconstructed fake images from Earth observation imagery [23]. The encoder–decoder architecture of Skip-GANomaly, in combination with the skip-connections, makes it efficient in recreating even complex remote sensing imagery.

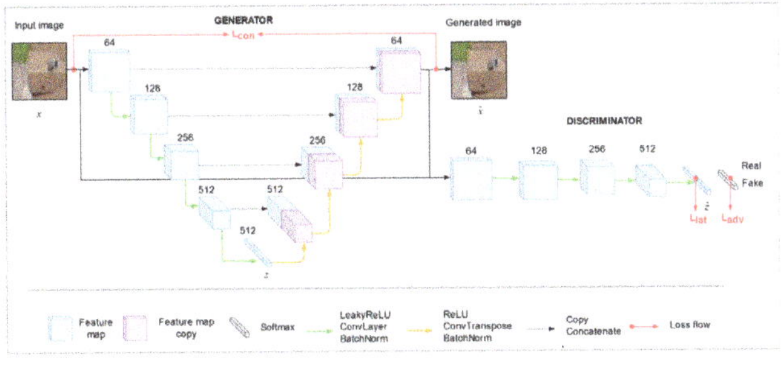

Figure 1. Skip-GANomaly architecture. Adapted from [24].

Skip-GANomaly makes use of three distinctive losses to guide its training, called the latent loss (L_{lat}), the adversarial loss (L_{adv}) and the contextual loss (L_{con}). L_{adv} accounts for the correctness of the classification (fake or real). L_{con} accounts for the generated image, and steers the model to create fake images that are contextually sound, i.e., images that look realistic. L_{lat} is a loss that steers the encoders inside the Generator and Discriminator to create similar representations of the image latent vector z [24]. Each loss contributes to the overall loss according to their corresponding weight (w). The losses are described in the following equations:

$$L_{adv} = \|f(x) - f(\hat{x})\|_2 \tag{1}$$

where,

$$f(.) = \mathbb{E}_{x \sim p_x} \left[\log D(.) \right] \tag{2}$$

$$L_{con} = \|x - \hat{x}\|_1 \tag{3}$$

$$L_{lat} = \|z - \hat{z}\|_2 \tag{4}$$

The overall loss is described as follows:

$$L = w_{adv}L_{adv} + w_{con}L_{con} + w_{lat}L_{lat} \tag{5}$$

Several hyper-parameters influence the performance of the model. Besides the general parameters such as batch size, learning rate or decay rate, model specific parameters include loss weights, the size of the latent vector z, and the number of encoder layers inside the Generator and Discriminator. Details on how these parameters are tuned can be found in Section 3.4.

A modification was made to the network. In the original network, after each epoch of training, the Area Under the Curve (AUC) score was calculated using the validation dataset. After training finished, a model for inference was selected based on the epoch in which it obtained the highest AUC score [24]. This makes the original implementation not a truly unsupervised approach, since a validation dataset is still required (i.e., examples of damage are still needed). Therefore, we choose to save the best performing model when the lowest Generator loss was found. This ensures that the model is chosen that is best able to generate fake images, which is the main principle of Skip-GANomaly. We verified that this approach yielded performance comparable to the original implementation, without the need of annotated test samples.

During inference, each image is classified as either damaged or undamaged by obtaining anomaly scores. Per-pixel anomaly scores are derived by simple image differencing between the input and the generated image. Each corresponding channel is subtracted from each other and averaged per pixel to obtain per-pixel anomaly scores. An image anomaly score is obtained by averaging the per-pixel anomaly scores. The closer to one, the higher the probability that the image is anomalous. After obtaining anomaly scores for all test samples, a classification threshold was determined in order to classify the images. This threshold is characterized as the intersection between the distribution of anomaly scores of normal and abnormal samples. Any sample with an anomaly score below the threshold was classified as normal and any value above the threshold as abnormal. Ideally, a model with a high descriptive value should result in non-overlapping distributions of the normal and abnormal samples with a clear threshold.

Finally, alterations and additions were applied to Skip-GANomaly in an attempt to boost results for the satellite imagery dataset. First, with the idea of weighing the generation of building pixels more than other pixels, we attempted to direct the attention of Skip-GANomaly by adding building masks as input in an approach similar to the one described in [48]. Furthermore, with the idea of utilizing the building information in the multiple epochs, similar to the approach described in [16], we stacked pre- and post-imagery into a 6-channel image and implemented an early, late or full feature fusion approach. These additions only provided marginal improvements. Our findings for stacking pre-

and post-imagery were in line with those found in [29]. The goal of this study was to investigate the applicability of ADGANs for building damage detection. Considering that improvements of the model were beyond our scope of work and only marginal, these lines of investigation were not explored any further and the original implementation was maintained.

3.2. Data

As mentioned earlier, a satellite and an UAV dataset were used in this research. This section will describe both datasets.

3.2.1. xBD Dataset

We made use of the xBD satellite imagery dataset [49]. It was created with the aim of aiding the development of post-disaster damage and change detection models. It consists of 162.787 pre- and post-event RGB satellite images from a variety of disaster events around the globe. These include floods, (wild)fire, hurricane, earthquake, volcano and tsunami. The resolution of the images is 1024 × 1024, the GSD ranges from 1.25 m to 3.25 m and annotated building polygons were included. The original annotations contained both quantitative and qualitative labels: 0—no damage, 1—minor damage, 2—major damage and 3—destroyed [50]. The annotation and quality control process is described in [50]. The dataset contained neither structural building information nor disaster metrics such as flood levels or peak ground acceleration (PGA). Figure 2 shows example pre- and post-event images of a location where a volcanic eruption took place. The color of the building polygons indicates the building damage level. For our purpose, all labels were converted to binary labels. All images with label 0 received the new label 0—undamaged, and the ones with label 1, 2 or 3 received the label 1—damaged. We note that even though damage is labelled under the umbrella-label of the event that caused it, damage is most often induced by secondary events such as for example debris flow, pyroclastic flow or secondary fires. For the sake of clarity, we will refer to the umbrella-label when referring to induced damage. Example imagery of each event can be found in Figure 3a–j. This dataset was used in the Xview2 challenge where the objective was to localize and classify building damage [51]. The ranked top-3 submissions reported amongst others an overall F1-score of 0.738 using a multi-temporal fusion approach [29].

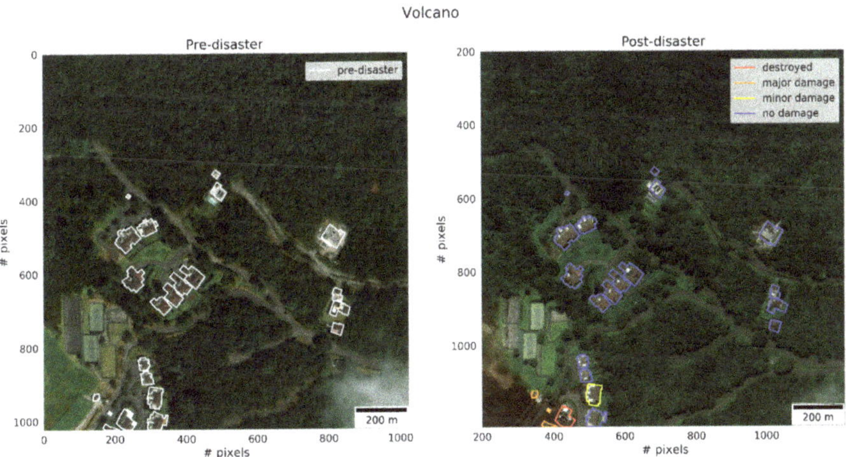

Figure 2. Example from the xBD dataset showing pre- and post-event satellite images from a location where a volcanic eruption took place. Several buildings and sport facilities are visible. The post-event image shows damage induced by volcanic activity. The buildings are outlined and the damage level is depicted by the polygon color. The scale bars are approximate.

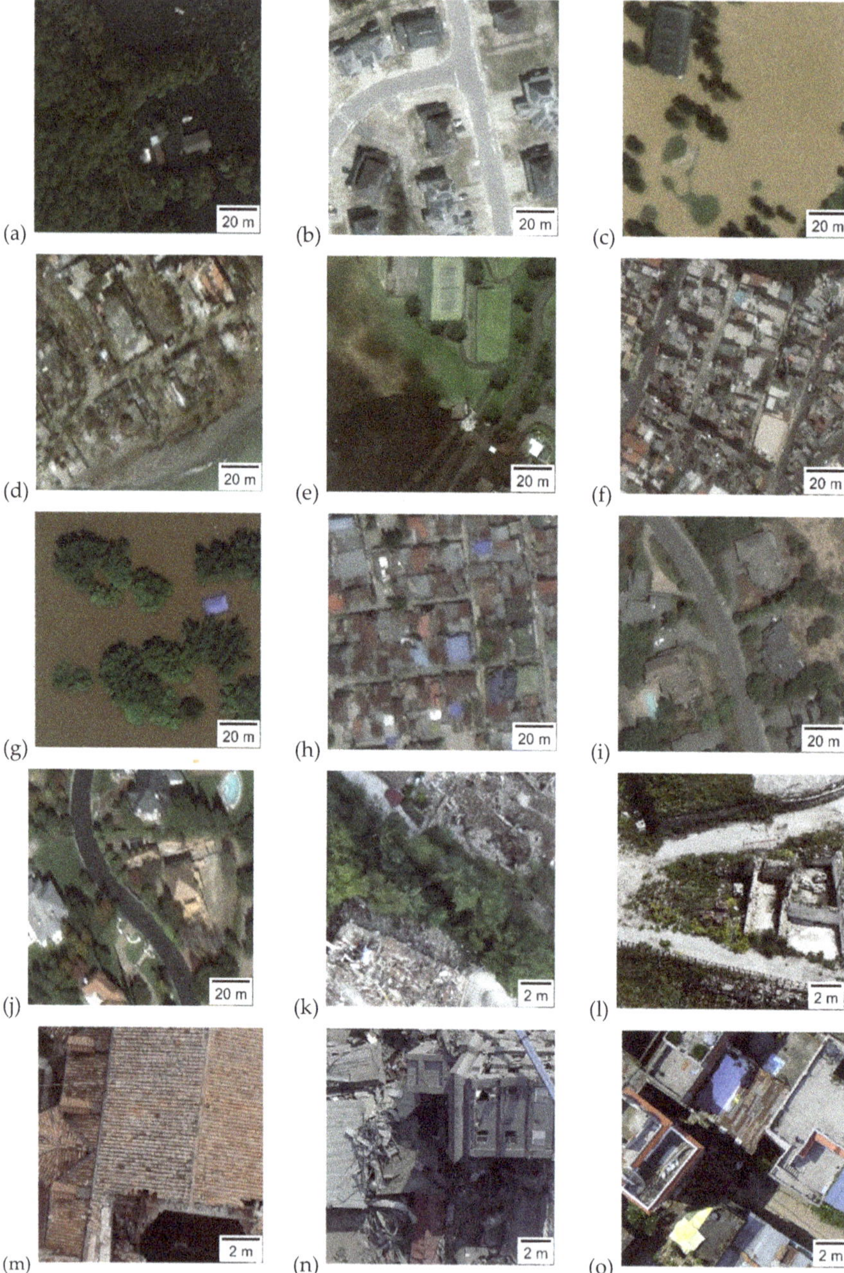

Figure 3. Examples of Satellite imagery used for testing: (**a**) Hurricane Florence (USA), (**b**) Hurricane Michael (USA), (**c**) Hurricane Harvey (USA), (**d**) Hurricane Mathew (Haiti), (**e**) Volcano (Guatemala), (**f**) Earthquake (Mexico), (**g**) Flood (Midwest), (**h**) Tsunami (Palu, Indonesia), (**i**) Wildfire (Santa-Rosa USA) and (**j**) Fire (Socal, USA). Examples of **UAV imagery** used for testing: (**k**) Earthquake (Pescara del Tronto, Italy), (**l**) Earthquake (L'Aquila, Italy), (**m**) Earthquake (Mirabello, Italy), (**n**) Earthquake (Taiwan) and (**o**), Earthquake (Nepal). The scale bars are approximate.

3.2.2. UAV Dataset

The UAV dataset was constructed manually from several datasets that depict the aftermath of several earthquake events. Examples can be found in Figure 3k–o. The UAV images were collected for different purposes and, therefore, the image resolution and the GSD vary and range around 6000 × 4000 pixels and from 0.02 to 0.06 m, respectively [13]. Moreover, the camera angle differed between nadir and oblique view. The UAV dataset contained no pre-event imagery and, therefore, the undamaged patches were obtained from undamaged sections in the images (see Section 3.3). Finally, the dataset contained neither structural building information nor PGA values.

3.3. Data Pre-Processing and Selection

Before the experiments were executed, the datasets were first treated to create different data-subsets. This section describes the different data treatments, while the next section describes how they were used in different experiments. The data treatments can be summarized into three categories: (i) varying patch size, (ii) removal of vegetation and shadows, and (iii) selection of data based on location or disaster type. For each dataset, we experimented with different cropping sizes. The rationale behind this step was that the larger the image, the more area is covered. Therefore, especially in satellite imagery where multiple objects are present, larger images often contain a high visual variety. As explained earlier, the Generator attempts to learn the image distribution, which is directly influenced by the visual variety contained in the images. When the learned image distribution is broad, a building damage has more chance to fall within this distribution, resulting in a reconstructed image that closely resembles the input image. The resulting distance between the input and generated images would be small and, therefore, the sample is expected to be misclassified as undamaged. We expected that restricting the patch size creates a more homogeneous and less visually varied scene. Especially cropping images around buildings would steer the Generator to learn mainly the image distribution of buildings. Therefore, any damage to buildings was expected to fall more easily outside the learned distribution, resulting in accurate damage detections and thus an increase in true positives.

The satellite imagery was cropped into patches of 256 × 256, 64 × 64 and 32 × 32 (Figure 4). By dividing the original image in a grid of four by four, patches of 256 × 256 could be easily obtained. However, the visual variety in these patches was likely still high. Smaller sizes of 64 × 64 or 32 × 32 would reduce this variety. However, simply dividing the original image systematically into patches of 64 × 64 or 32 × 32 resulted in a large amount of training patches that did not contain buildings. These patches did not contribute to learning the visual distribution of buildings. Therefore, the building footprints were used to construct 32 × 32 and 64 × 64 patches only around areas that contained buildings. To achieve this, the central point of each individual building polygon was selected and a bounding box of the correct size was constructed around this central point. We note that in real-world application, building footprints are not always available; however, this step is not necessarily required considering that it only intends to reduce the number of patches containing no buildings, even though there are various ways to derive building footprints. Open source repositories such as OpenStreetMap provide costless building footprints for an increasing number of regions, and supervised or unsupervised deep learning are proficient in extracting building footprints from satellite imagery [52–54]. Therefore, the proposed cropping strategy and subsequent training can be completely unsupervised and automated.

Figure 4. Illustration of different cropping strategies for the xBD dataset from the original patch size of 1024×1024 to 256×256, 64×64 and 32×32. The scale bars are approximate.

The UAV images were cropped in sizes of 512×512, 256×256 and 64×64. Larger patch sizes were chosen than for the xBD dataset to compensate for the difference in image resolution. More detail could be observed in larger sized UAV patches. Compare, for example, the amount of detail that can be observed in the smallest patches of Figures 4 and 5. Unlike for the xBD dataset, building footprints were not available. In general, building footprints for UAV imagery are difficult to obtain from open sources because, compared with satellite imagery, they are not necessarily georeferenced. Moreover, footprints would be difficult to visualize because of the varying perspectives and orientation of buildings in UAV imagery. Therefore, the 512×512 patches were extracted and labelled manually. Because of varying camera angles, patches displayed both facades and rooftops. Since no pre-event imagery was available, undamaged patches were obtained by extracting image patches from regions where no damage was visible. Binary labels were assigned to each image: 0—undamaged, or 1—damaged. The cropping strategy for the smaller sizes consisted of simply cropping around the center pixel (Figure 5).

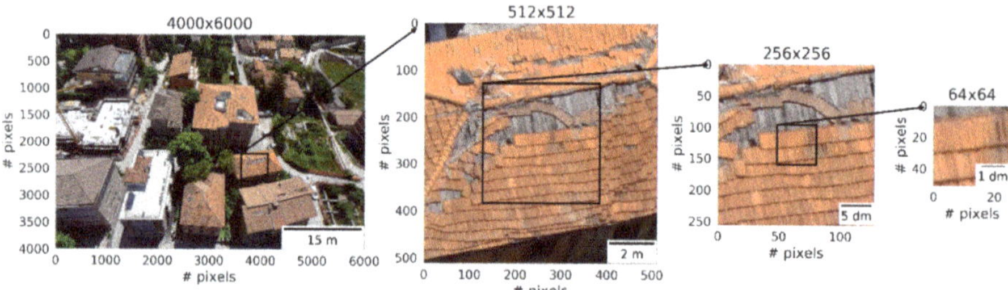

Figure 5. Illustration of cropping strategies for the UAV dataset from the original patch size of 4000×6000 to 512×512, 256×256 and 64×64. The scale bars are approximate and refer to the front of the scene.

Next, the cropped patches were pre-processed. In order to investigate how sensitive this method is against different pre-processing, images were removed from the dataset based on the presence of vegetation or shadows. Vegetation and shadows remain challenging in deep learning-based remote sensing applications. Shadows obscure objects of interest, but also introduce strong variation in illumination [55]. Depending on the varying lighting conditions, vegetation is prone to produce shadows and, therefore, varying Red, Green, Blue and illumination values [56]. Therefore, the image distribution learned by the Generator is expected to be broad. This means that any damage found on buildings is more likely to fall within this learned image distribution and, therefore, to be well reconstructed in the fake image. A well-reconstructed damage leads to lower anomaly scores, which is not the objective. We showed in [23] how removing these visually complex patches from the training set improve damage classification because the learned image distribution was expected to be narrower. Therefore, we created data subsets for training following the same procedure, using the

Shadow Index (SI; Equation (6)) and the Green–Red Vegetation Index (GRVI; Equation (7)) [57,58]. Images containing more than 75 or 10 percent vegetation and/or shadows, respectively, were removed from the original dataset. Using these datasets, we showed how increasingly stricter pre-processing, and thus decreasingly visually complex patches, influences performance. Removing images from a dataset is not ideal since it limits the practicality of the proposed methodology because it reduces the proportion of patches on which it can do inference (see Figure 6). The test dataset in the novegshad@10% data subset is 8 percent of the original test dataset. Therefore, we further experimented with masking the pixels that contain vegetation and shadow in an attention-based approach, as explained in Section 3.1. However, this was not considered as a further line of investigation since results did not improve.

$$SI = \sqrt{(256 - Blue) * (256 - Green)} \tag{6}$$

$$GRVI = \frac{\rho_{green} - \rho_{red}}{\rho_{green} + \rho_{red}} \tag{7}$$

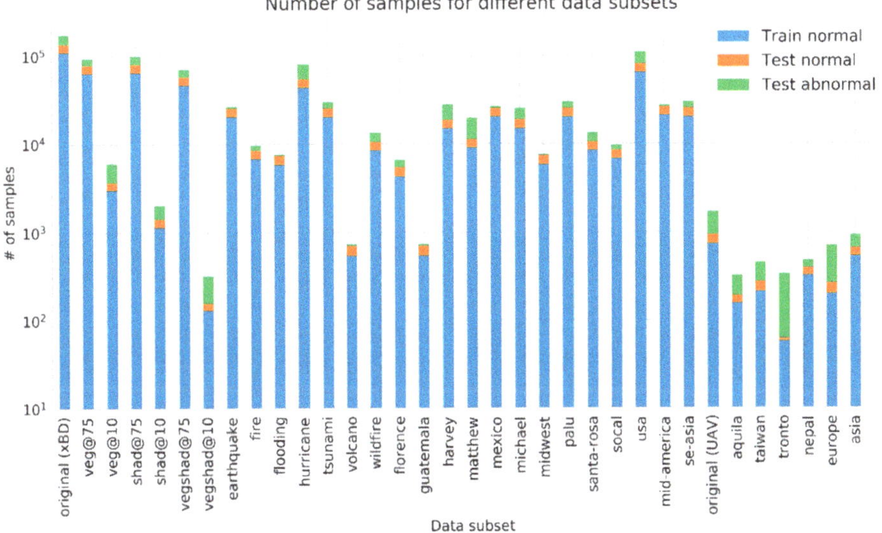

Figure 6. Number of samples in each data subset. Original refers to the complete un-preprocessed dataset. Y-axis is in log-scale.

Only the satellite patches of size 64 × 64 and 32 × 32 were pre-processed in this manner. Even though these sizes were already constrained to display maximum buildings and minimal surroundings using cropping techniques, some terrain and objects were often still present (see Figure 4). Satellite patches of size 256 × 256 were not pre-processed in this manner. Satellite images of this size usually contained more than 75 percent vegetation and/or shadow and, therefore, removing these images resulted in data subsets for training that were too small. UAV patches were also not pre-processed this way, since careful consideration was taken during manual patch extraction to ensure they do not contain vegetation or shadows.

Finally, selections of UAV and satellite patches were made based on the image location and the continent of the image location. Here the assumption was made that buildings were more similar in appearance if located in the same continent or country. Trained models were expected to transfer well to other locations if the buildings looked similar. Additionally, satellite patch selections were made based on the disaster type in order to investigate whether buildings affected by the same disaster type could yield a high performance. Here we consider that end-users might already possess a database of pre-event imagery of the same disaster of different locations around the globe, while they are not in possession of pre-event imagery of the country or continent that appears similar to the location of interest. Table 1 shows a summary of all the resulting satellite and UAV data subsets.

Table 1. Overview of the data subsets used in this research. Data subsets were created based on (1) resolutions and (2) data selections, which include pre-processing (removal vegetation and/or shadows), disaster-event location, disaster-event continent and disaster-type. * Not for satellite patches of size 256 × 256.

Dataset	Satellite (xBD)				UAV	
Resolutions	256 × 256 / 64 × 64 / 32 × 32				512 × 512 / 256 × 256 / 64 × 64	
Category	Data Pre-Processing	Data Selection: *Location*	Data Selection: *Continent*	Data Selection: *Disaster*	Data Selection: *Location*	Data Selection: *Continent*
	No vegetation (<75%) *	Guatemala (volcano)	North-America	Flood	Pescara del Tronto (Italy; earthquake)	Asia
	No vegetation (<10%) *	Florence (USA; hurricane)	Mid-America	Wildfire	Kathmandu (Nepal; earthquake)	Europe
	No shadow (<75%) *	Harvey (USA; hurricane)	South East Asia	Volcano	L'Aquila (Italy; earthquake)	South-America
	No shadow (<10%) *	Matthew (Haiti; hurricane)		Hurricane	Portoviejo (Ecuador; earthquake)	
← Category values	No vegetation and shadow (<75%) *	Michael (USA; hurricane)		Earthquake	Mirabello (Italy; earthquake)	
	No vegetation and shadow (<10%) *	Mexico City (Mexico; earthquake)		Tsunami	Taiwan (China; earthquake)	
		Midwest (USA; flood)				
		Palu (Indonesia; tsunami)				
		Santa-Rosa (USA; wildfire)				
		Socal (USA; fire)				

Each data subset was divided into a train and test set. Figure 6 shows the sample size of each subset. The train-set only consisted of undamaged images, and the test set contained both undamaged and damaged samples. For the satellite imagery, the undamaged samples in the train set came from the pre-event imagery, whereas the undamaged samples in the test set came from the post-event imagery. For the UAV imagery, the undamaged samples both came from the post-event imagery. The samples were divided over the train and test set in an 80 and 20 percent split. The original baseline dataset denotes the complete UAV or complete satellite dataset.

We note that the UAV dataset size was relatively low. However, the authors of [44] found that the ability of an ADGAN to reproduce normal samples was still high when trained on a low amount of training samples. We verified that low number of samples had no influence on the ability of Skip-GANomaly to produce realistic output imagery, and thus we conclude that the Generator was able to learn the image distribution well, which was the main goal. For the reasons explained above, these numbers of UAV samples were deemed acceptable.

3.4. Experiments

The experiments were divided into two parts. Part one showed whether the method is applicable and/or sensitive to preprocessing. The experiments consisted of training and evaluating Skip-GANomaly models using the different pre-processed data subsets from Table 1, described in Section 3.3. Part two showed whether the method can be transferred to different geographic locations or disasters. The experiments consisted of training and testing a Skip-GANomaly model on the different location, continent and disaster data subsets. Each trained model, including the ones trained in part one, was cross-tested on the test set of each other data subset.

The training procedure maintained for part one and part two can be described as follows: A Skip-GANomaly model was trained from scratch using the train-set. Before training, the model was tuned for the hyper-parameters, w_{adv}, w_{con}, w_{lat}, learning rate and batch size, using grid-search. When the best set of hyper-parameters was found, the model was retrained from scratch using these parameter values for 30 epochs, after which it did not improve further. As explained earlier, using a modification, during training the best performing model was saved based on the lowest generator loss value. After training was completed, the model was evaluated on the test set. Training and evaluation ran on a desktop with a dual Intel Xeon Gold (3.6GHz) 8-cores CPU and a Titan XP GPU (12GB). Training for 30 epochs took approximately 8 hours using patches of 64×64 and a batch size of 64. Inference on a single image of 64×64 took approximately 3.9 ms.

For part two of the experiments, the transferability was analyzed by testing each of the previously trained models on the test set of all other data subsets. For example, all trained UAV models were evaluated on the UAV-imagery of all patch sizes from all locations and continents. All trained satellite models were evaluated on satellite-imagery of all patch sizes, from all locations and continents and from all pre-processing manners. To deal with different patch sizes, the images were up- or down-sampled during testing. Finally, transferability was not only analyzed intra-platform, but also cross-platform. This means that all models trained on different subsets of satellite imagery were also evaluated on the test set of all different subsets of UAV imagery and vice versa.

The F1-score, recall, precision and accuracy were used to describe performance. A high recall is important, because it shows that most instances of damage are indeed recognized as damage. In practice, this means that it can be trusted that no damage goes unnoticed. A high precision is also important, because it shows that from all the recognized instances of damage, most are indeed damage. Moreover, this means that it can be trusted that all the selected instances of damage are indeed damaged, and no time has to be spent on manually filtering out false positives. The F1-score represents the balance between recall and precision.

3.5. Comparison Against State-of-the-Art

In order to investigate how close our results can get to those of supervised methods, we compared the results of our experiments against results obtained using supervised deep learning approaches. In earlier work, we showed how unsupervised methods drastically underperformed compared to our method and, therefore, unsupervised methods such as One Class Support Vector Machine are left out of the comparison [23]. In order to make a fair comparison, we considered approaches that made use of a single epoch and, ideally, datasets that resemble ours in GSD, resolution, location and disaster-types. Therefore, we compared the results obtained using satellite-based models against the xView2 baseline model, and ranked competitors in the xView2 competition [29]. The xView2 baseline model first trained a U-Net architecture to extract building polygons. Afterwards, they used a Resnet50 architecture pre-trained on ImageNet to classify different degrees of building classifications. The ranked contenders [29] used a multi-temporal approach where both localization and classification was learned simultaneously by feeding the pre- and post-disaster images into two architectures with shared weights. The architectures consisted of ResNet50, which was topped with Feature Pyramid Heads, and were pre-trained on ImageNet. Finally, we compared the results obtained using UAV-based models with results obtained by [13]. Here, comparable UAV-images were used from post-earthquake scenes to train an adapted DenseNet121 network with and without fine-tuning. The authors carried out several cross-validation tests where each time a different dataset was used for testing, to investigate the influence of training data on performance.

4. Results

This section will describe the performance of Skip-GANomaly to detect building damage from satellite and UAV imagery. We show the more interesting results to avoid lengthy descriptions of all tests that were carried out. In addition, we present a closer look at the cases in which the model succeeded or failed to detect damage, and show how anomaly scores could be used to map damage. Additionally, the cross-test results are presented, which offer insight into the transferability of this method. Finally, a comparison between our results and supervised method is presented.

4.1. Performance of Skip-GANomaly on Satellite Imagery

First, we examined the performance of Skip-GANomaly on satellite imagery when using different data pre-processing techniques on the baseline dataset (all disasters combined). The main result showed that, especially when strict pre-processing was applied, e.g., removing all patches that contained more than 10 percent of vegetation or shadow (novegshad@10%), performance improved compared to baseline, although it only reached a recall value of 0.4 (Figure 7). A similar trend was found for aerial imagery in an earlier work [23]. Their performance improved the most when the novegshad@10% rule was applied. Finally, contrary to expectations and excluding the performance of novegshad@10% on 32 × 32 patches, no clear trend was observed for specific patch sizes. In some cases, the smaller sizes performed well and the larger size did not, and vice versa.

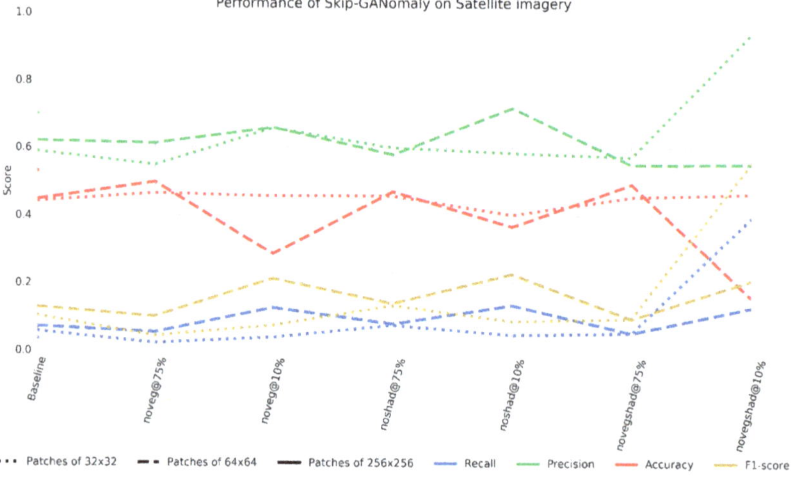

Figure 7. Performance of Skip-GANomaly on pre-processed satellite patches of size 256 × 256 (only baseline) 64 × 64 and 32 × 32.

Next, we examined the performance of Skip-GANomaly on satellite imagery when pre-selected by disaster type and without any pre-processing. Overall, we found that the performance of disaster-based models improved compared to baseline. Earlier, we found evidence that pre-processing improved performance. Therefore, in addition we tested the performance of disaster-based models when the training subsets of size 32×32 were pre-processed according to the novegshad@10% rule. The rule was not applied to subsets of size 256×256 or 64×64 because this resulted in subset sizes too small for training. The difference in performance is shown in Table 2. Again, we observed that performance improved for each individual disaster case.

Table 2. Difference in performance of Skip-GANomaly disaster-based models when trained on 32×32 satellite patches when (not) pre-processed based on the novegshad@10% rule. The grey background indicates the pre-processed values and bold values indicates which model performs best.

Model	Pre-processed	Recall	Precision	F1-score
Earthquake	No	0.110	**0.212**	0.022
	Yes	**0.333**	0.111	**0.167**
Flooding	No	0.455	**0.555**	0.500
	Yes	**0.500**	0.500	**0.500**
Hurricane	No	0.143	0.643	0.234
	Yes	**0.325**	**0.662**	**0.436**
Tsunami	No	0.040	0.365	0.073
	Yes	**0.141**	**0.926**	**0.245**
Wildfire	No	0.321	0.855	0.467
	Yes	**0.778**	**0.989**	**0.871**

We noted interesting differences between the performances of different disaster-based models (Table 2). Because a secondary damage induced by hurricanes is floods, it was expected that the performance for flood- and hurricane-based models would be comparable. However, this was not the case. In fact, it was observed that for the disaster types Hurricane and Tsunami (Table 2) and for the corresponding locations in Table 3, recall tended to be low compared to precision. We argue that this can be attributed to several reasons related to context, which will be explained in Section 4.3.

Finally, we examined the performance of Skip-GANomaly on satellite imagery when pre-selected based on location or continent location. In addition, the performance was examined when pre-processing according to the novegshad@10% rule was applied to patches of size 32×32. Again, we found that pre-processing improved the performance in a similar way, as was shown for the disaster-based models (Table 3).

4.2. Performance of Skip-GANomaly on UAV Imagery

Figure 8 shows the performance of UAV-based models. The main results show that the performance of UAV-based models was generally higher than that of the satellite-based models. Moreover, similar to the findings for satellite location-based models, we observed that the performance of UAV location-based models improved compared to baseline (all UAV-patches combined), with the exception of Asian locations (Nepal and Taiwan). Europe obtained a recall, precision and F1-score of 0.591, 0.97 and 0.735, respectively. As expected, UAV location-based models with similar building characteristics performed comparably. For example, models trained on location in Italy performed similarly (L'Aquila and Pescara del Tronto). This time, we did observe a pattern in performance of different patch sizes. Generally, models trained using the larger images size of 512×512 performed poorly, compared to models trained on smaller patch sizes. Especially for the Asian location-based models, the smaller sizes perform better. In the next section, we explain why context is likely the biggest influencer for the difference in performances.

Table 3. Difference in performance of Skip-GANomaly location-based models when trained on 32×32 satellite patches when (not) pre-processed based on the novegshad@10% rule. Locations that are not listed did not have sufficient training samples. The grey background indicates the pre-processed values and bold values indicates which model performs best.

Model	Pre-processed	Recall	Precision	F1-score
Harvey (USA; hurricane)	No	0.019	0.719	0.036
	Yes	**0.198**	**0.800**	**0.317**
Matthew (Haiti; hurricane)	No	**0.144**	0.625	**0.234**
	Yes	0.053	**1.00**	0.100
Michael (USA; hurricane)	No	**0.291**	**0.800**	**0.427**
	Yes	0.286	0.421	0.340
Mexico City (Mexico; earthquake)	No	0.055	0.002	0.005
	Yes	**0.333**	**0.111**	**0.167**
Midwest (USA; flood)	No	0.470	0.570	0.515
	Yes	**0.750**	**0.600**	**0.667**
Palu (Indonesia; tsunami)	No	0.099	0.393	0.158
	Yes	**0.141**	**0.926**	**0.245**
Santa-Rosa (USA; wildfire)	No	0.303	0.856	0.448
	Yes	**0.684**	**0.985**	**0.807**
Socal (USA; fire)	No	0.087	0.329	0.137
	Yes	**0.538**	**0.667**	**0.596**
North-America	No	0.099	0.718	0.175
	Yes	**0.652**	**0.970**	**0.780**
Mid-America	No	0.162	0.024	0.041
	Yes	**0.333**	**0.100**	**0.154**
South East Asia	No	0.031	0.366	0.058
	Yes	**0.099**	**0.854**	**0.177**

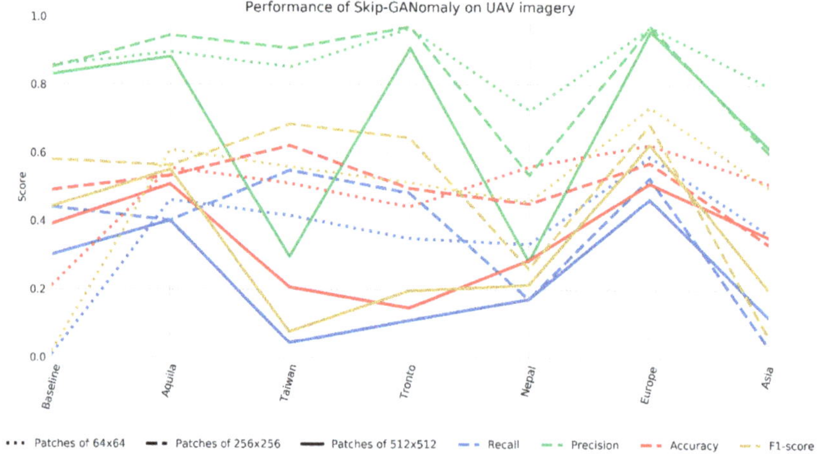

Figure 8. Performance of Skip-GANomaly on UAV imagery of size 512×512, 256×256 and 64×64 for different locations.

4.3. The Importance of Context

We investigated whether the characteristics of the different satellite data sources, especially for different disaster events, could explain why the method worked better for some disasters than the other. Certain disasters, such as floods or fires, induced both building damage and damage to their surroundings. Other disasters such as earthquakes mainly induced damage to buildings only. Large-scale damage can be better detected from satellite imagery than small-scale damage, because satellite imagery contains inherently coarse resolutions. Most likely, the ADGAN is more efficient in detecting building damage from large-scale disasters by considering the surroundings of the building.

To investigate this idea, we aimed at understanding how the anomaly score distribution corresponds to the large-scale or small-scale damage pattern, by plotting the anomaly scores over the original satellite image. High anomaly scores on pixels indicate that the model considered these pixels to be the most anomalous. These pixels weigh more towards classification.

Figure 9 shows multiple houses destroyed by a wildfire. The image was correctly classified as damaged for all patch sizes. However, it seems that context contributed to the correct classification for the larger scale image (256 × 256), because the burned building surroundings resulted in high anomaly scores, whereas the buildings itself obtained lower scores. This suggested that, for this particular patch size, the surroundings have more discriminative power to derive correct classifications than the building itself. For smaller patch sizes, as explained in Section 3.3, the assumption was made that the smaller the patch size, the more adept the ADGAN would be in learning the image distribution of the building characteristics, instead of its surroundings. For the example in Figure 9, this seemed to hold true. In the patches of size 64 × 64 and 32 × 32, high anomaly scores were found all throughout the image, including the damaged building itself. This suggested that our assumption was correct. In short, the large-scale damage pattern of wildfire, plus the removal of vegetation, resulted in a high performing model. This example suggests that our method is capable of recognizing and mapping large-scale disaster induced damage.

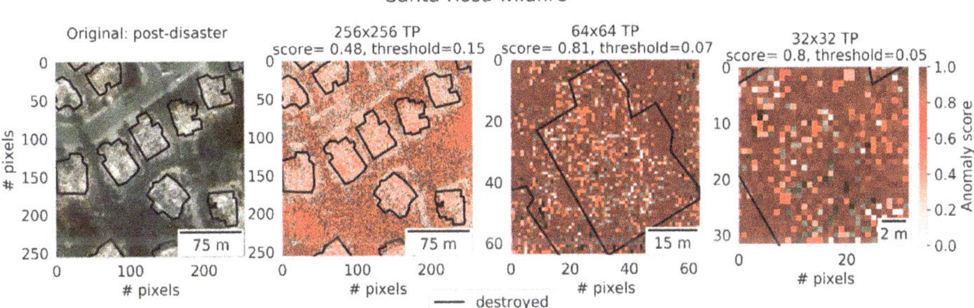

Figure 9. Post-wildfire satellite imagery from the USA showing multiple damaged buildings overlaid with anomaly scores and building polygon. The classification, classification threshold and anomaly scores are indicated (TP = True positive). The scale bars are approximate. High anomaly scores on burnt building surroundings and smaller patch sizes lead to correct classifications.

Another example of the influence of context can be seen in Figure 10. Here, a building was completely inundated during a flood event in the Midwest (USA), meaning that its roof and outline were not visible from the air. Even though, for all patch sizes, the image was correctly classified as damaged. We noticed how the anomaly scores were mostly located in the regions that were inundated. The anomaly scores on the unperceivable building itself naturally did not stand out. This suggests that the correct classifications were mostly derived from contextual information. No evidence of a better understanding of building damage could be observed for the smaller patch sizes (unlike for the wildfire example), because the building and its outline were not visible. Still, a correct classification was made due to the contextual information in these patch sizes. This example shows that the model was still adept in making correct inferences on buildings obscured by a disaster. Although removing vegetation generally improved classifications (Figure 7), for floods in particular performance only changed marginally (Table 2). Most of the flooded regions coincided with vegetated regions, which suggests that the large-scale damage patterns induced by floods are of strong descriptive power and weigh up against the negative descriptive power of vegetation.

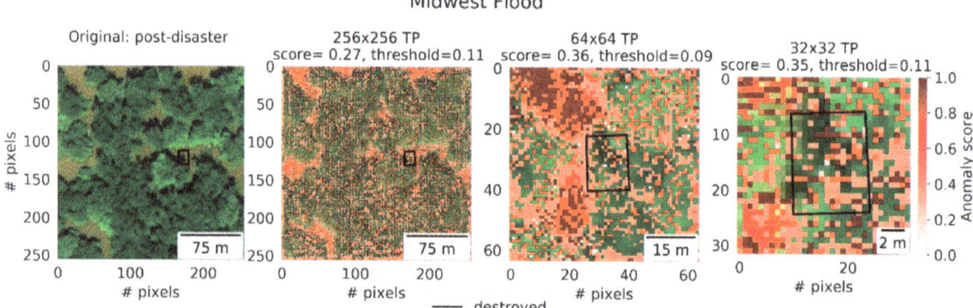

Figure 10. Post-flood satellite imagery from the USA showing a damaged building, overlaid with anomaly scores and building polygons. The classification, classification threshold and anomaly scores are indicated (TP = True positive). The scale bars are approximate. High anomaly scores on flooded areas resulted in correct classifications, regardless of patch size.

The examples above showed how context contributed to correct building damage classifications in cases where disasters clearly induced large-scale damage or changes to the surroundings. However, in cases where no clear damage was induced in the surroundings, context was shown to contribute negatively to classifications. Figure 11 shows an example of the Mexico earthquake event that spared multiple buildings. The patches of all sizes were wrongly classified as damaged, even though no large-scale damage was visible in its surroundings. The high anomaly scores in the 256×256 image were shown to exist all throughout the image, and to a lesser degree on the buildings itself. This example suggests that the context was visually too varied, which resulted in many surrounding pixels to obtain a high anomaly score. Moreover, unlike for the flood example, no clear damage pattern is present to counterbalance the visual variety of vegetation. As explained in Section 3.3, the variance could stem from vegetation. However, removing vegetation resulted in only modest improvements (Table 2). Therefore, the variation is also suspected to result from the densely built-up nature of the location. See for example Figure 3f, where a densely built-up region in Mexico without vegetation is visible. Even at the smaller patch sizes, the surroundings are varied, making it difficult to understand what is damaged and what is not. This example showed that context is less useful to derive earthquake-induced damage.

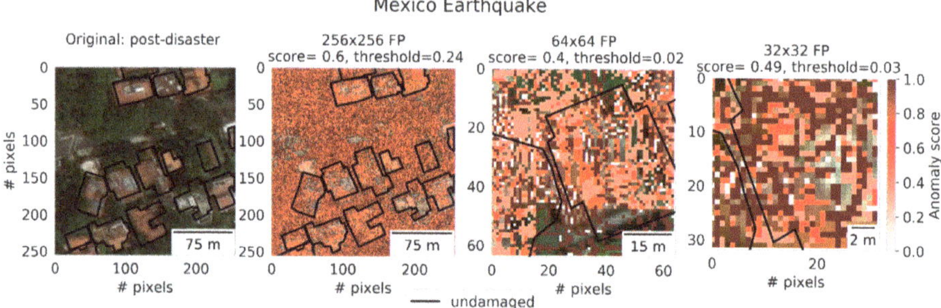

Figure 11. Post-earthquake satellite scene from Mexico showing undamaged buildings overlaid with anomaly scores and building polygons. The classification, classification threshold and anomaly scores are indicated (FP = False positive). The scale bars are approximate. High anomaly scores induced by varying building surroundings resulted in false positives, regardless of the patch size.

We argue that the examples shown so far could explain why recall for hurricane- and tsunami-based models was low (Tables 2 and 3). Where for flood events damage largely coincided with homogeneous flood patterns, the damage pattern for hurricanes and tsunamis was heterogeneous (Figure 3a–d,h). Large-scale floods and small-scale debris are the main indicators of damage. In addition, the locations suffer from a mixture between dense and less dense built-up areas. We suspect that when a mixture of damage patterns and inherently heterogeneous built-up area is present, lower performance and therefore a low recall value can be expected.

The examples shown above for satellite imagery can be used to illustrate the observed difference in performance for UAV location-based models, where smaller patch sizes were shown to perform better, particularly for the Asian locations (Nepal and Taiwan). Similar to the Mexico earthquake, Nepal and Taiwan are densely built-up. Moreover, the earthquakes in Nepal and Taiwan did not induce large-scale damage in the surroundings, meaning that the surroundings do not carry any descriptive power. However, unlike the Mexican imagery, due to the larger GSDs retained in the UAV imagery, reducing the patch size does result in less visual variety retained in smaller images. Therefore, similar to the wildfire event, and according to our previously stated assumption, smaller patch sizes result in a better understanding of the image distribution of buildings. Therefore, smaller images obtained higher anomaly scores on the building itself, instead of its surroundings, leading to performance improvements. See, for example, Figure 12.

Figure 12. Earthquake UAV scene from Nepal showing a damaged building overlaid with anomaly scores. The classification, classification threshold and anomaly scores are indicated (TP = True positive, FN = False Negative). The scale bars are approximate. The 64 × 64 patch yielded a correct classification due to a better understanding of the building image distribution.

In summary, the presented examples showed how context is important for damage detections. Particularly, we found that both the scale of the induced damage and the GSD of the imagery decide whether context plays a significant role. One added benefit observed from these results is that in cases where context does play a role, the derived anomaly maps can be used to map damage. These damage maps are useful to first responders in order to quickly allocate relief resources.

4.4. Cross-Tests

Finally, we evaluated the performance of trained UAV- and satellite-based models on the test set of other data subsets to find out how transferable our proposed method is.

First, we highlight a couple of general findings for satellite-based cross-tests. In general, a heterogeneous pattern in performance was found when a trained satellite-based model was tested on the test-set of another satellite data subset. In some cases performance increased compared to the dataset on which it was trained, while for others it decreased. Overall, no noteworthy improvements in performance were observed. Second, we observed that satellite-based models did not transfer well to image datasets that had different patch sizes than the ones on which they were trained. For example, satellite-based models trained on patches of size 32 × 32 performed worse when tested on patches of other sizes. This could be caused by context and GSD, as explained in Section 4.3. Finally, in some cross-tests, instead of the model, the dataset seemed to be the driving factor behind systematically high performances. For example, the flooding data subset yielded on average a F1-score of 0.62 for all patch sizes combined. This was systematically higher than for all other datasets. We expect that these results were driven by the flood-induced damage pattern, as explained in Section 4.3.

Next, we examined the cross-test results for UAV-based models. Again, general findings are highlighted. First, we observed that UAV-based models trained on specific patch sizes transferred well to patches of other sizes. A model trained on 32 × 32 patches from L'Aquila performed equally well on 64 × 64 patches from L'Aquila. We expect that this can be explained by the level of detail that is retained in UAV patches when the image is up- or down sampled during inference. Second, contrary to the findings for satellite imagery, no dataset seemed to drive the performance of datasets in a specific way.

Finally, we examined cross-platform transferability where we observed the performance of trained UAV or satellite models that were tested against the test set of either UAV or satellite imagery. The immediate observation is that the proposed method cannot be transferred cross-platform.

In summary, the models showed to transfer well if the test buildings look similar to the ones on which they were trained. Contrary to the desired outcome, transferability to buildings that are different in style was not unilaterally shown. We argue that the range of building appearances is yet too large for the Generator to learn an image distribution that is narrow enough to distinguish damaged buildings. Moreover, it is learned that no assumption can be made on the visual likeliness of buildings based on geographic appearance. However, we argue that transferability is less of an issue compared to traditional supervised methods considering that our method can be trained using pre-event data for the location of interest, which is often easily achieved.

4.5. Comparison Against State-of-the-Art

Table 4 shows how our proposed method compares against other state-of-of the art methods. Before comparing results for satellite-based models, we note that the xView2 baseline and the ranked contenders scores in the first and second row were based on the combined localization and multi-class classification F1-scores [29]. The third row shows the classification F1-score of the ranked contenders [29]. Our score is based on the binary classification F1-score. Because the supervised approaches considered all disasters at once, our reported F1-score is the average of F1-scores obtained using the pre-processed methods using patches of 32 × 32, which were reported in Table 2. Our method improved on the xView2 baseline but performed lower than the supervised method of the ranked contenders. Considering that our method is unsupervised, uses only pre-event imagery, makes no use of building footprints during damage detection training and is not pre-trained to boost performance, our results are encouraging and show that reasonable results can be obtained with minimal effort.

Table 4. Performance differences between our proposed method and comparable supervised CNN approaches.

	Method	Recall	Precision	F1-score
Satellite	Supervised—*localization and classification* (xView2 baseline [29])	-	-	0.265
	Supervised—*localization and classification* (Ranked contenders [29])	-	-	0.741
	Supervised—*classification* (Ranked contenders [29])	-	-	0.697
	Ours—*classification*	-	-	0.444
UAV	Supervised—*no fine-tuning* [13]	0.538–0.814	0.741–0.934	0.623–0.826
	Supervised—*fine-tuning* [13]	0.651–0.969	0.803–0.933	0.725–0.915
	Ours	0.591	0.97	0.735

Before comparing results for UAV-based models, we note that the supervised results are reported as ranges, considering how different cross validation experiments were carried and reported in the original paper [13]. Our results are derived from our best performing European-based model using patches of 64 × 64 (Figure 8). Our results perform on par with the supervised approach without fine-tuning. In fact, our precision value exceeds the highest precision value found by the supervised methods. Our F1-score is constantly within the range of values found for the supervised method. Our recall score comes close to the lower range of recall values obtained using supervised methods. This suggested that, as can be expected, supervised methods outperform our unsupervised method. Moreover, it suggests that having examples of damage of interest are of benefit to recall values. Nonetheless, our method is more practical considering that it is unsupervised, and only requires single epoch imagery and, therefore, can be directly applied in post-disaster scenarios.

Finally, we note how the recall values for both the supervised and our method were always lower than the precision values, suggesting that building damage detection remains a difficult task despite using supervised methods.

5. Discussion

5.1. Applicability and Sensitivity of Skip-GANomaly

First, the applicability and sensitivity of satellite-based models are discussed. Satellite imagery is generally a difficult data type for unsupervised damage detection tasks due to their visual complexity, which is subject to temporal variety, and due to large GSDs, which make it difficult to detect detailed damage. This difficulty is reflected by the baseline results obtained using our proposed method. However, we showed how performance can be improved. In line with the results found for aerial imagery [23], reducing the complexity of the baseline dataset, e.g., removing vegetation and shadowed patches, improved performances, especially for the 32 × 32 novegshad@10% based model. Cropping the training patch sizes in order to reduce the visual complexity did not always yield better performances. We argued in Section 4.3 that context is suspected to play a role. Comparing the performance of our best scoring pre-processed model, a F1-score of 0.556, to the F1-score of 0.697 obtained by ranked contenders in the Xview2 contest for damage classification, our results are encouraging [29]. This is especially so considering that our method was unsupervised and only used a single epoch, whereas their methodology was supervised and multi-epoch.

We conclude that satellite-based ADGANs are sensitive to pre-processing, and reducing the complexity of the training data by applying pre-processing helps to improve performance. This finding is not necessarily novel. However, our specific findings on how reducing the complexity for specific disaster-types influences performance has provided insight on the importance of context, and allowed us to define practical guidelines that can be applied by end users. As an example, for disasters such as floods and fires, pre-processing is not strictly necessary to obtain good results. However, for other disasters, a downside of this method is that, once stricter pre-processing rules are applied, the numbers of samples on which the model can conduct inference declines. Future research can look into other ways to deal with vegetation or shadows. The idea of weighing these objects differently during training can be the focus, which, as explained in Section 3.1, was explored in early phases of this research.

Next, the applicability of UAV-based models will be discussed. We found that the UAV-based baseline model performed generally better than satellite-based baseline models. Location-based UAV models surpassed the performance of all satellite-based models, with the F1-score reaching 0.735. These results are satisfactory when compared to the F1-score of 0.931 obtained by [13], who used a supervised CNN to detect building damage from similar post-event UAV datasets. Again, considering our unsupervised and single-epoch approach, which only makes use of undamaged buildings for training, our method showed to be promising.

The importance of contextual information was explained in Section 4.3. We showed how flood or fire-induced building damage was likely deduced from contextual information, rather than from the building itself. The contextual information has a negative influence towards the classification in disaster events where the damage is small-scale and the affected area is densely built-up. These findings suggest that, in practice, each event has to be approached case by case. However, we are able to provide broad practical guidelines: when the disaster has the characteristic of inducing large-scale damage to the terrain such as floods, the image training size can be 256×256.

Finally, we showed in Section 4.3 how detailed damage maps can be derived using simple image differencing between the original and generated image. As of yet, we are not aware of any other method that can produce both image classifications and damage segmentations without explicitly working towards both these tasks using dedicated supervised deep learning architectures and training schemes. Our method, therefore, shows a practical advantage compared to other methods.

Future work can focus consider the following: the rationale behind our sensitivity analysis was that reducing the visual information being fed to the Generator steers the Generators inability to recreate damaged scenes, which in turn helps the Discriminator distinguish fake from real. As an extra measure, future work can focus on strengthening the discriminative power of the Discriminator earlier on in the training process, by allowing it to train more often than the Generator, thus increasing its strength while not limiting the reconstructive capabilities of the Generator. Future work can also investigate the potential of alternative image distancing methods to obtain noiseless anomaly scores. The log-ratio operator for example, often used to difference synthetic aperture radar (SAR) imagery, considers the neighborhood pixels to determine whether a pixel has changed [59]. It is expected that such a differencing method lead to a decrease of noisy anomaly scores, and thus a better ability to distinguish between anomalous and normal samples.

5.2. Transferability

In general, a heterogeneous performance was observed for satellite-based models when tested on test-sets of other satellite sub-sets. Performance fluctuated for the different datasets and regardless of whether the model tested well on the dataset for which it was trained. This suggests that satellite-based models do not transfer well to other geographic locations or other typologies of disasters. This finding is in contrast with one specific finding from our preliminary work. There, we found that aerial-based models, trained on patches that were pre-processed according to the novegshad@10% rule, transferred well to other datasets [23]. We did not observe the same for the satellite-based model novegshad@10%. A possible explanation can be that the novegshad@10% model was not able to find the normalcy within other datasets, because the amount of training samples is small (see Figure 6). Therefore, the learned image distribution is too narrow. This could have led to an overestimation of false positives once this model was tested on other datasets.

Contrastingly, a homogeneous performance was observed for UAV-based models when tested on test-sets of other UAV sub-sets. Consistent performance was observed when models were tested on different datasets or different patch sizes. In addition, the model performance stayed high if the performance was high for the dataset on which it was trained. Specifically, we found that models transferred well if the buildings on which the model was tested looked similar to the buildings on which it was trained. For example, locations in Italy (L'Aquila and Pescara del Tronto) looked similar and were shown to transfer well. Locations in Asia (Taiwan and Nepal) looked very dissimilar in appearance and did not transfer well. Similar conclusions for the transferability of Italian locations were found in [13]. In line with the conclusion drawn in [13], we agree that the transferability of a model depends on whether the test data resemble the data on which it was trained. A model that cannot find the normalcy in other datasets is likely to overestimate damage in this dataset. Therefore, our previously stated assumption that buildings in the same geographic region look alike is not always valid. In future approaches, attention should be given to how geographic regions are defined. Categorizing buildings not based on the continent in which they are located, but on lower geographic units such as municipalities or provinces, might lead to a better approximation by the AGDAN of what constitutes a normal building in that geographic unit.

5.3. Practicality in Real-world Operations

The general conclusion is drawn that ADGANs can be used for damage detection from satellite images on the condition that the imagery is pre-processed to contain minimal vegetation and shadows. Considering how pre-processing is largely automated, this step is not a limitation. Nonetheless, cases were found where models yielded high performance, regardless of the presence of vegetation and shadows. The performance of satellite-based models trained on original imagery from flood and fire disasters was high and, therefore, these datasets do not have to be pre-processed, thus saving time.

We showed that damage maps could be constructed in the cases where context provides a significant contribution. These show, in detail, where damage is located. During inference, these maps can be created instantly, and they can therefore provide valuable information in the post-disaster response and recovery phase.

As stated in the introduction, a main limitation of UAV-based models is that UAV-based imagery needs to be collected in the pre-event stage. Considering how UAV-imagery collection is still a human-driven task, this might be difficult to achieve. However, the advantage is that data acquisitions can take place any time during the pre-event stage. Therefore, practical advice to end-users who wish to apply this methodology is to collect UAV-imagery of buildings in the pre-event stage in advance.

A final note of consideration is the following: the assumption is made that the normal dataset is free of anomalies. However, day-to-day activities such as constructions can result in visual deviations from normal that are not strictly damage [45]. In practice, care has to be taken to make the distinction between what is damaged and what is simply an anomaly.

6. Conclusions

In this paper, we proposed the use of a state of the art ADGAN, Skip-GANomaly, for unsupervised post-disaster building damage detection, using only imagery of undamaged buildings from the pre-epoch phase. The main advantage of this method is that it can be developed in the pre-event stage and deployed in the post-event stage, thus aiding disaster response and recovery. Special attention was given to the transferability of this method to other geographic regions or other typologies of damage. Additionally, several Earth observation platforms were considered, since they offer different advantages for data variety and data availability. Specifically, we investigated (1) the applicability of ADGANs to detect post-disaster building damage from different remote sensing platforms, (2) the sensitivity of this method against different types of pre-processing or data selections, and (3) the generalizability of this method over different typologies of damage or locations.

In line with earlier findings, we found that the satellite-based models were sensitive against the removal of objects that contained a high visual variety: vegetation and shadows. Removing these objects resulted in an increase in performance compared to the baseline. Additionally, in order to reduce the visual variety in the original images, experiments were conducted with varying image patch sizes. No clear difference in performance of different patch sizes was observed. UAV-based models yielded high performance when detecting earthquake-induced damage. Contrary to satellite-based models, UAV-based models trained on smaller patches obtained higher scores.

UAV imagery contained small GSDs and showed damage in high detail. Therefore, models based on UAV-imagery transferred well to other locations, which is in line with earlier findings. Models based on satellite-imagery did not transfer well to other locations. The results made it evident that image characteristics (patch size and GSD), and the characteristics of the disaster induced damage (large-scale and small-scale), play a role in the ability of satellite-based models to transfer to other locations.

Compared to supervised approaches, the obtained results are good achievements, especially considering the truly unsupervised and single-epoch nature of the proposed method. Moreover, the limited time needed for training in the pre-event stage and for inference in the post-event stage (see Section 3.4) make this method automatic and fast, which is essential for its practical application in post-disaster scenarios.

Author Contributions: Conceptualization, S.T. and F.N.; Analysis, S.T.; Methodology, S.T. and F.N.; Supervision, F.N., N.K. and G.V.; Original draft, S.T. and F.N.; Review and editing, S.T., F.N., N.K. and G.V. All authors have read and agreed to the published version of the manuscript.

Funding: Financial support has been provided by the Innovation and Networks Executive Agency (INEA) under the powers delegated by the European Commission through the Horizon 2020 program "PANOPTIS–Development of a decision support system for increasing the resilience of transportation infrastructure based on combined use of terrestrial and airborne sensors and advanced modelling tools", Grant Agreement number 769129.

Conflicts of Interest: The authors declare no conflict of interest.

References

1. Eguchi, R.T.; Huyck, C.K.; Ghosh, S.; Adams, B.J.; Mcmillan, A. Utilizing New Technologies in Managing Hazards and Disasters. In *Geospatial Techniques in Urban Hazard and Disaster Analysis*; Showalter, P.S., Lu, Y., Eds.; Springer: Dordrecht, The Netherlands, 2009; pp. 295–323.

2. Ji, M.; Liu, L.; Buchroithner, M. Identifying collapsed buildings using post-earthquake satellite imagery and convolutional neural networks: A case study of the 2010 Haiti Earthquake. *Remote Sens.* **2018**, *10*, 1689. [CrossRef]

3. Miura, H.; Aridome, T.; Matsuoka, M. Deep learning-based identification of collapsed, non-collapsed and blue tarp-covered buildings from post-disaster aerial images. *Remote Sens.* **2020**, *12*, 1924. [CrossRef]

4. Sublime, J.; Kalinicheva, E. Automatic post-disaster damage mapping using deep-learning techniques for change detection: Case study of the Tohoku tsunami. *Remote Sens.* **2019**, *11*, 1123. [CrossRef]

5. Kerle, N.; Nex, F.; Duarte, D.; Vetrivel, A. Uav-based structural damage mapping-results from 6 years of research in two european projects. *Int. Arch. Photogramm. Remote Sens. Spat. Inf. Sci.* **2019**, 187–194. [CrossRef]

6. Duarte, D.; Nex, F.; Kerle, N.; Vosselman, G. Satellite Image Classification of Building Damages using Airborne and Satellite Image Samples in a Deep Learning Approach. In Proceedings of the ISPRS Annals of Photogrammetry, Remote Sensing and Spatial Information Sciences, Riva del Garda, Italy, 4–7 June 2018; Volume IV-2, pp. 89–96.

7. Nex, F.; Duarte, D.; Steenbeek, A.; Kerle, N. Towards real-time building damage mapping with low-cost UAV solutions. *Remote Sens.* **2019**, *11*, 287. [CrossRef]

8. Gerke, M.; Kerle, N. Automatic structural seismic damage assessment with airborne oblique Pictometry© imagery. *Photogramm. Eng. Remote Sens.* **2011**, *77*, 885–898. [CrossRef]

9. Copernicus Emergency Management Service. Available online: https://emergency.copernicus.eu/ (accessed on 24 September 2020).

10. Zou, Q.; Zhang, Z.; Li, Q.; Qi, X.; Wang, Q.; Wang, S. DeepCrack: Learning hierarchical convolutional features for crack detection. *IEEE Trans. Image Process.* **2019**, *28*, 1498–1512. [CrossRef]

11. Ghaffarian, S.; Kerle, N.; Pasolli, E.; Arsanjani, J.J. Post-disaster building database updating using automated deep learning: An integration of pre-disaster OpenStreetMap and multi-temporal satellite data. *Remote Sens.* **2019**, *11*, 2427. [CrossRef]

12. Pi, Y.; Nath, N.D.; Behzadan, A.H. Convolutional neural networks for object detection in aerial imagery for disaster response and recovery. *Adv. Eng. Inform.* **2020**, *43*, 101009. [CrossRef]

13. Nex, F.; Duarte, D.; Tonolo, F.G.; Kerle, N. Structural Building Damage Detection with Deep Learning: Assessment of a State-of-the-Art CNN in Operational Conditions. *Remote Sens.* **2019**, *11*, 2765. [CrossRef]

14. Vetrivel, A.; Gerke, M.; Kerle, N.; Nex, F.; Vosselman, G. Disaster damage detection through synergistic use of deep learning and 3D point cloud features derived from very high resolution oblique aerial images, and multiple-kernel-learning. *ISPRS J. Photogramm. Remote Sens.* **2018**, *140*, 45–59. [CrossRef]

15. Fernandez Galarreta, J.; Kerle, N.; Gerke, M. UAV-based urban structural damage assessment using object-based image analysis and semantic reasoning. *Nat. Hazards Earth Syst. Sci.* **2015**, *15*, 1087–1101. [CrossRef]

16. Duarte, D.; Nex, F.; Kerle, N.; Vosselman, G. Detection of seismic façade damages with multi- temporal oblique aerial imagery. *GIScience Remote Sens.* **2020**, *57*, 670–686. [CrossRef]

17. Akçay, S.; Atapour-Abarghouei, A.; Breckon, T.P. GANomaly: Semi-supervised Anomaly Detection via Adversarial Training. In Proceedings of the Asian Conference on Computer Vision, Perth, Australia, 2–6 December 2018; Springer: Cham, The Netherlands, 2018; pp. 622–637.

18. See, L.; Mooney, P.; Foody, G.; Bastin, L.; Comber, A.; Estima, J.; Fritz, S.; Kerle, N.; Jiang, B.; Laakso, M.; et al. Crowdsourcing, citizen science or volunteered geographic information? The current state of crowdsourced geographic information. *ISPRS Int. J. Geo-Inf.* **2016**, *5*, 55. [CrossRef]

19. Alom, Z.; Taha, T.M.; Yakopcic, C.; Westberg, S.; Sidike, P.; Nasrin, M.S.; Van Essen, B.C.; Awwal, A.A.S.; Asari, V.K. The History Began from AlexNet: A Comprehensive Survey on Deep Learning Approaches. *arXiv* **2018**, arXiv:1803.

20. Celik, T. Unsupervised Change Detection in Satellite Images Using Principal Component Analysis and k-means Clustering. *IEEE Geosci. Remote Sens. Lett.* **2009**, *6*, 772–776. [CrossRef]

21. Ghaffarian, S.; Kerle, N.; Filatova, T. Remote sensing-based proxies for urban disaster risk management and resilience: A review. *Remote Sens.* **2018**, *10*, 1760. [CrossRef]

22. Scholkopf, B.; Platt, J.C.; Shawe-Taylor, J.; Smola, A.J.; Williamson, R.C. Estimating the Support of a High-Dimensional Distribution. *Neural Comput.* **2001**, *13*, 1443–1471. [CrossRef]

23. Tilon, S.M.; Nex, F.; Duarte, D.; Kerle, N.; Vosselman, G. Infrastructure degradation and post-disaster damage detection using anomaly detecting Generative Adversarial Networks. *ISPRS Ann. Photogramm. Remote. Sens. Spat. Inf. Sci.* **2020**, *V*, 573–582. [CrossRef]

24. Akçay, S.; Atapour-Abarghouei, A.; Breckon, T.P. Skip-GANomaly: Skip Connected and Adversarially Trained Encoder-Decoder Anomaly Detection. In Proceedings of the 2019 International Joint Conference on Neural Networks (IJCNN), Budapest, Hungary, 14–19 July 2019; pp. 1–8.

25. Dumagpi, J.K.; Jung, W.Y.; Jeong, Y.J. A new GAN-based anomaly detection (GBAD) approach for multi-threat object classification on large-scale x-ray security images. *IEICE Trans. Inf. Syst.* **2020**, *E103D*, 454–458. [CrossRef]

26. Fujioka, T.; Kubota, K.; Mori, M.; Kikuchi, Y.; Katsuta, L. Efficient Anomaly Detection with Generative Adversarial Network for Breast Ultrasound Imaging. *Diagnostics* **2020**, *10*, 456. [CrossRef]

27. Schlegl, T.; Seeböck, P.; Waldstein, S.M.; Schmidt-Erfurth, U.; Langs, G. Unsupervised anomaly detection with generative adversarial networks to guide marker discovery. In Proceedings of the International Conference on Information Processing in Medical Imaging, Boone, NC, USA, 25–30 June 2017; Springer: Cham, The Netherlands, 2017; pp. 146–147.

28. Duarte, D.; Nex, F.; Kerle, N.; Vosselman, G. Multi-resolution feature fusion for image classification of building damages with convolutional neural networks. *Remote Sens.* **2018**, *10*, 1636. [CrossRef]

29. Weber, E.; Kané, H. Building Disaster Damage Assessment in Satellite Imagery with Multi-Temporal Fusion. *arXiv* **2020**, arXiv:2004.05525.

30. Wheeler, B.J.; Karimi, H.A. Deep learning-enabled semantic inference of individual building damage magnitude from satellite images. *Algorithms* **2020**, *13*, 195. [CrossRef]

31. Gupta, A.; Watson, S.; Yin, H. Deep Learning-based Aerial Image Segmentation with Open Data for Disaster Impact Assessment. *arXiv* **2020**, arXiv:2006.05575.

32. Jiao, Z.; Zhang, Y.; Xin, J.; Mu, L.; Yi, Y.; Liu, H.; Liu, D. A Deep Learning Based Forest Fire Detection Approach Using UAV and YOLOv3. In Proceedings of the 2019 1st International Conference on Industrial Artificial Intelligence (IAI), Shenyang, China, 23–27 July 2019; pp. 1–5.

33. Goodfellow, I.J.; Pouget-Abadie, J.; Mirza, M.; Xu, B.; Warde-Farley, D.; Ozair, S.; Courville, A.; Bengio, Y. Generative Adversarial Nets. In Proceedings of the Advances in Neural Information Processing Systems, Montreal, Canada, 8–13 December 2014; pp. 2672–2680.

34. Douzas, G.; Bacao, F. Effective data generation for imbalanced learning using conditional generative adversarial networks. *Expert Syst. Appl.* **2018**, *91*, 464–471. [CrossRef]

35. Liu, D.; Long, C.; Zhang, H.; Yu, H.; Dong, X.; Xiao, C. ARShadowGAN: Shadow Generative Adversarial Network for Augmented Reality in Single Light Scenes. In Proceedings of the 2020 IEEE/CVF Conference on Computer Vision and Pattern Recognition (CVPR), Seatttle, WA, USA, 29 October 2020; pp. 8136–8145.

36. Antoniou, A.; Storkey, A.; Edwards, H. Data Augmentation Using Generative Adversarial Network. *arXiv* **2018**, arXiv:1711.04340v3.

37. Wang, L.; Chen, W.; Yang, W.; Bi, F.; Yu, F.R. A State-of-the-Art Review on Image Synthesis with Generative Adversarial Networks. *IEEE Access* **2020**, *8*, 63514–63537. [CrossRef]

38. Frid-Adar, M.; Diamant, I.; Klang, E.; Amitai, M.; Goldberger, J.; Greenspan, H. GAN-based synthetic medical image augmentation for increased CNN performance in liver lesion classification. *Neurocomputing* **2018**, *321*, 321–331. [CrossRef]

39. Maeda, H.; Kashiyama, T.; Sekimoto, Y.; Seto, T.; Omata, H. Generative adversarial network for road damage detection. *Comput. Civ. Infrastruct. Eng.* **2020**, 1–14. [CrossRef]

40. Mao, Q.; Lee, H.Y.; Tseng, H.Y.; Ma, S.; Yang, M.H. Mode seeking generative adversarial networks for diverse image synthesis. In Proceedings of the IEEE Computer Society Conference on Computer Vision and Pattern Recognition, Long Beach, CA, USA, 15–21 June 2019; pp. 1429–1437.

41. Ngo, P.C.; Winarto, A.A.; Kou, C.K.L.; Park, S.; Akram, F.; Lee, H.K. Fence GAN: Towards better anomaly detection. In Proceedings of the International Conference on Tools with Artificial Intelligence, ICTAI, Portland, OR, USA, 4–6 November 2019; pp. 141–148.

42. Zenati, H.; Foo, C.S.; Lecouat, B.; Manek, G.; Chandrasekhar, V.R. Efficient GAN-Based Anomaly Detection. *arXiv* **2018**, arXiv:1802.06222.

43. Schlegl, T.; Seeböck, P.; Waldstein, S.M.; Langs, G.; Schmidt-Erfurth, U. f-AnoGAN: Fast unsupervised anomaly detection with generative adversarial networks. *Med. Image Anal.* **2019**, *54*, 30–44. [CrossRef]

44. Tang, T.W.; Kuo, W.H.; Lan, J.H.; Ding, C.F.; Hsu, H.; Young, H.T. Anomaly detection neural network with dual auto-encoders GAN and its industrial inspection applications. *Sensors* **2020**, *20*, 3336. [CrossRef]

45. Berg, A.; Felsberg, M.; Ahlberg, J. Unsupervised adversarial learning of anomaly detection in the wild. In Proceedings of the Frontiers in Artificial Intelligence and Applications, Santiago de Compostela, Spain, 29 August–8 September 2020; Volume 325, pp. 1002–1008.

46. Zhong, J.; Xie, W.; Li, Y.; Lei, J.; Du, Q. Characterization of Background-Anomaly Separability With Generative Adversarial Network for Hyperspectral Anomaly Detection. *IEEE Trans. Geosci. Remote Sens.* **2020**, 1–12. [CrossRef]

47. Ronneberger, O.; Fischer, P.; Brox, T. U-net: Convolutional networks for biomedical image segmentation. In Proceedings of the International Conference on Medical Image Computing and Computer-Assisted Intervention, Munich, Germany, 5–9 October 2015; pp. 234–241.

48. Eppel, S. Classifying a Specific Image Region Using Convolutional Nets With an ROI Mask as Input. *arXiv* **2018**, arXiv:1812.00291.

49. Gupta, R.; Hosfelt, R.; Sajeev, S.; Patel, N.; Goodman, B.; Doshi, J.; Heim, E.; Choset, H.; Gaston, M. XBD: A dataset for assessing building damage from satellite imagery. *arXiv* **2019**, arXiv:1911.09296v1.

50. Gupta, R.; Goodman, B.; Patel, N.; Hosfelt, R.; Sajeev, S.; Heim, E.; Doshi, J.; Lucas, K.; Choset, H.; Gaston, M. Creating xBD: A Dataset for Assessing Building Damage from Satellite Imagery. In Proceedings of the IEEE Conference on Computer Vision and Pattern Recognition Workshops, Long Beach, CA, USA, 16–20 June 2019; pp. 10–17.

51. XView2: Assess Building Damage. Available online: xview2.org (accessed on 24 September 2020).

52. Brovelli, M.A.; Zamboni, G. A new method for the assessment of spatial accuracy and completeness of OpenStreetMap building footprints. *ISPRS Int. J. Geo-Inf.* **2018**, *7*, 289. [CrossRef]

53. Milosavljević, A. Automated processing of remote sensing imagery using deep semantic segmentation: A building footprint extraction case. *ISPRS Int. J. Geo-Inf.* **2020**, *9*, 486. [CrossRef]

54. Ma, J.; Wu, L.; Tang, X.; Liu, F.; Zhang, X.; Jiao, L. Building extraction of aerial images by a global and multi-scale encoder-decoder network. *Remote Sens.* **2020**, *12*, 2350. [CrossRef]

55. Ma, L.; Jiang, B.; Jiang, X.; Tian, Y. Shadow removal in remote sensing images using features sample matting. In Proceedings of the International Geoscience and Remote Sensing Symposium (IGARSS), Milan, Italy, 26–31 July 2015; pp. 4412–4415.

56. Suh, H.K.; Hofstee, J.W.; van Henten, E.J. Improved vegetation segmentation with ground shadow removal using an HDR camera. *Precis. Agric.* **2018**, *19*, 218–237. [CrossRef]

57. Motohka, T.; Nasahara, K.N.; Oguma, H.; Tsuchida, S. Applicability of Green-Red Vegetation Index for remote sensing of vegetation phenology. *Remote Sens.* **2010**, *2*, 2369–2387. [CrossRef]

58. Azizi, Z.; Najafi, A.; Sohrabi, H. Forest Canopy Density Estimating, Using Satellite Images. In Proceedings of the International Archives of the Photogrammetry, Remote Sensing and Spatial Information Sciences, Beijing, China, 3–11 July 2008; Volume XXXVII, pp. 1127–1130.

59. Zhao, J.; Gong, M.; Liu, J.; Jiao, L. Deep learning to classify difference image for image change detection. In Proceedings of the 2014 International Joint Conference on Neural Networks (IJCNN), Beijing, China, 6–11 July 2014; pp. 411–417.

Remote Sens. **2020**, *12*, 4193

remote sensing

MDPI

Article

Aerial Imagery Feature Engineering Using Bidirectional Generative Adversarial Networks: A Case Study of the Pilica River Region, Poland

Maciej Adamiak [1,*], Krzysztof Będkowski [2] and Anna Majchrowska [3]

[1] SoftwareMill, 02-791 Warsaw, Poland
[2] Faculty of Geographical Sciences, Institute of Urban Geography, Tourism and Geoinformation, University of Lodz, 90-139 Łódź, Poland; krzysztof.bedkowski@geo.uni.lodz.pl
[3] Department of Physical Geography, Faculty of Geographical Sciences, University of Lodz, 90-139 Łódź, Poland; anna.majchrowska@geo.uni.lodz.pl
* Correspondence: maciej.adamiak@softwaremill.com

Abstract: Generative adversarial networks (GANs) are a type of neural network that are characterized by their unique construction and training process. Utilizing the concept of the latent space and exploiting the results of a duel between different GAN components opens up interesting opportunities for computer vision (CV) activities, such as image inpainting, style transfer, or even generative art. GANs have great potential to support aerial and satellite image interpretation activities. Carefully crafting a GAN and applying it to a high-quality dataset can result in nontrivial feature enrichment. In this study, we have designed and tested an unsupervised procedure capable of engineering new features by shifting real orthophotos into the GAN's underlying latent space. Latent vectors are a low-dimensional representation of the orthophoto patches that hold information about the strength, occurrence, and interaction between spatial features discovered during the network training. Latent vectors were combined with geographical coordinates to bind them to their original location in the orthophoto. In consequence, it was possible to describe the whole research area as a set of latent vectors and perform further spatial analysis not on RGB images but on their lower-dimensional representation. To accomplish this goal, a modified version of the big bidirectional generative adversarial network (BigBiGAN) has been trained on a fine-tailored orthophoto imagery dataset covering the area of the Pilica River region in Poland. Trained models, precisely the generator and encoder, have been utilized during the processes of model quality assurance and feature engineering, respectively. Quality assurance was performed by measuring model reconstruction capabilities and by manually verifying artificial images produced by the generator. The feature engineering use case, on the other hand, has been presented in a real research scenario that involved splitting the orthophoto into a set of patches, encoding the patch set into the GAN latent space, grouping similar patches latent codes by utilizing hierarchical clustering, and producing a segmentation map of the orthophoto.

Keywords: machine learning; generative adversarial networks; feature engineering; orthophoto; unsupervised segmentation

Citation: Adamiak, M.; Będkowski, K.; Majchrowska, A. Aerial Imagery Feature Engineering Using Bidirectional Generative Adversarial Networks: A Case Study of the Pilica River Region, Poland. *Remote Sens.* **2021**, *13*, 306. https://doi.org/10.3390/rs13020306

Received: 20 December 2020
Accepted: 14 January 2021
Published: 17 January 2021

1. Introduction

There is no doubt that aerial imagery is a source of valuable information about geographical space. The rapid development of remote sensing technology supported by a significant improvement in access to remote sensing imagery [1] led to an increased interest in the potential use of the collected material among academia, government, and private sector representatives in areas such as urban planning, agriculture, transport, etc. Substantial quantities of image data have become available in recent years thanks to opening public access to images acquired by satellites such as Landsat 8 [2], Sentinel-2

A/B [3], and Pléiades [4]. Furthermore, due to the epidemiological situation in Poland, the government decided to open access to national orthophoto resources [5]. Access to high-quality and properly curated image repositories undoubtedly promotes the development of new ideas and contributes to the emergence of various methods and techniques for analyzing collected data.

The use of aerial and satellite images in the basic task of remote sensing that deals with land cover and land use classification is indisputable. At an early stage of remote sensing development, the possibility of distinguishing certain spatial units by interpreting the spectral, textural, and structural features of the image was indicated. Olędzki postulated extracting homogenous fragments of satellite images called photomorphic units. These units were similar in terms of structure and texture and originated from natural processes and man-made transformations of the environment [6,7]. Descriptive definitions of image features were soon replaced by mathematical formulas [8]. Further development has led to the introduction of object classification procedures, in which, in addition to the brightness parameters of the image pixels, their neighborhood and the shape and size of the distinguished objects were taken into account [9]. Object-oriented analysis is based on databases and fuzzy logic. Probably the most popular implementation of this paradigm in remote sensing applications is in the one originally developed in eCognition [10]. These techniques have been successfully applied to research on landscape structure and forestry [11]. Referring to division units used in physico-geographic regionalization [12,13], Corine Land Cover [14,15], or for the purposes of ecological research [16,17], an additional meaning and hierarchical structure [18,19] can be also given to units distinguished from the landscape. At the same time, it is important to properly take care of the appropriate adjustments of the scale of the study and data relevant for the analyzed problem, so as not to overlook important features of the area that may affect the reliability of the analysis, i.e., to mitigate the issues connected with spatial object scaling and the scale problem [20]. What is important is that the separation of landscape units cannot be based only on image data [21]. It is also necessary to take into account data on lithology, morphogenesis, terrain, geodiversity [22], water, and vegetation. The latter is interesting due to the fact that vegetation is represented in remote sensing imagery to the greatest extent both in terms of properties and structure. Therefore, it has great potential for being utilized in landscape quantification [23]. It should be mentioned that the process of identifying landscape units is also affected by human activity and created by the cultural landscape. Another major achievement in remote sensing image classification is the introduction of algorithms based on neural networks [24].

The influence of machine learning and deep learning on contemporary remote sensing techniques and their support in geographical space analysis is undeniable [25]. There are multiple fascinating applications of machine learning (ML) and deep learning (DL) in the remote sensing domain like land use classification [26], forest area semantic segmentation [27], species detection [28], recognition of patches of alpine vegetation [29,30], classification of urban areas [31], roads detection [32], etc.

What a significant part of these studies have in common is the focus on utilizing convolutional neural network (CNN) architectures capable of solving problems that can be brought down to traditional computer vision (CV) tasks like semantic segmentation, instance segmentation, or classification. This is directly associated with the underlying mechanism that enables the network to encode complex image features. CNN's convolutional filters are gradually trained to gain the ability to detect the presence of specific patterns. Frequently, the training routine is performed in a supervised manner. The model is presented with target data and uses it to learn the solution. Supervised learning is capable of achieving extraordinary results but at the same time relies on access to manually labeled data. Another incredibly interesting approach is to train the neural network without any pre-existing labels to let it discover the patterns on its own. Although unsupervised learning algorithms like clustering are well-known among remote sensing researchers, utilizing convolutional neural networks is still to gain trust. The way of training a neural network can be even more intriguing when you exchange human supervision with machine supervision, and let multiple neural networks control their learning progress and work like adversaries.

Generative adversarial networks (GANs) are constructed from at least one discriminator and one generator network. The main goal of these two networks is to compete with each other in the form of a two-player minimax game [33]. The generator tries to deceive the discriminator by producing artificial samples, and the discriminator assesses whether it is dealing with real or generator-originating samples. The generator network is producing samples from a specified data distribution by transforming vectors of noise [34]. This technique was successfully applied in multiple remote sensing activities from upsampling satellite imagery [35], deblurring [36] to artificial sample generation [37]. GANs' artificial data creation capabilities are not the only aspect that makes them interesting for remote sensing. When exploring the theory behind GANs, one should observe that, to perform its work, the generator retains all the information needed to produce a complex sample using only a much simpler representation called the latent code [33]. In terms of spatial analysis, this means that the network is able to produce a realistic image of an area using only a handful of configuration parameters as input. In the classic approach to GANs, this image recipe is reserved only for artificially generated samples. It was the introduction of bidirectional GANs and adversarial feature learning [38] that allowed to extract the latent code from ground truth (real) samples. The novelty of this approach when applied to aerial imagery is that it allows performing advanced spatial analysis using lower-dimensional representations of the orthophoto computed by a state-of-the-art neural network rather than utilizing raw image data. This method resembles algorithms like principal component analysis (PCA) but, instead of treating the image on the pixel level, it operates on the spatial features level and, therefore, offers a richer analysis context. The projection, a latent vector, serves as a lightweight representation of the image and holds information about the strength, occurrence, and interaction between spatial features discovered during the network training. This interesting capability opens up new possibilities for geographical space interpretation such as

- extracting features to fit in a variety of machine learning and spatial analysis algorithms like geographically weighted regression, support vector machines, etc.;
- minimizing resource consumption when processing large areas;
- discovering new features of analyzed areas by carefully exploring the network latent space.

The principal goal of our study is to evaluate the potential of bidirectional generative adversarial networks in remote sensing feature engineering activities and unsupervised segmentation. Therefore, the following hypotheses have been defined:

1. The image reconstruction process is strong enough to produce artificial images that closely resemble the original;
2. Similar orthophoto patches can produce latent space codes that are close to each other in the network latent space, therefore, preserving the similarity after encoding;
3. Latent codes enhanced by geographical coordinates can serve as artificial features used during geographical space interpretation by classical algorithms such as agglomerative clustering.

2. Materials and Methods

Figure 1 presents an overview of the proposed procedure composed of the following steps: preparing an orthophoto patches dataset, training the big bidirectional generative adversarial network (BigBiGAN), utilizing the network encoding module to convert orthophoto patches to their latent codes, enriching the data with geographical coordinates, and performing geospatial clustering on enriched latent codes.

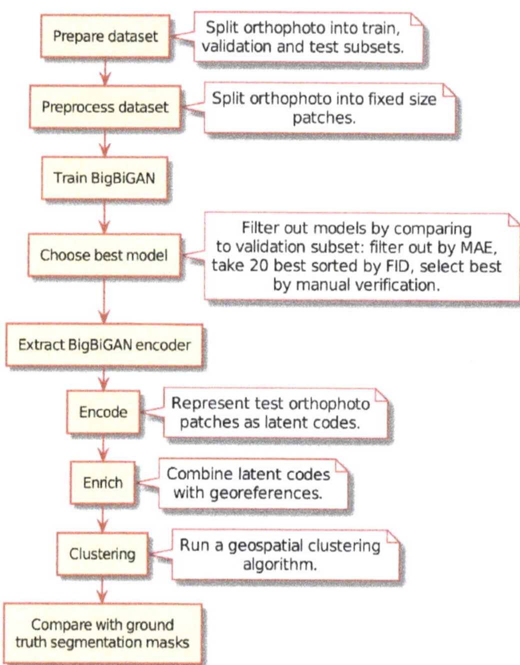

Figure 1. Investigation overview.

2.1. Research Area

To be able to produce precise results, generative models need to be trained on high-quality datasets. The dataset needs to be large enough to cover the variety of spatial features that the encoder will be able to utilize when interpreting the input image. The authors decided to utilize RGB orthophotos of the Pilica River and Sulejowski Reservoir regions in Poland. The area from which the samples have been obtained includes the Plica River valley between Maluszyn and Tomaszów Mazowiecki together with adjacent areas (see Figure 2).

According to the physico-geographical regionalization of Poland [13], the southern and eastern parts of the area are located in the province of Polish Uplands, the macroregion of the Przedbórz Upland, the mesoregions of Włoszczowa Basin, Radomsko Hills, Przedbórz-Małogoszcz Range, and Opoczno Hills. The northwestern part is located in the mesoregions of the Piotrków Plain and the Białobrzegi Valley, which are part of the South Mazovian Hills macroregion in the Central European Lowland Province. The consequence of the location in the border zone of the Polish Uplands and the Central European Lowland is the interpenetration of features characteristic of both provinces and the relative diversification of the natural environment of the area.

According to the tectonic regionalization [39], a fragment of the area located south of Przedbórz includes the Szczecin–Miechów Synclinorium, constructed mainly from Cretaceous rock formations. The rest of the area belongs to the Mid-Polish Anticlinorium, dominated by Jurassic carbonate rocks.

The axis of the selected area is the Pilica River valley. The chosen section of the valley is in a natural condition. The Pilica River flows in an unregulated, sinuous to a meandering channel that is not embanked along its entire length from Maluszyn to the vicinity of Sulejów. There, it reaches the Smardzewice dam waters of the Sulejowski Reservoir. The valley floor descends from 211 m above sea level in the south to 154 m above sea level in the north, and the stream gradient equals 0.51%. The width of the valley varies from about 300 m in the vicinity of Sulejów and Przedbórz to over 3 km near Łęg Ręczyński. It reaches its greatest width in places of well-formed levels of over-flood terraces occurring on both sides of the floodplain.

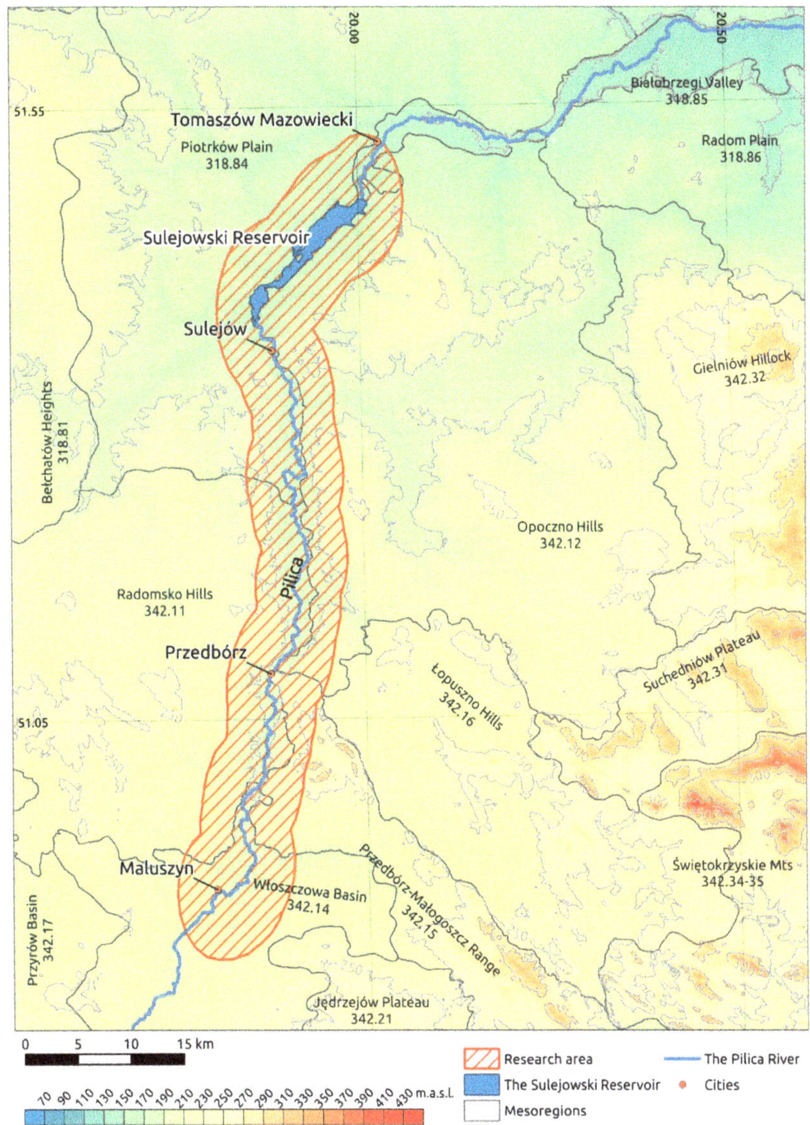

Figure 2. Research area, source: own elaboration based on physico-geographical regionalization of Poland [13] and Head Office of Geodesy and Cartography data [40]. Mesoregions codes are consistent with those introduced in the referenced paper and therefore enable precise localization of the research area within Polish mesoregions.

The valley cuts down the adjacent plain and undulating moraine uplands to about 20–25 m. These landforms were formed in the Quaternary, mainly during the Pleistocene glaciations of the Middle Polish Complex. Within the uplands on both sides of the Pilica River, the thickness of Quaternary sediments decreases from the north to the south. The surface area of Mesozoic outcrops increases, which is a result of the weakening of the landforming capacity of the ice sheets as they entered the uplands. Absolute heights of culminations, in the form of isolated hills built of Mesozoic Jurassic and Cretaceous rocks within the Radomsko and Opoczno Hills, are also increasing, e.g., Diabla Góra 272 m, Czartoria Range 267 m, Bąkowa Góra 282 m, the form of ridges in the Przedbórz–Małogoszcz Range exceeding 300 m above sea level, or Bukowa Góra 336 m.

The varied topography and the near-surface geological structures, in addition to the humidity conditions, shape the mosaic of land cover types. The area is poorly urbanized. The ground moraine plateaus are dominated by arable land, fluvioglacial plains, and other sandy areas largely occupied by forests and abandoned arable lands. In the valley, the over-flood terraces are characterized by complex systems of arable land, fallow land, forests, and meadows. The floodplain is dominated by meadows and pastures, in many places overgrown with shrubs and trees after their agricultural use ceased [41].

2.2. Dataset

The Pilica River region dataset covers the area of 691.86 km^2 and was generated using 138 orthophoto sheets that intersect with a 4 km buffer around the Pilica River from Sulejów to Maluszyn and Sulejowski Reservoir in Łódź Voivodeship in Poland. All orthophotos were acquired using GEOPORTAL2 [42] and possess three channels—R-red, G-green, and B-blue with 25 cm pixel ground sample distance (see Figure 3).

Figure 3. Examples of the Pilica River region dataset samples (512 px × 512 px patches). From the upper left corner: forest and a barely visible forest path outline, farmlands cut by a sandy road planted with trees, forest with an adjacent land abandonment, forest, farmlands, river valley, forest and an overgrown meadow, forest with clearly outlined shadow and a dirt road, farmlands, water reservoir, forest cut by a road, recently plowed farmland with balks, and a young forest.

During the preprocessing phase each image was split into 128 px × 128 px, 256 px × 256 px, 512 px × 512 px, and 1024 px × 1024 px patches. This step was crucial for electing the optimal image size and resizing approach to satisfy the requirements of the chosen neural network architecture and its internal complexity. The choice of image size directly influences hardware requirements, the ability of the neural network to learn image features needed during the assessment of the reconstruction process, and, important from a GAN perspective, overall dataset size. It was highly important to utilize the patches large enough to be described by complete and interpretable spatial features. Image size also affects the size of the input and output tensors and the handful of technical parameters that a processing unit can handle. The authors decided that, during the research, a single GPU (Nvidia Titan RTX in US) and CPU (AMD Threadripper 1950 in US) will be used, and all computations have to fit their representative capacity. This is due to ensuring that the results can be reproduced without using a multi-GPU cluster. In consequence, the authors decided to utilize:

- A series of 256 px × 256 px patches for encoder input, which were resized from 512 px × 512 px patches using bilinear interpolation. Although 256 px × 256 px is either the network nor hardware limit, it gives the opportunity to choose a larger batch, which significantly affects BigBiGAN performance. Furthermore, 512 px × 512 px (128 m × 128 m) patch is richer in spatial information;
- A series of 128 px × 128 px patches for image discriminator input, explicitly defined by BigBiGAN architecture for 256 px × 256 px encoder input;
- A series of 128 px × 128 px patches for generator output, which were the minimal interpretable patch size.

Geographical references of each patch and source image metadata have been preserved to enable reprojecting the results back to their original location. Patches acquired from 137 images were divided, in accordance with the established practice of machine learning, into two subsets of 512 px × 512 px images in the following proportions 0.95 and 0.05, forming a training set (39826 patches) and validation set (2096 patches). Remaining images were also processed and formed two test sets—one containing 1224 (256 px × 256 px) patches and another containing 306 (512 px × 512 px) patches. The authors introduced an additional test set of 256 px × 256 px patches that were smaller than the defined training size to verify whether the solution is capable of handling input material potentially containing less spatial information than it was trained on.

Afterward, a data augmentation procedure was defined to increase the diversity of managed datasets by applying specified image transformations in a random manner. Augmentation of the dataset is important from the point of view of GAN because the network has a higher chance to adapt to different conditions such as lighting or spatial feature shape changes, and at the same time, less data is needed for the network to converge. The authors decided to utilize basic post-processing computer vision techniques, such as adding or subtracting a random value to all pixels in an image, blurring the image using gaussian kernels, applying random four-point perspective transformations, rotating with clipping, or flipping the image. What is important, each transformation was applied only during the training phase and the decision of whether to apply it was random. Finally, a TensorFlow data processing pipeline(US; Mountain View; California) was implemented to ensure that reading and augmenting the data would efficiently utilize all computational resources. The main goal was to support the GPU with constant data flow orchestrated by the CPU and enable shuffling across batches, which turned out to be crucial when working with complex network architectures and utilizing a relatively small batch size, i.e., below 128 samples.

2.3. Generative Adversarial Network

The authors decided to use the bidirectional generative neural network (BiGAN) [38] architecture as a starting point and gradually updated its elements to end up with the final solution closely resembling BigBiGAN. An interesting, proven property of these architectures is the ability to perform the inverse mapping from input data to the latent representation. This makes BiGAN and BigBiGAN great candidates to address the research problem, i.e., finding a transformation capable of mapping a multichannel image to a fixed size vector representation. BigBiGAN can be used to shift a real image to the latent space using the encoder network.

The resulting latent space code can be then utilized as generator input to reconstruct an image similar to the original encoder input. Achieving the same input and output is hard or even impossible due to the fact that the pixel-wise reconstruction quality is not even a task for bidirectional GANs, and therefore, there is no loss function assigned to assess it. One can think of reconstruction as a process of enabling a mechanism of lossy image compression and decompression that operates—not on pixel level—but feature level. The similarity measure can be chosen arbitrarily but has to have sufficient power to reliably score the resemblance of the input and output images passed through the encoder and generator. A high-quality encoder is powerful enough to store information regarding crucial spatial features of the input image, thus making it a great candidate for the main module in an automatic feature engineering mechanism to automatically generate large numbers of candidate properties and selecting the best by their information gain [43].

To avoid recreating an existing solution, the authors decided to focus on reusing the BigBiGAN design and adjusting it to processing orthophoto images (see Figure 4). BigBi-GAN consists of five neural networks—a generator and an encoder, which are accompanied by discriminators that assess their performance in producing respectively artificial images and latent codes. Results from both intermediate discriminators are then combined by the main discriminator. In the research, a modification of BigBiGAN was utilized to tackle the problem of encoding orthophoto patches to the network underlying latent space. Although the generator and main discriminator architectures have been preserved, the encoder and intermediate discriminators went through a minor modification. As suggested in a study on the BigBiGAN [44], the RevNet model was simplified to reduce the number of parameters needed to train the encoder. Intermediate discriminators contained fewer multilayer perceptron modules (MLP), which were composed of smaller numbers of neurons. In consequence, this enabled the use of slightly bigger batches and, therefore, yielded better results at the cost of a training time increase. The final architecture was implemented in TensorFlow 2 and Keras. Figure 4 presents the final model training sequence blueprint.

2.4. Hierarchical Clustering

Latent space code is a 120-dimensional vector of real numbers produced by applying a GAN encoder on an orthophoto patch. Such code contains information regarding spatial features present in the scope of the encoded patch. Each part of the code controls the strength and occurrence of one or more spatial features discovered during the neural network training. One of the important features of the latent space is that codes that are closer to each other in terms of the Euclidean distance (L2 norm) are more similar in terms of the represented features, i.e., two forest area patches will be closer in the latent space than a forest area and farmland patches [45].

Furthermore, each patch holds information regarding its georeferences. To simplify further analyses, georeferences were expressed as the location of the patch center. Patch center geographical coordinates were preserved during the computation and combined with corresponding latent codes. This opened the possibility to describe a larger area, composed of multiple patches, in the form of a 120-dimensional point cloud where each point holds the information regarding its original location. The combination of georeferences and latent space code is called a georeferenced latent space for the purpose of this research (see Figure 5).

The similarity between patches, precisely between their encodings, and information regarding geographical location can serve as input for methods and techniques of geospatial clustering. During their research, the authors focused on utilizing hierarchical clustering to discover a predefined number of clusters in a patch dataset describing a single test orthophoto. Hierarchical clustering is a general family of clustering algorithms that build nested clusters by successively merging or splitting them [46]. The metric used for the merge strategy is determined by the linkage strategy. For the purpose of clustering the georeferenced latent space, Ward's linkage method [47] was used. Ward's method minimizes the sum of squared differences within all clusters. It is a variance-minimizing approach and, in this sense, is similar to the k-means objective function but tackled with an agglomerative hierarchical approach. The connectivity matrix has been calculated using the k-nearest neighbors algorithm (*k*-NN).

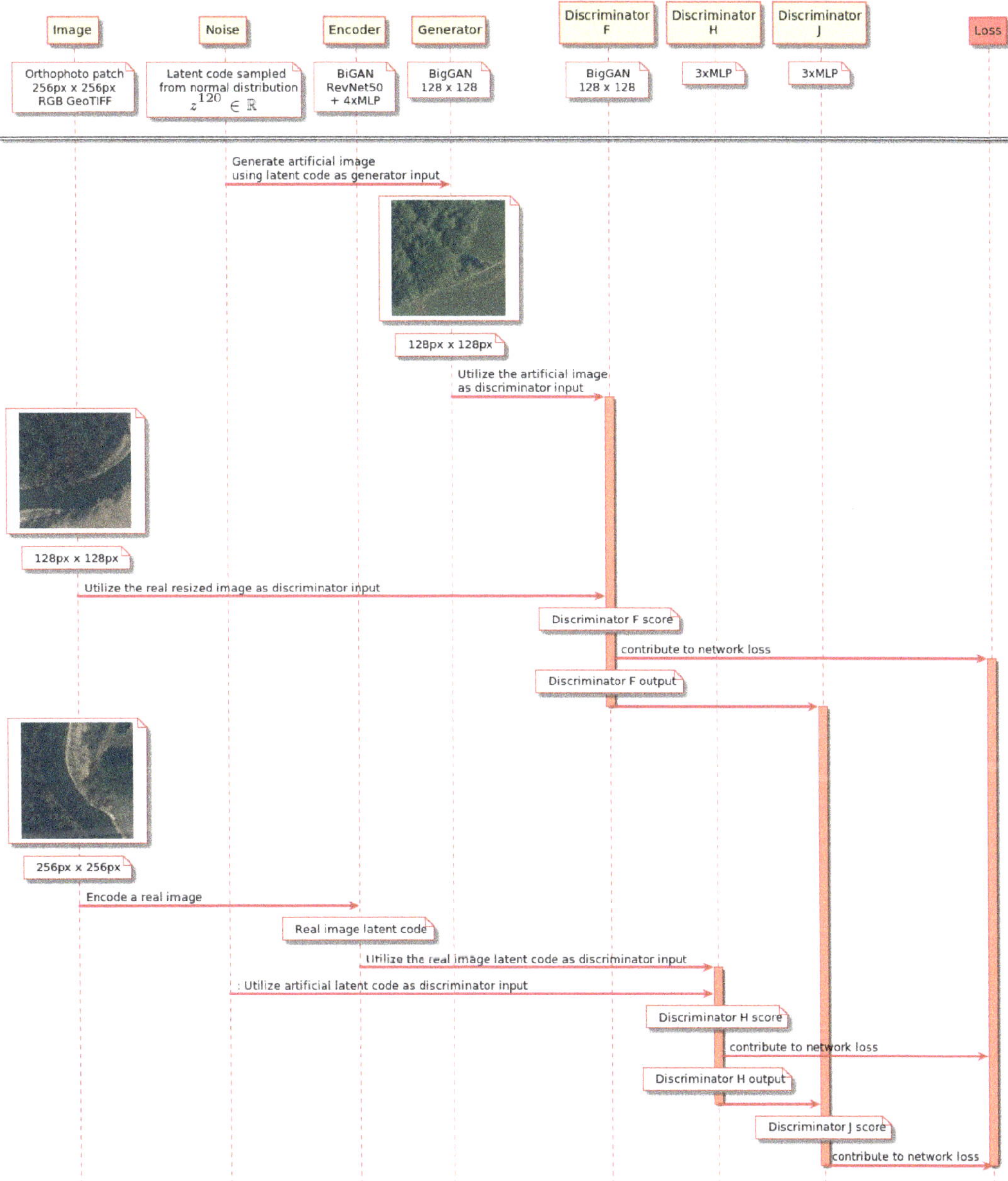

Figure 4. Big bidirectional generative adversarial network (BigBiGAN) training process.

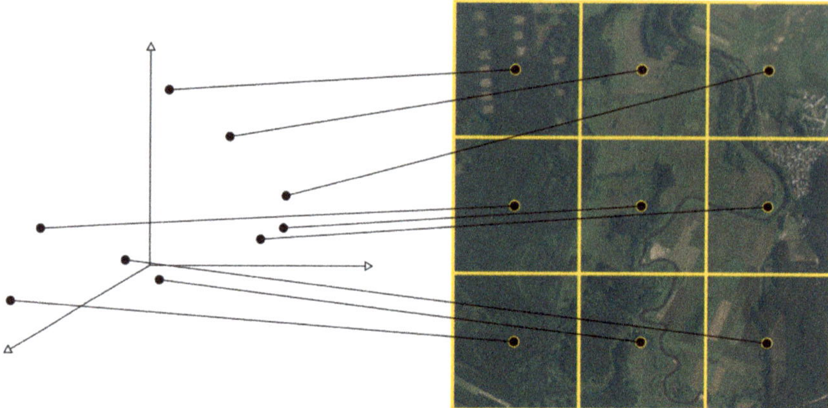

Figure 5. Latent space enhanced with geographical coordinates. The simplified image presents an example of encoding of the nine orthophoto patches (right) to a three-dimensional latent space (left) illustrated as a three-dimensional Cartesian coordinate system. Each latent code (a point in the latent space) carries additional information regarding georeferences that enables tracking its origin. Points that are closer in the three-dimensional latent space refer to patches that are visually similar. In the research, a 120-dimension latent space was trained.

3. Results

3.1. Model Training

Multiple model training session revealed that tackling the objective of training an orthophoto patch encoder is inseparably related to preparing generator and discriminator neural networks that are complex enough to learn all the features present in the input orthophoto. The networks have to be able to produce high-quality artificial images and determine whether the image is artificially generated or not, respectively. This directly influences the following:

- The overall size of the neural network, which is crucial due to GPU memory limitations and affects training duration;
- Maximum patch size that can be used as input during training and inference phases and is related to the level of detail offered in the processed dataset;
- Batch size, which has a significant influence on the stability and quality of generative adversarial models [44].

Initially chosen BiGAN architecture utilizes many concepts from previously designed networks such as deep convolutional GAN (DCGAN) [48] that, due to their simplicity, are not suitable for processing complex or large images. Therefore, their usefulness in the analysis of aerial imagery is limited. Although BiGAN offered all of the required earlier features, it was not capable of processing an orthophoto patch of size exceeding 32 pixels in both dimensions. This was a huge limitation due to the fact that, with a given 25 cm pixel ground sample distance, this method covered roughly the area of 64 m^2. In consequence, the processed patch did not carry enough details to allow a reliable assessment of the similarity between real and artificial images. Attempts to increase the maximum processed input size led to swapping default BiGAN generator and discriminator models with other network types based on deep residual blocks [49] and inception modules [50]. The overall architecture of the generator and discriminator pair resembled BigGAN [51].

After multiple experiments, the authors confirmed that, despite the ability to generate images up to 512 px $\times$ 512 px, the network was not capable of learning a reliable bidirectional mapping between the image and the latent space. This was due to the fact that the encoder architecture was lacking in comparison with its powerful counterparts. This problem has been addressed and mitigated in the paper describing large adversarial features learning and the big bidirectional generative adversarial network (BigBiGAN) [44] by introducing intermediate discriminators and proposing a stronger encoder model (Supplementary Materials).

3.2. Reconstruction

BigBiGAN neural network was trained for 200,000 steps with a batch containing 32 randomly picked patches from the training set. The trained model was saved during each reconstruction period that occurred every 1000 steps. During this period, patches from the validation set were fed, in inference time, to the encoder and generator to measure their power in creating artificial samples and close to the inversible encoding in terms of spatial features. Three types of metrics were calculated for each saved model to evaluate the reconstruction quality—pixel-wise mean absolute error (MAE) of image values normalized between –1 and 1, Fréchet inception distance (FID) [52] on a pre-trained InceptionV3 model, and by performing perceptual evaluation similar to that presented in the Human Eye Perceptual Evaluation (HYPE) paper [53]. MAE above 0.5 was used to discard low-quality models that were not able to effectively reconstruct input images in the early stages of the training. Then, FID values of all preserved models were compared and 20 with the highest score were selected. The average FID score was equal to 86.36 $\pm$ 7.28 in contrast to the state-of-the-art BigBiGAN baseline FID, which was equal to 31.19 $\pm$ 0.37.

The final model was selected by comparing the results of human evaluation of 21 arbitrarily chosen samples from the validation dataset with their reconstructed counterparts created by the network for each model. The human reader had an objective to assess whether each of the 42 images is real or artificial. The last verification phase resulted in selecting the model from the 170th reconstruction period, which yielded the least accuracy during human perceptual evaluation (accuracy: 59.5%, f-score: 0.6663). Samples and their reconstruction results are presented in Figure 6.

The overall quality of the reconstruction was assessed as sufficient during both quantitative and qualitative verification. For the selected model evaluated on non-scaled images (pixel values between 0 and 255), MAE was 27.213, structural similarity index (SSIM) [54] was 0.942, and peak signal-to-noise ratio (PSNR) [55] was equal to 42.731. From the analysis of human reader' misclassifications, it was clear that the chosen model is exceptionally good in reproducing areas like forests, land abandonment, and farmlands. The characteristic spatial features are preserved after encoding. Shadows cast by trees are consistent and natural. In the majority of cases, artificial and real images are indistinguishable. Mediocre results were achieved for urbanized areas. Reconstructed roads keep their linear character and surface type information. Although the model is capable of generating buildings, due to the high variety of housing types present in the research area and possible undersampling, the results are far from realistic. It is interesting that the link between residential areas and roads was maintained in multiple samples. Unfortunately, the generator is not capable of serving samples that contain water areas such as rivers or lakes. From all analyzed images from the training and validation set only a few presented water, which indicates weak encoding capabilities. Furthermore, all were significantly disrupted. The authors confirmed that this is related to undersampling and the insufficient information present in the RGB orthophoto. To tackle this issue, access to rich, multispectral imagery or digital terrain model (DTM) is required, or the model itself needs to be enriched to utilize additional class embeddings that could be derived from existing thematic maps or projects like Geoportal TBD [56].

Figure 6. Reconstruction result of 21 validation samples. Ground truth is represented by real tile images placed on the left. Images on the right were reconstructed by the generator from real images latent codes acquired through the encoder.

3.3. Feature Engineering

BigBiGAN encoder possesses an interesting capability that enables it to shift the input image into the latent space constructed during network training. The encoding, a 120-dimensional vector, should be considered simultaneously a compressed version of the input orthophoto and a recipe for generating a similar artificial image in terms of spatial features. The latter phenomenon is called representation learning. What is important, due to the nature of latent space, similar data points, i.e., those that were encoded from similar images, are closer to each other. This opens an interesting possibility to understand the structural similarity between images by performing the analysis not on the raw image input but only using latent codes.

In the research, the authors utilized the trained encoder to perform inference on a set of 256 px × 256 px test patches (see Figure 7). The 1224 test patches were converted into their latent space codes and represented as a geopandas [57] data frame containing 1224 rows, 120 encoding value columns, identifier, and a geometry column. Afterward, distance weights between patch centroids were calculated utilizing the *k*-NN algorithm [58]. The data frame and distance weights served as input parameters to the agglomerative clustering algorithm. Figure 8 represents the results for a specified number of clusters.

Figure 7. Test area 72961_840430_M-34-40-B-a-2-3 [42].

Simultaneously, ground truth segmentation masks were prepared by manually dividing the test image into a fixed number of regions. For the number of clusters between 2 and 10, there was an average of 17.97% ± 8.7% patch-wise difference between ground truth and the unsupervised approach results. The more clusters were predicted the difference was larger. Figure 9 represents the best result, which was acquired for six clusters where the unsupervised approach misclassified 6% of patches.

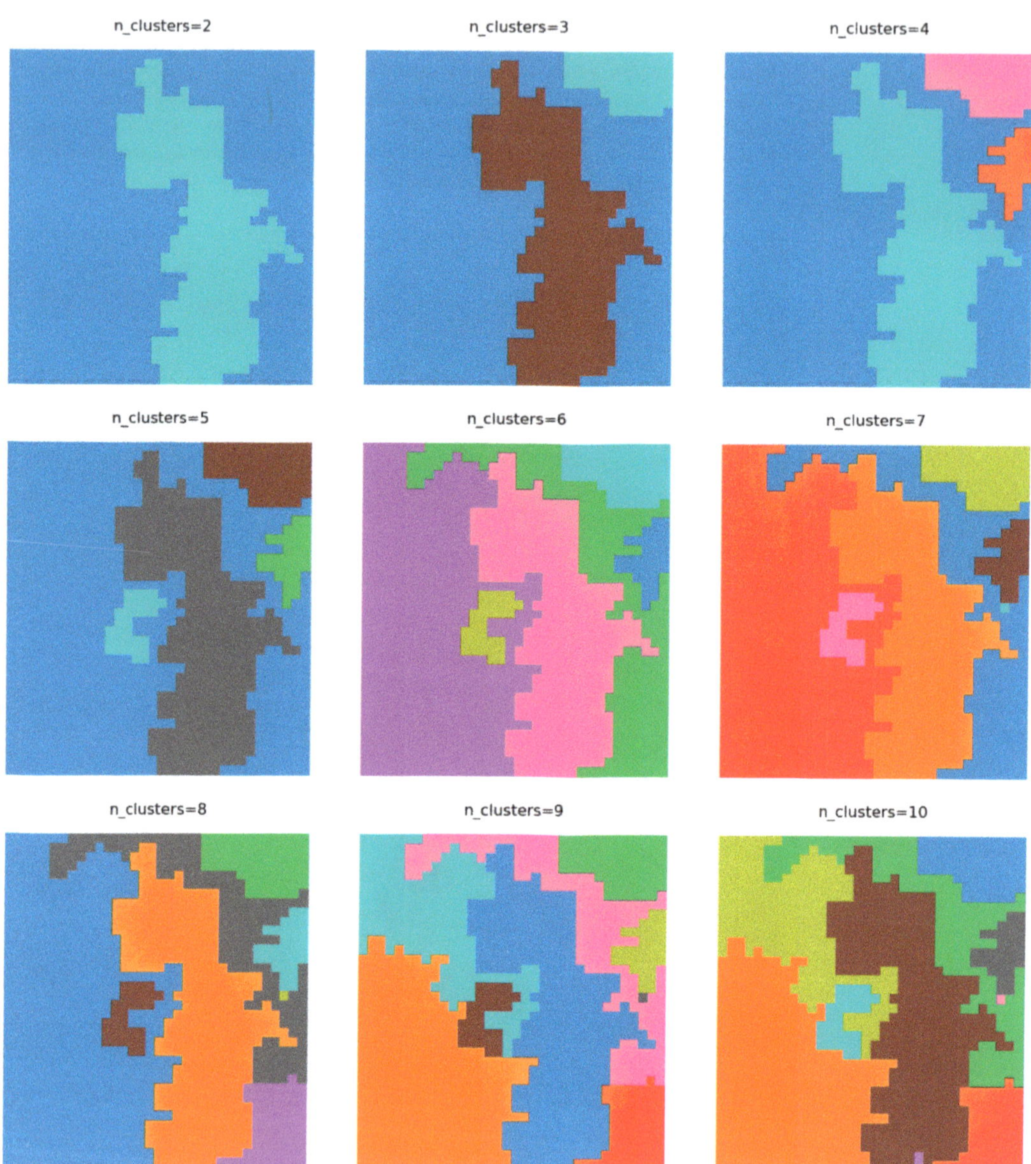

Figure 8. Agglomerative clustering of 72961_840430_M-34-40-B-a-2-3 sample encoded patches using different cluster numbers.

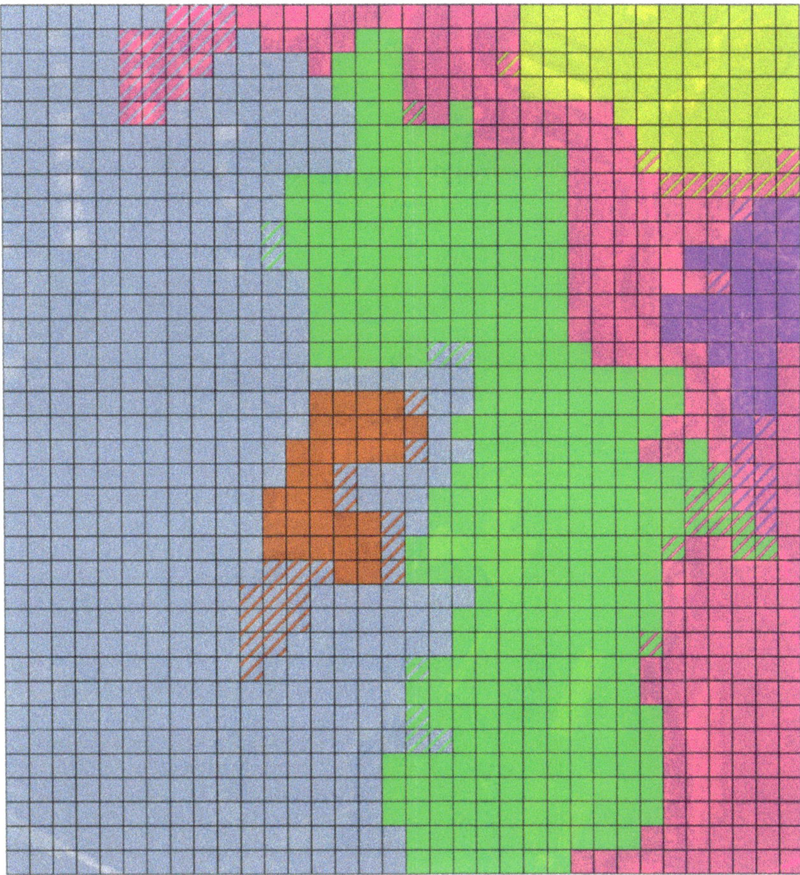

Figure 9. Clustering results of sample orthophoto (72961_840430_M-34-40-B-a-2-3 [42]) patches for fixed number of clusters (*n* clusters = 6). Each color represents a different cluster. Filled squares are the result of latent space clustering. The line pattern indicates the difference between the latent space clustering results and ground truth prepared by manual annotation.

4. Discussion

Utilizing a neural network as a key element of a feature engineering pipeline is a promising idea. The concept of learning the internal representation of data is not new and was extensively studied after the introduction of autoencoders (AE) [59]. Unlike regular autoencoders, bidirectional GANs do not make assumptions about the structure or distribution of the data making them agnostic to the domain of the data [38]. This makes them perfectly suited for use beyond working with RGB images and opens the opportunity to apply them in remote sensing where processing hyperspectral imagery is a standard use case.

One of the main challenges when utilizing a GAN is determining how big a research dataset is needed to feed the network to obtain the required result. The performance of the generator and, therefore, the overall quality of the reconstruction process and network encoding capabilities are tightly coupled with the input data. To be able to properly encode an image, BigBiGAN needs to learn different types of spatial features and discover how they interact with each other. In the early stages of the research, we identified that the size of the dataset had a positive influence on reconstruction quality. We initially worked with around 10% of the final dataset in order to rapidly prototype the solution. The results were not satisfying, i.e., we were not able to produce artificial samples that resembled ground truth data. This prompted us to gradually increase the dataset's size. Authors are far from estimating the correct size of the dataset that could yield the best possible result for a specific research area. We are sure that addressing this issue will be important in the future development of this method.

The method of measuring the training progression of generative models still remains a problematic issue. The standard approach of monitoring the loss value during training and validation is not applicable due to the fact that all GAN components interact with each other, and the loss value is calculated against a specific point in time during the training process and, therefore, is ephemeral and incomparable with previous epochs. There are multiple ways of controlling how the training should progress, e.g., by using Wasserstein loss [60], applying gradient penalty [61], or spectral normalization [62]. Nevertheless, it is difficult to make a clear statement of what loss value identifies a perfectly trained network. Furthermore, applying GAN to tackle the problems within the remote sensing domain is still a novelty. It is difficult to find references in the scientific literature or open-source projects that could be helpful in determining the proper course of model training.

Although nontrivial, measuring the quality of bidirectional GAN image reconstruction capabilities seems to be a valid approach to the task of model quality assurance. An encoder, by design, always yields a result. It is just as true for a state-of-the-art model and its poorly trained counterparts. Encoder output cannot be directly interpreted, which makes it hard to evaluate its quality. The generator, on the other hand, produces a visible result that can be measured. According to the assumptions of bidirectional models, the encoding and decoding process should to some extent be reversible [38]. Hence, the artificially produced image should resemble, in terms of features, its reconstruction origin, i.e., the real image in which latent code was used to create an artificial sample. In other words, checking generator results operating on strictly defined latent codes determines the quality of the entire GAN.

A naive method of verification of the degree to which an orthophoto generated image looks realistic would be to directly compare it to its reconstruction origin. Pixel-wise mean absolute error (MAE) or a similar metric can give the researchers insight, to a limited extent, regarding the quality of produced samples. Unfortunately, this technique only allows getting rid of obvious errors such as a significant mistake in the overall color of the land cover. This is due to MAE not promoting textural and structural correctness, which may lead to poor diagnostic quality in some conditions [63]. One can approach a similar problem when using PNSR. To some extent, SSIM addresses the issue of measuring absolute errors by analyzing structural information. On the other hand, this method is not taking into account the location of spatial features. BigBiGAN reconstruction process only preserves features and their interaction not their specific placement in the analyzed image. Inception score (IS) and Fréchet inception distance (FID) address this problem by measuring the quality of the artificial sample by scoring the GAN capability to produce realistic features [34]. The main drawback of the IS is that it can be misinterpreted in case of mode collapse [64], i.e., the generator is able to produce only a single sample despite the latent code used as input. FID is much stronger in terms of assessing the quality of the generator. What is important, both metrics utilize a pre-trained Inception classifier [50] to capture relevant image features and therefore are dependent on its quality. There are multiple pre-trained models of Inception available. Many of them were created using large datasets such as ImageNet [65]. The authors are not aware of whether a similar dataset for aerial imagery exists. The use of FID is advisable and, as confirmed during the research, it is valuable in proving the capabilities of the generator, but it needs an Inception network trained on a dedicated aerial imagery dataset to be reliable. This way, the score calculated would depend on real spatial features existing in the geographical space. What is more, this approach is only applicable to RGB images. To perform FID calculation for hyperspectral images, a fine-tailored classifier should be trained. Not surprisingly, one of the most effective ways of verifying the quality of artificial images is through human judgment. This takes on even greater importance when approaching the research subject requires specialized knowledge and skills, as exemplified by the analysis of aerial or satellite imagery. Unfortunately, qualitative verification is time-consuming and has to be supported by a quantitative method, which can aid in preselecting potentially good samples.

BigBiGAN accompanied by hierarchical clustering can be effectively used as a building block of an unsupervised orthophoto segmentation pipeline. The results of performing this procedure on a test orthophoto (see Figure 9) proves that the solution is powerful enough to divide the area into a meaningful predefined number of regions. Particularly noteworthy is the precise separation of forests, arable lands, and build-up areas. There is also room for improvement. Currently, the network is not capable of segmenting out tree felling areas located in the northwest and the river channel, which would be very beneficial from the point of view of landscape analysis. Furthermore, it also incorrectly combined pastures and arable lands. The main drawback of this method is the need to predefine the number of clusters. What is more, when increasing the number of clusters, artifacts started to occur, and the algorithm predicted small areas that were not identified as distinct regions in the ground truth image (Figure 8, n clusters = 7–10). Further analysis of latent codes and features that they represent is needed to understand the origin of this issue.

BigBiGAN clustering procedure results resemble, to some extent, the segmentation of the area performed during the Corine Land Cover project in 2018 (Figure 10). It is interesting that the proposed GAN procedure shows a better fit with the boundaries of individual areas than CLC. Nevertheless, CLC has a great advantage over the result generated using GAN, i.e., each tile possesses information about the land cover types that it represents. CLC land cover codes are consistent across all areas involved in the study, which makes this dataset very useful in terms of even sophisticated analysis. This does not mean, however, that the GAN cannot be rearmed to carry information about the land cover types. In the initial BigGAN paper, the authors proposed a solution to enrich each part of the neural network with a mechanism that would enable working with class embeddings [44]. The authors did not use the aforementioned solution to maintain the unsupervised nature of the procedure. An interesting solution would be to compare the latent codes of patches located within different regions to check how similar they are and use this information to join similar, distant regions. To achieve this, a more advanced dataset is needed to cover a larger area and prevent undersampling of occurring less frequently but spatially significant features. Comparison with CLC is also interesting due to the differences in the creation of both sets. CLC is prepared using a semi-supervised procedure that involves multiple different information sources. In contrast, the GAN approach utilizes only orthophotos and is fully unsupervised. Another interesting approach would be to utilize Corine Land Cover (CLC) as the source of model labels and retrain the network to also possess the notion of land cover types. This way, we would gain an interesting solution that would offer a way of producing CLC-like annotations in different precision levels and using different data sources.

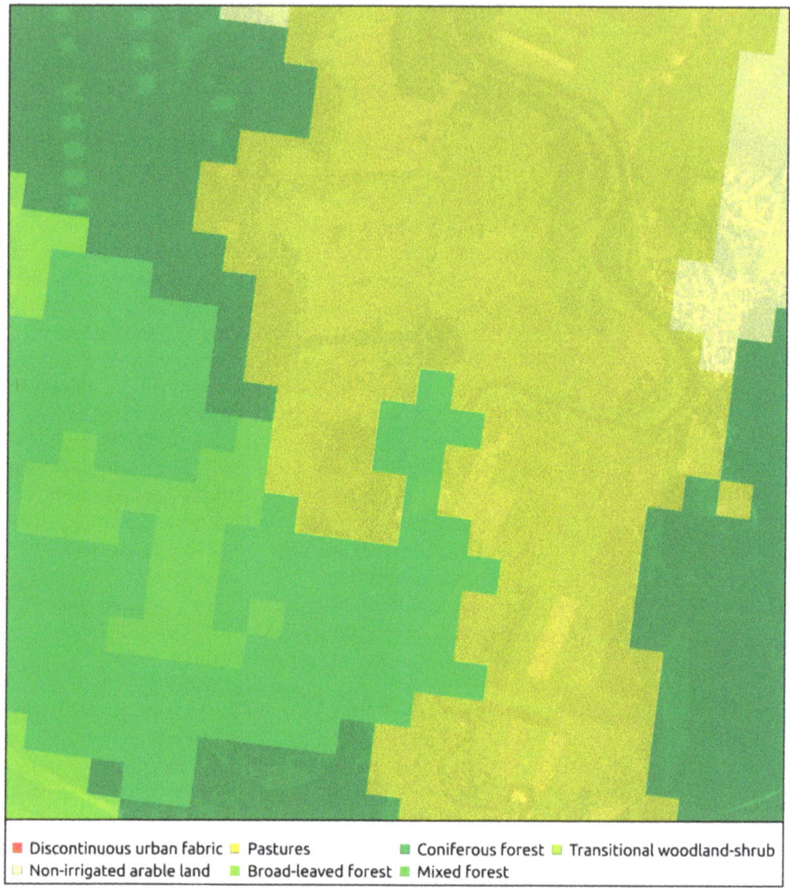

Figure 10. Orthophoto sheet 72961_840430_M-34-40-B-a-2-3 and Corine Land Cover Project (2018) segmentation [66].

5. Conclusions

Generative adversarial networks are a powerful tool that definitely found their place in both geographical information systems (GIS) and machine learning toolboxes. In the case of remote sensing imagery processing, they provide a data augmentation mechanism of creating decent quality artificial data samples, enhancing, or even fixing existing images, and also can actively participate in feature extraction. The latter gives the researchers access to new information encoded in the latent space. During the research, authors confirmed that the bidirectional generative adversarial network (BigBiGAN) encoder module can be successfully used to compress RGB orthophoto patches to lower-dimensional latent vectors.

The encoder performance was assessed indirectly by evaluating the network reconstruction capabilities. Pixel-wise comparison between ground truth and reconstruction output yielded the following results: mean absolute error (MAE) 27.213, structural similarity index (SSIM) 0.942, peak signal-to-noise ratio (PSNR) 42.731, and Fréchet inception distance (FID) 86.36 $\pm$ 7.28. Furthermore, the encoder was tested by utilizing output latent vectors to perform geospatial clustering of a chosen area from the Pilica River region (94% patch-wise accuracy against manually prepared segmentation mask). The case study proved that orthophoto latent vectors, combined with georeferences, can be used during spatial analysis, e.g., in region delimitation or by producing reliable segmentation masks.

The main advantage of the proposed procedure is that the whole training process is unsupervised. The utilized neural network is capable of discovering even complex spatial features and code them in the network underlying latent space. In addition, handling relatively lightweight latent vectors during analysis rather than raw orthophoto proved to significantly facilitate the study. During processing and analysis, there was no need to possess a real image (37MB) but only a recipe to compute in on the fly (3MB). The authors think this feature has great potential in the commercial application of the procedure to lower disk space and network transfer requirements when processing large remote sensing datasets.

On the other hand, the presented method is substantially difficult to implement, configure, and train; it is prone to errors and is demanding in terms of computation costs. To achieve a decent result, one must be ready for a long run of trials and errors mainly related to tuning the model and estimating the required dataset size. Regarding latent vectors, authors have identified a major flaw related to the lack of possibility to precisely describe the meaning of each dimension. The main disadvantage of the proposed procedure is that the majority of steps during the evaluation of the model involves human engagement.

The authors are certain that utilizing BigBiGAN on a more robust and rich dataset, like multispectral imagery, backed by digital terrain model (DTM) and at the same time working on reducing the internal complexity of the network to enable processing larger patches will result in a handful of valuable discoveries. The main focus of the research team in the future will be the verification of the proposed method on a greater scale. Future work will involve performing geospatial clustering of latent codes acquired for all Polish geographic regions and presenting the comparison between classically distinguished regions and their automatically generated counterparts.

Supplementary Materials: The following are available online at https://www.mdpi.com/2072-4292/13/2/306/s1. Encoder model in h5 format with sample data is available on github.com (maciej-adamiak/bigbigan-feature-engineering).

Author Contributions: Conceptualization, M.A.; Methodology, M.A..; Software, M.A; Validation, K.B. and A.M.; Formal analysis, M.A.; Investigation, M.A.; Resources, M.A.; Data curation, M.A.; Writing—original draft preparation, M.A., K.B, and A.M.; Writing—review and editing, M.A., K.B, and A.M.; Visualization, M.A.; Supervision, K.B and A.M.; Project administration, M.A. All authors have read and agreed to the published version of the manuscript.

Funding: This research received no external funding.

Institutional Review Board Statement: Not applicable.

Informed Consent Statement: Not applicable.

Data Availability Statement: Publicly available datasets were analyzed in this study. This data can be found here: https://www.geoportal.gov.pl/.

Acknowledgments: We would like to thank Mikołaj Koziarkiewicz, Maciej Opała, Kamil Rafałko and Tomasz Napierała for helpful remarks and an additional linguistic review.

Conflicts of Interest: The authors declare no conflict of interest.

References

1. Kussul, N.; Lavreniuk, M.; Skakun, S.; Shelestov, A. Deep Learning Classification of Land Cover and Crop Types Using Remote Sensing Data. *IEEE Geosci. Remote Sens. Lett.* **2017**, *14*, 778–782. [CrossRef]
2. Landsat 8—Landsat Science. Available online: https://landsat.gsfc.nasa.gov/landsat-8/ (accessed on 24 October 2020).
3. Sentinel-2—ESA Operational EO Missions—Earth Online—ESA. Available online: https://earth.esa.int/web/guest/missions/esa-operational-eo-missions/sentinel-2 (accessed on 24 October 2020).
4. Pleiades—eoPortal Directory—Satellite Missions. Available online: https://earth.esa.int/web/eoportal/satellite-missions/p/pleiades (accessed on 24 October 2020).
5. Dziennik Ustaw 2020 r. poz. 1086. Available online: https://www.dziennikustaw.gov.pl/DU/2020/1086 (accessed on 24 October 2020).

6. Olędzki, J.R. *Geographical Conditions of the Diversity of the Satellite Image of Poland and Its Division into Photomorphic Units*; Wydawnictwa Uniwersytetu Warszawskiego: Warszawa, Poland, 1992.
7. Olędzki, J.R. Geographical regions of Poland. *Teledetekcja Środowiska* **2001**, *38*, 302.
8. Haralick, R.M.; Shanmugam, K.; Dinstein, I. Textural Features for Image Classification. *IEEE Trans. Syst. Man Cybern.* **1973**, 610–621. [CrossRef]
9. Haralick, R.M.; Shapiro, L.G. Image segmentation techniques. *Comput. Vis. Graph. Image Process.* **1985**, *29*, 100–132. [CrossRef]
10. eCognition | Trimble Geospatial. Available online: https://geospatial.trimble.com/products-and-solutions/ecognition (accessed on 12 November 2020).
11. Wężyk, P.; De Kok, R.; Kozioł, K. Application of the object-based image analysis of VHR satellite images in land-use classification. *Rocz. Geomatyki-Ann. Geomat.* **2006**, *4*, 227–238.
12. Kondracki, J. *Physical and Geographic Regions of Poland*; Wydawnictwa Uniwersytetu Warszawskiego: Warszawa, Poland, 1977.
13. Solon, J.; Borzyszkowski, J.; Bidłasik, M.; Richling, A.; Badora, K.; Balon, J.; Brzezińska-Wójcik, T.; Chabudziński, Ł.; Dobrowolski, R.; Grzegorczyk, I.; et al. Physico-geographical mesoregions of Poland: Verification and adjustment of boundaries on the basis of contemporary spatial data. *Geogr. Pol.* **2018**, *91*, 143–170. [CrossRef]
14. Lewiński, S. Identification of land cover and land use forms on landsat ETM+ satellite image using the method of object-oriented classification. *Rocz. Geomatyki-Ann. Geomat.* **2006**, *4*, 139–150.
15. Lewiński, S. Comparison of object-oriented classification to traditional pixel-based classification with reference to automation of the process of land cover and land use data base creation. *Rocz. Geomatyki-Ann. Geomat.* **2007**, *5*, 63–70.
16. Kosiński, K. Application of Region Growing procedure to meadow classification based on Landsat ETM+ images. *Rocz. Geomatyki-Ann. Geomat.* **2005**, *3*, 69–76.
17. Kosiński, K. Application of structural features in the Landsat ETM+ image in object classification of landscape-vegetation complexes. *Archiwum Fotogram. Kartografii i Teledetekcji* **2007**, *17a*, 385–394.
18. Lang, S.; Burnett, C.; Blaschke, T. Multiscale object-based image analysis—A key to the hierarchical organisation of landscapes. *Ekológia* **2004**, *23*, 148–156.
19. Adamczyk, J.; Będkowski, K. Object-based analysis as a method of improving classification quality. *Rocz. Geomatyki-Ann. Geomat.* **2006**, *4*, 37–46.
20. Adamczyk, J. The effect of scaling methods on the calculation of environmental indices. *Ecol. Quest.* **2013**, *17*, 9–23. [CrossRef]
21. Białousz, S.; Chmiel, J.; Fijałkowska, A.; Różycki, S. Application of satellite images and GIS technology for updating of soil-landscape units—Examples for small scales mapping. *Archiwum Fotogram. Kartografii i Teledetekcji* **2010**, *21*, 21–32.
22. Kot, R. Application of the geodiversity index for defining the relief's diversity based on the example of the Struga Toruńska representative basin, Chełmno Lakeland. *Probl. Ekol. Kraj.* **2012**, *33*, 87–96.
23. Solon, J. Assessment of diversity of landscape on the basis of analysis of spatial structure of vegetation. *Pr. Geogr.* **2002**, *185*, 193–209.
24. Atkinson, P.M.; Tatnall, A.R.L. Introduction Neural networks in remote sensing. *Int. J. Remote Sens.* **1997**, *18*, 699–709. [CrossRef]
25. Chen, Y.; Lin, Z.; Zhao, X.; Wang, G.; Gu, Y. Deep Learning-Based Classification of Hyperspectral Data. *IEEE J. Sel. Top. Appl. Earth Obs. Remote Sens.* **2014**, *7*, 2094–2107. [CrossRef]
26. Luus, F.P.S.; Salmon, B.P.; van den Bergh, F.; Maharaj, B.T.J. Multiview Deep Learning for Land-Use Classification. *IEEE Geosci. Remote Sens. Lett.* **2015**, *12*, 2448–2452. [CrossRef]
27. Adamiak, M.; Biczkowski, M.; Leśniewska-Napierała, K.; Nalej, M.; Napierała, T. Impairing Land Registry: Social, Demographic, and Economic Determinants of Forest Classification Errors. *Remote Sens.* **2020**, *12*, 2628. [CrossRef]
28. Cabezas, M.; Kentsch, S.; Tomhave, L.; Gross, J.; Caceres, M.L.L.; Diez, Y. Detection of Invasive Species in Wetlands: Practical DL with Heavily Imbalanced Data. *Remote Sens.* **2020**, *12*, 3431. [CrossRef]
29. Sobczak, M.; Folbrier, A.; Kozłowska, A.; Pabjanek, P.; Wrzesien, M.; Zagajewski, B. Assessment of the potential of hyperspectral data and techniques for mountain vegetation analysis. In *Imaging Spectroscopy: New Quality in Environmental Studies*; EARSeL & Warsaw University: Warsaw, Poland, 2005; pp. 763–780.
30. Zagajewski, B. Assessment of neural networks and Imaging Spectroscopy for vegetation classification of the High Tatras. *Teledetekcja Środowiska* **2010**, *43*, 1–113.
31. Iwaniak, A.; Krówczyńska, M.; Paluszyński, W. Applying neural networks to urban area classification in satellite images. *Acta Scientiarum Polonorum. Geodesia et Descriptio Terrarum* **2002**, *1*, 5–13.
32. Krawiec, K.; Wyczałek, I. Supervised road detection using machine learning methodology. *Archiwum Fotogram. Kartografii Teledetekcji* **2006**, *16*, 361–371.
33. Goodfellow, I.J.; Pouget-Abadie, J.; Mirza, M.; Xu, B.; Warde-Farley, D.; Ozair, S.; Courville, A.; Bengio, Y. Generative Adversarial Networks. *arXiv* **2014**, arXiv:1406.2661.
34. Salimans, T.; Goodfellow, I.; Zaremba, W.; Cheung, V.; Radford, A.; Chen, X. Improved Techniques for Training GANs. *arXiv* **2016**, arXiv:1606.03498.
35. Dong, R.; Li, C.; Fu, H.; Wang, J.; Li, W.; Yao, Y.; Gan, L.; Yu, L.; Gong, P. Improving 3-m Resolution Land Cover Mapping through Efficient Learning from an Imperfect 10-m Resolution Map. *Remote Sens.* **2020**, *12*, 1418. [CrossRef]
36. Burdziakowski, P. A Novel Method for the Deblurring of Photogrammetric Images Using Conditional Generative Adversarial Networks. *Remote Sens.* **2020**, *12*, 2586. [CrossRef]

37. Zhao, W.; Chen, X.; Chen, J.; Qu, Y. Sample Generation with Self-Attention Generative Adversarial Adaptation Network (SaGAAN) for Hyperspectral Image Classification. *Remote Sens.* **2020**, *12*, 843. [CrossRef]

38. Donahue, J.; Krähenbühl, P.; Darrell, T. Adversarial Feature Learning. *arXiv* **2017**, arXiv:1605.09782.

39. Żelaźniewicz, A.; Aleksandrowski, P.; Buła, Z.; Karnkowski, P.H.; Konon, A.; Ślączka, A.; Żaba, J.; Żytko, K. *Tectonic Regionalization of Poland*; Komitet Nauk Geologicznych PAN: Wrocław, Poland, 2011; ISBN 978-83-63377-01-4.

40. Head Office of Geodesy and Cartography. Data for the Numerical Terrain Model with a Grid Interval of at Least 100 m. Available online: http://www.gugik.gov.pl/pzgik/dane-bez-oplat/dane-dotyczace-numerycznego-modelu-terenu-o-interwale-siatki-co-najmniej-100-m-nmt_100 (accessed on 9 January 2021).

41. Krysiak, S. Ecological aspects of land use changes in chosen types of natural landscapes in Central Poland. *Probl. Ekol. Kraj.* **2008**, *21*, 299–310.

42. Geoportal.gov.pl. Available online: http://geoportal.gov.pl (accessed on 11 November 2020).

43. Domingos, P. A few useful things to know about machine learning. *Commun. ACM* **2012**, *55*, 78–87. [CrossRef]

44. Donahue, J.; Simonyan, K. Large Scale Adversarial Representation Learning. *arXiv* **2019**, arXiv:1907.02544.

45. Mukherjee, S.; Asnani, H.; Lin, E.; Kannan, S. ClusterGAN: Latent Space Clustering in Generative Adversarial Networks. *arXiv* **2019**, arXiv:1809.03627. [CrossRef]

46. 2.3. Clustering—Scikit-Learn 0.23.2 Documentation. Available online: https://scikit-learn.org/stable/modules/clustering.html#hierarchical-clustering (accessed on 24 October 2020).

47. Ward, J.H. Hierarchical Grouping to Optimize an Objective Function. *J. Am. Stat. Assoc.* **1963**, *58*, 236–244. [CrossRef]

48. Radford, A.; Metz, L.; Chintala, S. Unsupervised Representation Learning with Deep Convolutional Generative Adversarial Networks. *arXiv* **2016**, arXiv:1511.06434.

49. He, K.; Zhang, X.; Ren, S.; Sun, J. Deep Residual Learning for Image Recognition. *arXiv* **2015**, arXiv:1512.03385.

50. Szegedy, C.; Liu, W.; Jia, Y.; Sermanet, P.; Reed, S.; Anguelov, D.; Erhan, D.; Vanhoucke, V.; Rabinovich, A. Going Deeper with Convolutions. *arXiv* **2014**, arXiv:1409.4842.

51. Brock, A.; Donahue, J.; Simonyan, K. Large Scale GAN Training for High Fidelity Natural Image Synthesis. *arXiv* **2019**, arXiv:1809.11096.

52. Landau, B.V. The Fréchet distance between multivariate normal distributions. *J. Multivar. Anal.* **1982**, *12*, 450–455. [CrossRef]

53. Zhou, S.; Gordon, M.L.; Krishna, R.; Narcomey, A.; Fei-Fei, L.; Bernstein, M.S. HYPE: A Benchmark for Human eYe Perceptual Evaluation of Generative Models. *arXiv* **2019**, arXiv:1904.01121.

54. Wang, Z.; Bovik, A.C.; Sheikh, H.R.; Simoncelli, E.P. Image quality assessment: From error visibility to structural similarity. *IEEE Trans. Image Process.* **2004**, *13*, 600–612. [CrossRef] [PubMed]

55. Korhonen, J.; You, J. Peak signal-to-noise ratio revisited: Is simple beautiful? In Proceedings of the 2012 Fourth International Workshop on Quality of Multimedia Experience, Yarra Valley, Australia, 5–7 July 2012; pp. 37–38.

56. Head Office of Geodesy and Cartography Integrated Copies of Databases of Topographic Objects. Available online: http://www.gugik.gov.pl/pzgik/zamow-dane/baza-danych-obiektow-topograficznych-bdot-10k (accessed on 11 November 2020).

57. GeoPandas 0.8.0—GeoPandas 0.8.0 Documentation. Available online: https://geopandas.org/ (accessed on 24 October 2020).

58. sklearn.neighbors.KNeighborsClassifier—Scikit-Learn 0.23.2 Documentation. Available online: https://scikit-learn.org/stable/modules/generated/sklearn.neighbors.KNeighborsClassifier.html (accessed on 24 October 2020).

59. Kingma, D.P.; Welling, M. An Introduction to Variational Autoencoders. *FNT Mach. Learn.* **2019**, *12*, 307–392. [CrossRef]

60. Frogner, C.; Zhang, C.; Mobahi, H.; Araya-Polo, M.; Poggio, T. Learning with a Wasserstein Loss. *arXiv* **2015**, arXiv:1506.05439.

61. Gulrajani, I.; Ahmed, F.; Arjovsky, M.; Dumoulin, V.; Courville, A. Improved Training of Wasserstein GANs. *arXiv* **2017**, arXiv:1704.00028.

62. Miyato, T.; Kataoka, T.; Koyama, M.; Yoshida, Y. Spectral Normalization for Generative Adversarial Networks. *arXiv* **2018**, arXiv:1802.05957.

63. Deora, P.; Vasudeva, B.; Bhattacharya, S.; Pradhan, P.M. Structure Preserving Compressive Sensing MRI Reconstruction using Generative Adversarial Networks. *arXiv* **2020**, arXiv:1910.06067.

64. Thanh-Tung, H.; Tran, T. On Catastrophic Forgetting and Mode Collapse in Generative Adversarial Networks. *arXiv* **2020**, arXiv:1807.04015.

65. ImageNet: A Large-Scale Hierarchical Image Database—IEEE Conference Publication. Available online: https://ieeexplore.ieee.org/document/5206848 (accessed on 26 October 2020).

66. Forslund, L. CLC 2018—Copernicus Land Monitoring Service. Available online: https://land.copernicus.eu/pan-european/corine-land-cover/clc2018 (accessed on 11 November 2020).

MDPI

St. Alban Anlage 66

4052 Basel

Switzerland

Tel. +41 61 683 77 34

Fax +41 61 302 89 18

www.mdpi.com

Remote Sensing Editorial Office

E-mail: remotesensing@mdpi.com

www.mdpi.com/journal/remotesensing